Nucleic Acids and Molecular Biology

Volume 30

Series Editor
Janusz M. Bujnicki
International Institute of Molecular
and Cell Biology
Laboratory of Bioinformatics and
Protein Engineering
Trojdena 4
02-109 Warsaw
Poland

For further volumes:
http://www.springer.com/series/881

Katsuhiko S. Murakami • Michael A. Trakselis
Editors

Nucleic Acid Polymerases

Springer

Editors
Katsuhiko S. Murakami
Dept. of Biochem. and Mol. Biology
The Pennsylvania State University
University Park
Pennsylvania
USA

Michael A. Trakselis
Department of Chemistry
University of Pittsburgh
Pittsburgh
Pennsylvania
USA

ISSN 0933-1891 ISSN 1869-2486 (electronic)
ISBN 978-3-642-39795-0 ISBN 978-3-642-39796-7 (eBook)
DOI 10.1007/978-3-642-39796-7
Springer Heidelberg New York Dordrecht London

Printed on acid-free paper

Springer is part of Springer Science+Business Media (www.springer.com)

Preface

More than any other class of enzymes, nucleic acid polymerases are directly responsible for an overabundance of enzymatic, regulatory, and maintenance activities in the cell. DNA polymerases accurately replicate copies of genomes in all forms of life as well as have specialized roles in DNA repair and immune response. RNA polymerases are most noted for their active roles in controlling gene expression during transcription but can also be utilized in self-replicating ribozymes and viral replication. Although the general sequence homology, structural architecture, and mechanism are conserved, they have evolved to incorporate deoxynucleotides (dNTPs) or ribonucleotides (rNTPs) explicitly. Various nucleic acid polymerases have specificities for RNA or DNA templates, incorporate dNTPs or rNTPs, and can be template dependent or independent. Here, we provide examples on the latest understanding of each class of nucleic acid polymerase, their structural and kinetic mechanisms, and their respective roles in the central dogma of life.

This book provides a catalog and description of the multitude of polymerases (both DNA and RNA) that contribute to genomic replication, maintenance, and gene expression. Evolution has resulted in tremendously efficient enzymes capable of repeated extremely rapid syntheses that have captivated researchers' interests for decades. We are inspired by work that started over 60 years ago and is actively pursued today for a fundamental understanding of life, contributions to human health and disease, and current and future biotechnology applications. Nucleic acid polymerases are fascinating on a number of levels, yet still continue to surprise us with novel modes of action revealed through ongoing and future studies described within this volume.

We wish to thank all the authors for their specific expertise and willingness to participate in this comprehensive review of nucleic acid polymerases. We are also grateful to the many investigators before us (including our research mentors: Stephen Benkovic and Akira Ishihama) who began and continue this important line of research. We believe this book will be useful for a wide range of researchers in both the early and later stages of their careers. We would be thrilled if this volume becomes the go-to resource for nucleic acid polymerase structure, function, and mechanism for years to come.

Pittsburgh, PA Michael A. Trakselis
University Park, PA Katsuhiko S. Murakami

Contents

1 **Introduction to Nucleic Acid Polymerases: Families, Themes, and Mechanisms** 1
Michael A. Trakselis and Katsuhiko S. Murakami

2 **Eukaryotic Replicative DNA Polymerases** 17
Erin Walsh and Kristin A. Eckert

3 **DNA Repair Polymerases** 43
Robert W. Sobol

4 **Eukaryotic Y-Family Polymerases: A Biochemical and Structural Perspective** 85
John M. Pryor, Lynne M. Dieckman, Elizabeth M. Boehm, and M. Todd Washington

5 **DNA Polymerases That Perform Template-Independent DNA Synthesis** 109
Anthony J. Berdis

6 **Archaeal DNA Polymerases: Enzymatic Abilities, Coordination, and Unique Properties** 139
Michael A. Trakselis and Robert J. Bauer

7 **Engineered DNA Polymerases** 163
Roberto Laos, Ryan W. Shaw, and Steven A. Benner

8 **Reverse Transcriptases** 189
Stuart F.J. Le Grice and Marcin Nowotny

9 **Telomerase: A Eukaryotic DNA Polymerase Specialized in Telomeric Repeat Synthesis** 215
Andrew F. Brown, Joshua D. Podlevsky, and Julian J.-L. Chen

10 **Bacteriophage RNA Polymerases** 237
Ritwika S. Basu and Katsuhiko S. Murakami

11 Mitochondrial DNA and RNA Polymerases 251
Y. Whitney Yin

12 Eukaryotic RNA Polymerase II 277
David A. Bushnell and Roger D. Kornberg

13 Plant Multisubunit RNA Polymerases IV and V 289
Thomas S. Ream, Jeremy R. Haag, and Craig S. Pikaard

14 Structure, Dynamics, and Fidelity of RNA-Dependent RNA Polymerases 309
David D. Boehr, Jamie J. Arnold, Ibrahim M. Moustafa, and Craig E. Cameron

Index 335

Chapter 1
Introduction to Nucleic Acid Polymerases: Families, Themes, and Mechanisms

Michael A. Trakselis and Katsuhiko S. Murakami

Keywords Polymerase • Mechanism • Structure • Function • Catalysis

Abbreviations

CPD	Cyclobutane pyrimidine dimers
E. coli	*Escherichia coli*
FDX	Fidaxomicin
FILS	Facial dysmorphism, immunodeficiency, livedo, and short statures
kDa	Kilodaltons
pol	Polymerase
Pol I	*E. coli* DNA polymerase I
RdRp	RNA-dependent RNA polymerase
Rif	Rifampicin
rRNA	Ribosomal RNA
TLS	Translesion synthesis
UV	Ultraviolet light
XPD	*Xeroderma pigmentosum*

M.A. Trakselis (✉)
Department of Chemistry, University of Pittsburgh, Pittsburgh, PA 15260, USA
e-mail: mtraksel@pitt.edu

K.S. Murakami (✉)
Department of Biochemistry and Molecular Biology, Pennsylvania State University, University Park, PA 16802, USA

The Center for RNA Molecular Biology, Pennsylvania State University, University Park, PA 16802, USA
e-mail: kum14@psu.edu

K.S. Murakami and M.A. Trakselis (eds.), *Nucleic Acid Polymerases*, Nucleic Acids and Molecular Biology 30, DOI 10.1007/978-3-642-39796-7_1,

1.1 Introduction/Discovery/Classification

Template-dependent and template-independent nucleotidyl transfer reactions are fundamentally important in the maintenance of the genome as well as for gene expression in all organisms and viruses. These reactions are conserved and involve the condensation of an incoming nucleotide triphosphate at the 3′ hydroxyl of the growing oligonucleotide chain with concomitant release of pyrophosphate. DNA polymerase I (Pol I) isolated from *E. coli* extracts was initially characterized in in vitro reactions well over 50 years ago by the seminal work of Arthur Kornberg's laboratory (Kornberg 1957; Lehman et al. 1958; Bessman et al. 1958). Inspired by this work, the discovery of a DNA-dependent RNA polymerase quickly followed in 1960 from a variety of researchers including Samuel Weiss (Weiss and Gladstone 1959), Jerald Hurwitz (Hurwitz et al. 1960), Audrey Stevens (Stevens 1960), and James Bonner (Huang et al. 1960). These early enzymatic characterizations of DNA-dependent deoxyribonucleotides and ribonucleotide incorporations gave credibility both to Watson and Crick's DNA double helix model (Watson and Crick 1953) and the transcription operon model proposed by François Jacob and Jacques Monod (Jacob and Monod 1961).

Prior to 1990, few DNA polymerase members were known. Pol I, Pol II, and Pol III from bacteria defined the initial A, B, and C families of polymerases (Braithwaite and Ito 1993), respectively. Eukaryotic polymerases adopted a Greek letter nomenclature (Weissbach et al. 1975) and included cellular B-family polymerases pol α, pol β, pol δ, pol ε, and pol ζ, and the mitochondrial A-family polymerase pol γ. Rapid progress in genome sequencing, search algorithms, and further biochemical analysis identified other putative DNA polymerases both in bacteria and eukaryotes prompting the expansion of the Greek letter nomenclature. Families D, X, and Y were created to classify unique polymerase in archaea as well as those with specialized functions in DNA repair (Burgers et al. 2001; Ishino et al. 1998; Ohmori et al. 2001). After inclusion of reverse transcriptase enzymes that are RNA-dependent DNA polymerases including telomerase, the DNA polymerase families now number seven (Table 1.1). Although the number of human DNA polymerases stands at 16 members, a recently characterized human archaeo-eukaryotic (AEP) DNA primase (Prim-Pol) has both RNA and DNA synthesis abilities (L. Blanco, personal communication) suggesting that other uncharacterized enzymes may have additional unidentified roles in DNA synthesis. This chapter introduces and highlights chapters within this series and puts the DNA and RNA polymerase families, structures, and mechanisms in context.

1.1.1 DNA Polymerase Families and Function

Most DNA-dependent DNA polymerases have a single catalytic subunit (Fig. 1.1). These single subunits are generally active on their own but are regulated with

Table 1.1 Model DNA polymerase family members

Family	Viral	Bacterial	Archaeal	Eukaryotic
A	T7 gp5	Pol I		Pol γ (mito)
		Klenow[a]		Pol θ
		Taq Pol		Pol ν
B	T4/RB69 gp43	Pol II	PolB1 (B2 and B3)[b]	Pol α
	phi29 Pol			Pol δ
				Pol ε
				Pol ζ
C		Pol III		
D			Pol D[c]	
X				Pol β
				Pol λ
				Pol μ
				TdT
Y		Pol IV	Pol Y	Pol η
		Pol V		Pol ι
				Pol κ
				Rev1
RT[d]	RT			Telomerase

[a]Klenow is the C-terminal truncation of *E. coli* Pol I
[b]Crenarchaea generally have thee Pol B enzymes, while euryarchaea have one
[c]Pol D is only found in euryarchaea phyla of archaea
[d]Reverse transcriptase (RT) is RNA-dependent DNA polymerase

regard to function through various accessory proteins that direct and restrain catalysis to specific DNA substrates. For the most part, Y-family DNA repair polymerases adopt a slightly more open active site to accommodate base damage and are devoid of any proofreading exonuclease domains (Chap. 4). These structural features are required for replication past a variety of lesions in the template strand during DNA replication to maintain the integrity of the fork. DNA repair polymerases (X and Y family) are actively involved in maintaining our genome under intense DNA-damaging stressors. The ability to prevent mutagenesis is their main cellular role, but changes in expression levels and disruptions of DNA repair pathways are common in promoting cancer and tumorigenesis (Chap. 3). Viral, bacterial, and some archaeal DNA polymerases (A and B families) are primarily single-subunit enzymes. They are held at the replication fork through dynamic interactions with accessory proteins to maintain high local concentrations during active replication (Chap. 6). The DNA replication polymerase enzymatic accuracy (fidelity) is unprecedented and is primarily responsible for maintaining stable genomes of all organisms. In bacteria and eukaryotes, the DNA replication polymerases (B and C families) have evolved to contain additional subunits that are almost always associated with the catalytic subunit as holoenzyme complexes (Chap. 2). The recently discovered D-family replication polymerases from certain archaea are also multisubunit enzymes and are presumably ancestral precursors to their eukaryotic homologs (Chap. 6). The polymerase accessory subunits have a variety of roles that are only just being identified including maintaining structural

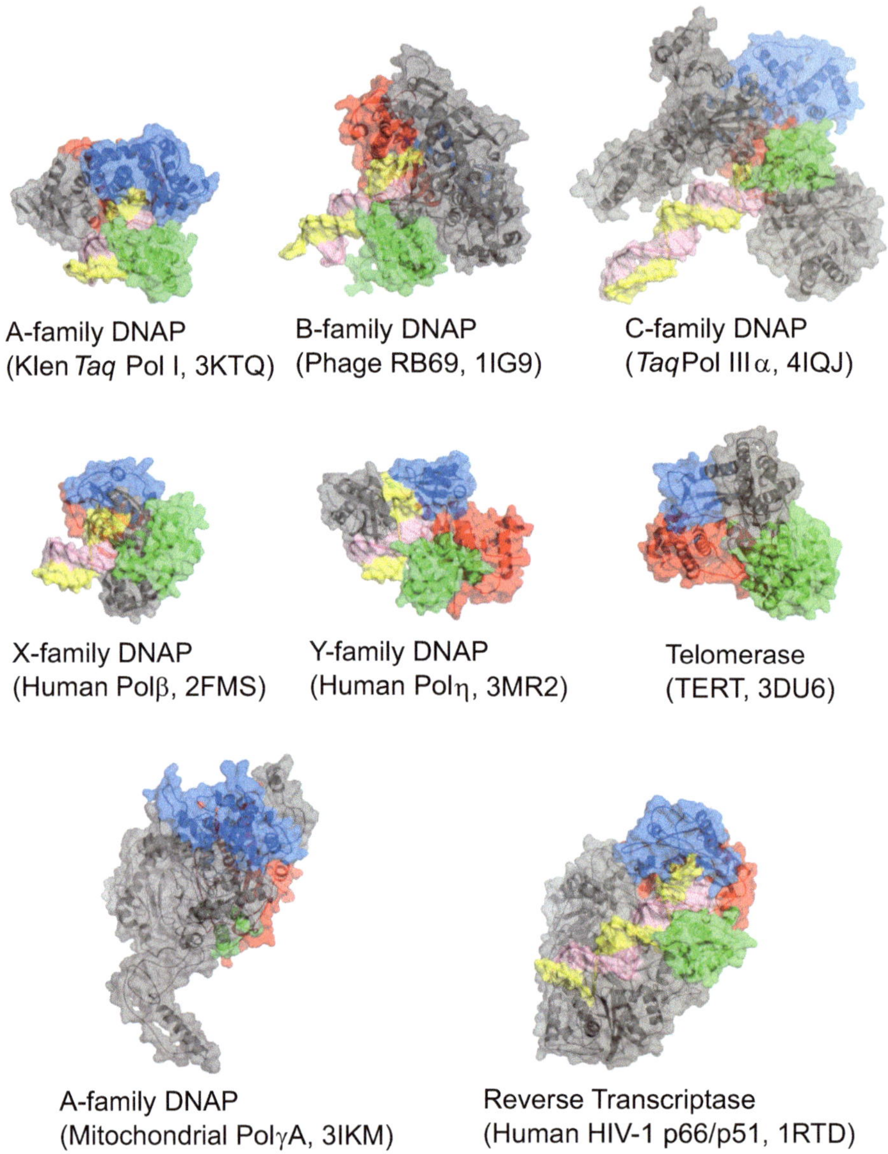

Fig. 1.1 Gallery of DNA polymerases. All polymerases and DNA are shown as cartoon models with partially transparent molecular surfaces. Their names of family, sources, and PDB accession codes are indicated. Protein structures are colored *gray* and key subdomains are colored (thumb, *green*; palm, *red*; fingers, *blue*). Nucleic acids are colored *yellow* for the template DNA and *pink* for the primer DNA. All polymerases are depicted using the same scale in this figure and also in Fig. 1.2 for direct comparison of their sizes

geometries, additional enzymatic activities, and interactions with other polymerase accessory proteins.

Although DNA polymerases are generally template-dependent enzymes and follow Watson and Crick base-pairing rules, there is a subclass of polymerases (primarily X family) that are template independent (Chap. 5). These polymerases are involved in aspects of DNA repair where DNA strands have lost connectivity and require additional nucleotide additions at the ends to facilitate repair. They also have a unique biological role contributing to random incorporations and corresponding diversity of antibodies required for immunological responses. A similar type of DNA extension is also required at the ends of chromosomes to maintain their length during DNA replication, but instead of random nucleotide additions, the enzyme telomerase uses an RNA template strand as a cofactor for sequence-specific DNA repeat additions called telomeres (Chap. 9). This RNA-dependent DNA polymerase, telomerase, is unique to organisms with linear genomes and is also implicated in a variety of human diseases and aging.

1.1.2 RNA Polymerase Families and Function

All cellular organisms including bacteria, archaea, and eukaryotes use multi-subunit DNA-dependent RNA polymerase for transcribing most of the RNAs in cells (Fig. 1.2) (Werner and Grohmann 2011). Bacterial RNA polymerase is the simplest form of this family (composed of the minimum five subunits), whereas archaeal and eukaryotic RNA polymerases possess additional polypeptides to form ~11–17 subunit complexes. Bacteria and archaea use a single type of RNA polymerase for transcribing all genes, whereas eukaryotes have three different enzymes, Pol I, Pol II, and Pol III, and synthesize the large ribosomal RNA (rRNA) precursor, messenger RNA (mRNA), and short untranslated RNAs including 5S rRNA and transfer RNA (tRNA), respectively (Chap. 12). In plant, there are two additional 12-subunit RNA polymerases, Pol IV and Pol V, that play important roles in RNA-mediated gene-silencing pathways (Chap. 13). The archaeal transcription system has been characterized as a hybrid of eukaryotic and bacterial transcription systems; the archaeal basal transcription apparatus is very similar to that of eukaryote, but its transcriptional regulatory factors are similar to those of bacteria (Hirata and Murakami 2009; Jun et al. 2011).

Bacteriophage encodes single-unit RNA polymerase of ~100 kDa molecular weight, which expresses bacteriophage genes on host bacterial cells for generating progeny phage particles (Chap. 10). Although bacteriophage RNA polymerase is about four times smaller than the cellular RNA polymerases (Fig. 1.2), it is able to carry out almost all functions in transcription cycle observed in cellular RNA polymerases. Its primary and three-dimensional structures are similar to A-family polymerase, which also includes mitochondrial RNA polymerase expressing genes from mitochondrial DNA (Fig. 1.2) (Chap. 11).

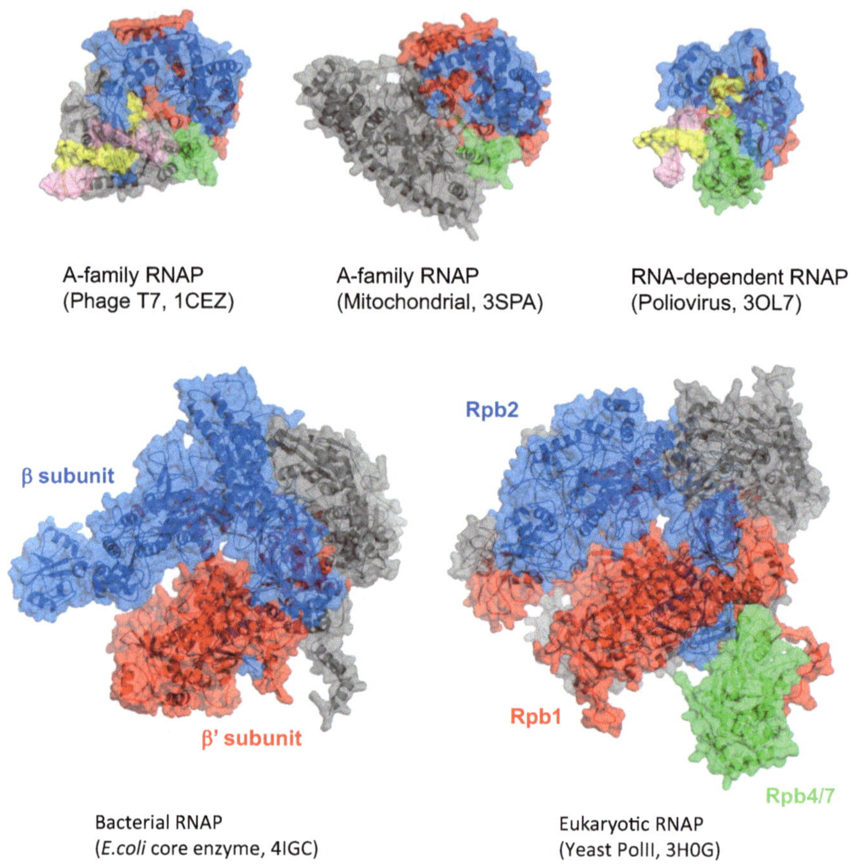

Fig. 1.2 Gallery of RNA polymerases. All polymerases and DNA are shown as cartoon models with partially transparent molecular surfaces. Their names of family, sources, and PDB accession codes are indicated. Protein structures are colored *gray* and key subdomains are colored (thumb, *green*; palm, *red*; fingers, *blue* for the A-family bacteriophage-type RNA polymerase and RdRp; largest subunit, *red*; second largest subunit, *blue*; protruding stalk, *green* for the cellular RNA polymerases). Nucleic acids are colored *yellow* for the template DNA and *pink* for the non-template DNA. All polymerases are depicted using the same scale in this figure and as in Fig. 1.1 for direct comparison of their sizes

RNA viruses including influenza, rhinovirus, hepatitis C, and poliovirus have RNA-dependent RNA polymerases (RdRps) that are responsible for replicating their RNA genomes and expressing their genes (Fig. 1.2). The RdRps are targets for antiviral therapies, but their higher mutation rates due to lack of proofreading endonuclease activity generate resistant variants to compromise antiviral therapies (Chap. 14).

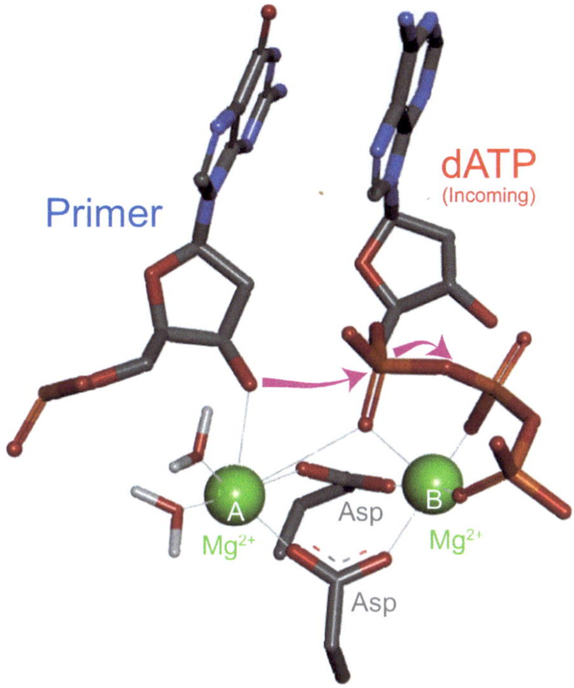

Fig. 1.3 Highlights the two-metal-ion mode of catalysis for DNA and RNA polymerases. Two conserved aspartates coordinate metals A and B in the active site. Metal A activates the 3′-OH for attack on the 5′ α-phosphate of the incoming nucleotide (either dATP or ATP) with release of β-γ pyrophosphate. Metal B neutralizes the negative charge on the phosphates as well as buildup in the transition state

1.2 Conserved Polymerase Structures

The original structure of the C-terminal fragment of *E. coli* Pol I (Klenow fragment) identified the general architecture of DNA polymerases to resemble a right hand with subdomains similar to fingers, thumb, and palm regions (Fig. 1.1) (Ollis et al. 1985). Although sequence homology from different DNA polymerase families has diverged quite significantly, the general organization of all polymerase structures is very similar (Figs. 1.1 and 1.2), suggesting that they may have evolved from a common ancestor. In fact, both the DNA and RNA polymerases catalyze essentially the same chemical reaction with subtle differences ensuring accurate incorporation of their respective nucleotides (Fig. 1.3) (Steitz 1993).

1.2.1 DNA Polymerase Structural and Kinetic Mechanisms

The two most important and conserved residues are aspartates contained within the palm domain that act to coordinate two metal ions (Mg^{2+}) for catalysis (Fig. 1.3). Metal A lowers the bonding potential of the hydrogen at the 3′-OH, activating the 3′-O$^-$ for attack at the α-phosphate of the incoming nucleotide. Metal B aids pyrophosphate leaving and stabilizes structures of the pentacovalent transition state. The mechanism was originally proposed based on the 3′–5′ removal of nucleotides in the exonuclease site of DNA Pol I (Beese et al. 1993). This two-metal-ion mechanism for phosphoryl transfer is identical for DNA and RNA polymerases and extremely similar to analogous reactions involving RNA-catalyzed reactions including splicing (Steitz and Steitz 1993). It is hypothesized that this mechanism was the basis of catalysis in the RNA world and has maintained its core features with all modern polynucleotide polymerases. Interesting, this basic two-metal-ion mechanism has recently been challenged by the observation of a third metal ion in the active site of pol η that acts to neutralize the negative charge buildup in the transition state and protonates the leaving group pyrophosphate (Nakamura et al. 2012). It will be interesting to see if this transient third metal ion also exists in other polymerases suggesting a common theme and expansion of the traditional two-metal-ion phosphoryl transfer mechanisms. The two other domains (fingers and thumb) have diverged significantly throughout the polymerase families but contain functionally analogous elements. The fingers domain acts to correctly position the incoming nucleotide with the template, while the thumb domain aids in DNA binding and successive nucleotide additions (processivity).

To increase the fidelity (accuracy) of continuous nucleotide incorporation, some DNA polymerases from the A, B, and C families have a separate exonuclease (3′–5′) domain which verifies correct incorporation and removes an incorrectly incorporated base. For bacterial and archaeal family A and B polymerases, the exonuclease activity is included in a separate domain within the contiguous polypeptide sequence. In the *E. coli* Pol III holoenzyme as well as eukaryotic B-family polymerase, the exonuclease activity is contained within a separate polypeptide. The first structure of DNA bound in the exonuclease domain was with the Klenow fragment and suggested a common two-metal-ion catalysis mechanism for removal of nucleotides as well (Beese and Steitz 1991). The exonuclease site was defined as having three conserved carboxylate residues coordinating both metal ions, binding to the DNA, and activating catalysis and removal of an incorrectly incorporate base.

In addition to exonuclease activity, high-fidelity DNA polymerases also maintain accuracy through kinetic checkpoints ensuring accurate base pairing (Fig. 1.4). The general consensus is that the polymerase domain alone accounts for fidelity values of 10^{-5} to 10^{-6} and inclusion of the exonuclease proofreading domain contributes another 10^{-2} for total fidelity values of 10^{-7} to 10^{-8} (1 error in every 100 million or 99.999999 % accurate) (Kunkel 2004). DNA polymerases from other families including X and Y have significantly lower fidelity values (10^{-2} to 10^{-5}) accounted by the more frequent error rates, lesion bypass abilities, and absent exonuclease domains (Chaps. 3–5).

For the majority of A-, B-, as well as some Y-family polymerases, a slow step prior to chemistry (step 3) ensures correct base pairing before phosphodiester bond formation. Based on the fusion of structural and kinetic data, it was originally postulated that an open-to-closed transition in the fingers domain was the slow step in the mechanism. More recently, the open-to-closed transition was measured directly using fluorescence and found to be fast relative to step 3, prompting the

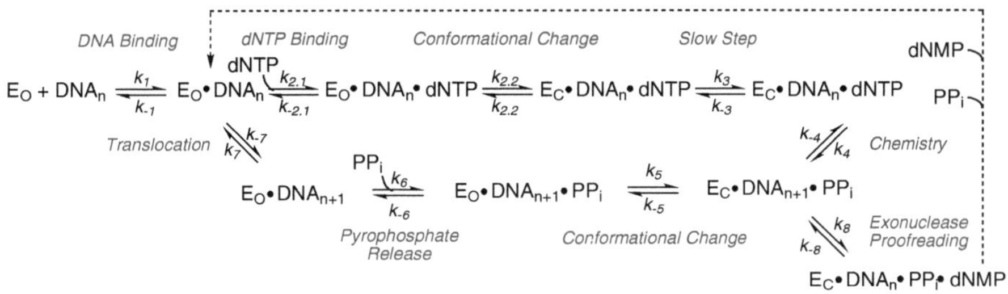

Fig. 1.4 General kinetic mechanism for high-fidelity DNA polymerases. The enzyme undergoes a fast open (E_O)-to-closed (E_C) transition (step 2.2) after binding DNA and nucleotide. Kinetic checkpoints include the slow step prior to chemistry (step 3) as well steps 5 and 8 after chemistry to activate the proofreading function if necessary

inclusion of steps 2a and 2b into the mechanism (Joyce et al. 2008; Johnson 2010). The identity of the slow step 3 is still unknown and may instead be associated with a change in metal ion coordination of either metal B or an incoming metal C in preparation of moving forward through the transition state towards chemistry (Nakamura et al. 2012).

The kinetic checkpoints themselves ensure that correct nucleotides are optimally positioned in the active site over incorrect ones to promote catalysis. Prevention of rNTP binding in DNA polymerase active sites is restricted by a steric gate towards the 2′-OH (step 2.1) (Delucia et al. 2006) as well as reduced rate of fingers closing (step 2.1) limiting their incorporation (Joyce et al. 2008). For Klenow and T7 pol, incorrect dNTP incorporation is prevented by a slower chemistry step 4 than for correct dNTPs defining polymerase fidelity (Dahlberg and Benkovic 1991). Step 5 following phosphoryl transfer is also considered a kinetic slow step and is important for increasing the probability for proofreading (step 8) in the case of a misincorporated base (Kuchta et al. 1988). Although this is not an absolute kinetic mechanism for nucleotide selection in all DNA polymerases (Fig. 1.4), the basic principles explain a number of the mechanistic facets required for maintaining high nucleotide fidelity. Whether or not this complex scheme holds as a general mechanism for all DNA polymerases remains to be determined, but it is likely to be accurate for high-fidelity DNA polymerases in particular.

Fast, successive, and accurate nucleotide additions require that the polymerase remains associated with the template after a translocation step (step 7) for multiple rounds of catalysis or processivity. DNA polymerases by themselves are not highly processive and are not able to incorporate more than 20–50 successive nucleotides in a single binding event. The exception seems to be the B-family DNA polymerase from phi29 which has extremely robust strand displacement activity and processivity of replication of several thousand bases (Blanco et al. 1989; Kamtekar et al. 2006). Phi29 Pol has a specific insertion called the terminal protein region 2 (TPR2) that acts with the palm and thumb subdomains to encircle and close around the DNA template limiting dissociation. Increased processivity has also been seen after oligomerization of some archaeal DNA polymerases effectively encircling the template (Chap. 6). In both of these examples, the polymerases use a topological linkage of the protein to DNA to remain bound to the template for efficient and successive incorporations. More commonly, interactions of DNA polymerases with toroidal accessory factors (clamp proteins) achieve the same result of increased processivity by coupling the DNA polymerase with the template, limiting dissociation (Trakselis and Benkovic 2001; Bloom 2009). These clamp proteins (PCNA, in particular) have specific interaction domains that bind consensus sequences in DNA polymerases and other genomic maintenance proteins that act to recruit and retain enzymes at the replication fork (Moldovan et al. 2007).

1.2.2 RNA Polymerase Structural Mechanism

For the nucleotidyl transfer reaction by RNA polymerase, a two-metal-ion catalytic mechanism has been proposed, which is common in the DNA polymerase, as the enzyme possesses two divalent catalytic and nucleotide-binding metal cations (Mg^{2+}) chelated by two or three Asp residues at the enzyme active site (Fig. 1.3). Both metal ions are proposed to have octahedral coordination at physiological Mg^{2+} concentrations.

For transcribing RNA using DNA template, DNA-dependent RNA polymerase including cellular RNA polymerases (Chaps. 12 and 13), bacteriophage RNA polymerase (Chap. 10), and mitochondrial RNA polymerase (Chap. 11) unwinds a small region of double-stranded DNA to the single-stranded form and synthesizes RNA as a complementary sequence of the template. The unwound DNA region is called the transcription bubble that contains a DNA-RNA hybrid of ~8 base pairs. For synthesis of RNA, nucleotide substrate and catalytically essential divalent metals in addition to the single-stranded template DNA must be accommodated at the active site. One of four ribonucleotide triphosphates—ATP, GTP, CTP, and UTP—forms a Watson–Crick base pair with a DNA template base, and its α-phosphate group is attached by a 3′-hydroxyl of the growing end of the RNA. As a result, a linear RNA polymer is built in the 5′ to 3′ direction.

The overall shape of cellular RNA polymerases including bacterial, archaeal, and all types of eukaryotic enzymes is crab claw-like with a wide internal channel for double-stranded DNA binding (Fig. 1.2, Chaps. 12 and 13). The enzyme active site is located on the back wall of the channel, where an essential Mg^{2+} ion is chelated by three Asp of the absolutely conserved NADFDGD motif in the largest subunit. Compared to the bacterial RNA polymerase, archaeal and all eukaryotic RNA polymerases possess a characteristic protruding stalk that is formed by a heterodimer, and their relative positioning of the main body and stalk is also highly conserved.

The structure of bacteriophage-type RNA polymerases including mitochondrial enzyme resembles cupped right hand with palm, fingers, and thumb subdomains and a cleft that can accommodate double-stranded DNA (Fig. 1.2). Not only the overall structure of polymerases but also the secondary structures of subdomains in the bacteriophage-type RNA polymerase are highly conserved in the A-family DNA polymerase. The enzyme active site is located on the palm subdomain, where an essential Mg^{2+} ion is chelated by two Asp of the conserved motifs A and C. The conserved motif B is in the mobile finger subdomain, which changes its position during the nucleotide addition cycle and plays an important role in the nucleotide selection (Chaps. 10 and 11).

The overall shape of RdRps is similar to other nucleic acid polymerases, having "cupped right hand" structure and fingers, thumb, and palm subdomains (Fig. 1.2, Chap. 14). Because of its extension of the fingers, RdRp has more fully enclosed the active site, which may enhance their protein stability and enzyme processivity for their genomic RNA replication function.

1.3 Implications in Disease/Therapy

Although accurate DNA synthesis is a hallmark of high-fidelity DNA replication polymerases, a number of other polymerases have been implicated in various diseases and aging underlying their importance for further study. The best known example involves telomerase. When normal somatic cells replicate in the absence of telomerase, they undergo successive shortening of their telomeric ends, termed the end replication problem (Allsopp and Harley 1995). The shortening of telomeres acts as a clock determining the life of a cell ultimately causing senescence and cell death. However, in cancer cells, telomerase is upregulated preventing telomeric shortening and increasing cellular survival giving rise to immortal cells found in tumors. Although telomerase deficiency is most notable in the genetic disorder, *dyskeratosis congenita*, mutations in telomerase are also associated with anemia, other bone marrow-related diseases, and lung fibrosis. The unifying diagnostic indicator in all cases is short telomeres (Armanios 2009). The mechanism of RNA-mediated DNA telomeric synthesis by telomerase will be discussed in great detail in Chap. 9.

Translesion DNA polymerases are specialized low-fidelity DNA polymerases that can insert bases opposite a lesion, bypassing the damage, while potentially inducing point mutations. It is hypothesized that potential mutagenesis is favored over complete replication arrest and fork collapse. Translesion synthesis (TLS) is regarded as being responsible for the large increase in point mutations found in cancer genomes (Bielas et al. 2006). These Y-family DNA polymerases generally have much less fidelity and more open active sites accommodating a variety of DNA template lesions including oxidations, deaminations, abasic sites, methylations, and a host of environmental mutagens and are described in detail in Chap. 4.

Mutations in the Y-family Pol η account for the inheritable genetic disease, *xeroderma pigmentosum* (XPD) (Masutani et al. 1999). This disease sensitizes cells to UV light, significantly increasing the risk of skin carcinomas. Pol η is known to bypass thymine cyclobutane pyrimidine dimers (CPD) caused by UV cross-linking of adjacent residues (Johnson et al. 2000). Genetic mutations in Pol η associated with XPD disrupt the contacts with the DNA limiting its activity. Therefore, this translesion DNA polymerase has evolved a specific role in replication over UV-induced damage, and mutations in Pol η are responsible for replication fork collapse, double-strand breaks, and chromosomal breaks. The only other DNA polymerase found to be associated with an inheritable genetic disease is Pol ε where mutations in the large subunit give rise to splicing changes that cause decreased expression and predicted truncated protein products in FILS (facial dysmorphism, immunodeficiency, livedo, and short statures) syndrome patients (Pachlopnik Schmid et al. 2012).

The X-family base excision repair DNA polymerase (Pol β) (Chap. 3) has been found to have sporadic mutations in human tumors (Starcevic et al. 2004). Increased expression of Pol β has also been measured in a number of cancers interfering with normal DNA replication and causing mutations (Albertella et al. 2005; Tan et al. 2005). Several small-molecule inhibitors have been found to increase sensitivity to chemotherapeutic agents by blocking action of Pol β and seem to be a viable avenue for cancer therapy (Goellner et al. 2012). Other X-family DNA polymerases including terminal deoxytransferase (TdT) (Chap. 5) have been implicated in leukemia and carcinomas through altered expression levels. New nucleoside analogs have been shown to be specific towards TdT controlling expression levels and sensitizing cancer cells to conventional treatment.

Therefore, it seems there is an opportunity for targeted X- and Y-family DNA polymerase inhibition by either controlling expression levels or using as adjuvants with DNA-damaging radiation or chemotherapy (Lange et al. 2011). The challenges will be to avoid toxicity issues common with previous inhibitors, selectively target cancer cells, and act specifically on one of the 16 human DNA polymerases. Not an easy task, but with preliminary successes for Pol β and Tdt, the opportunity also exists for other DNA polymerases. The more we can emphasize structural differences in the active sites, identify allosteric regions, or detect novel mechanistic features, the better a position we are in to develop novel therapeutic agents. Success will require understanding the balance of DNA polymerase actions in a variety of cell types and developing screening methods to simultaneously measure effects on multiple DNA polymerases.

RNA polymerase is an essential enzyme in bacteria and virus and, as such, is a proven target for antibiotics and antiviral drugs (Chaps. 11 and 14). Fidaxomicin (FDX) is an inhibitor of bacterial RNA polymerase and is one of the newest antibiotics approved by the US Food and Drug Administration (FDA) for treatment of *Clostridium difficile*-associated diarrhea. The best characterized antibiotic against bacterial RNA polymerase is Rifampicin (Rif), which has been used as the first-line drug for infectious bacteria treatment, including tuberculosis, over four decades. However, a high incidence of Rif-resistant bacterial strains with RNA polymerase mutations is one of our public health challenges. Although many Rif-resistant *Mycobacterium tuberculosis* with RNA polymerase mutants have been derived in laboratory, only three residues with specific amino acid substitution account for ~85 % of *M. tuberculosis* Rif-resistant strains found in clinical isolates.

Therefore, structures of these three Rif-resistant RNAPs can move one step forward the structure-based discovery of improved Rif for tuberculosis treatment.

1.4 Remaining Questions and Future Directions

DNA and RNA polymerases have evolved naturally over millions of years to be highly accurate enzymes for incorporating the four available deoxyribonucleotides for DNA synthesis and ribonucleotides for RNA synthesis to faithfully maintain genomes and to express genes. Recent research efforts have focused on evolving these high-fidelity enzymes to have altered enzymatic properties or nucleotide specificities required for a variety of biotechnology applications (Chap. 7). Some of the goals are to amplify ancient genomes, incorporate alternative genetic alphabets, and replicate chemically and environmentally modified templates. In addition to traditional biotechnology development, these engineered polymerases have the potential to revolutionize synthetic biology by creating safe artificial living systems that incorporate unnatural DNA analogs for the creation of anything from drugs/metabolites to energy.

Although the basic mechanisms of incorporation, proofreading, and fidelity are well characterized for a number of DNA polymerases, there are many remaining questions on how polymerases function within the context of the replisome during normal DNA replication or repair. For example, do the kinetics or fidelities change drastically when accessory proteins are interacting with the polymerase? How does the unwinding rate of the DNA helicase control DNA polymerase kinetics? How is high-fidelity synthesis coordinated with error-prone lesion bypass when multiple polymerases are available? Answers to these questions will require the ability to both assemble and test more components of the replisome simultaneously in vitro and probe the kinetics within the context of an actively replicating cell.

The expression levels of DNA polymerases in various cancer cell types and stem cells are also an exciting avenue for study. Stem cells in particular need to maintain highly stable genomes. In these cells, the distribution of polymerases should favor high-fidelity enzymes and may even include suppressors against X- and Y-family polymerases. On the other hand, cancer cells are active mutators, and it would not be surprising to find inactivating mutations, loss, or rearrangements of DNA polymerases as more individual cancer cell sequencing results are available. In addition, DNA polymerases may be inactivated through alterations in DNA methylation patterns or RNAi changes. Pol θ in particular has been shown to have a significant difference in expression levels in breast tumors over non-tumor cells (Lemee et al. 2010). Therefore, it will also be important to assess any expression level deviations for DNA polymerases in individual cells to understand equilibrium changes that may be occurring at the replication fork and their resulting consequences on genomic stability. If DNA polymerase distributions can be determined first, it is conceivable that targeted DNA polymerase therapies will better sensitize cells to radiation or chemotherapy.

References

Albertella MR, Lau A, O'Connor MJ (2005) The overexpression of specialized DNA polymerases in cancer. DNA Repair (Amst) 4(5):583–593

Allsopp RC, Harley CB (1995) Evidence for a critical telomere length in senescent human fibroblasts. Exp Cell Res 219(1):130–136

Armanios M (2009) Syndromes of telomere shortening. Annu Rev Genomics Hum Genet 10:45–61

Beese LS, Steitz TA (1991) Structural basis for the 3′-5′ exonuclease activity of *Escherichia coli* DNA polymerase I: a two metal ion mechanism. EMBO J 10(1):25–33

Beese LS, Derbyshire V, Steitz TA (1993) Structure of DNA polymerase I Klenow fragment bound to duplex DNA. Science 260(5106):352–355

Bessman MJ, Lehman IR, Simms ES, Kornberg A (1958) Enzymatic synthesis of deoxyribonucleic acid. II. General properties of the reaction. J Biol Chem 233(1):171–177

Bielas JH, Loeb KR, Rubin BP, True LD, Loeb LA (2006) Human cancers express a mutator phenotype. Proc Natl Acad Sci USA 103(48):18238–18242

Blanco L, Bernad A, Lazaro JM, Martin G, Garmendia C, Salas M (1989) Highly efficient DNA synthesis by the phage phi 29 DNA polymerase. Symmetrical mode of DNA replication. J Biol Chem 264(15):8935–8940

Bloom LB (2009) Loading clamps for DNA replication and repair. DNA Repair (Amst) 8(5):570–578

Braithwaite DK, Ito J (1993) Compilation, alignment, and phylogenetic relationships of DNA polymerases. Nucleic Acids Res 21(4):787–802

Burgers PM, Koonin EV, Bruford E, Blanco L, Burtis KC, Christman MF, Copeland WC, Friedberg EC, Hanaoka F, Hinkle DC, Lawrence CW, Nakanishi M, Ohmori H, Prakash L, Prakash S, Reynaud CA, Sugino A, Todo T, Wang Z, Weill JC, Woodgate R (2001) Eukaryotic DNA polymerases: proposal for a revised nomenclature. J Biol Chem 276(47):43487–43490

Dahlberg ME, Benkovic SJ (1991) Kinetic mechanism of DNA polymerase I (Klenow fragment): identification of a second conformational change and evaluation of the internal equilibrium constant. Biochemistry 30(20):4835–4843

Delucia AM, Chaudhuri S, Potapova O, Grindley ND, Joyce CM (2006) The properties of steric gate mutants reveal different constraints within the active sites of Y-family and A-family DNA polymerases. J Biol Chem 281(37):27286–27291

Goellner EM, Svilar D, Almeida KH, Sobol RW (2012) Targeting DNA polymerase as for therapeutic intervention. Curr Mol Pharmacol 5(1):68–87

Hirata A, Murakami KS (2009) Archaeal RNA polymerase. Curr Opin Struct Biol 19(6):724–731

Huang RC, Maheshwari N, Bonner J (1960) Enzymatic synthesis of RNA. Biochem Biophys Res Com 3(6):689–694

Hurwitz J, Bresler A, Diringer R (1960) The enzymic incorporation of ribonucleotides into polyribonucleotides and the effect of DNA. Biochem Biophys Res Com 3(1):15–19

Ishino Y, Komori K, Cann IKO, Koga Y (1998) A novel DNA polymerase family found in Archaea. J Bacteriol 180(8):2232–2236

Jacob F, Monod J (1961) Genetic regulatory mechanisms in the synthesis of proteins. J Mol Biol 3:318–356

Johnson KA (2010) The kinetic and chemical mechanism of high-fidelity DNA polymerases. Biochim Biophys Acta 1804(5):1041–1048

Johnson RE, Washington MT, Prakash S, Prakash L (2000) Fidelity of human DNA polymerase eta. J Biol Chem 275(11):7447–7450

Joyce CM, Potapova O, Delucia AM, Huang X, Basu VP, Grindley ND (2008) Fingers-closing and other rapid conformational changes in DNA polymerase I (Klenow fragment) and their role in nucleotide selectivity. Biochemistry 47(23):6103–6116

Jun SH, Reichlen MJ, Tajiri M, Murakami KS (2011) Archaeal RNA polymerase and transcription regulation. Crit Rev Biochem Mol Biol 46(1):27–40

Kamtekar S, Berman AJ, Wang J, Lazaro JM, de Vega M, Blanco L, Salas M, Steitz TA (2006) The phi29 DNA polymerase: protein-primer structure suggests a model for the initiation to elongation transition. EMBO J 25(6):1335–1343

Kornberg A (1957) Enzymatic synthesis of deoxyribonucleic acid. Harvey Lect 53:83–112

Kuchta RD, Benkovic P, Benkovic SJ (1988) Kinetic mechanism whereby DNA polymerase I (Klenow) replicates DNA with high fidelity. Biochemistry 27(18):6716–6725

Kunkel TA (2004) DNA replication fidelity. J Biol Chem 279(17):16895–16898

Lange SS, Takata K, Wood RD (2011) DNA polymerases and cancer. Nat Rev Cancer 11(2):96–110

Lehman IR, Bessman MJ, Simms ES, Kornberg A (1958) Enzymatic synthesis of deoxyribonucleic acid. I. Preparation of substrates and partial purification of an enzyme from Escherichia coli. J Biol Chem 233(1):163–170

Lemee F, Bergoglio V, Fernandez-Vidal A, Machado-Silva A, Pillaire MJ, Bieth A, Gentil C, Baker L, Martin AL, Leduc C, Lam E, Magdeleine E, Filleron T, Oumouhou N, Kaina B, Seki M, Grimal F, Lacroix-Triki M, Thompson A, Roche H, Bourdon JC, Wood RD, Hoffmann JS, Cazaux C (2010) DNA polymerase theta up-regulation is associated with poor survival in breast cancer, perturbs DNA replication, and promotes genetic instability. Proc Natl Acad Sci USA 107(30):13390–13395

Masutani C, Araki M, Yamada A, Kusumoto R, Nogimori T, Maekawa T, Iwai S, Hanaoka F (1999) *Xeroderma pigmentosum* variant (XP-V) correcting protein from HeLa cells has a thymine dimer bypass DNA polymerase activity. EMBO J 18(12):3491–3501

Moldovan GL, Pfander B, Jentsch S (2007) PCNA, the maestro of the replication fork. Cell 129 (4):665–679

Nakamura T, Zhao Y, Yamagata Y, Hua YJ, Yang W (2012) Watching DNA polymerase eta make a phosphodiester bond. Nature 487(7406):196–201

Ohmori H, Friedberg EC, Fuchs RP, Goodman MF, Hanaoka F, Hinkle D, Kunkel TA, Lawrence CW, Livneh Z, Nohmi T, Prakash L, Prakash S, Todo T, Walker GC, Wang Z, Woodgate R (2001) The Y-family of DNA polymerases. Mol Cell 8(1):7–8

Ollis DL, Brick P, Hamlin R, Xuong NG, Steitz TA (1985) Structure of large fragment of *Escherichia coli* DNA polymerase I complexed with dTMP. Nature 313(6005):762–766

Pachlopnik Schmid J, Lemoine R, Nehme N, Cormier-Daire V, Revy P, Debeurme F, Debre M, Nitschke P, Bole-Feysot C, Legeai-Mallet L, Lim A, de Villartay JP, Picard C, Durandy A, Fischer A, de Saint BG (2012) Polymerase epsilon1 mutation in a human syndrome with facial dysmorphism, immunodeficiency, livedo, and short stature ("FILS syndrome"). J Exp Med 209(13):2323–2330

Starcevic D, Dalal S, Sweasy JB (2004) Is there a link between DNA polymerase beta and cancer? Cell Cycle 3(8):998–1001

Steitz TA (1993) DNA-dependent and RNA-dependent DNA-polymerases. Curr Opin Struct Biol 3(1):31–38

Steitz TA, Steitz JA (1993) A general two-metal-ion mechanism for catalytic RNA. Proc Natl Acad Sci USA 90(14):6498–6502

Stevens A (1960) Incorporation of the adenine ribonucleotide into RNA by cell fractions from, E. coli B. Biochem Biophys Res Com 3(1):92–96

Tan XH, Zhao M, Pan KF, Dong Y, Dong B, Feng GJ, Jia G, Lu YY (2005) Frequent mutation related with overexpression of DNA polymerase beta in primary tumors and precancerous lesions of human stomach. Cancer Lett 220(1):101–114

Trakselis MA, Benkovic SJ (2001) Intricacies in ATP-dependent clamp loading: variations across replication systems. Structure 9(11):999–1004

Watson JD, Crick FH (1953) Molecular structure of nucleic acids. Nature 171:737–738

Weiss SB, Gladstone L (1959) A mammalian system for the incorporation of cytidine triphosphate into ribonucleic acid. J Am Chem Soc 81(15):4118–4119

Weissbach A, Baltimore D, Bollum F, Gallo R, Korn D (1975) Nomenclature of eukaryotic DNA polymerases. Science 190(4212):401–402

Werner F, Grohmann D (2011) Evolution of multisubunit RNA polymerases in the three domains of life. Nat Rev Microbiol 9(2):85–98

Chapter 2
Eukaryotic Replicative DNA Polymerases

Erin Walsh and Kristin A. Eckert

Abstract DNA replication is a dynamic process that requires the precise coordination of numerous cellular proteins. At the core of replication in eukaryotic cells are three DNA polymerases, Pol α, Pol δ, and Pol ε, which function cooperatively to ensure efficient and high-fidelity genome replication. These enzymes are members of the B family of DNA polymerases, characterized by conserved amino acid motifs within the polymerase active sites. Pol α is a DNA polymerase of moderate fidelity that lacks 3′→5′ exonuclease activity, while Pols δ and ε are processive, high-fidelity polymerases with functional 3′→5′ exonuclease activities. Each polymerase exists as a holoenzyme complex of a large polymerase catalytic subunit and several smaller subunits. The Pol α holoenzyme possesses primase activity, which is required for de novo synthesis of RNA–DNA primers at replication origins and at each new Okazaki fragment. In one model of eukaryotic DNA replication, Pol ε functions in leading strand DNA synthesis, while Pol δ functions primarily in lagging strand synthesis. This chapter discusses the biochemical properties of eukaryotic replicative polymerases and how biochemical properties shape their functional roles in replication initiation, replication fork elongation, and the checkpoint responses.

Keywords DNA replication fork • S phase checkpoint • DNA polymerase fidelity • primase • proofreading exonuclease • replisome • genome stability

---

E. Walsh • K.A. Eckert (✉)
Department of Pathology, Jake Gittlen Cancer Research Foundation, Pennsylvania State University College of Medicine, Hershey, PA 17033, USA
e-mail: kae4@psu.edu

K.S. Murakami and M.A. Trakselis (eds.), *Nucleic Acid Polymerases*, Nucleic Acids and Molecular Biology 30, DOI 10.1007/978-3-642-39796-7_2,

2.1 Biochemical Properties of Replicative Polymerases

2.1.1 Polymerase α-Primase

2.1.1.1 Overview

A formal nomenclature for eukaryotic DNA polymerases has been adopted, in which cellular DNA polymerases are given Greek letters in order of their historical discovery (Weissbach et al. 1975; Burgers et al. 1990). Accordingly, Pol α-primase was the first mammalian polymerase to be purified and characterized (Yoneda and Bollum 1965). Pol α-primase holoenzyme is a heterotetrameric protein complex in all eukaryotes studied (Table 2.1) (see Muzi-Falconi et al. 2003 for review). The p180 large subunit contains the DNA polymerase active site (Plevani et al. 1985; Wong et al. 1986). The B subunit has no known enzymatic activity, but performs a regulatory function, possibly linking the pol α holoenzyme to components of the replication fork (Collins et al. 1993). DNA polymerases lack the ability to carry out de novo DNA synthesis and, in eukaryotes, require a 3′-OH provided by an RNA primer in order to initiate DNA synthesis. Primase activity is contained within the tightly associated p49/p58 complex (Plevani et al. 1985; Nasheuer and Grosse 1988). The architecture and subunit arrangement of the *Saccharomyces cerevisiae* (*S. c.*) Pol α-primase holoenzyme have been examined by X-ray crystallography and electron microscopy. The B subunit is tethered to the p180 subunit through a structured, flexible linker (Klinge et al. 2009). The four-subunit holoenzyme exists as a dumbbell-shaped particle, with the catalytic primase and polymerase active sites present in distinct lobes of the complex, separated by ~100 Å (Nunez-Ramirez et al. 2011).

2.1.1.2 An Essential Polymerase and Primase

Pol α polymerase activity is essential for chromosomal replication. Genetic analyses of mutants in *S. c.* have demonstrated that both the catalytic and regulatory B subunits are required in vivo for viability, and mutants in either gene exhibit defects in DNA replication and progression through S-phase (Johnson et al. 1985; Foiani et al. 1994; Budd and Campbell 1987). In mammalian cells, Pol α-primase neutralizing antibodies inhibit DNA synthesis (Miller et al. 1985; Kaczmarek et al. 1986). Thermosensitive yeast *POL1* mutants display elevated genetic instability due to defects in replication (Gutiérrez and Wang 2003; Liu et al. 1999). The primase activity is also essential in eukaryotic cells. Disruption of either the *PRI1* or *PRI2* genes is lethal in *S. c.* due to replication defects (Foiani et al. 1989; Lucchini et al. 1987). Characterization of conditional *PRI1* and *PRI2* mutants demonstrates that loss of primase activity causes increased mitotic recombination and spontaneous mutation rates, possibly due to defects in replication and impaired meiosis (Longhese et al. 1993).

Table 2.1 Composition of the major replicative DNA polymerases in eukaryotes

		Designation (gene/protein)		
Subunit	Function	Human	*S. cerevisiae*	*S. pombe*
Pol α holoenzyme				
A	Polymerase	*POLA1*/p180	*POL1*	*pol1*
B	Regulatory	*POLA2*/p70	*POL12*	*pol12*
	Primase regulatory	*PRIM2*/p58	*PRI2*	*pri2*
	Primase catalytic	*PRIM1*/p49	*PRI1*	*pri1*
Pol δ holoenzyme				
A	Polymerase, 3′→5′ exonuclease	*POLD1*/p125	*POL3*	*pol3*
B	Regulatory	*POLD2*/p50	*POL31*	*cdc1*
C	Regulatory	*POLD3*/p68[a]	*POL32*	*cdc27*
D	Regulatory (DNA damage)	*POLD4*/p12	–	*cdm1*
Pol ε holoenzyme				
A	Polymerase, 3′→ 5′ exo	*POLE1*/p261	*POL2*	*cdc20*
B	Regulatory	*POLE2*/p59	*DPB2*	*dpb2*
C	Double-stranded DNA binding	*POLE3*/p17	*DPB3*	*dpb3*
D	Double-stranded DNA binding	*POLE4*/p12	*DPB4*	*dpb4*

[a]Also referred to as the p66 subunit

2.1.1.3 Coordinated Primase and Polymerase Activities

In eukaryotic cells, in vivo studies have demonstrated that RNA primers of Okazaki fragments are attached to short DNA chains (Kitani et al. 1984). The Pol α holoenzyme is a unique replicative protein complex, possessing two coupled catalytic activities (Hu et al. 1984). Primase initiates synthesis of RNA at polypyrimidine tracts within the DNA template, preferentially within a T-rich region, and this occurs at many sites along the DNA template (Yamaguchi et al. 1985; Kitani et al. 1984). The primase catalytic site maps to the p49 subunit (Nasheuer and Grosse 1988), in which a carboxylic triad is essential for its function in generating RNA primers (Copeland and Tan 1995). The *S. c.* Pol α/p49 subunit is sufficient for synthesis of RNA, but highly inefficient without the p58 subunit (Santocanale et al. 1993). Biochemically, p58 is necessary for synthesis of the first dinucleotide of the RNA primer and modulates the rate and stability of subsequent extension steps (Copeland and Wang 1993). GTP is preferentially incorporated into the first dinucleotide, possibly to stabilize the short primer (Sheaff and Kuchta 1993). Although RNA primers generated by purified primase can vary in length from ~2 to 10 nucleotides in the presence of dNTPs (Nasheuer and Grosse 1988), only those ≥ 7 nucleotides are extended by Pol α, and formation of an RNA primer of this length terminates primase activity (Kuchta et al. 1990). The p58 subunit regulates primase processivity and ensures the correct RNA primer length is synthesized (Zerbe and Kuchta 2002). The DNA polymerase activity of Pol α extends RNA primers with ~20–30 dNTPs to generate an RNA–DNA hybrid molecule (Thompson et al. 1995).

Mechanistically, the primase and polymerase activities of the Pol α holoenzyme are tightly coupled to ensure that new primers are not synthesized until the previous primer has been extended by polymerase (Sheaff et al. 1994). Functional coordination of the two activities is dependent upon the presence of dNTPs (Hu et al. 1984). An intramolecular transfer of the RNA primer from the primase active site to the polymerase active site occurs rapidly when dNTPs are present (Sheaff et al. 1994). Structurally, flexible tethering of Pol α holoenzyme lobes containing the primase and polymerase centers increases the efficiency of primer transfer between the two active sites (Nunez-Ramirez et al. 2011). Structural and biochemical studies suggest that the p58 subunit is involved in the switch from primase to polymerase through conformational changes (Arezi et al. 1999; Agarkar et al. 2011). Consistent with its role in synthesizing short RNA–DNA primers, the Pol α holoenzyme lacks the high processivity possessed by the other major replicative holoenzymes (see below), although it does possess the capacity to carry out robust DNA synthesis in vitro. The rate of DNA synthesis by calf thymus Pol α-primase is similar to that of Pols ε and δ (Weiser et al. 1991), and Pol α-primase displays an in vitro processivity of ~20–100 nucleotides (Hohn and Grosse 1987). Pol α-primase interacts with the p70 subunit of replication protein A (RPA), which stimulates both the polymerase activity and processivity of the holoenzyme (Braun et al. 1997), possibly by increasing the affinity for primer termini (Maga et al. 2001).

2.1.1.4 A Moderately Accurate Polymerase

The in vitro fidelity of the mammalian Pol α holoenzyme purified from multiple sources has been determined using several genetic assays (Kunkel and Loeb 1981; Kunkel 1985; Eckert et al. 1997). The majority of errors created by Pol α are single-base substitutions, followed by one-base deletion errors (Kunkel 1985; Eckert et al. 1997). The purified *S. c.* Pol α/p180 catalytic subunit and the *S. c.* Pol α holoenzyme have similar overall fidelities but display statistically significant differences in error rates within specific sequence contexts (Kunkel et al. 1989). The *S. c.* Pol α/p180 subunit exhibits an error rate of 1/9,900 and 1/12,000 per nucleotide synthesized, for single-base substitutions and deletions, respectively (Kunkel et al. 1991). The human Pol α/p180 subunit base substitution error rate can be as low as 1/42,000 nucleotides in low-pH buffer conditions (Eckert and Kunkel 1993). Direct comparison of purified mammalian Pol α-primase, Pol δ, and Pol ε preparations demonstrated that Pol α possesses the lowest fidelity of the three major eukaryotic replicative polymerases (Thomas et al. 1991).

One mechanism by which Pol α maintains its moderate fidelity is the low rate of extending mismatched DNA primers (Perrino and Loeb 1989). Some calf thymus Pol α holoenzyme pause sites correspond to sites of increased mismatched 3′-primer termini (Fry and Loeb 1992), consistent with a low rate of mispair extension. Pol α holoenzyme pausing within microsatellite sequences also is correlated with the rate of misalignment-based errors (Hile and Eckert 2004).

Replication auxiliary factors enhance the DNA fidelity of Pol α-primase. RPA decreases dNTP misincorporation rate by decreasing the affinity of the Pol α holoenzyme to mismatches while increasing the affinity to matched DNA primer termini, suggesting that it may function as a "fidelity clamp" that modulates the Pol α holoenzyme for faithful DNA synthesis (Maga et al. 2001). Terminal misincorporation events at Pol α pause sites are also avoided in the presence of RPA (Suzuki et al. 1994). Pol α-primase lacks 3′→5′ exonuclease activity due to amino acid changes at catalytic residues within the domain, which limits its intrinsic fidelity (Pavlov et al. 2006b). However, *S. c.* Pol δ can perform intermolecular proofreading of errors made by Pol α-primase (Pavlov et al. 2006a). Based on an estimation that ~4–8 × 10^4 RNA–DNA primers are synthesized by Pol α-primase during replication in human cells (Muzi-Falconi et al. 2003), such mechanisms of enhanced fidelity may be essential in maintaining genome stability.

In contrast to its DNA polymerase fidelity, the priming activity exhibits very low fidelity during RNA synthesis. Primase readily misincorporates NTPs during in vitro RNA synthesis, often polymerizing primers containing consecutive mismatches (Sheaff and Kuchta 1994). Importantly, after intramolecular transfer, such mismatched RNA primers are readily extended by the DNA polymerase activity in the presence of dNTP substrates.

2.1.1.5 Posttranslational Regulation

The Pol α holoenzyme exists as an assembled complex throughout the cell cycle (Ferrari et al. 1996), and its activity is regulated by cyclin-dependent kinases in a cell cycle-dependent manner (Voitenleitner et al. 1999). The p180 subunit is a phosphoprotein that becomes hyperphosphorylated in G_2/M phase, while the B subunit is phosphorylated only in G_2/M (Nasheuer et al. 1991). Pol α phosphorylation results in lowered single-stranded DNA binding affinity, lowered DNA synthesis activity, and an inhibition of DNA replication (Nasheuer et al. 1991; Voitenleitner et al. 1999).

2.1.2 DNA Polymerase δ

2.1.2.1 Overview

Pol δ was originally identified as a novel DNA polymerase purified from rabbit bone marrow that possessed a very active 3′→5′ exonuclease activity (Byrnes et al. 1976). Subsequently, *S. c.* POL3 was identified as the yeast homolog of Pol δ (Boulet et al. 1989). In all eukaryotes, Pol δ is a multimeric complex, wherein the largest subunit harbors the DNA polymerase and 3′→5′ exonuclease catalytic domains, and is tightly associated with a regulatory B subunit (Table 2.1). Mammalian Pol δ holoenzyme is a heterotetrameric protein consisting of two additional

C and D subunits, p68 and p12. The fission yeast *Schizosaccharomyces pombe* (*S. p.*) also encodes a heterotetrameric Pol δ, possessing the Cdm1 subunit which shows some sequence similarity to the mammalian p12 subunit. The budding yeast *S. c.* Pol δ is a heterotrimer of the A and B subunits together with a C subunit ortholog, Pol32.

The structure of a truncated form of the *S. c.* Pol δ/p125 subunit in ternary complex with DNA and dNTP substrates has been solved at 2 Å resolution (Swan et al. 2009). The polymerase domain has the characteristic right-hand structure of palm, fingers, and thumb subdomains. The 3′→5′ exonuclease domain is separated from the polymerase domain by 45 Å. Pol δ also possesses a novel N-terminal domain that interacts with 10–20 nucleotides of the ssDNA template, upstream of the polymerase active site. The structure of a truncated *S. c.* Pol δ holoenzyme as determined by small-angle X-ray scattering analysis consists of a globular catalytic core (A subunit) flexibly linked to an elongated tail comprised of the B and C subunits (Jain et al. 2009). The interaction of Pol3 (A subunit) with the Pol31 and Pol32 (B and C subunits) is stabilized by binding of a 4Fe-4S metal cofactor cluster to four conserved cysteine residues in the CysB motif of the Pol3 C-terminal domain (Netz et al. 2012).

2.1.2.2 An Essential Polymerase

Evidence that Pol δ is an essential component of the eukaryotic replication machinery has been derived from several models. Disruption of Pol δ is lethal in *S. p.*, and thermosensitive mutants of pol3 arrest in S-phase of the cell cycle (Francesconi et al. 1993). Deletion of the *S. p.* cdc27 gene (Pol δ/C subunit) results in cell cycle arrest (MacNeill et al. 1996). Although the orthologous *S. c POL32* gene is not essential, deletion mutants do display abnormal phenotypes (Gerik et al. 1998). *S. c. POL3* mutants harboring different substitutions at a catalytically essential residue, Leu^{612}, exhibit a wide range of phenotypic deficiencies (Venkatesan et al. 2006). Only 8 of 19 mutants are viable and display varying degrees of genotoxic sensitivity, cell cycle defects, and morphological abnormalities (Venkatesan et al. 2006). Replication studies in *Xenopus* egg extracts demonstrated that immunodepletion of Pol δ leads to a significant decrease in DNA synthesis and the accumulation of unreplicated, single-stranded DNA gaps (Fukui et al. 2004). Knockdown of Pol δ/p125 in HeLa cells causes an accumulation of cells in early, middle, and late S-phase, and cells do not enter mitosis (Bermudez et al. 2011). Homozygous disruption of *Pold1* in mice was shown to cause embryonic lethality between E4.5 and E7.5, due to significant defects in DNA synthesis (Uchimura et al. 2009). Profoundly, homozygous mutations of highly conserved amino acids within motif A of the Pol δ active site (L604G and L604K) are embryonic lethal in mice, and mice heterozygote for the L604K mutant exhibit reduced life spans and accelerated tumorigenesis (Venkatesan et al. 2007). Mouse embryonic fibroblasts heterozygous for these mutant proteins display elevated mutation rates and chromosomal instability relative to wild-type cells (Venkatesan et al. 2007). The severe replication defects associated with loss of Pol δ in each of these models were vital in uncovering the polymerase's role as a major component of the eukaryotic replication fork.

2.1.2.3 Efficient DNA Synthesis in the Presence of Replication Accessory Factors

Pol δ requires replication accessory proteins to achieve its greatest efficiency. Shortly after the discovery of Pol δ, an auxiliary protein was identified, which co-eluted with Pol δ during purification and significantly enhanced its in vitro DNA synthesis on several DNA templates (Tan et al. 1986). This protein was later identified as proliferating cell nuclear antigen (PCNA) and was shown to greatly enhance calf thymus Pol δ processivity in vitro (Prelich et al. 1987; Bravo et al. 1987). PCNA is now known as the eukaryotic sliding clamp protein, a homotrimer ring-shaped protein that encircles DNA and tethers replication proteins, allowing movement along the DNA template (Moldovan et al. 2007). Replication Factor C (RFC), a heteropentameric complex, is responsible for loading PCNA onto DNA through an ATP-dependent mechanism. In the presence of PCNA, RFC, and RPA, *S. c.* Pol δ processivity is enhanced from 6 nucleotides to >600 nucleotides (Chilkova et al. 2007). Human Pol δ holoenzyme activity increases >50-fold and processivity is stimulated in the presence of PCNA (Xie et al. 2002). However, the human Pol δ holoenzyme differs biochemically from *S. c.* Pol δ holoenzyme, as human Pol δ dissociates more readily from DNA templates, even in the presence of accessory factors (Bermudez et al. 2011).

Recombinant human Pol δ can be purified in several subassemblies, all of which retain DNA synthesis activity (Zhou et al. 2012; Podust et al. 2002). The Pol δ p125/p50 heterodimer (also referred to as the core dimer) is a tightly associated complex with low specific activity. The three subunit complex, Pol δ3 (core + p68 subunit), displays high DNA synthesis activity but is unstable in vitro. The p68 subunit is essential for synthesis of long DNA products by Pol δ. The p12 subunit increases stability of the holoenzyme and increases DNA synthesis activity. All four subunits of human Pol δ individually interact with PCNA, which may allow for flexibility during DNA replication, as many proteins functionally interact with PCNA during Okazaki fragment maturation (Zhang et al. 1999; Wang et al. 2011; Li et al. 2006b). PCNA interacting motifs also were identified within the C-termini of all three subunits of *S. c.* Pol δ, which are needed for efficient DNA synthesis (Acharya et al. 2011). PCNA stimulation of DNA synthesis activity differs quantitatively among the various human Pol δ subassemblies, in the order Pol δ4 > Pol δ3 (core + p68) > Pol δ2 core (Zhou et al. 2012). Kinetically, PCNA reduces the K_m for DNA template binding and increases the V_{max} of the calf thymus Pol δ catalytic core, suggesting that PCNA might stimulate Pol δ processivity by increasing its residence time on the DNA template and the rate of nucleotide incorporation (Ng et al. 1991).

2.1.2.4 An Accurate Polymerase

Replication of the genome requires accurate DNA synthesis in order to avoid the accumulation of deleterious mutations. As demonstrated using in vitro mutation assays, purified Pol δ generally possesses a high DNA synthesis fidelity (see Prindle and Loeb 2012 for review). The *S. c.* Pol δ holoenzyme incorporates less than one-base substitution error out of 80,000 nucleotides synthesized (Fortune et al. 2005). For human Pol δ, less than one-base misinsertion error occurred per 220,000 nucleotides (Schmitt et al. 2009). However, *S. c.* and human Pol δ holoenzymes do not exhibit high fidelity during in vitro synthesis of repetitive microsatellite DNA sequences (Hile et al. 2012; Abdulovic et al. 2011). At the heart of Pol δ's high fidelity is its intrinsic $3' \rightarrow 5'$ exonuclease activity (Simon et al. 1991), which enables proofreading upon incorporation of incorrect nucleotides during DNA synthesis (Kunkel et al. 1987). Both human and *S. c.* exonuclease-deficient Pol δ forms exhibit approximately 10-fold higher base substitution error rates than wild-type Pol δ in vitro (Fortune et al. 2005; Schmitt et al. 2009). In vivo, yeast strains carrying inactivating mutations within the *POL*3 exonuclease domain exhibit a 100-fold increased spontaneous mutation rate (Morrison and Sugino 1994). Furthermore, mice with homozygous substitutions at highly conserved residues within the proofreading domain of Pol δ exhibit a high incidence of cancer and decreased survival (Goldsby et al. 2002; Albertson et al. 2009).

Intrinsic kinetic properties of Pol δ also are key determinants of its high fidelity. *S. c.* Pol δ strongly favors incorporation of correct versus incorrect nucleotides during synthesis (Dieckman et al. 2010). A highly conserved leucine residue within the Pol δ active site has been extensively studied in yeast, mice, and humans (Leu-612, 604 and 606, respectively) for its role in ensuring high-fidelity DNA synthesis. Amino acid substitutions at this site cause allele-specific phenotypic effects in *S. c.* and mice (Venkatesan et al. 2006, 2007). In vitro, the human Pol δ holoenzyme L606G mutant is highly error prone (Schmitt et al. 2010). Interestingly, the L606K mutant exhibits higher fidelity than wild-type Pol δ but decreased bypass of DNA adducts (Schmitt et al. 2010), suggesting that the high fidelity conferred by this active site residue is balanced by the need for the wild-type enzyme to perform other activities, including translesion synthesis at the replication fork.

Efficient proofreading requires partitioning of DNA substrates containing 3′ terminal mispairs from the polymerase to the exonuclease active sites (Khare and Eckert 2002). An *S. c. POL*3 active site mutation that impairs this partitioning results in decreased DNA synthesis fidelity and increased spontaneous mutation rate (McElhinny et al. 2007). Purified *S. c.* Pol δ interacts with base pairs distant from the templating base, which may allow the polymerase to "sense" base mismatches (Swan et al. 2009).

Although PCNA stimulates Pol δ activity, it may decrease its fidelity. In vitro, PCNA increases the rate of nucleotide misincorporation by *S. c.* Pol δ, resulting in a significant reduction in fidelity (Hashimoto et al. 2003). Similarly, nucleotide misincorporations by calf thymus Pol δ increase ~27-fold in the presence of PCNA, and PCNA may enable Pol δ to extend mismatched base pairs by stabilizing the Pol δ-template-primer complex (Mozzherin et al. 1996). However, addition of both PCNA and RPA had no effect on *S. c.* Pol δ base substitution error rates, and the addition of PCNA and RPA to Pol δ DNA synthesis reactions contributed to a decreased rate of large deletion errors within directly repeated sequences (Fortune et al. 2006). Further studies are needed to fully understand the effect of accessory factors on Pol δ fidelity. Interestingly, the Werner syndrome protein (WRN), a RecQ family helicase and 3′→5′ exonuclease, can enable high-fidelity DNA synthesis by Pol δ through excision of primer-template mismatches prior to polymerase extension (Kamath-Loeb et al. 2012).

2.1.2.5 Posttranslational Regulation

Phosphorylation may represent an important mechanism of Pol δ regulation (see Lee et al. 2012 for review). Mammalian Pol δ protein levels peak at the G1/S border, and the p125 subunit is most actively phosphorylated during S-phase (Zeng et al. 1994). The B subunit (p50) is also phosphorylated in vivo and is an in vitro substrate of the cyclin A-Cdk2 cell cycle-dependent kinase (Li et al. 2006a). The C subunit (p68) can be phosphorylated by cyclin-dependent kinases (CycE-Cdk2, CycACdk1, or CycA-Cdk2) in vitro, and PCNA interferes with this phosphorylation (Ducoux et al. 2001). In addition, mammalian Pol δ/p125, p68, and p12 subunits can be phosphorylated by casein kinase 2 in vitro and subsequently dephosphorylated by protein phosphatase-1 (Gao et al. 2008), suggesting an additional regulatory circuit for regulation. Phosphorylation of Pol δ/p68 coincides with Pol δ association with chromatin at the start of S-phase (Lemmens et al. 2008). The Pol δ/p68 subunit also contains a phosphorylation site for protein kinase A, and phosphomimetic mutation of this residue decreases Pol δ affinity for PCNA and processivity (Rahmeh et al. 2011). Thus, phosphorylation may serve to regulate Pol δ activity by controlling its interaction with DNA and/or auxiliary proteins during replication.

2.1.3 DNA Polymerase ε

2.1.3.1 Overview

Purified Pol ε was first characterized as a larger form of Pol δ that was highly processive, but not stimulated by PCNA in vitro (Syvaoja and Linn 1989). The Pol ε holoenzyme exists as a heterotetrameric protein in all eukaryotes studied (Table 2.1) (Pursell et al. 2008). The large catalytic subunit contains the polymerase and 3′→5′ exonuclease active sites within the N-terminus, and binding domains for smaller subunits and PCNA within the C-terminus. The C and D subunits form a subcomplex that is important for double-stranded DNA binding (Tsubota et al. 2003). The *S. c.* Pol ε holoenzyme structure has been solved at 20 Å resolution using cryo-electron microscopy and single-particle analyses (Asturias et al. 2006). The three small subunits (B–D) form a discrete extended tail structure, separated from the large catalytic subunit by a flexible hinge. Similar to the *S. c.* Pol δ, a 4Fe-4S cluster is bound to four conserved cysteine residues (the CysB motif) within the C-terminal domain of Pol2 and may be essential for stabilizing the *S. c.* Pol ε holoenzyme (Netz et al. 2012).

2.1.3.2 An Essential Component of the Eukaryotic Replication Machinery

The Pol ε holoenzyme is essential for chromosomal replication. In *S. c.*, *POL2* reading frame disruptions are nonviable due to arrest of DNA replication; however, truncation mutations that maintain catalytic activity are viable with a slow growth phenotype (Morrison et al. 1990). Deletion of the *POL2* N-terminus, which encodes the polymerase catalytic domain, also is viable, although mutants display severe replication defects, including slow fork movement, prolonged S-phase, and shortened telomeres (Kesti et al. 1999; Feng and D'Urso 2001; Dua et al. 1999). Thermosensitive yeast mutants also demonstrate that nonfunctional Pol ε leads to defective chromosomal replication and the accumulation of short DNA fragments (Araki et al. 1992; Budd and Campbell 1993). Immunodepletion of Pol ε from *Xenopus* egg extracts significantly impedes elongation of nascent DNA strands and causes the accumulation of short replication intermediates (Waga et al. 2001). In HeLa cells, transient knockdown of Pol ε causes an accumulation of cells in early S-phase and a decreased rate of replication fork movement (Bermudez et al. 2011). Disruption of Pol ε in mice results in embryonic lethality (Menezes and Sweasy 2012).

2.1.3.3 A Highly Efficient and Processive Polymerase

Pol ε and Pol δ differ in their biochemical interactions with PCNA. PCNA binds the Pol ε p261 (N-terminus), p59, and p12 subunits and increases the rate of nucleotide incorporation by the holoenzyme (Bermudez et al. 2011). In the presence of PCNA, RFC, and RPA, *S. c.* Pol ε processivity is stimulated ~6-fold, less than the ~100-fold stimulation observed for Pol δ (Chilkova et al. 2007). *S. c.* Pol ε has a high affinity for DNA and low affinity for PCNA, while *S. c.* Pol δ displays the opposing affinities for each, suggesting that Pol ε might load onto DNA independently of PCNA, while Pol δ requires preloading of PCNA (Chilkova et al. 2007). However, *S. c.* and human Pol ε are more processive than Pol δ, even in the absence of accessory factors (Dua et al. 2002) (Bermudez et al. 2011). High *S. c.* Pol ε processivity requires a minimal primer duplex stem length of 40 base pairs, and structural studies suggest that the tail domain formed by the B–D subunits contributes directly to Pol ε processivity by binding double-stranded DNA, precluding the need for PCNA to enhance DNA affinity (Asturias et al. 2006). The processivities of both the *S. c.* Pol ε polymerase and exonuclease activities are reduced in subassemblies lacking the C and D subunits, relative to the holoenzyme form (Aksenova et al. 2010).

2.1.3.4 A Highly Accurate Polymerase

Pol ε is perhaps the most accurate eukaryotic DNA polymerase. Purified calf thymus Pol ε base substitution error rates determined in vitro ranged from 1/30,000 to 1/400,000 mutations per nucleotide synthesized, lower than either Pol δ or Pol α (Thomas et al. 1991). The *S. c.* Pol ε holoenzyme exhibits very low base substitution and single-base deletion error rates, $\leq 2 \times 10^{-5}$ and $\leq 5 \times 10^{-7}$ (Shcherbakova et al. 2003), respectively. Error rates of human Pol ε have not yet been determined because they are close to background rates for in vitro mutation assays (Korona et al. 2011). Although the in vitro fidelity of purified *S. c.* Pol ε lacking the C and D subunits is the same as the Pol ε holoenzyme form, deletion of *DPB3* and *DPB4* results in a slightly increased spontaneous mutation rate in vivo (Aksenova et al. 2010).

The high fidelity of Pol ε is due, in part, to its intrinsic 3′→5′ exonuclease activity. Amino acid substitutions of a conserved motif within the exonuclease active site result in a ~20-fold increase in the *S. c.* spontaneous mutation rate (Morrison et al. 1991). Purified exonuclease-deficient *S. c.* Pol ε exhibits single-base substitution and deletion errors rates that are ~10- and 100-fold higher, respectively, than wild-type Pol ε (Shcherbakova et al. 2003). Kinetically, *S. c.* Pol ε mutants harboring a C1089Y substitution within the polymerase active site exhibit an increased rate of base misincorporation that may result from the inability to perform proper DNA shuffling between the polymerase and exonuclease domains (Shimizu et al. 2002). Human exonuclease-deficient Pol ε is ~5-fold more accurate than wild-type *S. c.* Pol ε for both base substitutions and single-base deletions (Korona et al. 2011). Homozygous loss of Pol ε exonuclease activity in mice results in an elevated spontaneous mutation rate and tumor incidence (Albertson et al. 2009). Specific polymerase active site residues also play a role in determining Pol ε fidelity. For instance, an M644F substitution within the *S. c.* Pol ε active site reduces fidelity due to an increased base misincorporation rate (Pursell et al. 2007a). The Pol ε M644 active site residue may modulate fidelity by maintaining proper geometry of the substrate binding pocket (Pursell et al. 2007a).

2.1.3.5 Posttranslational Modification

The posttranslational regulation of Pol ε has not been extensively studied. However, the *S. c.* B subunit, required for chromosomal replication (Araki et al. 1991), is phosphorylated during late G1 phase in a cell cycle-dependent manner (Kesti et al. 2004).

2.2 Functions at the Eukaryotic Replication Fork

2.2.1 Replication Initiation

The process of replication initiation takes place as two steps (see Masai et al. 2010 for review). The pre-RC complex (ORC; Cdc6; Cdt1; MCM10 and MCM2-6 helicase complex) is first loaded onto chromosomes at origins of replication. In the second step, pre-RCs are activated to generate replication fork through the kinase activities of Cdc7-Dbf4 and CDK, which allows loading of other proteins required for replication. Loading of the GINS complex, a heterotetrameric complex essential for replication (Takayama et al. 2003), leads to activation of the MCM helicase complex, which enables unwinding of duplex DNA at the replication fork. All three *S. c.* replicative polymerases (α, δ, ε) bind to origins in early S-phase (Hiraga et al. 2005). Pol ε is a component of the preloading complex (Pre-LC), which also contains GINS, Sld2, and Dpb11 (Muramatsu et al. 2010), suggesting that Pol ε may play an active role in replication initiation (Fig. 2.1). Indeed, loss of the Pol ε catalytic subunit in *S. p.* prevents loading of GINS and several other initiation proteins to origins (Handa et al. 2012). GINS function is required for recruitment of Pol ε to chromatin (Pai et al. 2009) and can stimulate Pol ε activity (Bermudez et al. 2011).

Pol α-primase interacts with several protein players in DNA replication initiation (Fig. 2.1). The B subunit of Pol α-primase interacts directly with ORC for its recruitment to initiation sites (Uchiyama and Wang 2004). MCM10 binds to the Pol α/p180 subunit and physically interacts with the Pol α holoenzyme in vitro and in vivo (Fien et al. 2004; Ricke and Bielinsky 2004; Warren et al. 2009). MCM10 maintains steady-state levels of the Pol α/p180 subunit (Ricke and Bielinsky 2004), loads Pol α holoenzyme onto chromatin (Zhu et al. 2007), and facilitates Pol α holoenzyme binding to primed DNA templates (Fien et al. 2004). Additionally, the GINS complex interacts with the primase subunits and stimulates DNA synthesis activity (De Falco et al. 2007).

2.2.2 Leading and Lagging Strand Replication

Upon activation of replication origins, Pol α-primase is responsible for generating RNA–DNA primers that initiate DNA synthesis on the leading and lagging strands of the replication fork. Temporal studies using *Xenopus* extracts demonstrated that Pol α-primase is loaded onto DNA after helicase-mediated unwinding has been

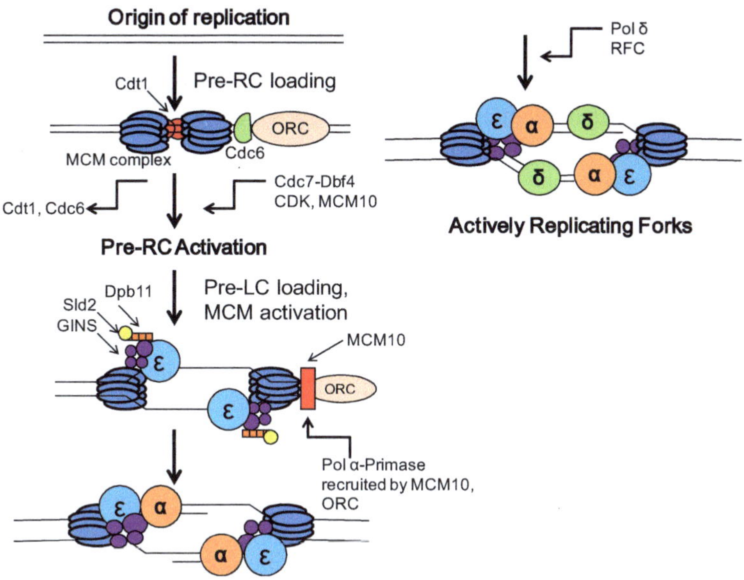

Fig. 2.1 Simplified schematic of replication initiation in eukaryotes. Initiation occurs at origins of replication, at which a pre-RC complex, consisting of the MCM complex, Cdc6, and Cdt1, is first loaded and then activated by the Cdc7-Dbf4 and CDK kinases. This leads to the recruitment of the pre-LC complex, consisting of GINS, Sld2, Dpb11, and Pol ε. Pol α-primase is recruited through interactions with MCM10 and also directly interacts with GINS. Mechanisms leading to Pol δ recruitment are currently unknown. Upon recruitment of all three replicative Pols, an active, bidirectional replication fork is generated, in which Pol ε is modeled as the leading strand Pol, and Pol δ is modeled as the lagging strand Pol. For simplicity, some replication-initiating factors and protein–protein interactions are not displayed. The reader is referred to Masai et al. 2010 and Araki 2010 for further details

initiated (Walter and Newport 2000). Switching from Pol α to Pol δ or Pol ε involves RFC. In vitro, RFC inhibits Pol α activity once a sufficient RNA–DNA primer is generated, by decreasing its affinity for the DNA template, thus coordinating a switch from Pol α to Pol δ (Mossi et al. 2000; Maga et al. 2000).

Several studies have suggested that Pols δ and ε have distinct responsibilities at the replication fork (Kunkel and Burgers 2008). In HeLa cells, Pol ε is more active in early S-phase while Pol δ activity increases in late S-phase (Rytkonen et al. 2006). Early biochemical evidence that Pol δ functions as the lagging strand Pol came from studies demonstrating its role in Okazaki fragment maturation, in which Pol δ functions together with FEN1, a 5′ flap endonuclease; DNA ligase 1; and PCNA (Burgers 2009). As replication takes place on the lagging strand, Pol δ encounters RNA primers within downstream Okazaki fragments every ~100–200 nucleotides, which must be removed in order to prevent deleterious duplication mutations (Burgers 2009). Yeast genetic analyses and in vitro studies demonstrated that Pol δ carries out strand displacement of one to two nucleotides within the RNA primer, followed by idling through its 3′→5′ exonuclease activity, which allows for FEN1 to remove the resulting ribonucleotide flap. This process occurs in several cycles until all ribonucleotides are removed and a DNA–DNA nick is generated, which is sealed by DNA ligase 1 (Garg et al. 2004; Jin et al. 2003; Burgers 2009). In contrast, Pol ε does not undergo idling at downstream primers or interact with FEN1 to generate ligatable nicks (Garg et al. 2004), evidence that is not consistent with a role in lagging strand replication.

Early studies of exonuclease-deficient *S. c.* Pol δ and Pol ε strains provided genetic evidence that the two polymerases act on opposite DNA strands during replication (Morrison and Sugino 1994; Shcherbakova and Pavlov 1996; Karthikeyan et al. 2000). More recently, yeast DNA polymerases δ and ε variants that have distinctive error signatures in vitro have been used to infer the strand preference for DNA synthesis (e.g., leading versus lagging) during in vivo DNA replication. Such genetic studies of Pol ε and Pol δ mutants are consistent with roles as the leading and lagging strand polymerase, respectively. The asymmetric *S. c.* Pol ε mutant (M644G) exhibits a high rate of T-dTMP mismatches relative to complementary A-dAMP mismatches in vitro and an elevated rate of T-dTMP mismatches in an origin orientation-dependent manner in vivo, consistent with its role in replication on the leading strand (Pursell et al. 2007b). Conversely, the *pol3-L612M* is an asymmetric Pol δ mutator that increases the rate of dT-dGMP mismatches but has a low rate for generating the complementary dA-dCMP mismatch in vitro (McElhinny et al. 2007), and genetic analysis of the *pol3-L612M* mutant showed that its signature errors occurred in vivo primarily on the lagging strand (Nick McElhinny et al. 2008). Deep DNA sequencing and analyses of L612M mutant genomes revealed the Pol error signature genome wide, suggesting that Pol δ is responsible for lagging strand replication across the *S. c.* genome (Larrea et al. 2010). Although such evidence supports the model in which Pol δ functions as the lagging strand Pol, alternative models have been proposed in which Pol δ may also contribute to replication on the leading strand (as reviewed in Kunkel and Burgers 2008). Importantly, it remains to be determined whether this model applies to replication in mammalian cells and to the fork after replication restart.

2.2.3 Replicative Polymerase Functions in Checkpoint Responses

Replicative polymerases are integral components of pathways monitoring progression of the mitotic cell cycle (Fig. 2.2). Pol α-primase is a central player in checkpoint signaling that ensures DNA replication is complete prior to progression into mitosis. Checkpoint activation of the Chk1 kinase in *Xenopus* extracts requires RNA synthesis by primase, but not DNA synthesis (Michael et al. 2000), and yeast mutants defective for primase activity are unable to activate the Chk1 and Rad53 kinases (Griffiths et al. 2001; Marini et al. 1997). The Cds1 protein kinase interacts directly with Pol α-primase, possibly to signal downstream cell cycle checkpoint

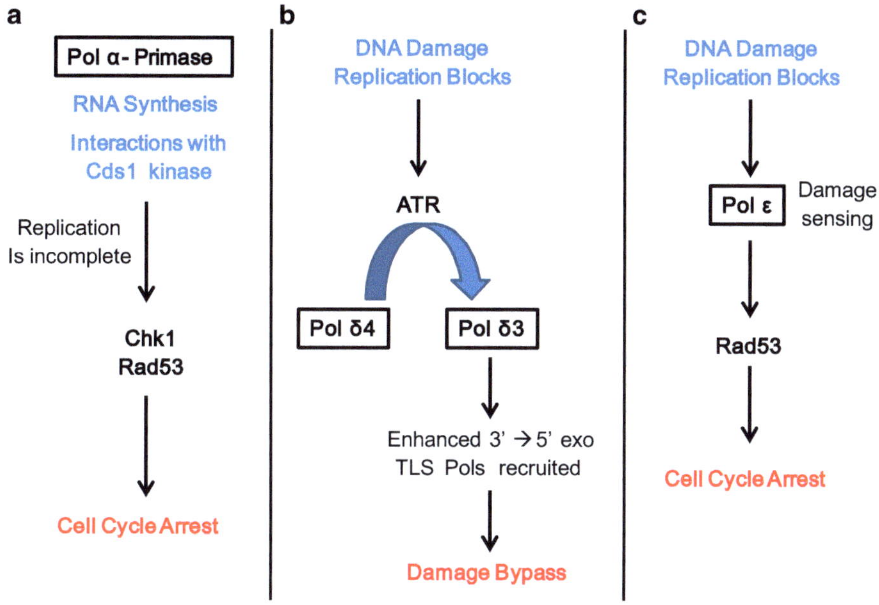

Fig. 2.2 Eukaryotic DNA polymerases in cellular checkpoint responses. (**a**) During S-phase of the cell cycle, Pol α-primase activity results in RNA synthesis, which serves as a signal that replication is incomplete and prevents cell cycle progression through activation of Chk1 and Rad53. Interactions between Pol α-primase and the Cds1 kinase also act as an upstream signal for Chk1 and Rad53 activation. (**b**) DNA damage and replication blocks act as a signal for ATR activation in cells, which leads to degradation of the p12 subunit of Pol δ4 and the formation of Pol δ3. The three-subunit form of Pol δ is proposed to play a role in damage bypass. (**c**) DNA damage and replication blocks act an upstream signal for Pol ε, which then triggers signaling for activation of Rad53 and cell cycle arrest under these conditions

responses (Murakami and Okayama 1995). Thus, the presence of RNA primers synthesized by primase during S-phase acts as a signal that replication is not yet complete in order to prevent cell cycle progression.

Over the past several years, a three subunit form of Pol δ, devoid of the p12 subunit, has been extensively studied for its role in regulating the DNA damage response in human cells (reviewed in Lee et al. 2012). Upon treatment of cells with various DNA damage-inducing agents, including UV, methyl methanesulfonate, hydroxyurea, and aphidicolin, the p12 subunit undergoes ubiquitylation-dependent degradation to form Pol δ3 (core + p68 subunit). Under conditions of low UV doses, this is dependent on activation of ATR, a major checkpoint response pathway in eukaryotic cells (Zhang et al. 2007). Interestingly, Pol δ3 displays increased exonuclease partitioning and decreased potential for bypass of various DNA lesions (Lee et al. 2012). These findings led to a model in which Pol δ3 is responsible for slowing replication progression at sites of DNA damage, which might allow for switching to a translesion synthesis polymerase (Lee et al. 2012).

Pol ε also is an essential component of cell cycle checkpoint responses. Pol ε functions upstream of Rad53 in the DNA damage response pathway (Navas et al. 1996). The C-terminus of *S. c.* POL2 has an essential role in initiating the S-phase checkpoint response, but is not involved in the G1 or G2/M checkpoints (Navas et al. 1995; Dua et al. 1998). In response to replication stress, Pol ε is required for one of two independent pathways leading to checkpoint activation (Puddu et al. 2011).

2.3 Perspective

Intensive basic research over the past 4 decades, using biochemical and genetic approaches and model systems from yeast to humans, has demonstrated that DNA polymerases cooperate to ensure efficient and accurate eukaryote genome replication. This chapter has provided a glimpse into how the biochemical properties of Pols α, δ, and ε shape their functional roles at the replication fork. The multisubunit holoenzyme structure of each polymerase provides a mechanism for regulating polymerase activities, allowing proper cell cycle progression and cellular responses to replication roadblocks, such as DNA damage. Studies of budding yeast have provided strong evidence that a fourth polymerase holoenzyme complex is present at the eukaryotic replication fork, namely, Pol ζ, whose biochemical activities are critical for translesion DNA replication (see Pavlov and Shcherbakova 2010 for review). Intriguingly, very recent papers have shown that the B and C subunits of Pol δ are shared with Pol ζ, suggesting that the catalytic domains of replicative polymerases can be exchanged within a holoenzyme complex (Baranovskiy et al. 2012; Johnson et al. 2012). This new evidence highlights the very dynamic nature of DNA polymerases at the replication fork to efficiently process all aspects of genomic maintenance as they arise during replication. Clearly, elucidating exactly how polymerases are regulated during DNA replication in eukaryotic cells is an emerging area of research but one that is already providing new and unexpected twists to the current paradigm.

References

Abdulovic AL, Hile SE, Kunkel TA, Eckert KA (2011) The *in vitro* fidelity of yeast DNA polymerase delta and polymerase epsilon holoenzymes during dinucleotide microsatellite DNA synthesis. DNA Repair 10(5):497–505. doi: 10.1016/j.dnarep.2011.02.003

Acharya N, Klassen R, Johnson RE, Prakash L, Prakash S (2011) PCNA binding domains in all three subunits of yeast DNA polymerase delta modulate its function in DNA replication. Proc Natl Acad Sci USA 108(44):17927–17932. doi:10.1073/pnas.1109981108

Agarkar VB, Babayeva ND, Pavlov YI, Tahirov TH (2011) Crystal structure of the C-terminal domain of human DNA primase large subunit: implications for the mechanism of the primase-polymerase alpha switch. Cell Cycle 10(6):926–931. doi: 4161/cc.10.6.15010

Aksenova A, Volkov K, Maceluch J, Pursell ZF, Rogozin IB, Kunkel TA, Pavlov YI, Johansson E (2010) Mismatch repair-independent increase in spontaneous mutagenesis in yeast lacking non-essential subunits of DNA polymerase epsilon. PLoS Genet 6(11):e1001209. doi:10.1371/journal.pgen.1001209

Albertson TM, Ogawa M, Bugni JM, Hays LE, Chen Y, Wang Y, Treuting PM, Heddle JA, Goldsby RE, Preston BD (2009) DNA polymerase epsilon and delta proofreading suppress discrete mutator and cancer phenotypes in mice. Proc Natl Acad Sci USA 106 (40):17101–17104. doi:10.1073/pnas.0907147106

Araki H, Hamatake RK, Johnston LH, Sugino A (1991) DPB2, the gene encoding DNA polymerase II subunit B, is required for chromosome replication in Saccharomyces cerevisiae. Proc Natl Acad Sci USA 88(11):4601–4605

Araki H, Ropp PA, Johnson AL, Johnston LH, Morrison A, Sugino A (1992) DNA polymerase II, the probable homolog of mammalian DNA polymerase epsilon, replicates chromosomal DNA in the yeast Saccharomyces cerevisiae. EMBO J 11(2):733–740

Araki H (2010) Cyclin-dependent kinase-dependent initiation of chromosomal DNA replication. Curr Opin Cell Biol 22(6):766–771. doi: 10.1016/jceb.2010.07.015

Arezi B, Kirk BW, Copeland WC, Kuchta RD (1999) Interactions of DNA with human DNA primase monitored with photoactivatable cross-linking agents: implications for the role of the p58 subunit. Biochemistry 38(39):12899–12907. doi:10.1021/bi9908991

Asturias FJ, Cheung IK, Sabouri N, Chilkova O, Wepplo D, Johansson E (2006) Structure of Saccharomyces cerevisiae DNA polymerase epsilon by cryo-electron microscopy. Nat Struct Mol Biol 13(1):35–43. doi:10.1038/nsmb1040, nsmb1040 [pii]

Baranovskiy AG, Lada AG, Siebler HM, Zhang Y, Pavlov YI, Tahirov TH (2012) DNA polymerase delta and zeta switch by sharing accessory subunits of DNA polymerase delta. J Biol Chem 287(21):17281–17287. doi:10.1074/jbc.M112.351122

Bermudez VP, Farina A, Raghavan V, Tappin I, Hurwitz J (2011) Studies on human DNA polymerase epsilon and GINS complex and their role in DNA replication. J Biol Chem 286 (33):28963–28977. doi:10.1074/jbc.M111.256289

Boulet A, Simon M, Faye G, Bauer GA, Burgers PM (1989) Structure and function of the Saccharomyces cerevisiae CDC2 gene encoding the large subunit of DNA polymerase III. EMBO J 8(6):1849–1854

Braun KA, Lao Y, He Z, Ingles CJ, Wold MS (1997) Role of protein–protein interactions in the function of replication protein A (RPA): RPA modulates the activity of DNA polymerase alpha by multiple mechanisms. Biochemistry 36(28):8443–8454. doi:10.1021/bi970473r

Bravo R, Frank R, Blundell PA, Macdonald-Bravo H (1987) Cyclin/PCNA is the auxiliary protein of DNA polymerase-delta. Nature 326(6112):515–517. doi:10.1038/326515a0

Budd M, Campbell JL (1987) Temperature-sensitive mutations in the yeast DNA polymerase I gene. Proc Natl Acad Sci USA 84(9):2838–2842

Budd ME, Campbell JL (1993) DNA polymerases delta and epsilon are required for chromosomal replication in Saccharomyces cerevisiae. Mol Cell Biol 13(1):496–505. doi:10.1128/mcb.13.1.496

Burgers PMJ (2009) Polymerase dynamics at the eukaryotic DNA replication fork. J Biol Chem 284(7):4041–4045. doi:10.1074/jbc.R800062200

Burgers PM, Bambara RA, Campbell JL, Chang LM, Downey KM, Hubscher U, Lee MY, Linn SM, So AG, Spadari S (1990) Revised nomenclature for eukaryotic DNA polymerases. Eur J Biochem 191(3):617–618

Byrnes JJ, Downey KM, Black VL, So AG (1976) A new mammalian DNA polymerase with 3′ to 5′ exonuclease activity: DNA polymerase delta. Biochemistry 15(13):2817–2823. doi:10.1021/bi00658a018

Chilkova O, Stenlund P, Isoz I, Stith CM, Grabowski P, Lundström E-B, Burgers PM, Johansson E (2007) The eukaryotic leading and lagging strand DNA polymerases are loaded onto primer-ends via separate mechanisms but have comparable processivity in the presence of PCNA. Nucleic Acids Res 35(19):6588–6597. doi:10.1093/nar/gkm741

Collins KL, Russo AA, Tseng BY, Kelly TJ (1993) The role of the 70 kDa subunit of human DNA polymerase alpha in DNA replication. EMBO J 12(12):4555–4566

Copeland WC, Tan X (1995) Active site mapping of the catalytic mouse primase subunit by alanine scanning mutagenesis. J Biol Chem 270(8):3905–3913

Copeland WC, Wang TS (1993) Enzymatic characterization of the individual mammalian primase subunits reveals a biphasic mechanism for initiation of DNA replication. J Biol Chem 268 (35):26179–26189

De Falco M, Ferrari E, De Felice M, Rossi M, Hubscher U, Pisani FM (2007) The human GINS complex binds to and specifically stimulates human DNA polymerase alpha-primase. EMBO Rep 8(1):99–103. doi:10.1038/sj.embor.7400870, 7400870 [pii]

Dieckman LM, Johnson RE, Prakash S, Washington MT (2010) Pre-steady state kinetic studies of the fidelity of nucleotide incorporation by yeast DNA polymerase delta. Biochemistry 49 (34):7344–7350. doi:10.1021/bi100556m

Dua R, Levy DL, Campbell JL (1998) Role of the putative zinc finger domain of Saccharomyces cerevisiae DNA polymerase epsilon in DNA replication and the S/M checkpoint pathway. J Biol Chem 273(45):30046–30055. doi:10.1074/jbc.273.45.30046

Dua R, Levy DL, Campbell JL (1999) Analysis of the essential functions of the C-terminal protein/protein interaction domain of Saccharomyces cerevisiae pol epsilon and its unexpected ability to support growth in the absence of the DNA polymerase domain. J Biol Chem 274 (32):22283–22288. doi:10.1074/jbc.274.32.22283

Dua R, Levy DL, Li CM, Snow PM, Campbell JL (2002) In vivo reconstitution of Saccharomyces cerevisiae DNA polymerase epsilon in insect cells. J Biol Chem 277(10):7889–7896. doi:10.1074/jbc.M108546200

Ducoux M, Urbach S, Baldacci G, Hubscher U, Koundrioukoff S, Christensen J, Hughes P (2001) Mediation of proliferating cell nuclear antigen (PCNA)-dependent DNA replication through a conserved p21(Cip1)-like PCNA-binding motif present in the third subunit of human DNA polymerase delta. J Biol Chem 276(52):49258–49266. doi:10.1074/jbc.M106990200, M106990200 [pii]

Eckert KA, Kunkel TA (1993) Fidelity of DNA synthesis catalyzed by human DNA polymerase alpha and HIV-1 reverse transcriptase: effect of reaction pH. Nucleic Acids Res 21 (22):5212–5220

Eckert KA, Hile SE, Vargo PL (1997) Development and use of an in vitro HSV-Tk forward mutation assay to study eukaryotic DNA polymerase processing of DNA alkyl lesions. Nucleic Acids Res 25(7):1450–1457. doi:10.1093/nar/25.7.1450

Feng W, D'Urso G (2001) Schizosaccharomyces pombe cells lacking the amino-terminal catalytic domains of DNA polymerase epsilon Are viable but require the DNA damage checkpoint control. Mol Cell Biol 21(14):4495–4504. doi:10.1128/mcb.21.14.4495-4504.2001

Ferrari M, Lucchini G, Plevani P, Foiani M (1996) Phosphorylation of the DNA polymerase alpha-primase B subunit is dependent on its association with the p180 polypeptide. J Biol Chem 271 (15):8661–8666

Fien K, Cho YS, Lee JK, Raychaudhuri S, Tappin I, Hurwitz J (2004) Primer utilization by DNA polymerase alpha-primase is influenced by its interaction with Mcm10p. J Biol Chem 279 (16):16144–16153. doi:10.1074/jbc.M400142200, M400142200 [pii]

Foiani M, Santocanale C, Plevani P, Lucchini G (1989) A single essential gene, PRI2, encodes the large subunit of DNA primase in Saccharomyces cerevisiae. Mol Cell Biol 9(7):3081–3087. doi:10.1128/mcb.9.7.3081

Foiani M, Marini F, Gamba D, Lucchini G, Plevani P (1994) The B subunit of the DNA polymerase alpha-primase complex in Saccharomyces cerevisiae executes an essential function at the initial stage of DNA replication. Mol Cell Biol 14(2):923–933

Fortune JM, Pavlov YI, Welch CM, Johansson E, Burgers PMJ, Kunkel TA (2005) Saccharomyces cerevisiae DNA Polymerase delta. J Biol Chem 280(33):29980–29987. doi:10.1074/jbc.M505236200

Fortune JM, Stith CM, Kissling GE, Burgers PMJ, Kunkel TA (2006) RPA and PCNA suppress formation of large deletion errors by yeast DNA polymerase delta. Nucleic Acids Res 34 (16):4335–4341. doi:10.1093/nar/gkl403

Francesconi S, Park H, Wang TS-F (1993) Fission yeast with DNA polymerase delta temperature-sensitive alleles exhibits cell division cycle phenotype. Nucleic Acids Res 21(16):3821–3828. doi:10.1093/nar/21.16.3821

Fry M, Loeb LA (1992) A DNA polymerase alpha pause site is a hot spot for nucleotide misinsertion. Proc Natl Acad Sci USA 89(2):763–767

Fukui T, Yamauchi K, Muroya T, Akiyama M, Maki H, Sugino A, Waga S (2004) Distinct roles of DNA polymerases delta and epsilon at the replication fork in Xenopus egg extracts. Genes Cells 9(3):179–191. doi:10.1111/j.1356-9597.2004.00716.x

Gao Y, Zhou Y, Xie B, Zhang S, Rahmeh A, H-s H, Lee MYWT, Lee EYC (2008) Protein phosphatase-1 is targeted to DNA polymerase delta via an interaction with the p68 subunit. Biochemistry 47(43):11367–11376. doi:10.1021/bi801122t

Garg P, Stith CM, Sabouri N, Johansson E, Burgers PM (2004) Idling by DNA polymerase delta maintains a ligatable nick during lagging-strand DNA replication. Genes Dev 18 (22):2764–2773. doi:10.1101/gad.1252304

Gerik KJ, Li X, Pautz A, Burgers PM (1998) Characterization of the two small subunits of Saccharomyces cerevisiae DNA polymerase delta. J Biol Chem 273(31):19747–19755

Goldsby RE, Hays LE, Chen X, Olmsted EA, Slayton WB, Spangrude GJ, Preston BD (2002) High incidence of epithelial cancers in mice deficient for DNA polymerase delta proofreading. Proc Natl Acad Sci USA 99(24):15560–15565. doi:10.1073/pnas.232340999

Griffiths DJF, Liu VF, Nurse P, Wang TSF (2001) Role of fission yeast primase catalytic subunit in the replication checkpoint. Mol Biol Cell 12(1):115–128. doi: 10.1091/mbc.12.1.115

Gutiérrez PJA, Wang TS-F (2003) Genomic instability induced by mutations in Saccharomyces cerevisiae POL1. Genetics 165(1):65–81

Handa T, Kanke M, Takahashi TS, Nakagawa T, Masukata H (2012) DNA polymerization-independent functions of DNA polymerase epsilon in assembly and progression of the replisome in fission yeast. Mol Biol Cell 23(16):3240–3253. doi:10.1091/mbc.E12-05-0339

Hashimoto K, Shimizu K, Nakashima N, Sugino A (2003) Fidelity of DNA polymerase delta holoenzyme from Saccharomyces cerevisiae: the sliding clamp proliferating cell nuclear antigen decreases its fidelity. Biochemistry 42(48):14207–14213. doi:10.1021/bi0348359

Hile SE, Eckert KA (2004) Positive correlation between DNA polymerase alpha-primase pausing and mutagenesis within polypyrimidine/polypurine microsatellite sequences. J Mol Biol 335 (3):745–759. doi: 10.1016/j.jmb.2003.10.075

Hile SE, Wang X, Lee MYWT, Eckert KA (2012) Beyond translesion synthesis: polymerase kappa fidelity as a potential determinant of microsatellite stability. Nucleic Acids Res 40 (4):1636–1647. doi: 10.1093/nar/gkr889

Hiraga S, Hagihara-Hayashi A, Ohya T, Sugino A (2005) DNA polymerases alpha, delta, and epsilon localize and function together at replication forks in Saccharomyces cerevisiae. Genes Cells 10(4):297–309. doi:10.1111/j.1365-2443.2005.00843.x, GTC843 [pii]

Hohn KT, Grosse F (1987) Processivity of the DNA polymerase alpha-primase complex from calf thymus. Biochemistry 26(10):2870–2878. doi:10.1021/bi00384a031

Hu SZ, Wang TS, Korn D (1984) DNA primase from KB cells. Evidence for a novel model of primase catalysis by a highly purified primase/polymerase-alpha complex. J Biol Chem 259 (4):2602–2609

Jain R, Hammel M, Johnson RE, Prakash L, Prakash S, Aggarwal AK (2009) Structural insights into yeast DNA polymerase delta by small angle X-ray scattering. J Mol Biol 394(3):377–382. doi:10.1016/j.jmb.2009.09.066, S0022-2836(09)01216-9 [pii]

Jin YH, Ayyagari R, Resnick MA, Gordenin DA, Burgers PMJ (2003) Okazaki fragment maturation in yeast. J Biol Chem 278(3):1626–1633. doi:10.1074/jbc.M209803200

Johnson LM, Snyder M, Chang LMS, Davis RW, Campbell JL (1985) Isolation of the gene encoding yeast DNA polymerase I. Cell 43(1):369–377

Johnson RE, Prakash L, Prakash S (2012) Pol31 and Pol32 subunits of yeast DNA polymerase delta are also essential subunits of DNA polymerase zeta. Proc Natl Acad Sci USA 109 (31):12455–12460. doi:10.1073/pnas.1206052109

Kaczmarek L, Miller MR, Hammond RA, Mercer WE (1986) A microinjected monoclonal antibody against human DNA polymerase-alpha inhibits DNA replication in human, hamster, and mouse cell lines. J Biol Chem 261(23):10802–10807

Kamath-Loeb AS, Shen J-C, Schmitt MW, Loeb LA (2012) The Werner syndrome exonuclease facilitates DNA degradation and high fidelity DNA polymerization by human DNA polymerase delta. J Biol Chem 287(15):12480–12490. doi:10.1074/jbc.M111.332577

Karthikeyan R, Vonarx EJ, Straffon AFL, Simon M, Gr F, Kunz BA (2000) Evidence from mutational specificity studies that yeast DNA polymerases delta and epsilon replicate different DNA strands at an intracellular replication fork. J Mol Biol 299(2):405–419. doi: 10.1006/jmbi.200.3744

Kesti T, Flick K, Keränen S, Syväoja JE, Wittenberg C (1999) DNA polymerase μ catalytic domains Are dispensable for DNA replication, DNA repair, and cell viability. Mol Cell 3 (5):679–685. doi: 10.1016/S1097-2765(00)80361-5

Kesti T, McDonald WH, Yates JR, Wittenberg C (2004) Cell cycle-dependent phosphorylation of the DNA polymerase epsilon subunit, Dpb2, by the Cdc28 cyclin-dependent protein kinase. J Biol Chem 279(14):14245–14255

Khare V, Eckert KA (2002) The proofreading 3′→5′ exonuclease activity of DNA polymerases: a kinetic barrier to translesion DNA synthesis. Mutat Res 510 (1–2):45–54. doi: 10.1016/S0027-5107(02)00251-8

Kitani T, Yoda K, Okazaki T (1984) Discontinuous DNA replication of Drosophila melanogaster is primed by octaribonucleotide primer. Mol Cell Biol 4(8):1591–1596

Klinge S, Nunez-Ramirez R, Llorca O, Pellegrini L (2009) 3D architecture of DNA Pol alpha reveals the functional core of multi-subunit replicative polymerases. EMBO J 28 (13):1978–1987. doi:10.1038/emboj.2009.150, emboj2009150 [pii]

Korona DA, LeCompte KG, Pursell ZF (2011) The high fidelity and unique error signature of human DNA polymerase epsilon. Nucleic Acids Res 39(5):1763–1773. doi:10.1093/nar/gkq1034

Kuchta RD, Reid B, Chang LM (1990) DNA primase. Processivity and the primase to polymerase alpha activity switch. J Biol Chem 265(27):16158–16165

Kunkel TA (1985) The mutational specificity of DNA polymerases-alpha and -gamma during in vitro DNA synthesis. J Biol Chem 260(23):12866–12874

Kunkel TA, Burgers PM (2008) Dividing the workload at a eukaryotic replication fork. Trends Cell Biol 18(11):521–527. doi: 10.1016/j.tcb.2008.08.005

Kunkel TA, Loeb LA (1981) Fidelity of mammalian DNA polymerases. Science 213 (4509):765–767. doi:10.2307/1686608

Kunkel TA, Sabatino RD, Bambara RA (1987) Exonucleolytic proofreading by calf thymus DNA polymerase delta. Proc Natl Acad Sci USA 84(14):4865–4869

Kunkel TA, Hamatake RK, Motto-Fox J, Fitzgerald MP, Sugino A (1989) Fidelity of DNA polymerase I and the DNA polymerase I-DNA primase complex from Saccharomyces cerevisiae. Mol Cell Biol 9(10):4447–4458

Kunkel TA, Roberts JD, Sugino A (1991) The fidelity of DNA synthesis by the catalytic subunit of yeast DNA polymerase alpha alone and with accessory proteins. Mutat Res 250:175–182

Larrea AA, Lujan SA, Nick McElhinny SA, Mieczkowski PA, Resnick MA, Gordenin DA, Kunkel TA (2010) Genome-wide model for the normal eukaryotic DNA replication fork. Proc Natl Acad Sci USA 107(41):17674–17679. doi:10.1073/pnas.1010178107, 1010178107 [pii]

Lee MYWT, Zhang S, Lin SHS, Chea J, Wang X, LeRoy C, Wong A, Zhang Z, Lee EYC (2012) Regulation of human DNA polymerase delta in the cellular responses to DNA damage. Environ Mol Mutagen 53(9):683–698. doi:10.1002/em.21743

Lemmens L, Urbach S, Prudent R, Cochet C, Baldacci G, Hughes P (2008) Phosphorylation of the C subunit (p66) of human DNA polymerase delta. Biochem Biophys Res Commun 367 (2):264–270. doi:10.1016/j.bbrc.2007.12.083, S0006-291X(07)02692-7 [pii]
Li H, Xie B, Rahmeh A, Zhou Y, Lee MY (2006a) Direct interaction of p21 with p50, the small subunit of human DNA polymerase delta. Cell Cycle 5 (4):428–436
Li H, Xie B, Zhou Y, Rahmeh A, Trusa S, Zhang S, Gao Y, Lee EYC, Lee MYWT (2006b) Functional roles of p12, the fourth subunit of human DNA polymerase delta. J Biol Chem 281 (21):14748–14755. doi:10.1074/jbc.M600322200
Liu VF, Bhaumik D, Wang TS-F (1999) Mutator phenotype induced by aberrant replication. Mol Cell Biol 19(2):1126–1135
Longhese MP, Jovine L, Plevani P, Lucchini G (1993) Conditional mutations in the yeast DNA primase genes affect different aspects of DNA metabolism and interactions in the DNA polymerase alpha-primase complex. Genetics 133(2):183–191
Lucchini G, Francesconi S, Foiani M, Badaracco G, Plevani P (1987) Yeast DNA polymerase–DNA primase complex; cloning of PRI 1, a single essential gene related to DNA primase activity. EMBO J 6(3):737–742
MacNeill SA, Moreno S, Reynolds N, Nurse P, Fantes PA (1996) The fission yeast Cdc1 protein, a homologue of the small subunit of DNA polymerase delta, binds to Pol3 and Cdc27. EMBO J 15(17):4613–4628
Maga G, Stucki M, Spadari S, Hubscher U (2000) DNA polymerase switching: I. Replication factor C displaces DNA polymerase alpha prior to PCNA loading. J Mol Biol 295(4):791–801. doi: 10.1006/jmbi.1999.3394
Maga G, Frouin I, Spadari S, Hubscher U (2001) Replication protein A as a "fidelity clamp" for DNA polymerase alpha. J Biol Chem 276(21):18235–18242. doi:10.1074/jbc.M009599200
Marini F, Pellicioli A, Paciotti V, Lucchini G, Plevani P, Stern DF, Foiani M (1997) A role for DNA primase in coupling DNA replication to DNA damage response. EMBO J 16(3):639–650
Masai H, Matsumoto S, You Z, Yoshizawa-Sugata N, Oda M (2010) Eukaryotic chromosome DNA replication: where, when, and how? Annu Rev Biochem 79(1):89–130. doi:10.1146/annurev.biochem.052308.103205
McElhinny SAN, Stith CM, Burgers PMJ, Kunkel TA (2007) Inefficient proofreading and biased error rates during inaccurate DNA synthesis by a mutant derivative of Saccharomyces cerevisiae DNA polymerase delta. J Biol Chem 282(4):2324–2332. doi:10.1074/jbc.M609591200
Menezes MR, Sweasy JB (2012) Mouse models of DNA polymerases. Environ Mol Mutagen 53 (9):645–665. doi:10.1002/em.21731
Michael WM, Ott R, Fanning E, Newport J (2000) Activation of the DNA replication checkpoint through RNA synthesis by primase. Science 289(5487):2133–2137. doi: 10.1126/science.289.5487.2133
Miller MR, Ulrich RG, Wang TS, Korn D (1985) Monoclonal antibodies against human DNA polymerase-alpha inhibit DNA replication in permeabilized human cells. J Biol Chem 260 (1):134–138
Moldovan G-L, Pfander B, Jentsch S (2007) PCNA, the maestro of the replication fork. Cell 129 (4):665–679. doi: 10.1016/j.cell.2007.05.003
Morrison A, Sugino A (1994) The 3′→5′ exonucleases of both DNA polymerases delta and epsilon participate in correcting errors of DNA replication in Saccharomyces cerevisiae. Mol Gen Genet 242(3):289–296
Morrison A, Araki H, Clark AB, Hamatake RK, Sugino A (1990) A third essential DNA polymerase in S. Cerevisiae. Cell 62(6):1143–1151
Morrison A, Bell JB, Kunkel TA, Sugino A (1991) Eukaryotic DNA polymerase amino acid sequence required for 3′–5′ exonuclease activity. Proc Natl Acad Sci USA 88(21):9473–9477
Mossi R, Keller RC, Ferrari E, Hubscher U (2000) DNA polymerase switching: II. Replication factor C abrogates primer synthesis by DNA polymerase alpha at a critical length. J Mol Biol 295(4):803–814. doi: 10.1006/jmbi.1999.3395

Mozzherin DJ, McConnell M, Jasko MV, Krayevsky AA, Tan C-K, Downey KM, Fisher PA (1996) Proliferating cell nuclear antigen promotes misincorporation catalyzed by calf thymus DNA polymerase delta. J Biol Chem 271(49):31711–31717. doi:10.1074/jbc.271.49.31711

Murakami H, Okayama H (1995) A kinase from fission yeast responsible for blocking mitosis in S phase. Nature 374(6525):817–819. doi: 10.1038/374817a0

Muramatsu S, Hirai K, Tak Y-S, Kamimura Y, Araki H (2010) CDK-dependent complex formation between replication proteins Dpb11, Sld2, Pol Epsilon, and GINS in budding yeast. Genes Dev 24(6):602–612. doi:10.1101/gad.1883410

Muzi-Falconi M, Giannattasio M, Foiani M, Plevani P (2003) The DNA polymerase alpha-primase complex: multiple functions and interactions. ScientificWorldJournal 3:21–33. doi:10.1100/tsw.2003.05

Nasheuer HP, Grosse F (1988) DNA polymerase alpha-primase from calf thymus. Determination of the polypeptide responsible for primase activity. J Biol Chem 263(18):8981–8988

Nasheuer HP, Moore A, Wahl AF, Wang TS (1991) Cell cycle-dependent phosphorylation of human DNA polymerase alpha. J Biol Chem 266(12):7893–7903

Navas TA, Zhou Z, Elledge SJ (1995) DNA polymerase epsilon links the DNA replication machinery to the S phase checkpoint. Cell 80(1):29–39

Navas TA, Sanchez Y, Elledge SJ (1996) RAD9 and DNA polymerase epsilon form parallel sensory branches for transducing the DNA damage checkpoint signal in Saccharomyces cerevisiae. Genes Dev 10(20):2632–2643. doi:10.1101/gad.10.20.2632

Netz DJ, Stith CM, Stumpfig M, Kopf G, Vogel D, Genau HM, Stodola JL, Lill R, Burgers PM, Pierik AJ (2012) Eukaryotic DNA polymerases require an iron-sulfur cluster for the formation of active complexes. Nat Chem Biol 8(1):125–132. doi:10.1038/nchembio.721, nchembio.721 [pii]

Ng L, Tan CK, Downey KM, Fisher PA (1991) Enzymologic mechanism of calf thymus DNA polymerase delta. J Biol Chem 266(18):11699–11704

Nick McElhinny SA, Gordenin DA, Stith CM, Burgers PMJ, Kunkel TA (2008) Division of labor at the eukaryotic replication fork. Mol Cell 30(2):137–144. doi: 10.1016.j.mol.cell.2008.02.022

Nunez-Ramirez R, Klinge S, Sauguet L, Melero R, Recuero-Checa MA, Kilkenny M, Perera RL, Garcia-Alvarez B, Hall RJ, Nogales E, Pellegrini L, Llorca O (2011) Flexible tethering of primase and DNA Pol alpha in the eukaryotic primosome. Nucleic Acids Res 39 (18):8187–8199. doi:10.1093/nar/gkr534, gkr534 [pii]

Pai CC, Garcia I, Wang SW, Cotterill S, MacNeill SA, Kearsey SE (2009) GINS inactivation phenotypes reveal Two pathways for chromatin association of replicative alpha and epsilon DNA polymerases in fission yeast. Mol Biol Cell 20(4):1213–1222. doi:10.1091/mbc.E08-04-0429

Pavlov YI, Shcherbakova PV (2010) DNA polymerases at the eukaryotic fork-20 years later. Mutat Res 685(1–2):45–53

Pavlov YI, Frahm C, McElhinny SAN, Niimi A, Suzuki M, Kunkel TA (2006a) Evidence that errors made by DNA polymerase alpha are corrected by DNA polymerase delta. Curr Biol 16 (2):202–207

Pavlov YI, Shcherbakova PV, Rogozin IB, Kwang WJ (2006b) Roles of DNA polymerases in replication, repair, and recombination in eukaryotes. Int Rev Cytol 255:41–132. doi: 10.1016/S0074-7696(06)55002-8

Perrino FW, Loeb LA (1989) Differential extension of 3′ mispairs is a major contribution to the high fidelity of calf thymus DNA polymerase-alpha. J Biol Chem 264(5):2898–2905

Plevani P, Foiani M, Valsasnini P, Badaracco G, Cheriathundam E, Chang LM (1985) Polypeptide structure of DNA primase from a yeast DNA polymerase-primase complex. J Biol Chem 260 (11):7102–7107

Podust VN, Chang L-S, Ott R, Dianov GL, Fanning E (2002) Reconstitution of human DNA polymerase delta using recombinant baculoviruses: the p12 subunit potentiates DNA polymerizing activity of the four-subunit enzyme. J Biol Chem 277(6):3894–3901. doi: 10.1074.jbc.M109684200

Prelich G, Tan C-K, Kostura M, Mathews MB, So AG, Downey KM, Stillman B (1987) Functional identity of proliferating cell nuclear antigen and a DNA polymerase-[delta] auxiliary protein. Nature 326(6112):517–520

Prindle MJ, Loeb LA (2012) DNA polymerase delta in DNA replication and genome maintenance. Environ Mol Mutagen 53(9):666–682. doi:10.1002/em.21745

Puddu F, Piergiovanni G, Plevani P, Muzi-Falconi M (2011) Sensing of replication stress and Mec1 activation act through two independent pathways involving the 9-1-1 complex and DNA polymerase epsilon. PLoS Genet 7(3):e1002022. doi: 10.137/journal.pgen.1002022

Pursell ZF, Isoz I, Lundstrom E-B, Johansson E, Kunkel TA (2007a) Regulation of B family DNA polymerase fidelity by a conserved active site residue: characterization of M644W, M644L and M644F mutants of yeast DNA polymerase epsilon. Nucleic Acids Res 35(9):3076–3086. doi:10.1093/nar/gkm132

Pursell ZF, Isoz I, Lundstrom E-B, Johansson E, Kunkel TA (2007b) Yeast DNA polymerase epsilon participates in leading-strand DNA replication. Science 317(5834):127–130. doi:10.1126/science.1144067

Pursell ZF and Kunkel TA (2008) Chapter 4 DNA polymerase epsilon: a polymerase of unusual size (and complexity). In: Progress in nucleic acid research and molecular biology, vol 82. Academic Press, London, UK; P. Michael Conn, Editor, pp 101–145. doi: 10.1016/S0079-6603(08)00004-4

Rahmeh AA, Zhou Y, Xie B, Li H, Lee EYC, Lee MYWT (2011) Phosphorylation of the p68 subunit of Pol delta acts as a molecular switch to regulate its interaction with PCNA. Biochemistry 51(1):416–424. doi:10.1021/bi201638e

Ricke RM, Bielinsky A-K (2004) Mcm10 regulates the stability and chromatin association of DNA polymerase-alpha. Mol Cell 16(2):173–185. doi: 10.1016/jmolcell.2004.09.017

Rytkonen AK, Vaara M, Nethanel T, Kaufmann G, Sormunen R, Laara E, Nasheuer HP, Rahmeh A, Lee MY, Syvaoja JE, Pospiech H (2006) Distinctive activities of DNA polymerases during human DNA replication. FEBS J 273(13):2984–3001. doi:10.1111/j.1742-4658.2006.05310.x, EJB5310 [pii]

Santocanale C, Foiani M, Lucchini G, Plevani P (1993) The isolated 48,000-dalton subunit of yeast DNA primase is sufficient for RNA primer synthesis. J Biol Chem 268(2):1343–1348

Schmitt MW, Matsumoto Y, Loeb LA (2009) High fidelity and lesion bypass capability of human DNA polymerase delta. Biochimie 91(9):1163–1172. doi: 10.1016/j.biochi.2009.06.007

Schmitt MW, Venkatesan RN, Pillaire M-J, Hoffmann J-S, Sidorova JM, Loeb LA (2010) Active site mutations in mammalian DNA polymerase delta alter accuracy and replication fork progression. J Biol Chem 285(42):32264–32272. doi:10.1074/jbc.M110.147017

Shcherbakova PV, Pavlov YI (1996) $3' \rightarrow 5'$ Exonucleases of DNA Polymerases ε and δ Correct Base Analog Induced DNA Replication Errors on Opposite DNA Strands in Saccharomyces cerevisiae. Genetics 142(3):717–726

Shcherbakova PV, Pavlov YI, Chilkova O, Rogozin IB, Johansson E, Kunkel TA (2003) Unique error signature of the four-subunit yeast DNA polymerase epsilon. J Biol Chem 278(44):43770–43780. doi:10.1074/jbc.M306893200

Sheaff RJ, Kuchta RD (1993) Mechanism of calf thymus DNA primase: slow initiation, rapid polymerization, and intelligent termination. Biochemistry 32(12):3027–3037. doi:10.1021/bi00063a014

Sheaff RJ, Kuchta RD (1994) Misincorporation of nucleotides by calf thymus DNA primase and elongation of primers containing multiple noncognate nucleotides by DNA polymerase alpha. J Biol Chem 269(30):19225–19231

Sheaff RJ, Kuchta RD, Ilsley D (1994) Calf thymus DNA polymerase alpha-primase: "communication" and primer-template movement between the two active sites. Biochemistry 33(8):2247–2254. doi:10.1021/bi00174a035

Shimizu K, Hashimoto K, Kirchner JM, Nakai W, Nishikawa H, Resnick MA, Sugino A (2002) Fidelity of DNA polymerase epsilon holoenzyme from budding yeast Saccharomyces cerevisiae. J Biol Chem 277(40):37422–37429. doi:10.1074/jbc.M204476200

Simon M, Giot L, Faye G (1991) The 3′ to 5′ exonuclease activity located in the DNA polymerase delta subunit of Saccharomyces cerevisiae is required for accurate replication. EMBO J 10 (8):2165–2170

Suzuki M, Izuta S, Yoshida S (1994) DNA polymerase alpha overcomes an error-prone pause site in the presence of replication protein-A. J Biol Chem 269(14):10225–10228

Swan MK, Johnson RE, Prakash L, Prakash S, Aggarwal AK (2009) Structural basis of high-fidelity DNA synthesis by yeast DNA polymerase [delta]. Nat Struct Mol Biol 16(9):979–986. doi: 10.1038/nsmb.1663

Syvaoja J, Linn S (1989) Characterization of a large form of DNA polymerase delta from HeLa cells that is insensitive to proliferating cell nuclear antigen. J Biol Chem 264(5):2489–2497

Takayama Y, Kamimura Y, Okawa M, Muramatsu S, Sugino A, Araki H (2003) GINS, a novel multiprotein complex required for chromosomal DNA replication in budding yeast. Genes Dev 17(9):1153–1165. doi:10.1101/gad.1065903

Tan CK, Castillo C, So AG, Downey KM (1986) An auxiliary protein for DNA polymerase-delta from fetal calf thymus. J Biol Chem 261(26):12310–12316

Thomas DC, Roberts JD, Sabatino RD, Myers TW, Tan CK, Downey KM, So AG, Bambara RA, Kunkel TA (1991) Fidelity of mammalian DNA replication and replicative DNA polymerases. Biochemistry 30(51):11751–11759. doi:10.1021/bi00115a003

Thompson HC, Sheaff RJ, Kuchta RD (1995) Interactions of calf thymus DNA polymerase alpha with primer/templates. Nucleic Acids Res 23(20):4109–4115

Tsubota T, Maki S, Kubota H, Sugino A, Maki H (2003) Double-stranded DNA binding properties of Saccharomyces cerevisiae DNA polymerase epsilon and of the Dpb3p-Dpb4p subassembly. Genes Cells 8 (11):873–888. doi: 10.1046/j.1365-2443.2003.00683

Uchimura A, Hidaka Y, Hirabayashi T, Hirabayashi M, Yagi T (2009) DNA polymerase delta is required for early mammalian embryogenesis. PLoS One 4(1):e4184. doi: 10.1371/journal.pone.0004184

Uchiyama M, Wang TSF (2004) The B-subunit of DNA polymerase alpha-primase associates with the origin recognition complex for initiation of DNA replication. Mol Cell Biol 24 (17):7419–7434. doi: 10.1128/mcb.24.17.7419-7434.2004

Venkatesan RN, Hsu JJ, Lawrence NA, Preston BD, Loeb LA (2006) Mutator phenotypes caused by substitution at a conserved motif a residue in eukaryotic DNA polymerase delta. J Biol Chem 281(7):4486–4494. doi:10.1074/jbc.M510245200

Venkatesan RN, Treuting PM, Fuller ED, Goldsby RE, Norwood TH, Gooley TA, Ladiges WC, Preston BD, Loeb LA (2007) Mutation at the polymerase active site of mouse DNA polymerase delta increases genomic instability and accelerates tumorigenesis. Mol Cell Biol 27 (21):7669–7682. doi:10.1128/mcb.00002-07

Voitenleitner C, Rehfuess C, Hilmes M, O'Rear L, Liao PC, Gage DA, Ott R, Nasheuer HP, Fanning E (1999) Cell cycle-dependent regulation of human DNA polymerase alpha-primase activity by phosphorylation. Mol Cell Biol 19(1):646–656

Waga S, Masuda T, Takisawa H, Sugino A (2001) DNA polymerase epsilon is required for coordinated and efficient chromosomal DNA replication in Xenopus egg extracts. Proc Natl Acad Sci USA 98(9):4978–4983. doi:10.1073/pnas.081088798

Walter J, Newport J (2000) Initiation of eukaryotic DNA replication: origin unwinding and sequential chromatin association of Cdc45, RPA, and DNA polymerase alpha. Mol Cell 5 (4):617–627. doi: 10.1016/S1097-2765(00)80241-5

Wang Y, Zhang Q, Chen H, Li X, Mai W, Chen K, Zhang S, Lee EYC, Lee MYWT, Zhou Y (2011) P50, the small subunit of DNA polymerase delta, is required for mediation of the interaction of polymerase delta subassemblies with PCNA. PLoS One 6(11):e27092. doi: 10.1371/journal.pone.0027092

Warren EM, Huang H, Fanning E, Chazin WJ, Eichman BF (2009) Physical interactions between Mcm10, DNA, and DNA polymerase alpha. J Biol Chem 284(36):24662–24672. doi: 10.1074/jbc.M109.020438

Weiser T, Gassmann M, Thommes P, Ferrari E, Hafkemeyer P, Hubscher U (1991) Biochemical and functional comparison of DNA polymerases alpha, delta, and epsilon from calf thymus. J Biol Chem 266(16):10420–10428

Weissbach A, Baltimore D, Bollum F, Gallo R, Korn D (1975) Nomenclature of eukaryotic DNA polymerases. Science 190(4212):401–402

Wong SW, Paborsky LR, Fisher PA, Wang TS, Korn D (1986) Structural and enzymological characterization of immunoaffinity-purified DNA polymerase alpha.DNA primase complex from KB cells. J Biol Chem 261(17):7958–7968

Xie B, Mazloum N, Liu L, Rahmeh A, Li H, Lee MYWT (2002) Reconstitution and characterization of the human DNA polymerase delta four-subunit holoenzyme. Biochemistry 41 (44):13133–13142. doi:10.1021/bi0262707

Yamaguchi M, Hendrickson EA, DePamphilis ML (1985) DNA primase-DNA polymerase alpha from simian cells: sequence specificity of initiation sites on simian virus 40 DNA. Mol Cell Biol 5(5):1170–1183. doi:10.1128/mcb.5.5.1170

Yoneda M, Bollum FJ (1965) Deoxynucleotide-polymerizing enzymes of calf thymus gland. I. Large scale purification of terminal and replicative deoxynucleotidyl transferases. J Biol Chem 240:3385–3391

Zeng XR, Hao H, Jiang Y, Lee MY (1994) Regulation of human DNA polymerase delta during the cell cycle. J Biol Chem 269(39):24027–24033

Zerbe LK, Kuchta RD (2002) The p58 subunit of human DNA primase is important for primer initiation, elongation, and counting. Biochemistry 41(15):4891–4900. doi:10.1021/bi016030b

Zhang P, Mo J-Y, Perez A, Leon A, Liu L, Mazloum N, Xu H, Lee MYWT (1999) Direct interaction of proliferating cell nuclear antigen with the p125 catalytic subunit of mammalian DNA polymerase delta. J Biol Chem 274(38):26647–26653. doi:10.1074/jbc.274.38.26647

Zhang S, Zhou Y, Trusa S, Meng X, Lee EYC, Lee MYWT (2007) A novel DNA damage response. J Biol Chem 282(21):15330–15340. doi:10.1074/jbc.M610356200

Zhou Y, Meng X, Zhang S, Lee EY, Lee MY (2012) Characterization of human DNA polymerase delta and its subassemblies reconstituted by expression in the MultiBac system. PLoS One 7 (6):e39156. doi:10.1371/journal.pone.0039156, PONE-D-12-08472 [pii]

Zhu W, Ukomadu C, Jha S, Senga T, Dhar SK, Wohlschlegel JA, Nutt LK, Kornbluth S, Dutta A (2007) Mcm10 and And-1/CTF4 recruit DNA polymerase alpha to chromatin for initiation of DNA replication. Genes Dev 21(18):2288–2299. doi: 10.1101/gad.1585607

Chapter 3
DNA Repair Polymerases

Robert W. Sobol

Abstract The nuclear and mitochondrial genomes are under constant attack from endogenous (metabolic) and exogenous genotoxins. The resulting genomic insults include damaged bases and nucleotides, deoxyribo- and ribonucleotide misincorporation, intra-strand and interstrand DNA cross-links, and single-strand and double-strand DNA breaks. As expected, efficient recognition and removal of these genotoxic lesions is critical to begin the repair process and restore genome integrity. With the exception of direct reversal mechanisms, repair of both the nuclear and mitochondrial genomes requires DNA synthesis to replace the nucleotides or DNA strands removed during the repair process. Whereas some DNA repair pathways co-opt replicative DNA polymerases to synthesize the DNA in the "repair patch," other DNA repair pathways have dedicated DNA polymerase enzymes. This chapter will detail the DNA polymerases central to the major mammalian DNA repair pathways and, where applicable, highlight the unique roles these DNA polymerases may play in protecting normal cells from mutagenic or genotoxic agents and in providing resistance to genotoxic chemotherapeutic treatments.

Keywords DNA repair • DNA polymerase • Mutation • Genome stability

R.W. Sobol (✉)
Department of Pharmacology & Chemical Biology, University of Pittsburgh Cancer Institute, Hillman Cancer Center, Research Pavilion, 5117 Centre Avenue, Suite 2.6a, Pittsburgh, PA 15213-1863, USA

University of Pittsburgh School of Medicine, Pittsburgh, PA, USA

Department of Human Genetics, University of Pittsburgh School of Public Health, Pittsburgh, PA, USA
e-mail: rws9@pitt.edu

K.S. Murakami and M.A. Trakselis (eds.), *Nucleic Acid Polymerases*, Nucleic Acids and Molecular Biology 30, DOI 10.1007/978-3-642-39796-7_3,

Abbreviations

5′dRP	5′-deoxyribose phosphate
8-oxodG	8-oxo-7,8-dihydro-2′-deoxyguanosine
AP	Apurinic/apyrimidinic
APE1	Apurinic/apyrimidinic endonuclease
BER	Base excision repair
dsDNA	Double-stranded DNA
FEN1	Structure-specific flap endonuclease 1
HR	Homologous recombination
KO	Knockout
LigI	DNA ligase I
LigIII	DNA ligase III
MEF	Mouse embryonic fibroblast
MGMT	O^6-methylguanine-DNA methyltransferase
MMR	Mismatch repair
NER	Nucleotide excision repair
NHEJ	Nonhomologous end joining
PARP1	Poly(ADP-ribose)polymerase-1
PARP2	Poly(ADP-ribose)polymerase-2
PCNA	Proliferating cell nuclear antigen
Polκ	DNA polymerase kappa
Polλ	DNA polymerase lambda
Polβ	DNA polymerase beta
Polα	DNA polymerase alpha
Polγ	DNA polymerase gamma
Polδ	DNA polymerase delta
Polη	DNA polymerase eta
Polθ	DNA polymerase theta
Polι	DNA polymerase iota
Polκ	DNA polymerase kappa
Polμ	DNA polymerase mu
RFC	Replication factor C
RNS	Reactive nitrogen species
ROS	Reactive oxygen species
SSBs	Single-strand breaks
ssDNA	Single-stranded DNA
TdT	Terminal deoxynucleotidyltransferase
UV	Ultraviolet
WRN	Werner syndrome protein helicase

3.1 Introduction

Human cells repair thousands of DNA lesions per day to prevent the accumulation of DNA mutations or genome aberrations that can impact cellular survival and genomic integrity (Lindahl 1993). These lesions manifest as base (e.g., deamination of cytosine to uracil) or nucleotide modifications (e.g., thymine–thymine dimers), deoxyribo- and ribonucleotide misincorporation (mismatches), intra-strand or inter-strand DNA cross-links, and single-strand or double-strand DNA breaks. These numerous and diverse types of DNA lesions derive from both endogenous and exogenous sources. Base damage can be induced by a variety of reactive oxygen species (ROS), reactive nitrogen species (RNS), and sources of DNA alkylation (Svilar et al. 2011). Such molecules stem from endogenous sources via cellular metabolism and exogenous sources mediated by environmental exposure. Additional modifications include simple and complex DNA adducts (Marnett et al. 2003; Marnett 2000; Knutson et al. 2007, 2009; Otteneder et al. 2006), ultraviolet light-induced pyrimidine dimers (Setlow and Setlow 1962), replication errors that give rise to mutations, deletions, or insertions (Kunkel and Erie 2005), as well as gross modifications such as DNA cross-links (Deans and West 2011) and DNA double-strand breaks (DSBs) from environmental, genetic, and endogenous sources (Friedberg et al. 2006).

To facilitate the repair of these lesions, cells have multiple DNA damage response and DNA repair mechanisms that signal the presence of lesions and promote DNA repair (Jackson and Bartek 2009; Friedberg et al. 2006; Wood et al. 2001, 2005; Hoeijmakers 2001). A general overview of mammalian DNA repair pathways and the lesions each repairs is depicted in Fig. 3.1. With the exception of direct reversal (DR) repair processes, each pathway utilizes one or more DNA polymerases as an integral part of the overall DNA repair pathway. The major DR pathway proteins include O^6-methylguanine DNA methyltransferase (MGMT) and the α-ketoglutarate-dependent dioxygenase enzymes: AlkB homologues 1, 2, and 3 (ALKBH1, ALKHB2, ALKBH3). These proteins directly reverse the damage to the DNA base without the requirement of new DNA synthesis (Fu et al. 2012; Yi and He 2013), each with unique lesion specificity. Further detail on this mode of direct reversal DNA repair can be found elsewhere (Fu et al. 2012; Yi and He 2013).

The remaining DNA repair pathways depicted in Fig. 3.1 utilize DNA polymerases to replace the excised lesion-containing nucleotides. Base lesions and DNA SSBs are primarily repaired by the base excision repair (BER) pathway (Almeida and Sobol 2005, 2007; Svilar et al. 2011). As shown and as will be discussed, BER utilizes specific DNA polymerases depending on the initiating lesion, the subcellular location (nuclear vs. mitochondria), and the BER sub-pathway. Similarly, nonhomologous end joining (NHEJ), a pathway for repair of DNA DSBs, also utilizes specialized DNA polymerases (Lieber 2008; Lange et al. 2011; Ramsden 2011; Ramsden and Asagoshi 2012). Another class of specialized DNA polymerases, translesion DNA polymerases, are discussed in

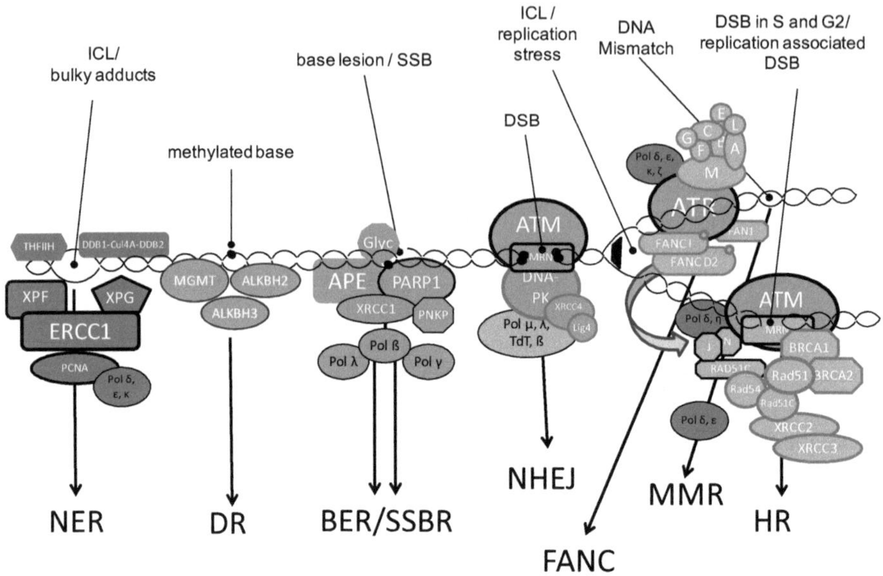

Fig. 3.1 Schematic representation of the mammalian DNA repair pathways. This figure depicts the mammalian DNA repair pathways, the major proteins within each pathway and highlights (*black lettering*) the DNA polymerases involved in each pathway. Adapted from Vens and Sobol (2013)

Chap. 4. The nucleotide excision repair (NER) pathway is a multi-protein, highly complex DNA repair pathway that plays an important role in the repair of DNA lesions induced by many genotoxins and facilitates the removal of bulky DNA adducts that grossly distort the DNA double helix and those that cause a block to transcription (Hoeijmakers 2001; de Laat et al. 1999; Wood 1996; Shuck et al. 2008). As depicted in Fig. 3.1, NER utilizes primarily replicative DNA polymerases but also uses DNA polymerase kappa (Polκ) separate from its role in lesion bypass or translesion DNA synthesis (Ogi et al. 2010). The remaining pathways for the repair of DNA mismatches (MMR), DNA DSBs via homologous recombination (HR), or DNA intra-strand cross-links via the FANC pathway either co-opt replicative DNA polymerases or use specialized polymerases to synthesize DNA after lesion removal or to replicate DNA from the homologous template. The following sections will provide an overview of these DNA repair pathways, emphasizing the role of the DNA polymerases specific to each pathway. Where appropriate, each section will also include relevant discussion on the alterations in these DNA polymerases in cancer since defects in these DNA repair pathways can promote tumorigenesis and are common in human cancers (Hanahan and Weinberg 2011; Harper and Elledge 2007; Curtin 2012; O'Driscoll 2012; Hoeijmakers 2009).

3.2 DNA Polymerases in Base Excision Repair

The proteins of the base excision repair (BER) pathway participate in the repair of dozens of base modifications that result from alkylating agents, reactive nitrogen species, and reactive oxygen species (oxidative DNA damage), among others (Svilar et al. 2011; Almeida and Sobol 2005, 2007). Such damage can arise from numerous exogenous and endogenous sources, resulting in a multitude of detrimental cellular effects, including mutations, genome rearrangements, altered gene expression, and the onset of cell death or senescence (Hoeijmakers 2001; Baute and Depicker 2008; Hegde et al. 2011). The BER pathway model shown in Fig. 3.2 is initiated by a DNA glycosylase such as MYH, a unique glycosylase with specificity for a normal base (adenine) when paired opposite the ROS modified form of deoxyguanosine, 8-oxo-7,8-dihydro-2′-deoxyguanosine (8-oxodG) (Svilar et al. 2011; David et al. 2007). The remaining ten DNA glycosylases are specific for many types of base lesions, as reviewed elsewhere (Almeida and Sobol 2007; Svilar et al. 2011; Fu et al. 2012). Once the base lesion is removed, the product, an abasic or apurinic/apyrimidinic (AP) site, is a substrate for an endonuclease specific for AP sites, the AP endonucleases APE1 or APE2 (although the majority activity results from APE1) (Almeida and Sobol 2007). There is general consensus that the resulting DNA single-strand break after APE1 (or APE2) cleavage forms a nucleation site for scaffold proteins such as PARP1 and XRCC1 followed by recruitment of the proteins needed to complete repair (not shown in this figure) (Almeida and Sobol 2007). Either DNA polymerase β (Polβ) or DNA polymerase lambda (Polλ) can be recruited to conduct end-trimming and DNA synthesis. Polβ is considered the major end-trimming (5′dRP lyase activity) and DNA polymerase enzyme in BER although, as will be detailed below, Polλ plays a significant role in oxidative damage repair. Alternate DNA polymerases have also been suggested to participate in BER, depending on the base lesion and the subcellular location (nuclear vs. mitochondrial), as will be discussed below. The short-patch BER pathway (Fig. 3.2, left panel) likely contributes 90 % of the repair mediated by BER, but if the 5′ end of the gap is blocked such that end-trimming (5′dRP lyase activity) is attenuated, both Polβ and DNA polymerase δ (Polδ) or DNA polymerase ε (Polε) can extend the repair patch to 2–12 bases, completing a form of BER known as long-patch BER (Fig. 3.2, right panel). Finally, the repair gap is sealed or ligated by either DNA ligase III (LigIII) or DNA ligase I (LigI). Recently, it was suggested that LigI functions as the primary BER DNA ligase in the nucleus with LigIII playing a predominant role in the mitochondria (Gao et al. 2011; Simsek et al. 2011).

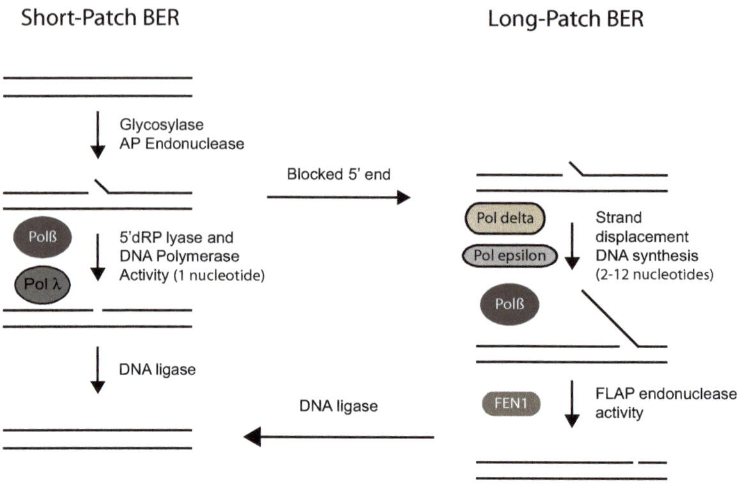

Fig. 3.2 Schematic for short-patch and long-patch BER. Simplified diagram depicting the two sub-pathways for BER: short-patch and long-patch. In short-patch BER, the cleaved AP site can be further processed by the 5′dRP lyase activity of Polβ or Polλ, followed by DNA synthesis and ligation. However, if the 5′ end of the downstream DNA is blocked and cannot be processed, strand-displacement DNA synthesis can proceed. Processing requires FEN1 to remove the 2–12 base flap, followed by DNA ligation

3.2.1 DNA Polβ as the Primary BER Polymerase

DNA polymerase β (Polβ) is a member of the X-family of DNA polymerases (Burgers et al. 2001; Bebenek and Kunkel 2004) and is an essential BER protein, considered the major or primary BER DNA polymerase. At 335 amino acids (39 kDa), Polβ is the smallest of the human DNA polymerases (Beard and Wilson 2006; Lange et al. 2011; Sobol et al. 1996). Polβ has two active sites. The 5′dRP lyase activity is restricted to the 8 kDa N-terminal domain and requires the active site residue K72, whereas the nucleotidyl transferase or DNA polymerase activity resides within the C-terminal domain and requires the aspartate triad D190, D192, and D256, as depicted in Fig. 3.3a. Structurally, the enzyme contains four domains (8K, Fingers, Palm, and Thumb), with the 8K and Fingers domain comprising the dRP lyase activity and the Palm and Thumb domains comprising the majority of the nucleotidyl transferase activity. As depicted in the diagram and structural representation (Fig. 3.3b), the single-stranded DNA (ssDNA) and double-stranded DNA (dsDNA) binding domains reside mostly in the N-terminus, with the enzyme inducing a bend in the DNA upon binding and nucleotide incorporation (Batra et al. 2006).

Since its initial discovery (Weissbach 1977; Weissbach et al. 1975a, b), Polβ was found to be unique in its enzymatic properties (Ono et al. 1979; Tanabe et al. 1979; Yoshida et al. 1979) as compared to the other newly characterized mammalian DNA polymerases alpha (Polα), gamma (Polγ), and delta (Polδ) (Byrnes

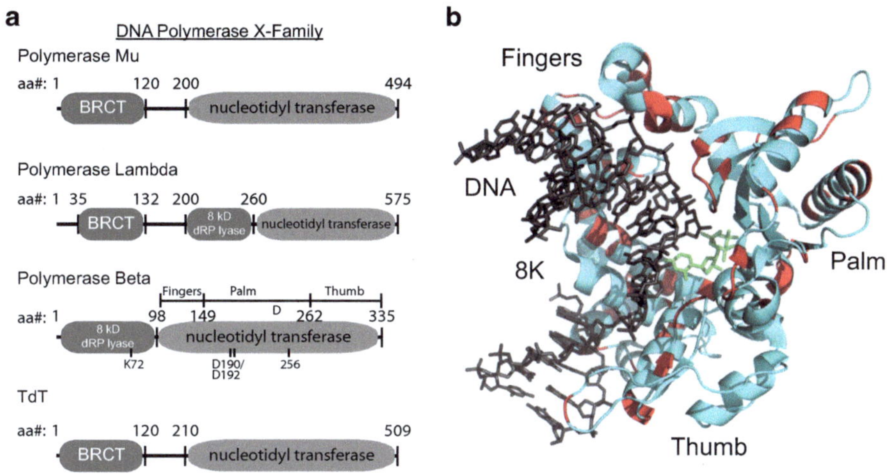

Fig. 3.3 Structural and functional details for DNA Polβ. (**a**) Cartoon depiction of DNA polymerases μ, λ, β, and TdT. For each, the amino acid length is indicated, as well as the domains for protein binding (BRCT), lyase activity (8 kDa lyase), and DNA polymerase activity (nucleotidyl transferase). The amino acid residues spanning each domain are also indicated. A linear depiction of the amino acid residues (1–335) of Polβ is shown, indicating the structural sub-domains as determined by crystallographic analysis (8K, fingers, palm, and thumb), the functional domains (dRP lyase and nucleotidyl transferase), and the essential active site residues (K72 and D190/192/256). (**b**) Ternary structure of DNA Polβ with DNA substrate and incoming nucleotide—structure (pdb2fms) depicting DNA Polymerase β (Polβ) with a gapped DNA substrate and dUMPNPP with magnesium in the catalytic site (Batra et al. 2006). The fingers, palm, and thumb domains of Polβ are indicated. The 8K domain is at the back of the structure facing away from the plane of the image and is shown behind the DNA in this orientation. Amino acids altered by germline or somatic mutations are colored *red* (Sobol 2012b; Donigan et al. 2012)

et al. 1976; Weissbach 1977). Of the four eukaryotic DNA polymerases identified by 1977, Polβ was considered to be "the" DNA repair polymerase (Hubscher et al. 1979; Siedlecki et al. 1980; Waser et al. 1979; Wawra and Dolejs 1979). These early studies defined a role for Polβ in repair using isolated nuclei or nuclear extracts, monitoring the incorporation of radioactive nucleosides following DNA damage (Hubscher et al. 1979; Siedlecki et al. 1980; Waser et al. 1979; Wawra and Dolejs 1979; Mosbaugh and Linn 1983). Although it was subsequently shown that Polα can also carry out gap-filling DNA synthesis in a base excision repair (BER) reaction (Mosbaugh and Linn 1984), the evidence continued to mount in support of Polβ acting as "the" DNA repair polymerase in the nucleus. Studies continued to identify a role for Polβ in the repair of damage induced by many different DNA damaging agents, including bleomycin (Seki and Oda 1986; DiGiuseppe and Dresler 1989), cigarette smoke (Cui et al. 2012), arsenic (Lai et al. 2011), UV-radiation (Orlando et al. 1988), benzo[a]pyrene (Ishiguro et al. 1987), methylmethane sulfonate (Park et al. 1991), ionizing radiation (Price 1993), G-T mis-pairs (Wiebauer and Jiricny 1990), and uracil (Dianov et al. 1992; Nealon et al. 1996; Singhal et al. 1995; Singhal and Wilson 1993). Interestingly, a truncated version of Polβ expressed in MEFs mediates a dependence of the cells on homologous recombination (HR) and sensitizes cells to radiation (Neijenhuis et al. 2009, 2010). Several groups reported complete BER in vitro with Polβ and additional purified proteins (Singhal et al. 1995; Nealon et al. 1996; Kubota et al. 1996). Like many DNA repair proteins, Polβ has been reported to be modified by and/or regulated by posttranslational modification, including phosphorylation (Tokui et al. 1991; Guo et al. 2008; Phosphosite 2010), acetylation (Hasan et al. 2002), methylation (El-Andaloussi et al. 2006, 2007), as well ubiquitinylation (Parsons et al. 2008, 2009; Sobol 2008). It is not yet clear how these modifications impact overall BER or the role of Polβ in BER as in many cases these observations have been limited to in vitro studies (Sobol 2008; Goellner et al. 2012).

Although it was demonstrated in heterologous systems (*Escherichia coli* and *Saccharomyces cerevisiae*) that Polβ can conduct DNA replication and repair in vivo (Blank et al. 1994; Ohnishi et al. 1990), it was not until a mouse gene knockout (KO) (Gu et al. 1994) was made that the specificity of the repair conducted by Polβ was defined (Sobol et al. 1996).

Characterization of the Polβ KO mouse (Gu et al. 1994; Sugo et al. 2000) and mouse embryonic fibroblasts (MEFs) deficient in Polβ (Sobol et al. 1996) clearly demonstrated a requirement for Polβ in repair of alkylation and oxidative DNA damage (Sobol et al. 1996; Horton et al. 2002) and provided a valuable resource to explore additional functions of Polβ (Esposito et al. 2000; Gonda et al. 2001), to evaluate the impact of Polβ on mutagenesis (Niimi et al. 2005; Cabelof et al. 2003; Sobol et al. 2002; Bennett et al. 2001; Poltoratsky et al. 2005) and mechanisms of genotoxin-induced cell death (Ochs et al. 1999, 2002; Horton et al. 2003, 2005; Le Page et al. 2003; Sobol et al. 2003; Trivedi et al. 2005; Cabelof et al. 2004; Tomicic et al. 2001), to investigate alternate or compensatory repair pathways in the absence of Polβ (Biade et al. 1998; Fortini et al. 1998, 1999; Stucki et al. 1998; Dianov et al. 1999; Braithwaite et al. 2005b), to address structure–function relationships or protein partners of Polβ in vivo (Kedar et al. 2002; Niimi et al. 2005; Sobol et al. 2000) and most recently to evaluate changes of gene expression in response to Polβ depletion (Li et al. 2012a). The most definitive and reproducible endpoint that has been used to evaluate Polβ participation in repair in vivo is survival following DNA damage such as exposure to alkylating agents (Sobol et al. 1996, 2000). Unfortunately, the Polβ mouse knockout is lethal just after birth (neonatal lethality) (Sugo et al. 2000; Gu et al. 1994), complicating efforts to evaluate the role of Polβ in an animal model. Surprisingly, it is the 5′dRP lyase function of Polβ (Matsumoto and Kim 1995) that appears to be essential and sufficient for alkylating agent resistance (Sobol et al. 2000). In the absence of Polβ (in MEFs), cells are unable to efficiently repair the highly toxic 5′dRP moiety and therefore are hypersensitive to different types of alkylating agents such as methylmethane sulfonate, *N*-methyl-*N*-nitrosourea, and *N*-methyl-*N*′-nitro-*N*-nitrosoguanidine (Sobol et al. 1996, 2000, 2002, 2003; Trivedi et al. 2005), the thymidine analog 5-hydroxymethyl-2′-deoxyuridine (Horton et al. 2003), as well as the therapeutic agent temozolomide (Trivedi et al. 2005; Horton et al. 2003) and radiation (Neijenhuis et al. 2005, 2009, 2010; Vens and Begg 2010). In human cells, Polβ is suggested to play a role in the cellular response to cisplatin (Kothandapani et al. 2011) but plays a significant role in the repair of the DNA lesions induced by the clinical alkylating agent temozolomide (Trivedi et al. 2008; Tang et al. 2010, 2011; Goellner et al. 2011; Stachelek et al. 2010), prompting a robust series of investigations to identify specific Polβ inhibitors (Goellner et al. 2012; Wilson et al. 2010; Barakat et al. 2012; Jaiswal et al. 2009).

In mice, an intact 5′dRP lyase domain appears to rescue the neonatal lethality of the Polβ mouse knockout yet does not appear to be sufficient to rescue all of the defects associated with the loss of Polβ in the mouse (Senejani et al. 2012). By expression of a mutant of Polβ deficient in polymerase activity (Y265C) in a Polβ KO background, it was demonstrated that the mice (homozygous for the Y265C mutant Polβ) were born at the expected Mendelian ratios, but loss of the polymerase function in vivo led to the accumulation of repair intermediates and less than 40 % survived 24 h after birth, remaining smaller than the WT littermates even 3 weeks after birth (Senejani et al. 2012).

Dozens of somatic or germline mutations of Polβ have been identified and characterized (Sobol 2012b; Nemec et al. 2012; Donigan et al. 2012), prompting the suggestion that Polβ may be mutated in as much as 30 % of human tumors (Starcevic et al. 2004; Sweasy et al. 2006; An et al. 2011). As depicted in Fig. 3.3b, these mutations are not limited to a single domain or active site and are found throughout the Polβ open reading frame (note the red-shaded regions in the cartoon depicting locations of somatic and germline mutations in the structure of Polβ). In many cases, these mutations show little or no effect but several mutations have significant impact on DNA polymerase activity or 5′dRP lyase activity. For example, the E295K mutant, first identified in gastric cancer (Iwanaga et al. 1999), is defective in nucleotidyl transferase activity, and the resulting protein is defective in BER, inducing cellular transformation when over-expressed (Lang et al. 2007; Li et al. 2012b). Conversely, the L22P cancer mutant is reported to be defective only in 5′dRP lyase activity (Dalal et al. 2008). It has been suggested that tumor-specific defects in BER such as a defect in Polβ may be exploited for selective therapeutic options (Neijenhuis et al. 2010), and so it remains to be determined if the presence of these Polβ mutants can be exploited clinically.

3.2.2 Emerging Role of Polλ in BER of Oxidative Damage

DNA polymerase lambda (Polλ) was first isolated and characterized as a beta-like (Polβ-like) polymerase (Aoufouchi et al. 2000; Garcia-Diaz et al. 2000; Nagasawa et al. 2000). Overall, Polλ is a 575 amino acid enzyme that participates in both BER and nonhomologous end joining (NHEJ). Unique to Polλ is an N-terminal BRCT domain that is essential for its role in NHEJ (see Sect. 3.2 below). Similar to Polβ, Polλ is an X-family polymerase with multiple domains, including both the nucleotidyl transferase activity domain and an 8K domain that contributes the 5′dRP lyase activity important for complete BER (Garcia-Diaz et al. 2001)

(Fig. 3.3a). Although its primary role is likely in NHEJ, the presence of the 5′dRP lyase activity (Garcia-Diaz et al. 2001) has prompted continued investigation into the role of Polλ in BER, even suggesting that in some cases, both Polβ and Polλ compete for repair of the same lesions but may have nonredundant roles in vivo depending on cellular state (Garcia-Diaz et al. 2002). However, it is clear that whether it is a backup or competing player in BER, Polλ has a significant role in BER (Lebedeva et al. 2005; Braithwaite et al. 2005a, b).

Both Polβ and Polλ participate in BER in DT40 cells (Tano et al. 2007) and have redundant as well as independent BER roles in MEFs (Braithwaite et al. 2010). Further, Polλ-deficient cells are sensitive to radiation (Vermeulen et al. 2007a), similar to that found by the same group for Polβ (Vermeulen et al. 2007b). However, the most prominent role for Polλ in BER appears to be in MYH-initiated BER, as depicted in the short-patch BER model shown in Fig. 3.2a. This unique BER process requires the removal of the normal adenine base when opposite the ROS lesion 8-oxodG and insertion of a cytidine opposite 8-oxodG to allow a second round of BER initiated by OGG1 (Sobol 2012a; David et al. 2007). Interestingly, both Polβ and Polλ can fill the gap opposite the 8-oxodG lesion (Brown et al. 2007) although Polλ may be more error prone (Brown et al. 2011). A role for Polλ in MYH-mediated repair has been shown in cells and with purified proteins (van Loon and Hubscher 2009).

Multiple structural studies have been completed for Polλ, specific for the lyase domain (DeRose et al. 2003) or the catalytic core (Garcia-Diaz et al. 2004), providing significant insight with regard to structure and function (Garcia-Diaz et al. 2005). The enzyme is phosphorylated by cdk2 (Frouin et al. 2005; Wimmer et al. 2008) and its stability is regulated by ubiquitylation (Markkanen et al. 2011). Further, the involvement of Polλ in BER of oxidative lesions is regulated by both cdk2-mediated phosphorylation and MULE-mediated ubiquitylation (Markkanen et al. 2012). Most interestingly, there is functional cross talk between these two PTMs in that phosphorylation of Thr553 on Polλ prevents ubiquitylation and proteasome-mediated degradation (Wimmer et al. 2008). More recently, it is suggested that long-patch DNA repair synthesis mediated by Polλ is enhanced by binding to the Werner syndrome protein helicase (WRN) (Kanagaraj et al. 2012).

Only one cancer mutant of Polλ has been characterized, but it appears to have a defect in NHEJ as opposed to BER (Terrados et al. 2009). Polλ inhibitors have also been developed and are suggested to have tenfold greater specificity to Polλ as compared to Polβ (Strittmatter et al. 2011). It remains to be determined if these novel tools can advance our understanding on the role of Polλ in BER.

3.2.3 Alternate or Backup Polymerases in BER

The most relevant or obvious backup DNA polymerases that function in BER (besides Polλ) are those that have been found to participate in the long-patch sub-pathway of BER (Fig. 3.2b). Long-patch BER is initiated similarly to short-patch BER to produce a nicked DNA intermediate but appears to have different DNA polymerase requirements. Repair completion requires a 3′OH moiety for proper nucleotidyl transfer and chain elongation. In cases where the 5′ moiety within the gap is refractory to Polβ lyase activity (Gary et al. 1999), Polδ, Polε, or Polβ, coupled with proliferating cell nuclear antigen (PCNA) and a variety of other proteins including the flap structure-specific endonuclease 1 (FEN1), poly (ADP-ribose)polymerase 1 (PARP1), and LigI, synthesizes DNA to fill the gap, resulting in a displaced DNA flap of 2–12 bases in length (Fortini et al. 1998; Stucki et al. 1998; Gary et al. 1999; Parlanti et al. 2002; Pascucci et al. 1999; Podlutsky et al. 2001; Matsumoto et al. 1999). DNA synthesis and strand displacement by Polβ is stimulated by the combined presence of FEN1 and PARP1 (Prasad et al. 2000, 2001) or RPA (DeMott et al. 1998). WRN is also observed to stimulate strand-displacement activities of Polβ (Harrigan et al. 2003) to facilitate long-patch BER (Harrigan et al. 2006). FEN1 then catalyzes the removal of the ensuing DNA flap, leaving a nick that has been transferred 2–12 nucleotides downstream of the original damage site. Finally, the intact DNA strand is restored by ligation mediated by LigI (Fig. 3.2b).

Recent studies with purified proteins or in cells (DT40 KO cells, MEF KO cells, or human cells following RNA interference) have implicated additional DNA polymerases that may participate in nuclear BER. DNA polymerase iota (Polι) is a Y-family polymerase and encodes a 5′dRP lyase activity (Bebenek et al. 2001) located in the 40-kDa domain spanning residues M79 to M445 (Prasad et al. 2003). Although Polι is shown to protect cells from oxidative stress suggesting a more prominent role in BER of oxidative damage (Petta et al. 2008), Polι appears to play little or no role in the repair of alkylation damage (Poltoratsky et al. 2008; Sobol 2007). Efforts are continuing to uncover the most significant biological role for Polι (Vidal and Woodgate 2009).

DNA polymerase theta (Polθ), an A-family polymerase, has also been suggested to be involved in BER (Ukai et al. 2006). As with the other BER DNA polymerases, Polθ contains a 5′dRP lyase domain (Prasad et al. 2009) and Polθ KO cells are sensitive to oxidative damage (Goff et al. 2009; Yousefzadeh and Wood 2013), all supportive for a role for Polθ in BER, as well as a role in the response to radiation (Higgins et al. 2010b). Interestingly, Polθ is known to be upregulated in breast cancers and, when over-expressed, correlates with poor prognosis (Higgins et al. 2010a; Lemee et al. 2010; Begg 2010).

3.2.4 Mitochondrial BER

BER has a well-defined role in repair of the mitochondrial genome (Bogenhagen et al. 2001), although recently it has been suggested that other repair pathways function in mitochondria (Kazak et al. 2012). Several nuclear BER enzymes also encode mitochondrial isoforms, i.e., UNG1 (Slupphaug et al. 1993). Additional mitochondrial BER enzymes have been summarized elsewhere (Svilar et al. 2011;

Liu and Demple 2010). The sole polymerase in mitochondria is DNA polymerase γ (Polγ), an A-family DNA polymerase essential for mitochondrial replication (Liu and Demple 2010) and implicated in mitochondrial BER (Stuart et al. 2005; Bogenhagen et al. 2001). The enzyme is comprised of the catalytic subunit (Polγ) and an accessory subunit (POLG2 or POLGB), reported to enhance the BER capacity of Polγ (Pinz and Bogenhagen 2006). As with the other BER DNA polymerases mentioned above, Polγ encodes a 5′dRP lyase activity domain (Longley et al. 1998), supporting its role in the short-patch sub-pathway of BER. Similar to that observed for nuclear BER, Polγ also supports a long-patch BER sub-pathway in mitochondria via both a FEN1-dependent (Liu et al. 2008) and a FEN1-independent (Szczesny et al. 2008) reaction, the latter possibly involving either DNA2 (Zheng et al. 2008) or EXOG (Tann et al. 2011). Interestingly, over 40 disease mutations have been identified in the gene for Polγ (POLG), many of which lead to mitochondrial disorders (Longley et al. 2005). It has yet to be determined if these mutations impact the role of Polγ in BER.

3.3 Unique and Specialized DNA Polymerases in Nonhomologous End Joining

The majority of DNA DSBs are repaired in mammalian cells by the nonhomologous end joining (NHEJ) pathway (Fig. 3.4a) (Lieber 2008; Downs et al. 2007). Primarily, the requisite DNA synthesis associated with NHEJ-mediated repair is via X-family DNA polymerases (Fig. 3.3a). As will be described below, repair of DSBs in the G_2-phase of the cell cycle or during the latter part of the S-phase of the cell cycle is primarily handled by the homologous recombination (HR) pathway (Fig. 3.4b). It is suggested that in G_2-phase and late S-phase, there are numerous factors that contribute to DSB repair pathway choice between HR and NHEJ (Brandsma and Gent 2012). However, in G_0-phase, G_1-phase and in the early part of S-phase, DSBs are repaired primarily by the NHEJ pathway. As depicted in Fig. 3.4a, the DNA ends are recognized by the KU heterodimer (KU70/KU80), a large DNA-binding protein with significant binding affinity to DNA ends (Doherty and Jackson 2001). Bound and activated KU undergoes a conformational change, increasing its affinity (hence recruitment) to the other critical factors required for NHEJ (Lieber 2008), including the nuclease complex Artemis/DNA-PKcs (Gell and Jackson 1999; Rivera-Calzada et al. 2007), the DNA polymerases (primarily Polμ and Polλ) (Ramsden 2011; Ramsden and Asagoshi 2012), and the ligase complex XLF/XRCC4/LigIV (Costantini et al. 2007; Gell and Jackson 1999). This large protein complex subsequently processes the broken, modified DNA ends, resulting in relegation/joining to repair the DSB. The overall process has been extensively reviewed elsewhere (Lieber 2008; Brandsma and Gent 2012; Murray et al. 2012; Boboila et al. 2012; Chapman et al. 2012; Kass and Jasin 2010; Malu et al. 2012; Pawelczak et al. 2011). Note that the recruitment of the

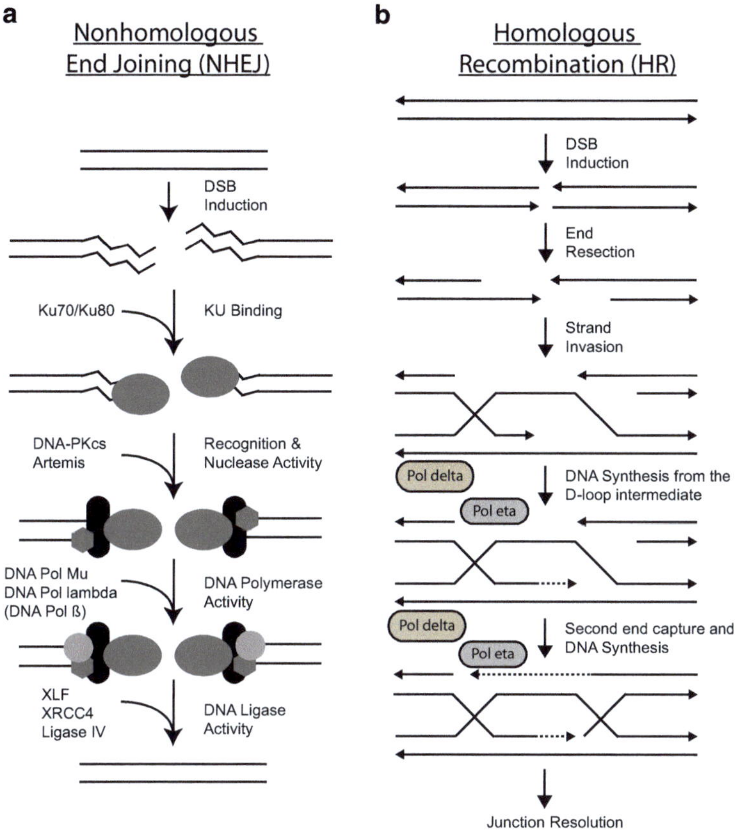

Fig. 3.4 DNA polymerases involved in double-strand break repair. (**a**) Schematic depicting the mechanism of NHEJ functioning in all phases of the cell cycle, showing KU binding to the broken ends of a DNA DSB, followed by nuclease activity to trim the ends, DNA polymerase activity for end processing, and DNA ligase activity to seal the DSB. (**b**) Classical scheme for HR-mediated repair of a DSB, showing DNA synthesis by either Polδ or Polη extending from the D-loop intermediate and followed by second-end capture and coordinated DNA synthesis of the opposite strand

essential polymerases for NHEJ primarily occurs via the BRCT domain of the polymerases (Fig. 3.3a) (DeRose et al. 2007; Matsumoto et al. 2012; Mueller et al. 2008; Gell and Jackson 1999). Details for each of the polymerases involved in mammalian NHEJ are described below.

3.3.1 DNA Polymerase Mu

DNA polymerase μ (Polμ) is an X-family DNA polymerase (Nick McElhinny and Ramsden 2003) with homology to TdT (Dominguez et al. 2000; Ruiz et al. 2001) containing both a BRCT domain in the N-terminus and the nucleotidyl transferase activity in the C-terminus (Fig. 3.3a). A role for Polμ in NHEJ was suggested following the discovery that Polμ interacts with KU and LigIV (Mahajan et al. 2002; Paull 2005). There are known NHEJ-related functional differences between Polμ and the other X-family polymerases (Bertocci et al. 2006). For example, mice deficient for Polμ are defective for immunoglobulin kappa chain rearrangement (Bertocci et al. 2003) but not Ig gene hypermutation (Bertocci et al. 2002). Further, over-expression of Polμ can impact the rate of somatic hypermutation (Ruiz et al. 2004). In addition, Polμ-deficient mice have a defect in hematopoiesis (Lucas et al. 2009). Structural studies have contributed to an in-depth understanding for the role of Polμ in NHEJ, highlighting slight but important differences with other X-family DNA polymerase members that might explain some of the substrate specificity for Polμ as compared to Polλ and Polβ (Moon et al. 2007).

Analysis of Polμ KO MEFs clearly establishes a role for Polμ in DSB repair of a variety of NHEJ substrates (Chayot et al. 2010, 2012; Capp et al. 2007). Similar to that seen for Polλ (see below), gap-filling activities in the NHEJ process mediated by Polμ are dependent on XLF (Akopiants et al. 2009), likely via interaction with the BRCT domain of Polμ (Mueller et al. 2008; DeRose et al. 2007). Polμ is also known to conduct translesion DNA synthesis, as will be discussed elsewhere in this series (Chap. 4). Defects in Polμ with regard to NHEJ can give rise to an increase in genomic abnormalities (e.g., chromosome aberrations) and should be associated with an increase in cancer.

3.3.2 DNA Polymerase Lambda

As described in Sect. 3.1.2, Polλ has a high degree of similarity to Polβ (Garcia-Diaz et al. 2000) and has a significant role in BER, particularly for oxidative damage (Lebedeva et al. 2005; Braithwaite et al. 2005a, b; Markkanen et al. 2012; Kanagaraj et al. 2012). The enzymatic properties of Polλ also suggested a role for this polymerase in NHEJ (Fan and Wu 2004; Lee et al. 2004; Bebenek et al. 2003). As was described above and as shown in Fig. 3.3a, Polλ has an N-terminal BRCT domain that promotes its role in NHEJ (Mueller et al. 2008). Polλ interacts with the XRCC4/LigIV complex via residue R57 in the BRCT domain (Mueller et al. 2008). Polλ-mediated gap filling during NHEJ also requires XLF, a core protein in the NHEJ pathway (Ahnesorg et al. 2006; Buck et al. 2006; Cavero et al. 2007; Revy et al. 2006). By characterizing whole-cell extracts from XLF-deficient human cells, it was determined that XLF is essential for gap filling by both Polλ and Polμ (Akopiants et al. 2009). It is suggested that XLF may align the DNA ends in the repair reaction, in concert with XRCC4 (Andres et al. 2012), DNA ligase IV (Ropars et al. 2011), KU (Yano et al. 2008, 2011), and APLF (Grundy et al. 2013). Proper polymerase fidelity is also required for Polλ with regard to NHEJ-mediated DNA synthesis, as was discovered by characterizing the single-nucleotide polymorphism (SNP) mutant at codon 438 (R438W) (Terrados et al. 2009). This point mutant of Polλ retains nucleotidyl transferase activity and 5′dRP lyase activity but has a reduction in base substitution fidelity (Terrados et al. 2009). Interestingly, this mutant negatively impacts the role of Polλ in NHEJ, leading to an increase in chromosomal aberrations (Terrados et al. 2009).

3.3.3 TdT

Terminal deoxynucleotidyltransferase (TdT) participates in a very restricted capacity in NHEJ. Expression of TdT is limited to cells productive for V(D)J recombination (Benedict et al. 2000), suggesting that a role for TdT is limited to NHEJ during V(D)J recombination. This unique X-family polymerase catalyzes the addition of nucleotides by a template-independent mechanism, for example, at the junction of rearranged Ig heavy chain and T-cell receptor gene segments during B-cell and T-cell maturation. This activity, even with purified protein, is consistent with a role in NHEJ (Ma et al. 2004). Recently, it has also been shown that TdT can carry out non-template-mediated nucleotide addition at a DSB junction but only in the presence of KU80 and XRCC4 (Boubakour-Azzouz et al. 2012). TdT binds to the essential NHEJ protein KU (Mahajan et al. 1999) as well as the DSB repair protein hPso4 (Mahajan and Mitchell 2003), and its role in V(D)J recombination is suppressed by binding to PCNA (Ibe et al. 2001). As with other X-family polymerases involved in NHEJ, the N-terminal BRCT domain of TdT (Mueller et al. 2008) does have a positive effect on nucleotide addition activity (Repasky et al. 2004). Although TdT shares significant sequence homology with the other X-family polymerase members (Fig. 3.3a), there does not appear to be any significant overlapping function of TdT with Polλ or Polμ (Bertocci et al. 2006). Further activities of TdT are discussed in Chap. 5.

3.3.4 DNA Polymerase Beta

DNA polymerase β (Polβ) is genetically similar to TdT (Anderson et al. 1987) and exhibits strong similarity to Polμ (Ruiz et al. 2001) as well as structural (DeRose et al. 2003) and functional (Ramadan et al. 2003) similarity to Polλ (Fig. 3.3a). In this light, Polβ has long been suggested to have a role, albeit minor, in NHEJ. Although there is evidence of a genetic interaction between Polβ and the NHEJ protein DNA-PKs (Niimi et al. 2005), this by itself does not implicate Polβ in

NHEJ. Mice with a reconstituted lymphoid system using Polβ KO fetal liver cells showed normal patterns of somatic hypermutation, suggesting little role for Polβ in this process (Esposito et al. 2000). Further, the lack of a BRCT domain in Polβ (Fig. 3.3a) would negatively impact its role in NHEJ since this protein–protein interaction domain (Woods et al. 2012) in Polλ, Polμ, and TdT is important for interacting with NHEJ protein partners (Mueller et al. 2008). However, it is possible that Polβ may play a limited role in microhomology-mediated end joining (MMEJ) (Crespan et al. 2012), a sub-pathway of NHEJ that is independent from KU and DNA ligase 4/XRCC4 (McVey and Lee 2008).

3.4 DNA Polymerases Critical to Homologous Recombination Repair of DNA Double-Strand Breaks

The homologous recombination (HR) pathway participates in several critical biological processes, including DNA repair, the rescue of stalled/collapsed DNA replication forks, meiotic chromosome segregation, and telomere maintenance (Sung and Klein 2006; Friedberg et al. 2006; Hoeijmakers 2001). As with NER and MMR, much of the effort in recent years to characterize the proteins involved in HR has focused on the early steps in this pathway including lesion (DSB) recognition, HR protein regulation, DSB repair pathway choice (HR vs. NHEJ), strand exchange processes, as well as the proteins involved in the resolution of holiday junctions (Barzel and Kupiec 2008; Bordeianu et al. 2011; Krejci et al. 2012; Sung and Klein 2006; Symington and Gautier 2011). A classical schematic for the HR pathway is depicted in Fig. 3.4b. Upon recognition of the DSB, the ends are processed through an end resection step, allowing strand invasion of the homologous strand of the sister chromatid, providing the template for HR-directed DNA synthesis from the D-loop and subsequently after second-end capture. Defining the DNA polymerase in this process was first shown in yeast where it was demonstrated that Polδ is preferentially recruited to complete DNA synthesis for HR (Maloisel et al. 2008). There are in fact numerous genetic examples demonstrating that yeast Polδ is involved in HR (Giot et al. 1997; Lydeard et al. 2007; Maloisel et al. 2004, 2008; Wang et al. 2004; Fabre et al. 1991). More recently, using purified proteins, it was shown that yeast Polδ, together with PCNA, is essential for DNA synthesis from the D-loop during HR (Li et al. 2009). In a more recent study, both yeast Polδ and yeast polymerase eta (Polη) contributed equally to DNA synthesis to extend the D-loop (Sebesta et al. 2011).

However, in chicken DT40 cells it was demonstrated that Polη participates in both HR and TLS (Kawamoto et al. 2005). Simultaneously, using purified human proteins and cell lysates, it was shown that human Polη promoted DNA synthesis from the D-loop intermediate (Fig. 3.4b) but this DNA synthesis step could not be conducted by human Polδ or by human polymerase iota (Polι) (McIlwraith et al. 2005). Human Polη, but not human Polδ or human Polι, was also able to mediate the capture and annealing of the second end of the resected DSB, in concert with RAD52. This was subsequently followed by DNA synthesis from the captured "second" DNA end (McIlwraith and West 2008) (Fig. 3.4b). Of course, some aspects of HR may require a TLS step. This will be discussed elsewhere in this series (Chap. 4).

3.5 DNA Polymerases as Essential Components in Nucleotide Excision Repair

The nucleotide excision repair (NER) pathway plays an important role in the repair of DNA lesions (Kuper and Kisker 2012) induced by many genotoxins and chemotherapeutics including DNA cross-linking agents such as chloroethylating agents, cisplatin, carboplatin, and lesions induced by a host of environmental agents including cigarette smoke (Friedberg et al. 2006) and ultraviolet (UV) light (Wood 1996; de Vries et al. 1995; Yeh et al. 2012). Put simply, NER facilitates the removal of bulky DNA adducts that grossly distort the DNA double helix and those that cause a block to transcription. Overall, the pathway consists of two complementary sub-pathways that have some overlap. These two sub-pathways are referred to as global genome repair (GGR–NER) and transcription-coupled repair (TCR–NER) and facilitate lesion recognition/confirmation and the assembly of the pre-incision complex. Molecular details on the proteins involved in NER can be found in several excellent reviews (Hoeijmakers 2001; de Laat et al. 1999; Wood 1996; Shuck et al. 2008; Hanawalt and Spivak 2008; Gillet and Scharer 2006). The two sub-pathways are distinct regarding the lesion recognition step but converge and utilize the same proteins to remove the ~22–28 base oligonucleotide containing the lesion. Until recently, the molecular details of the later steps in the pathway, the DNA synthesis steps, were not fully characterized (Kunkel and Van Houten 2006). Although much has yet to be worked out, recent studies have provided compelling evidence that the DNA synthesis step of NER involves three DNA polymerases (Fig. 3.5), the replicative DNA polymerases delta (Polδ) and epsilon (Polε) as well as the Y-family DNA polymerase kappa (Polκ) (Kunkel and Van Houten 2006; Ogi et al. 2010; Ogi and Lehmann 2006).

3.5.1 *Replicative Polymerases Delta and Epsilon in NER*

DNA polymerases δ (Polδ) and ε (Polε) are both B-family DNA polymerases with primary roles in DNA replication (Kunkel and Burgers 2008). The involvement of replicative DNA polymerases in DNA synthesis in human NER evolved from earlier studies with human cell extracts that implicated a requirement for the replication cofactor PCNA (Shivji et al. 1992). Subsequent studies demonstrated

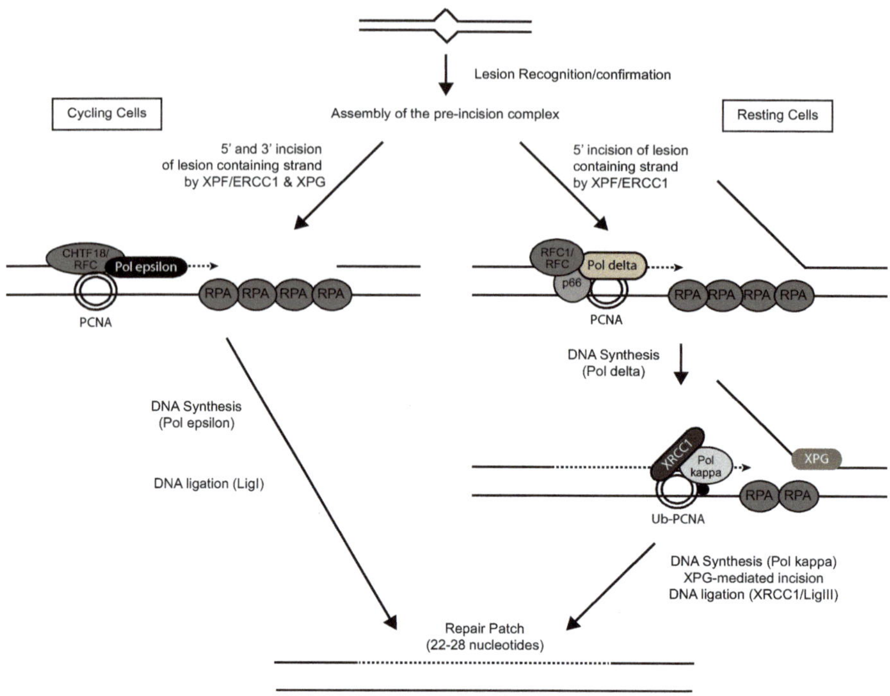

Fig. 3.5 DNA synthesis in NER by Polδ, Polε, and Polκ. Schematic depicting a role for Polδ, Polε, and Polκ in the DNA synthesis step of NER. The diagram shows the NER DNA polymerases in cycling and resting cells (Polδ, Polε, and Polκ) as well as the clamp loaders required for each polymerase (Ogi et al. 2010)

that DNA synthesis during NER was not affected by neutralizing antibodies to polymerase α (Polα) but was blocked by aphidicolin, suggesting a possible role for Polδ and/or Polε (Coverley et al. 1992). Subsequent elegant studies with purified human proteins clearly established the requirement for Polδ and/or Polε in DNA synthesis during NER (Shivji et al. 1995; Aboussekhra et al. 1995), ultimately defining a core set of proteins required for the repair of a cisplatin DNA adduct (Araujo et al. 2000). Replication factor C (RFC) was observed to be required for recruitment of Polδ (Overmeer et al. 2010).

As suggested above, the latest models suggest that Polδ, Polε, and Polκ (see below) are all involved in DNA synthesis during NER (Fig. 3.5) (Lehmann 2011; Ogi and Lehmann 2006; Ogi et al. 2010). Although it is not yet fully resolved as to the conditions or parameters that dictate polymerase choice in NER, several clues have emerged from biochemical studies (Fig. 3.5). Polδ is recruited by RFC1/p66 and loaded onto PCNA. The recruitment of Polκ does not require RFC1 but in fact is mediated by XRCC1 and is loaded onto ubiquitylated PCNA (see below). Conversely, Polε appears to be the polymerase of choice in cycling (dividing) cells in which CHTF18-RFC recruits Polε to load onto PCNA. Recruitment of Polε appears to favor dividing cells with high dNTPs and after dual incision by

XPF/ERCC1 and XPG, whereas Polδ requires RFC and PCNA for recruitment and likely favors nondividing cells (Lehmann 2011).

3.5.2 A Role for DNA Polymerase Kappa in NER Unrelated to Translesion Synthesis

DNA polymerase kappa (Polκ) is a Y-family DNA polymerase with a high error rate typical for this family of polymerases (Ohashi et al. 2000). Like other Y-family polymerases, Polκ can participate in DNA synthesis past bulky DNA lesions (translesion DNA synthesis, TLS) (Chap. 4) (Lange et al. 2011; Ziv et al. 2009) and would not be expected to participate in the DNA synthesis step of NER (Kunkel and Van Houten 2006). The low processivity and fidelity of Y-family polymerases (McCulloch and Kunkel 2008) (e.g., synthesis and incorporation of only one to five nucleotides before dissociation from the primer-template) would likely preclude Polκ from participation in NER to fill the gap of 22–28 nucleotides (Friedberg et al. 2006). However, in vitro studies have demonstrated that Polκ polymerizes up to 25 nucleotides before dissociation (Ohashi et al. 2000), supporting a possible role for Polκ in NER gap-filling DNA synthesis.

The first clue that Polκ may participate in NER gap-filling DNA synthesis was the demonstration that Polκ localized to repair foci with PCNA in a pattern that was unlike the other Y-family TLS polymerases eta (Polη) and iota (Polι) (Ogi et al. 2005). In a surprising finding using Polκ-KO MEFs, it was demonstrated that loss of Polκ reduced the level of NER following UV damage. Repair was not completely absent but was significantly reduced and clearly established a novel role for Polκ in NER (Ogi and Lehmann 2006). The same group followed this with a more detailed report implicating Polδ, Polε, and Polκ in NER (Ogi et al. 2010). As described above for Polδ and Polε and in Fig. 3.5, Polκ (in a complex with XRCC1) is recruited to complete DNA repair synthesis by ubiquitylated PCNA following repair DNA synthesis initiated by Polδ (Fig. 3.5, right side). Once repair is completed, the polymerase dissociates (Polκ) and XPG facilitates the 3′ incision to release the flap. This latter step is consistent with the observation that the 5′ incision by XPF/ERCC1 precedes the 3′ incision by XPG and that repair synthesis can proceed in the absence of XPG catalytic activity (Staresincic et al. 2009). Based on this recent model and available biochemical analysis (Fig. 3.5), recruitment and involvement of Polκ in NER requires XRCC1 and ubiquitylated PCNA for recruitment and likely favors low dNTPs and synthesis after 5′ incision by XPF/ERCC1 (Lehmann 2011).

3.6 The Mismatch Repair Pathway as a Replicative Polymerase Fidelity Factor

The DNA mismatch repair (MMR) pathway is involved in numerous processes involving DNA metabolism including repair of damage due to environmental or chemotherapeutic exposures, meiotic recombination, DNA damage signaling, and the correction or repair of numerous base–base mismatches and insertion/deletion loops (Fu et al. 2012; Li 2008; Wyatt and Pittman 2006; Modrich 2006; Jiricny 2006). The latter role of the MMR pathway functions to significantly improve DNA replication fidelity, as much as 1,000-fold, repairing errors made by Polα (Liberti et al. 2013; Nick McElhinny et al. 2010; Niimi et al. 2004), Polδ (Nick McElhinny et al. 2010; Lujan et al. 2012), and Polε (Lujan et al. 2012). Loss of MMR therefore promotes a mutator/genome instability phenotype that can predispose to an increase in mutations and cancer in eukaryotic model systems and humans (Arana and Kunkel 2010; Hubscher 2009; Kunkel 2009; Preston et al. 2010; Reha-Krantz 2010; Albertson et al. 2009).

3.6.1 A Reconstituted Human Mismatch Repair Pathway Utilizes DNA Polymerase Delta

A functional human MMR system has been reconstituted using recombinant proteins and artificial (plasmid) substrates (Modrich 2006). As depicted in the significantly simplified model shown in Fig. 3.6, mismatch recognition is primarily mediated by the heterodimers MUTSα (comprised of the proteins MSH2/MSH6 or MSH2/MSH3) and MUTLα (comprised of the proteins MLH1/PMS1) (Friedberg et al. 2006). Further details on mismatch recognition and MMR can be found in numerous reviews (Fu et al. 2012; Li 2008; Wyatt and Pittman 2006; Modrich 2006; Jiricny 2006). In an elegant series of biochemical studies, a completely reconstituted system was developed that was capable of supporting directional MMR ($3' \geq 5'$ or $5' \geq 3'$) that exploits a previously undiscovered latent endonuclease activity of MUTLα that is both ATP and mismatch dependent (Kadyrov et al. 2006). Together with EXO1, this in vitro system yields the proper substrate for DNA polymerase loading onto PCNA to allow DNA synthesis of the repair patch for MMR, estimated at 1,000 bases (Thomas et al. 1991) but can range from 200 to >2,000 base pairs, depending on the location of the mismatch and the cellular state (Modrich 2006). In this system, purified Polδ was utilized and found to be fully capable of supporting MMR DNA synthesis (Fig. 3.6).

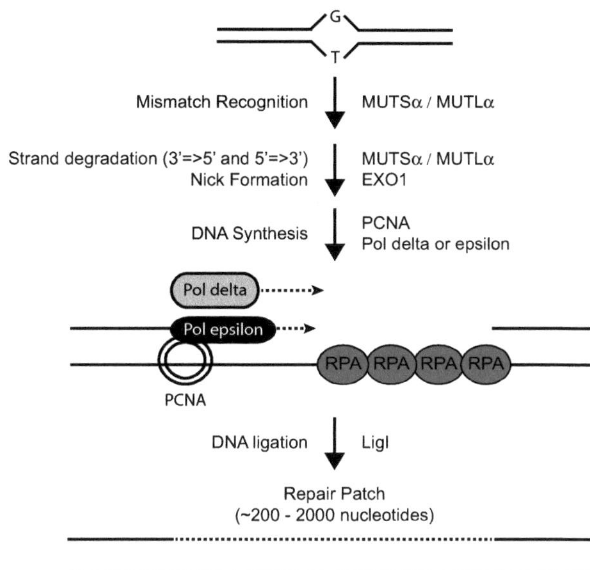

Fig. 3.6 DNA synthesis in human MMR. Schematic depicting a role for Polδ and Polε in the DNA synthesis step of MMR, resulting in a DNA repair patch ranging from 200 to 2,000 bases

3.6.2 Replicative Polymerases Delta and Epsilon in Eukaryotic MMR

Functional (in vivo) studies of eukaryotic MMR and the DNA polymerase requirements for MMR have been limited to *S. cerevisiae* or mouse model systems. As might be expected from the reconstituted system, both replicative polymerases (Polδ and Polε) likely play a role in MMR DNA synthesis. In this model system it is suggested that Polδ, together with Polα, uses the lagging strand as the template for DNA replication whereas Polε uses the leading strand as template (Larrea et al. 2010; Nick McElhinny et al. 2008). It is not yet established if the polymerase used in MMR is also strand specific. Although leading strand (Polε) and lagging strand (Polδ and Polα) fidelity differs, evidence is clear that MMR balances fidelity across both DNA strands (Lujan et al. 2012).

3.7 DNA Polymerase Involved in DNA Cross-link Repair

Characterizing the mechanism or mechanisms that mediate the repair of DNA interstrand cross-links (ICLs) has been a significantly challenging task, complicating the identification of DNA polymerases that may be involved in the repair process. Models have been proposed that depend on replication (Raschle et al. 2008) as well as those that are replication independent (Williams et al. 2012) and involve NER-related transcription-coupled repair or global genome repair and both models (replication dependent and replication independent) involve DNA translesion synthesis (Enoiu et al. 2012). It is generally accepted that complete repair of an ICL involves proteins from several pathways, including the FANC proteins for ICL recognition and signaling (Kim and D'Andrea 2012), HR and NER proteins for lesion processing (Hinz 2010; Wood 2010), as well as TLS polymerases to synthesize DNA across from the unhooked lesion (Enoiu et al. 2012; Ho et al. 2011; Ho and Scharer 2010; Klug et al. 2012; McHugh and Sarkar 2006; Nojima et al. 2005; Sharma and Canman 2012; Shen et al. 2006). One plausible model for the repair of ICLs is shown in Fig. 3.7. In this model, repair can be achieved by a replication-dependent (right panel) or replication-independent (left panel) process. In the left panel, the replication-independent process utilizes NER proteins to "unhook" the cross-link followed by a translesion DNA polymerase to synthesize DNA across the "lesion." This is followed by a second round of NER-mediated repair and DNA synthesis. The NER proteins involved in ICL repair may vary with the lesion. It was recently reported that cisplatin lesions are repaired in a replication-independent fashion utilizing TCR–NER proteins (Enoiu et al. 2012) whereas MMC and psoralen cross-links are suggested to utilize GGR–NER proteins for ICL repair (Hlavin et al. 2010; Wang et al. 2001; Muniandy et al. 2009). In some cases, BER proteins appear to play a role in ICL repair (Kothandapani et al. 2011; Kothandapani and Patrick 2013). The replication-dependent process utilizes FANC proteins to recognize the ICL and mediate unhooking and induce ICL-associated DSBs, in preparation for HR-mediated repair. Both processes rely heavily on translesion DNA polymerases to synthesize DNA opposite the "unhooked" DNA cross-link (Lange et al. 2011; Sharma and Canman 2012). A separate chapter in this series will discuss translesion DNA polymerases (Chap. 4).

3.8 Summary and Concluding Remarks

There are as many as 15 human DNA polymerases to facilitate DNA replication, DNA repair, and DNA lesion bypass (tolerance) in the nucleus and mitochondrial genomes (Burgers et al. 2001; Bebenek and Kunkel 2004). These are characterized by family or class based upon phylogenetic relationships, as described by Burgers et al. (2001). In most cases, the role of some human DNA polymerases in specific DNA repair pathways is expected, based either on data from *E. coli* or *S. cerevisiae* or on biochemical parameters. The high-processivity, high-fidelity replicative DNA polymerases (Polδ and Polε) are a likely option for synthesis of the longer repair patches needed for NER, MMR, and even long-patch BER, whereas the biochemical parameters of the X-family polymerases suggest they are well suited for DNA synthesis required for the short-patch or minimal DNA synthesis observed in BER and NHEJ. Yet, as the field advances, surprises still abound. For example, the Y-family human DNA polymerase Polκ participates in NER and human Polη but not Polδ or Polι is involved in HR-mediated DNA synthesis from the D-loop

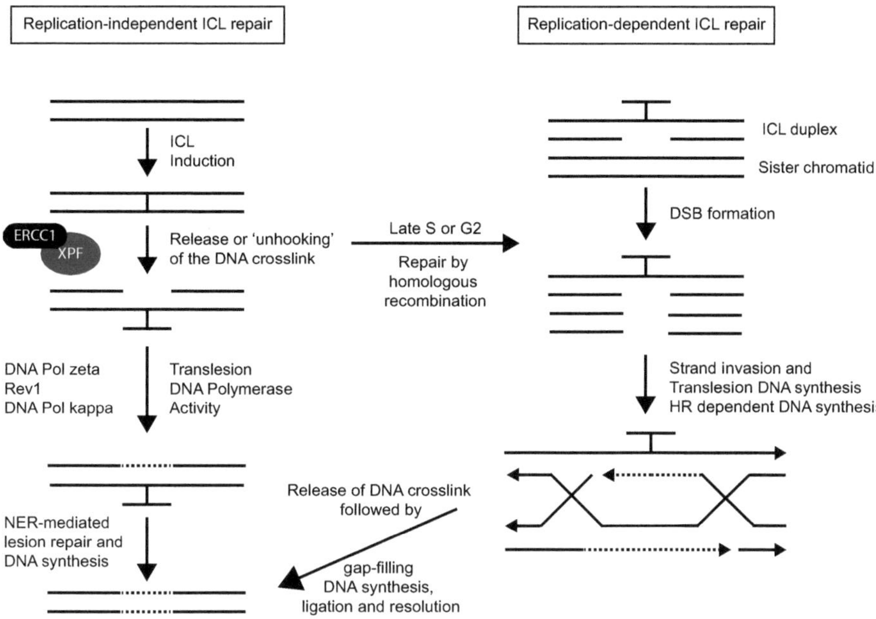

Fig. 3.7 Proposed mechanisms for repair of DNA cross-links. Schematic depicting a current model for the repair of DNA cross-links for cells in the G0/G1 phase of the cell cycle (*left*). Here, repair is mediated by a replication-independent mechanism. Following release or "unhooking" of the lesion (ICL), translesion DNA polymerases can fill the gap across from the lesion, followed by a second lesion removal step and DNA synthesis as in classical NER. On the *right* is a scheme for replication-dependent ICL repair. Here, the repair of DNA cross-links for cells in the late S-phase or G2 phase of the cell cycle would have the availability of the HR pathway to encode the information opposite the unhooked cross-link followed by a second round of DNA synthesis once the lesion (an unhooked cross-link) is removed

intermediate nor involved in second-end capture and the subsequent DNA synthesis step.

Considerable effort is still required to identify and characterize the DNA polymerases involved in many aspects of DNA repair and DNA metabolism. Mutations or defects in DNA polymerases affect response to DNA damaging agents (Sobol et al. 1996, 2000; Trivedi et al. 2005), antibody diversity (Seki et al. 2005), organism survival (Friedberg and Meira 2006), and overall genome maintenance (Prindle and Loeb 2012). The interrelationship between DNA synthesis fidelity and DNA repair is most evident by the cancer predisposition observed when replicative DNA polymerases are mutated in their proofreading domain (Palles et al. 2013). As more details emerge regarding the role of each DNA polymerase in DNA repair, it is expected that we will begin to understand the need for so many different DNA polymerases to maintain genome integrity as well as the multiple roles they may play in the diverse yet interrelated pathways for DNA repair.

References

Aboussekhra A, Biggerstaff M, Shivji MK, Vilpo JA, Moncollin V, Podust VN, Protic M, Hubscher U, Egly JM, Wood RD (1995) Mammalian DNA nucleotide excision repair reconstituted with purified protein components. Cell 80(6):859–868

Ahnesorg P, Smith P, Jackson SP (2006) XLF interacts with the XRCC4-DNA ligase IV complex to promote DNA nonhomologous end-joining. Cell 124(2):301–313. doi:10.1016/j.cell.2005.12.031

Akopiants K, Zhou RZ, Mohapatra S, Valerie K, Lees-Miller SP, Lee KJ, Chen DJ, Revy P, de Villartay JP, Povirk LF (2009) Requirement for XLF/Cernunnos in alignment-based gap filling by DNA polymerases lambda and mu for nonhomologous end joining in human whole-cell extracts. Nucleic Acids Res 37(12):4055–4062. doi:10.1093/nar/gkp283

Albertson TM, Ogawa M, Bugni JM, Hays LE, Chen Y, Wang Y, Treuting PM, Heddle JA, Goldsby RE, Preston BD (2009) DNA polymerase epsilon and delta proofreading suppress discrete mutator and cancer phenotypes in mice. Proc Natl Acad Sci USA 106(40):17101–17104. doi:10.1073/pnas.0907147106

Almeida KH, Sobol RW (2005) Increased specificity and efficiency of base excision repair through complex formation. In: Siede W, Doetsch PW, Kow YW (eds) DNA damage recognition. Marcel Dekker, New York, pp 33–64

Almeida KH, Sobol RW (2007) A unified view of base excision repair: lesion-dependent protein complexes regulated by post-translational modification. DNA Repair 6(6):695–711

An CL, Chen D, Makridakis NM (2011) Systematic biochemical analysis of somatic missense mutations in DNA polymerase beta found in prostate cancer reveal alteration of enzymatic function. Hum Mutat 32(4):415–423. doi:10.1002/humu.21465

Anderson RS, Lawrence CB, Wilson SH, Beattie KL (1987) Genetic relatedness of human DNA polymerase beta and terminal deoxynucleotidyltransferase. Gene 60(2–3):163–173

Andres SN, Vergnes A, Ristic D, Wyman C, Modesti M, Junop M (2012) A human XRCC4-XLF complex bridges DNA. Nucleic Acids Res 40(4):1868–1878. doi:10.1093/nar/gks022

Aoufouchi S, Flatter E, Dahan A, Faili A, Bertocci B, Storck S, Delbos F, Cocea L, Gupta N, Weill JC, Reynaud CA (2000) Two novel human and mouse DNA polymerases of the polX family. Nucleic Acids Res 28(18):3684–3693

Arana ME, Kunkel TA (2010) Mutator phenotypes due to DNA replication infidelity. Semin Cancer Biol 20(5):304–311. doi:10.1016/j.semcancer.2010.10.003

Araujo SJ, Tirode F, Coin F, Pospiech H, Syvaoja JE, Stucki M, Hubscher U, Egly JM, Wood RD (2000) Nucleotide excision repair of DNA with recombinant human proteins: definition of the minimal set of factors, active forms of TFIIH, and modulation by CAK. Genes Dev 14(3):349–359

Barakat KH, Gajewski MM, Tuszynski JA (2012) DNA polymerase beta (pol beta) inhibitors: a comprehensive overview. Drug Discov Today 17(15–16):913–920. doi:10.1016/j.drudis.2012.04.008

Barzel A, Kupiec M (2008) Finding a match: how do homologous sequences get together for recombination? Nat Rev Genet 9(1):27–37. doi:10.1038/nrg2224

Batra VK, Beard WA, Shock DD, Krahn JM, Pedersen LC, Wilson SH (2006) Magnesium-induced assembly of a complete DNA polymerase catalytic complex. Structure 14(4):757–766. doi:10.1016/j.str.2006.01.011

Baute J, Depicker A (2008) Base excision repair and its role in maintaining genome stability. Crit Rev Biochem Mol Biol 43(4):239–276. doi:10.1080/10409230802309905

Beard WA, Wilson SH (2006) Structure and mechanism of DNA polymerase beta. Chem Rev 106(2):361–382

Bebenek K, Kunkel TA (2004) Functions of DNA polymerases. Adv Protein Chem 69:137–165. doi:10.1016/S0065-3233(04)69005-X

Bebenek K, Tissier A, Frank EG, McDonald JP, Prasad R, Wilson SH, Woodgate R, Kunkel TA (2001) 5′-Deoxyribose phosphate lyase activity of human DNA polymerase iota in vitro. Science 291(5511):2156–2159

Bebenek K, Garcia-Diaz M, Blanco L, Kunkel TA (2003) The frameshift infidelity of human DNA polymerase lambda. Implications for function. J Biol Chem 278(36):34685–34690. doi:10.1074/jbc.M305705200

Begg A (2010) POLQ in breast cancer. Oncotarget 1(3):161–162

Benedict CL, Gilfillan S, Thai TH, Kearney JF (2000) Terminal deoxynucleotidyl transferase and repertoire development. Immunol Rev 175:150–157

Bennett SE, Sung JS, Mosbaugh DW (2001) Fidelity of uracil-initiated base excision DNA repair in DNA polymerase β-proficient and -deficient mouse embryonic fibroblast cell extracts. J Biol Chem 276(45):42588–42600

Bertocci B, De Smet A, Flatter E, Dahan A, Bories JC, Landreau C, Weill JC, Reynaud CA (2002) Cutting edge: DNA polymerases mu and lambda are dispensable for Ig gene hypermutation. J Immunol 168(8):3702–3706

Bertocci B, De Smet A, Berek C, Weill JC, Reynaud CA (2003) Immunoglobulin kappa light chain gene rearrangement is impaired in mice deficient for DNA polymerase mu. Immunity 19(2):203–211

Bertocci B, De Smet A, Weill JC, Reynaud CA (2006) Nonoverlapping functions of DNA polymerases mu, lambda, and terminal deoxynucleotidyltransferase during immunoglobulin V(D)J recombination in vivo. Immunity 25(1):31–41. doi:10.1016/j.immuni.2006.04.013

Biade S, Sobol RW, Wilson SH, Matsumoto Y (1998) Impairment of proliferating cell nuclear antigen-dependent apurinic/apyrimidinic site repair on linear DNA. J Biol Chem 273(2):898–902

Blank A, Kim B, Loeb LA (1994) DNA polymerase delta is required for base excision repair of DNA methylation damage in Saccharomyces cerevisiae. Proc Natl Acad Sci USA 91(19):9047–9051

Boboila C, Alt FW, Schwer B (2012) Classical and alternative end-joining pathways for repair of lymphocyte-specific and general DNA double-strand breaks. Adv Immunol 116:1–49. doi:10.1016/B978-0-12-394300-2.00001-6

Bogenhagen DF, Pinz KG, Perez-Jannotti RM (2001) Enzymology of mitochondrial base excision repair. Prog Nucleic Acid Res Mol Biol 68:257–271

Bordeianu G, Zugun-Eloae F, Rusu MG (2011) The role of DNA repair by homologous recombination in oncogenesis. Rev Med Chir Soc Med Nat Iasi 115(4):1189–1194

Boubakour-Azzouz I, Bertrand P, Claes A, Lopez BS, Rougeon F (2012) Terminal deoxynucleotidyl transferase requires KU80 and XRCC4 to promote N-addition at non-V(D)J chromosomal breaks in non-lymphoid cells. Nucleic Acids Res 40(17):8381–8391. doi:10.1093/nar/gks585

Braithwaite EK, Kedar PS, Lan L, Polosina YY, Asagoshi K, Poltoratsky VP, Horton JK, Miller H, Teebor GW, Yasui A, Wilson SH (2005a) DNA polymerase lambda protects mouse fibroblasts against oxidative DNA damage and is recruited to sites of DNA damage/repair. J Biol Chem 280(36):31641–31647

Braithwaite EK, Prasad R, Shock DD, Hou EW, Beard WA, Wilson SH (2005b) DNA polymerase lambda mediates a back-up base excision repair activity in extracts of mouse embryonic fibroblasts. J Biol Chem 280(18):18469–18475

Braithwaite EK, Kedar PS, Stumpo DJ, Bertocci B, Freedman JH, Samson LD, Wilson SH (2010) DNA polymerases beta and lambda mediate overlapping and independent roles in base excision repair in mouse embryonic fibroblasts. PLoS One 5(8):e12229

Brandsma I, Gent DC (2012) Pathway choice in DNA double strand break repair: observations of a balancing act. Genome Integr 3(1):9. doi:10.1186/2041-9414-3-9

Brown JA, Duym WW, Fowler JD, Suo Z (2007) Single-turnover kinetic analysis of the mutagenic potential of 8-oxo-7,8-dihydro-2′-deoxyguanosine during gap-filling synthesis catalyzed by

human DNA polymerases lambda and beta. J Mol Biol 367(5):1258–1269. doi:10.1016/j.jmb.2007.01.069
Brown JA, Pack LR, Sanman LE, Suo Z (2011) Efficiency and fidelity of human DNA polymerases lambda and beta during gap-filling DNA synthesis. DNA Repair (Amst) 10 (1):24–33. doi:10.1016/j.dnarep.2010.09.005
Buck D, Malivert L, de Chasseval R, Barraud A, Fondaneche MC, Sanal O, Plebani A, Stephan JL, Hufnagel M, le Deist F, Fischer A, Durandy A, de Villartay JP, Revy P (2006) Cernunnos, a novel nonhomologous end-joining factor, is mutated in human immunodeficiency with microcephaly. Cell 124(2):287–299. doi:10.1016/j.cell.2005.12.030
Burgers PM, Koonin EV, Bruford E, Blanco L, Burtis KC, Christman MF, Copeland WC, Friedberg EC, Hanaoka F, Hinkle DC, Lawrence CW, Nakanishi M, Ohmori H, Prakash L, Prakash S, Reynaud CA, Sugino A, Todo T, Wang Z, Weill JC, Woodgate R (2001) Eukaryotic DNA polymerases: proposal for a revised nomenclature. J Biol Chem 276(47):43487–43490
Byrnes JJ, Downey KM, Black VL, So AG (1976) A new mammalian DNA polymerase with 3′ to 5′ exonuclease activity: DNA polymerase delta. Biochemistry 15(13):2817–2823
Cabelof DC, Guo Z, Raffoul JJ, Sobol RW, Wilson SH, Richardson A, Heydari AR (2003) Base excision repair deficiency caused by polymerase β haploinsufficiency: accelerated DNA damage and increased mutational response to carcinogens. Cancer Res 63(18):5799–5807
Cabelof DC, Raffoul JJ, Nakamura J, Kapoor D, Abdalla H, Heydari AR (2004) Imbalanced base excision repair in response to folate deficiency is accelerated by polymerase beta haploinsufficiency. J Biol Chem 279(35):36504–36513
Capp JP, Boudsocq F, Besnard AG, Lopez BS, Cazaux C, Hoffmann JS, Canitrot Y (2007) Involvement of DNA polymerase mu in the repair of a specific subset of DNA double-strand breaks in mammalian cells. Nucleic Acids Res 35(11):3551–3560. doi:10.1093/nar/gkm243
Cavero S, Chahwan C, Russell P (2007) Xlf1 is required for DNA repair by nonhomologous end joining in Schizosaccharomyces pombe. Genetics 175(2):963–967. doi:10.1534/genetics.106.067850
Chapman JR, Taylor MR, Boulton SJ (2012) Playing the end game: DNA double-strand break repair pathway choice. Mol Cell 47(4):497–510. doi:10.1016/j.molcel.2012.07.029
Chayot R, Danckaert A, Montagne B, Ricchetti M (2010) Lack of DNA polymerase mu affects the kinetics of DNA double-strand break repair and impacts on cellular senescence. DNA Repair (Amst) 9(11):1187–1199. doi:10.1016/j.dnarep.2010.09.001
Chayot R, Montagne B, Ricchetti M (2012) DNA polymerase mu is a global player in the repair of non-homologous end-joining substrates. DNA Repair (Amst) 11(1):22–34. doi:10.1016/j.dnarep.2011.09.016
Costantini S, Woodbine L, Andreoli L, Jeggo PA, Vindigni A (2007) Interaction of the Ku heterodimer with the DNA ligase IV/Xrcc4 complex and its regulation by DNA-PK. DNA Repair (Amst) 6(6):712–722. doi:10.1016/j.dnarep.2006.12.007
Coverley D, Kenny MK, Lane DP, Wood RD (1992) A role for the human single-stranded DNA binding protein HSSB/RPA in an early stage of nucleotide excision repair. Nucleic Acids Res 20(15):3873–3880
Crespan E, Czabany T, Maga G, Hubscher U (2012) Microhomology-mediated DNA strand annealing and elongation by human DNA polymerases lambda and beta on normal and repetitive DNA sequences. Nucleic Acids Res 40(12):5577–5590. doi:10.1093/nar/gks186
Cui J, Zhao W, Xu X, Yang M, Ren Y, Zhang Z (2012) DNA polymerase beta is involved in the protection against the cytotoxicity and genotoxicity of cigarette smoke. Environ Toxicol Pharmacol 34(2):370–380. doi:10.1016/j.etap.2012.05.012
Curtin NJ (2012) DNA repair dysregulation from cancer driver to therapeutic target. Nat Rev Cancer 12(12):801–817. doi:10.1038/nrc3399
Dalal S, Chikova A, Jaeger J, Sweasy JB (2008) The Leu22Pro tumor-associated variant of DNA polymerase beta is dRP lyase deficient. Nucleic Acids Res 36(2):411–422
David SS, O'Shea VL, Kundu S (2007) Base-excision repair of oxidative DNA damage. Nature 447(7147):941–950. doi:10.1038/nature05978, nature05978 [pii]

de Laat WL, Jaspers NG, Hoeijmakers JH (1999) Molecular mechanism of nucleotide excision repair. Genes Dev 13(7):768–785

de Vries A, van Oostrom CT, Hofhuis FM, Dortant PM, Berg RJ, de Gruijl FR, Wester PW, van Kreijl CF, Capel PJ, van Steeg H, Verbeek SJ (1995) Increased susceptibility to ultraviolet-B and carcinogens of mice lacking the DNA excision repair gene XPA. Nature 377(6545): 169–173. doi:10.1038/377169a0

Deans AJ, West SC (2011) DNA interstrand crosslink repair and cancer. Nat Rev Cancer 11(7): 467–480. doi:10.1038/nrc3088

DeMott MS, Zigman S, Bambara RA (1998) Replication protein A stimulates long patch DNA base excision repair. J Biol Chem 273(42):27492–27498

DeRose EF, Kirby TW, Mueller GA, Bebenek K, Garcia-Diaz M, Blanco L, Kunkel TA, London RE (2003) Solution structure of the lyase domain of human DNA polymerase lambda. Biochemistry 42(32):9564–9574. doi:10.1021/bi034298s

DeRose EF, Clarkson MW, Gilmore SA, Galban CJ, Tripathy A, Havener JM, Mueller GA, Ramsden DA, London RE, Lee AL (2007) Solution structure of polymerase mu's BRCT Domain reveals an element essential for its role in nonhomologous end joining. Biochemistry 46(43):12100–12110. doi:10.1021/bi7007728

Dianov G, Price A, Lindahl T (1992) Generation of single-nucleotide repair patches following excision of uracil residues from DNA. Mol Cell Biol 12(4):1605–1612

Dianov GL, Prasad R, Wilson SH, Bohr VA (1999) Role of DNA polymerase β in the excision step of long patch mammalian base excision repair. J Biol Chem 274(20):13741–13743

DiGiuseppe JA, Dresler SL (1989) Bleomycin-induced DNA repair synthesis in permeable human fibroblasts: mediation of long-patch and short-patch repair by distinct DNA polymerases. Biochemistry 28(24):9515–9520

Doherty AJ, Jackson SP (2001) DNA repair: how Ku makes ends meet. Curr Biol 11(22): R920–R924

Dominguez O, Ruiz JF, Lain de Lera T, Garcia-Diaz M, Gonzalez MA, Kirchhoff T, Martinez AC, Bernad A, Blanco L (2000) DNA polymerase mu, homologous to TdT, could act as a DNA mutator in eukaryotic cells. EMBO J 19(7):1731–1742

Donigan KA, Sun KW, Nemec AA, Murphy DL, Cong X, Northrup V, Zelterman D, Sweasy JB (2012) Human POLB gene is mutated in high percentage of colorectal tumors. J Biol Chem 287 (28):23830–23839. doi:10.1074/jbc.M111.324947

Downs JA, Nussenzweig MC, Nussenzweig A (2007) Chromatin dynamics and the preservation of genetic information. Nature 447(7147):951–958. doi:10.1038/nature05980

El-Andaloussi N, Valovka T, Toueille M, Steinacher R, Focke F, Gehrig P, Covic M, Hassa PO, Schar P, Hubscher U, Hottiger MO (2006) Arginine methylation regulates DNA polymerase β. Mol Cell 22(1):51–62

El-Andaloussi N, Valovka T, Toueille M, Hassa PO, Gehrig P, Covic M, Hubscher U, Hottiger MO (2007) Methylation of DNA polymerase beta by protein arginine methyltransferase 1 regulates its binding to proliferating cell nuclear antigen. FASEB J 21(1):26–34

Enoiu M, Jiricny J, Scharer OD (2012) Repair of cisplatin-induced DNA interstrand crosslinks by a replication-independent pathway involving transcription-coupled repair and translesion synthesis. Nucleic Acids Res 40(18):8953–8964. doi:10.1093/nar/gks670

Esposito G, Texido G, Betz UA, Gu H, Muller W, Klein U, Rajewsky K (2000) Mice reconstituted with DNA polymerase β-deficient fetal liver cells are able to mount a T cell-dependent immune response and mutate their Ig genes normally. Proc Natl Acad Sci USA 97(3):1166–1171

Fabre F, Boulet A, Faye G (1991) Possible involvement of the yeast POLIII DNA polymerase in induced gene conversion. Mol Gen Genet 229(3):353–356

Fan W, Wu X (2004) DNA polymerase lambda can elongate on DNA substrates mimicking non-homologous end joining and interact with XRCC4-ligase IV complex. Biochem Biophys Res Commun 323(4):1328–1333. doi:10.1016/j.bbrc.2004.09.002

Fortini P, Pascucci B, Parlanti E, Sobol RW, Wilson SH, Dogliotti E (1998) Different DNA polymerases are involved in the short- and long-patch base excision repair in mammalian cells. Biochemistry 37(11):3575–3580

Fortini P, Parlanti E, Sidorkina OM, Laval J, Dogliotti E (1999) The type of DNA glycosylase determines the base excision repair pathway in mammalian cells. J Biol Chem 274(21): 15230–15236

Friedberg EC, Meira LB (2006) Database of mouse strains carrying targeted mutations in genes affecting biological responses to DNA damage Version 7. DNA Repair (Amst) 5(2):189–209

Friedberg EC, Walker GC, Siede W, Wood RD, Schultz RA, Ellenberger T (2006) DNA repair and mutagenesis, 2nd edn. ASM Press, Washington, DC

Frouin I, Toueille M, Ferrari E, Shevelev I, Hubscher U (2005) Phosphorylation of human DNA polymerase lambda by the cyclin-dependent kinase Cdk2/cyclin A complex is modulated by its association with proliferating cell nuclear antigen. Nucleic Acids Res 33(16):5354–5361

Fu D, Calvo JA, Samson LD (2012) Balancing repair and tolerance of DNA damage caused by alkylating agents. Nat Rev Cancer 12(2):104–120. doi:10.1038/nrc3185

Gao Y, Katyal S, Lee Y, Zhao J, Rehg JE, Russell HR, McKinnon PJ (2011) DNA ligase III is critical for mtDNA integrity but not Xrcc1-mediated nuclear DNA repair. Nature 471 (7337):240–244. doi:10.1038/nature09773

Garcia-Diaz M, Dominguez O, Lopez-Fernandez LA, de Lera LT, Saniger ML, Ruiz JF, Parraga M, Garcia-Ortiz MJ, Kirchhoff T, del Mazo J, Bernad A, Blanco L (2000) DNA polymerase lambda (Pol lambda), a novel eukaryotic DNA polymerase with a potential role in meiosis. J Mol Biol 301(4):851–867. doi:10.1006/jmbi.2000.4005

Garcia-Diaz M, Bebenek K, Kunkel TA, Blanco L (2001) Identification of an intrinsic 5-′-deoxyribose-5-phosphate lyase activity in human DNA polymerase lambda: a possible role in base excision repair. J Biol Chem 276(37):34659–34663

Garcia-Diaz M, Bebenek K, Sabariegos R, Dominguez O, Rodriguez J, Kirchhoff T, Garcia-Palomero E, Picher AJ, Juarez R, Ruiz JF, Kunkel TA, Blanco L (2002) DNA polymerase lambda, a novel DNA repair enzyme in human cells. J Biol Chem 277(15):13184–13191. doi:10.1074/jbc.M111601200

Garcia-Diaz M, Bebenek K, Krahn JM, Blanco L, Kunkel TA, Pedersen LC (2004) A structural solution for the DNA polymerase lambda-dependent repair of DNA gaps with minimal homology. Mol Cell 13(4):561–572

Garcia-Diaz M, Bebenek K, Gao G, Pedersen LC, London RE, Kunkel TA (2005) Structure-function studies of DNA polymerase lambda. DNA Repair (Amst) 4(12):1358–1367

Gary R, Kim K, Cornelius HL, Park MS, Matsumoto Y (1999) Proliferating cell nuclear antigen facilitates excision in long-patch base excision repair. J Biol Chem 274(7):4354–4363

Gell D, Jackson SP (1999) Mapping of protein-protein interactions within the DNA-dependent protein kinase complex. Nucleic Acids Res 27(17):3494–3502

Gillet LC, Scharer OD (2006) Molecular mechanisms of mammalian global genome nucleotide excision repair. Chem Rev 106(2):253–276. doi:10.1021/cr040483f

Giot L, Chanet R, Simon M, Facca C, Faye G (1997) Involvement of the yeast DNA polymerase delta in DNA repair in vivo. Genetics 146(4):1239–1251

Goellner EM, Grimme B, Brown AR, Lin YC, Wang XH, Sugrue KF, Mitchell L, Trivedi RN, Tang JB, Sobol RW (2011) Overcoming temozolomide resistance in glioblastoma via dual inhibition of NAD+ biosynthesis and base excision repair. Cancer Res 71(6):2308–2317. doi:10.1158/0008-5472.CAN-10-3213

Goellner EM, Svilar D, Almeida KH, Sobol RW (2012) Targeting DNA polymerase β for therapeutic intervention. Curr Mol Pharmacol 5(1):68–87

Goff JP, Shields DS, Seki M, Choi S, Epperly MW, Dixon T, Wang H, Bakkenist CJ, Dertinger SD, Torous DK, Wittschieben J, Wood RD, Greenberger JS (2009) Lack of DNA polymerase theta (POLQ) radiosensitizes bone marrow stromal cells in vitro and increases reticulocyte micronuclei after total-body irradiation. Radiat Res 172(2):165–174. doi:10.1667/RR1598.1

Gonda H, Sugai M, Katakai T, Sugo N, Aratani Y, Koyama H, Mori KJ, Shimizu A (2001) DNA polymerase β is not essential for the formation of palindromic (P) region of T cell receptor gene. Immunol Lett 78(1):45–49

Grundy GJ, Rulten SL, Zeng Z, Arribas-Bosacoma R, Iles N, Manley K, Oliver A, Caldecott KW (2013) APLF promotes the assembly and activity of non-homologous end joining protein complexes. EMBO J 32:112–125. doi:10.1038/emboj.2012.304

Gu H, Marth JD, Orban PC, Mossmann H, Rajewsky K (1994) Deletion of a DNA polymerase β gene segment in T cells using cell type-specific gene targeting. Science 265(5168):103–106

Guo A, Villen J, Kornhauser J, Lee KA, Stokes MP, Rikova K, Possemato A, Nardone J, Innocenti G, Wetzel R, Wang Y, MacNeill J, Mitchell J, Gygi SP, Rush J, Polakiewicz RD, Comb MJ (2008) Signaling networks assembled by oncogenic EGFR and c-Met. Proc Natl Acad Sci USA 105(2):692–697

Hanahan D, Weinberg RA (2011) Hallmarks of cancer: the next generation. Cell 144(5):646–674. doi:10.1016/j.cell.2011.02.013

Hanawalt PC, Spivak G (2008) Transcription-coupled DNA repair: two decades of progress and surprises. Nat Rev Mol Cell Biol 9(12):958–970. doi:10.1038/nrm2549

Harper JW, Elledge SJ (2007) The DNA damage response: ten years after. Mol Cell 28(5): 739–745. doi:10.1016/j.molcel.2007.11.015

Harrigan JA, Opresko PL, von Kobbe C, Kedar PS, Prasad R, Wilson SH, Bohr VA (2003) The Werner syndrome protein stimulates DNA polymerase β strand displacement synthesis via its helicase activity. J Biol Chem 278(25):22686–22695

Harrigan JA, Wilson DM 3rd, Prasad R, Opresko PL, Beck G, May A, Wilson SH, Bohr VA (2006) The Werner syndrome protein operates in base excision repair and cooperates with DNA polymerase beta. Nucleic Acids Res 34(2):745–754

Hasan S, El-Andaloussi N, Hardeland U, Hassa PO, Burki C, Imhof R, Schar P, Hottiger MO (2002) Acetylation regulates the DNA end-trimming activity of DNA polymerase β. Mol Cell 10(5):1213–1222

Hegde ML, Hegde PM, Rao KS, Mitra S (2011) Oxidative genome damage and its repair in neurodegenerative diseases: function of transition metals as a double-edged sword. J Alzheimers Dis 24(Suppl 2):183–198. doi:10.3233/JAD-2011-110281

Higgins GS, Harris AL, Prevo R, Helleday T, McKenna WG, Buffa FM (2010a) Overexpression of POLQ confers a poor prognosis in early breast cancer patients. Oncotarget 1(3):175–184

Higgins GS, Prevo R, Lee YF, Helleday T, Muschel RJ, Taylor S, Yoshimura M, Hickson ID, Bernhard EJ, McKenna WG (2010b) A small interfering RNA screen of genes involved in DNA repair identifies tumor-specific radiosensitization by POLQ knockdown. Cancer Res 70 (7):2984–2993. doi:10.1158/0008-5472.CAN-09-4040

Hinz JM (2010) Role of homologous recombination in DNA interstrand crosslink repair. Environ Mol Mutagen 51(6):582–603. doi:10.1002/em.20577

Hlavin EM, Smeaton MB, Noronha AM, Wilds CJ, Miller PS (2010) Cross-link structure affects replication-independent DNA interstrand cross-link repair in mammalian cells. Biochemistry 49(18):3977–3988. doi:10.1021/bi902169q

Ho TV, Scharer OD (2010) Translesion DNA synthesis polymerases in DNA interstrand crosslink repair. Environ Mol Mutagen 51(6):552–566. doi:10.1002/em.20573

Ho TV, Guainazzi A, Derkunt SB, Enoiu M, Scharer OD (2011) Structure-dependent bypass of DNA interstrand crosslinks by translesion synthesis polymerases. Nucleic Acids Res 39(17): 7455–7464. doi:10.1093/nar/gkr448

Hoeijmakers JH (2001) Genome maintenance mechanisms for preventing cancer. Nature 411(6835): 366–374

Hoeijmakers JH (2009) DNA damage, aging, and cancer. N Engl J Med 361(15):1475–1485. doi:10.1056/NEJMra0804615, 361/15/1475 [pii]

Horton JK, Baker A, Berg BJ, Sobol RW, Wilson SH (2002) Involvement of DNA polymerase β in protection against the cytotoxicity of oxidative DNA damage. DNA Repair (Amst) 1(4): 317–333

Horton JK, Joyce-Gray DF, Pachkowski BF, Swenberg JA, Wilson SH (2003) Hypersensitivity of DNA polymerase beta null mouse fibroblasts reflects accumulation of cytotoxic repair intermediates from site-specific alkyl DNA lesions. DNA Repair (Amst) 2(1):27–48

Horton JK, Stefanick DF, Naron JM, Kedar PS, Wilson SH (2005) Poly(ADP-ribose) polymerase activity prevents signaling pathways for cell cycle arrest following DNA methylating agent exposure. J Biol Chem 280(16):15773–15785

Hubscher U (2009) DNA replication fork proteins. Methods Mol Biol 521:19–33. doi:10.1007/978-1-60327-815-7_2

Hubscher U, Kuenzle CC, Spadari S (1979) Functional roles of DNA polymerases beta and gamma. Proc Natl Acad Sci USA 76(5):2316–2320

Ibe S, Fujita K, Toyomoto T, Shimazaki N, Kaneko R, Tanabe A, Takebe I, Kuroda S, Kobayashi T, Toji S, Tamai K, Yamamoto H, Koiwai O (2001) Terminal deoxynucleotidyltransferase is negatively regulated by direct interaction with proliferating cell nuclear antigen. Genes Cells 6(9):815–824

Ishiguro T, Otsuka F, Ochi T, Ohsawa M (1987) Involvement of DNA polymerases in the repair of DNA damage by benzo[a]pyrene in cultured Chinese hamster cells. Mutat Res 184(1):57–63

Iwanaga A, Ouchida M, Miyazaki K, Hori K, Mukai T (1999) Functional mutation of DNA polymerase β found in human gastric cancer: inability of the base excision repair in vitro. Mutat Res 435(2):121–128

Jackson SP, Bartek J (2009) The DNA-damage response in human biology and disease. Nature 461 (7267):1071–1078. doi:10.1038/nature08467

Jaiswal AS, Banerjee S, Panda H, Bulkin CD, Izumi T, Sarkar FH, Ostrov DA, Narayan S (2009) A novel inhibitor of DNA polymerase beta enhances the ability of temozolomide to impair the growth of colon cancer cells. Mol Cancer Res 7(12):1973–1983. doi:10.1158/1541-7786.MCR-09-0309

Jiricny J (2006) The multifaceted mismatch-repair system. Nat Rev Mol Cell Biol 7(5):335–346

Kadyrov FA, Dzantiev L, Constantin N, Modrich P (2006) Endonucleolytic function of MutLalpha in human mismatch repair. Cell 126(2):297–308. doi:10.1016/j.cell.2006.05.039

Kanagaraj R, Parasuraman P, Mihaljevic B, van Loon B, Burdova K, Konig C, Furrer A, Bohr VA, Hubscher U, Janscak P (2012) Involvement of Werner syndrome protein in MUTYH-mediated repair of oxidative DNA damage. Nucleic Acids Res 40(17):8449–8459. doi:10.1093/nar/gks648

Kass EM, Jasin M (2010) Collaboration and competition between DNA double-strand break repair pathways. FEBS Lett 584(17):3703–3708. doi:10.1016/j.febslet.2010.07.057

Kawamoto T, Araki K, Sonoda E, Yamashita YM, Harada K, Kikuchi K, Masutani C, Hanaoka F, Nozaki K, Hashimoto N, Takeda S (2005) Dual roles for DNA polymerase eta in homologous DNA recombination and translesion DNA synthesis. Mol Cell 20(5):793–799. doi:10.1016/j.molcel.2005.10.016

Kazak L, Reyes A, Holt IJ (2012) Minimizing the damage: repair pathways keep mitochondrial DNA intact. Nat Rev Mol Cell Biol 13(10):659–671. doi:10.1038/nrm3439

Kedar PS, Kim SJ, Robertson A, Hou E, Prasad R, Horton JK, Wilson SH (2002) Direct interaction between mammalian DNA polymerase β and proliferating cell nuclear antigen. J Biol Chem 277(34):31115–31123

Kim H, D'Andrea AD (2012) Regulation of DNA cross-link repair by the Fanconi anemia/BRCA pathway. Genes Dev 26(13):1393–1408. doi:10.1101/gad.195248.112

Klug AR, Harbut MB, Lloyd RS, Minko IG (2012) Replication bypass of N2-deoxyguanosine interstrand cross-links by human DNA polymerases eta and iota. Chem Res Toxicol 25(3):755–762. doi:10.1021/tx300011w

Knutson CG, Akingbade D, Crews BC, Voehler M, Stec DF, Marnett LJ (2007) Metabolism in vitro and in vivo of the DNA base adduct, M1G. Chem Res Toxicol 20(3):550–557. doi:10.1021/tx600334x

Knutson CG, Rubinson EH, Akingbade D, Anderson CS, Stec DF, Petrova KV, Kozekov ID, Guengerich FP, Rizzo CJ, Marnett LJ (2009) Oxidation and glycolytic cleavage of etheno and propano DNA base adducts. Biochemistry 48(4):800–809. doi:10.1021/bi801654j
Kothandapani A, Patrick SM (2013) Evidence for base excision repair processing of DNA interstrand crosslinks. Mutat Res 743–744:44–52. doi:10.1016/j.mrfmmm.2012.11.007
Kothandapani A, Dangeti VS, Brown AR, Banze LA, Wang XH, Sobol RW, Patrick SM (2011) Novel role of base excision repair in mediating cisplatin cytotoxicity. J Biol Chem 286(16): 14564–14574. doi:10.1074/jbc.M111.225375
Krejci L, Altmannova V, Spirek M, Zhao X (2012) Homologous recombination and its regulation. Nucleic Acids Res 40(13):5795–5818. doi:10.1093/nar/gks270
Kubota Y, Nash RA, Klungland A, Schar P, Barnes DE, Lindahl T (1996) Reconstitution of DNA base excision-repair with purified human proteins: interaction between DNA polymerase β and the XRCC1 protein. EMBO J 15(23):6662–6670
Kunkel TA (2009) Evolving views of DNA replication (in)fidelity. Cold Spring Harb Symp Quant Biol 74:91–101. doi:10.1101/sqb.2009.74.027
Kunkel TA, Burgers PM (2008) Dividing the workload at a eukaryotic replication fork. Trends Cell Biol 18(11):521–527. doi:10.1016/j.tcb.2008.08.005
Kunkel TA, Erie DA (2005) DNA mismatch repair. Annu Rev Biochem 74:681–710. doi:10.1146/annurev.biochem.74.082803.133243
Kunkel TA, Van Houten B (2006) Survival choices. Nat Cell Biol 8(6):547–549. doi:10.1038/ncb0606-547
Kuper J, Kisker C (2012) Damage recognition in nucleotide excision DNA repair. Curr Opin Struct Biol 22(1):88–93. doi:10.1016/j.sbi.2011.12.002
Lai Y, Zhao W, Chen C, Wu M, Zhang Z (2011) Role of DNA polymerase beta in the genotoxicity of arsenic. Environ Mol Mutagen 52(6):460–468. doi:10.1002/em.20643
Lang T, Dalal S, Chikova A, Dimaio D, Sweasy JB (2007) The E295K DNA polymerase beta gastric cancer-associated variant interferes with base excision repair and induces cellular transformation. Mol Cell Biol 27(15):5587–5596
Lange SS, Takata K, Wood RD (2011) DNA polymerases and cancer. Nat Rev Cancer 11(2): 96–110. doi:10.1038/nrc2998
Larrea AA, Lujan SA, Nick McElhinny SA, Mieczkowski PA, Resnick MA, Gordenin DA, Kunkel TA (2010) Genome-wide model for the normal eukaryotic DNA replication fork. Proc Natl Acad Sci USA 107(41):17674–17679. doi:10.1073/pnas.1010178107
Le Page F, Schreiber V, Dherin C, De Murcia G, Boiteux S (2003) Poly(ADP-ribose) polymerase-1 (PARP-1) is required in murine cell lines for base excision repair of oxidative DNA damage in the absence of DNA polymerase β. J Biol Chem 278(20):18471–18477
Lebedeva NA, Rechkunova NI, Dezhurov SV, Khodyreva SN, Favre A, Blanco L, Lavrik OI (2005) Comparison of functional properties of mammalian DNA polymerase lambda and DNA polymerase beta in reactions of DNA synthesis related to DNA repair. Biochim Biophys Acta 1751(2):150–158. doi:10.1016/j.bbapap.2005.05.012
Lee JW, Blanco L, Zhou T, Garcia-Diaz M, Bebenek K, Kunkel TA, Wang Z, Povirk LF (2004) Implication of DNA polymerase lambda in alignment-based gap filling for nonhomologous DNA end joining in human nuclear extracts. J Biol Chem 279(1):805–811. doi:10.1074/jbc.M307913200
Lehmann AR (2011) DNA polymerases and repair synthesis in NER in human cells. DNA Repair (Amst) 10(7):730–733. doi:10.1016/j.dnarep.2011.04.023
Lemee F, Bergoglio V, Fernandez-Vidal A, Machado-Silva A, Pillaire MJ, Bieth A, Gentil C, Baker L, Martin AL, Leduc C, Lam E, Magdeleine E, Filleron T, Oumouhou N, Kaina B, Seki M, Grimal F, Lacroix-Triki M, Thompson A, Roche H, Bourdon JC, Wood RD, Hoffmann JS, Cazaux C (2010) DNA polymerase theta up-regulation is associated with poor survival in breast cancer, perturbs DNA replication, and promotes genetic instability. Proc Natl Acad Sci USA 107(30):13390–13395. doi:10.1073/pnas.0910759107

Li GM (2008) Mechanisms and functions of DNA mismatch repair. Cell Res 18(1):85–98. doi:10.1038/cr.2007.115
Li X, Stith CM, Burgers PM, Heyer WD (2009) PCNA is required for initiation of recombination-associated DNA synthesis by DNA polymerase delta. Mol Cell 36(4):704–713. doi:10.1016/j.molcel.2009.09.036
Li J, Luthra S, Wang XH, Chandran UR, Sobol RW (2012a) Transcriptional profiling reveals elevated Sox2 in DNA polymerase ss null mouse embryonic fibroblasts. Am J Cancer Res 2 (6):699–713
Li Y, Gridley CL, Jaeger J, Sweasy JB, Schlick T (2012b) Unfavorable electrostatic and steric interactions in DNA polymerase beta E295K mutant interfere with the enzyme's pathway. J Am Chem Soc 134(24):9999–10010. doi:10.1021/ja300361r
Liberti SE, Larrea AA, Kunkel TA (2013) Exonuclease 1 preferentially repairs mismatches generated by DNA polymerase alpha. DNA Repair (Amst) 12:92–96. doi:10.1016/j.dnarep.2012.11.001
Lieber MR (2008) The mechanism of human nonhomologous DNA end joining. J Biol Chem 283(1):1–5. doi:10.1074/jbc.R700039200
Lindahl T (1993) Instability and decay of the primary structure of DNA. Nature 362(6422):709–715
Liu P, Demple B (2010) DNA repair in mammalian mitochondria: much more than we thought? Environ Mol Mutagen 51(5):417–426. doi:10.1002/em.20576
Liu P, Qian L, Sung JS, de Souza-Pinto NC, Zheng L, Bogenhagen DF, Bohr VA, Wilson DM 3rd, Shen B, Demple B (2008) Removal of oxidative DNA damage via FEN1-dependent long-patch base excision repair in human cell mitochondria. Mol Cell Biol 28(16):4975–4987
Longley MJ, Prasad R, Srivastava DK, Wilson SH, Copeland WC (1998) Identification of 5′-deoxyribose phosphate lyase activity in human DNA polymerase γ and its role in mitochondrial base excision repair in vitro. Proc Natl Acad Sci USA 95(21):12244–12248
Longley MJ, Graziewicz MA, Bienstock RJ, Copeland WC (2005) Consequences of mutations in human DNA polymerase gamma. Gene 354:125–131. doi:10.1016/j.gene.2005.03.029
Lucas D, Escudero B, Ligos JM, Segovia JC, Estrada JC, Terrados G, Blanco L, Samper E, Bernad A (2009) Altered hematopoiesis in mice lacking DNA polymerase mu is due to inefficient double-strand break repair. PLoS Genet 5(2):e1000389. doi:10.1371/journal.pgen.1000389
Lujan SA, Williams JS, Pursell ZF, Abdulovic-Cui AA, Clark AB, Nick McElhinny SA, Kunkel TA (2012) Mismatch repair balances leading and lagging strand DNA replication fidelity. PLoS Genet 8(10):e1003016. doi:10.1371/journal.pgen.1003016
Lydeard JR, Jain S, Yamaguchi M, Haber JE (2007) Break-induced replication and telomerase-independent telomere maintenance require Pol32. Nature 448(7155):820–823. doi:10.1038/nature06047
Ma Y, Lu H, Tippin B, Goodman MF, Shimazaki N, Koiwai O, Hsieh CL, Schwarz K, Lieber MR (2004) A biochemically defined system for mammalian nonhomologous DNA end joining. Mol Cell 16(5):701–713. doi:10.1016/j.molcel.2004.11.017
Mahajan KN, Mitchell BS (2003) Role of human Pso4 in mammalian DNA repair and association with terminal deoxynucleotidyl transferase. Proc Natl Acad Sci USA 100(19):10746–10751. doi:10.1073/pnas.1631060100
Mahajan KN, Gangi-Peterson L, Sorscher DH, Wang J, Gathy KN, Mahajan NP, Reeves WH, Mitchell BS (1999) Association of terminal deoxynucleotidyl transferase with Ku. Proc Natl Acad Sci USA 96(24):13926–13931
Mahajan KN, Nick McElhinny SA, Mitchell BS, Ramsden DA (2002) Association of DNA polymerase mu (pol mu) with Ku and ligase IV: role for pol mu in end-joining double-strand break repair. Mol Cell Biol 22(14):5194–5202
Maloisel L, Bhargava J, Roeder GS (2004) A role for DNA polymerase delta in gene conversion and crossing over during meiosis in Saccharomyces cerevisiae. Genetics 167(3):1133–1142. doi:10.1534/genetics.104.026260

Maloisel L, Fabre F, Gangloff S (2008) DNA polymerase delta is preferentially recruited during homologous recombination to promote heteroduplex DNA extension. Mol Cell Biol 28(4): 1373–1382. doi:10.1128/MCB.01651-07

Malu S, Malshetty V, Francis D, Cortes P (2012) Role of non-homologous end joining in V(D)J recombination. Immunol Res 54(1–3):233–246. doi:10.1007/s12026-012-8329-z

Markkanen E, van Loon B, Ferrari E, Hubscher U (2011) Ubiquitylation of DNA polymerase lambda. FEBS Lett 585(18):2826–2830. doi:10.1016/j.febslet.2011.03.069

Markkanen E, van Loon B, Ferrari E, Parsons JL, Dianov GL, Hubscher U (2012) Regulation of oxidative DNA damage repair by DNA polymerase lambda and MutYH by cross-talk of phosphorylation and ubiquitination. Proc Natl Acad Sci USA 109(2):437–442. doi:10.1073/pnas.1110449109

Marnett LJ (2000) Oxyradicals and DNA damage. Carcinogenesis 21(3):361–370

Marnett LJ, Riggins JN, West JD (2003) Endogenous generation of reactive oxidants and electrophiles and their reactions with DNA and protein. J Clin Invest 111(5):583–593. doi:10.1172/JCI18022

Matsumoto Y, Kim K (1995) Excision of deoxyribose phosphate residues by DNA polymerase β during DNA repair. Science 269(5224):699–702

Matsumoto Y, Kim K, Hurwitz J, Gary R, Levin DS, Tomkinson AE, Park MS (1999) Reconstitution of proliferating cell nuclear antigen-dependent repair of apurinic/apyrimidinic sites with purified human proteins. J Biol Chem 274(47):33703–33708

Matsumoto T, Go K, Hyodo M, Koiwai K, Maezawa S, Hayano T, Suzuki M, Koiwai O (2012) BRCT domain of DNA polymerase mu has DNA-binding activity and promotes the DNA polymerization activity. Genes Cells 17(9):790–806. doi:10.1111/j.1365-2443.2012.01628.x

McCulloch SD, Kunkel TA (2008) The fidelity of DNA synthesis by eukaryotic replicative and translesion synthesis polymerases. Cell Res 18(1):148–161. doi:10.1038/cr.2008.4

McHugh PJ, Sarkar S (2006) DNA interstrand cross-link repair in the cell cycle: a critical role for polymerase zeta in G1 phase. Cell Cycle 5(10):1044–1047

McIlwraith MJ, West SC (2008) DNA repair synthesis facilitates RAD52-mediated second-end capture during DSB repair. Mol Cell 29(4):510–516. doi:10.1016/j.molcel.2007.11.037

McIlwraith MJ, Vaisman A, Liu Y, Fanning E, Woodgate R, West SC (2005) Human DNA polymerase eta promotes DNA synthesis from strand invasion intermediates of homologous recombination. Mol Cell 20(5):783–792. doi:10.1016/j.molcel.2005.10.001

McVey M, Lee SE (2008) MMEJ repair of double-strand breaks (director's cut): deleted sequences and alternative endings. Trends Genet 24(11):529–538. doi:10.1016/j.tig.2008.08.007

Modrich P (2006) Mechanisms in eukaryotic mismatch repair. J Biol Chem 281(41):30305–30309

Moon AF, Garcia-Diaz M, Bebenek K, Davis BJ, Zhong X, Ramsden DA, Kunkel TA, Pedersen LC (2007) Structural insight into the substrate specificity of DNA polymerase mu. Nat Struct Mol Biol 14(1):45–53. doi:10.1038/nsmb1180

Mosbaugh DW, Linn S (1983) Excision repair and DNA synthesis with a combination of HeLa DNA polymerase beta and DNase V. J Biol Chem 258(1):108–118

Mosbaugh DW, Linn S (1984) Gap-filling DNA synthesis by HeLa DNA polymerase alpha in an in vitro base excision DNA repair scheme. J Biol Chem 259(16):10247–10251

Mueller GA, Moon AF, Derose EF, Havener JM, Ramsden DA, Pedersen LC, London RE (2008) A comparison of BRCT domains involved in nonhomologous end-joining: introducing the solution structure of the BRCT domain of polymerase lambda. DNA Repair (Amst) 7(8): 1340–1351. doi:10.1016/j.dnarep.2008.04.018

Muniandy PA, Thapa D, Thazhathveetil AK, Liu ST, Seidman MM (2009) Repair of laser-localized DNA interstrand cross-links in G1 phase mammalian cells. J Biol Chem 284(41): 27908–27917. doi:10.1074/jbc.M109.029025

Murray JM, Stiff T, Jeggo PA (2012) DNA double-strand break repair within heterochromatic regions. Biochem Soc Trans 40(1):173–178. doi:10.1042/BST20110631

Nagasawa K, Kitamura K, Yasui A, Nimura Y, Ikeda K, Hirai M, Matsukage A, Nakanishi M (2000) Identification and characterization of human DNA polymerase beta 2, a DNA polymerase beta -related enzyme. J Biol Chem 275(40):31233–31238. doi:10.1074/jbc.M004263200

Nealon K, Nicholl ID, Kenny MK (1996) Characterization of the DNA polymerase requirement of human base excision repair. Nucleic Acids Res 24(19):3763–3770

Neijenhuis S, Begg AC, Vens C (2005) Radiosensitization by a dominant negative to DNA polymerase beta is DNA polymerase beta-independent and XRCC1-dependent. Radiother Oncol 76(2):123–128. doi:10.1016/j.radonc.2005.06.020, S0167-8140(05)00249-5 [pii]

Neijenhuis S, Verwijs-Janssen M, Kasten-Pisula U, Rumping G, Borgmann K, Dikomey E, Begg AC, Vens C (2009) Mechanism of cell killing after ionizing radiation by a dominant negative DNA polymerase beta. DNA Repair (Amst) 8(3):336–346. doi:10.1016/j.dnarep.2008.11.008, S1568-7864(08)00398-4 [pii]

Neijenhuis S, Verwijs-Janssen M, van den Broek LJ, Begg AC, Vens C (2010) Targeted radiosensitization of cells expressing truncated DNA polymerase {beta}. Cancer Res 70(21): 8706–8714. doi:10.1158/0008-5472.CAN-09-3901

Nemec AA, Donigan KA, Murphy DL, Jaeger J, Sweasy JB (2012) Colon cancer-associated DNA polymerase beta variant induces genomic instability and cellular transformation. J Biol Chem 287(28):23840–23849. doi:10.1074/jbc.M112.362111

Nick McElhinny SA, Ramsden DA (2003) Polymerase mu is a DNA-directed DNA/RNA polymerase. Mol Cell Biol 23(7):2309–2315

Nick McElhinny SA, Gordenin DA, Stith CM, Burgers PM, Kunkel TA (2008) Division of labor at the eukaryotic replication fork. Mol Cell 30(2):137–144. doi:10.1016/j.molcel.2008.02.022

Nick McElhinny SA, Kissling GE, Kunkel TA (2010) Differential correction of lagging-strand replication errors made by DNA polymerases alpha and {delta}. Proc Natl Acad Sci USA 107 (49):21070–21075. doi:10.1073/pnas.1013048107

Niimi A, Limsirichaikul S, Yoshida S, Iwai S, Masutani C, Hanaoka F, Kool ET, Nishiyama Y, Suzuki M (2004) Palm mutants in DNA polymerases alpha and eta alter DNA replication fidelity and translesion activity. Mol Cell Biol 24(7):2734–2746

Niimi N, Sugo N, Aratani Y, Koyama H (2005) Genetic interaction between DNA polymerase β and DNA-PKcs in embryogenesis and neurogenesis. Cell Death Differ 12(2):184–191

Nojima K, Hochegger H, Saberi A, Fukushima T, Kikuchi K, Yoshimura M, Orelli BJ, Bishop DK, Hirano S, Ohzeki M, Ishiai M, Yamamoto K, Takata M, Arakawa H, Buerstedde JM, Yamazoe M, Kawamoto T, Araki K, Takahashi JA, Hashimoto N, Takeda S, Sonoda E (2005) Multiple repair pathways mediate tolerance to chemotherapeutic cross-linking agents in vertebrate cells. Cancer Res 65(24):11704–11711. doi:10.1158/0008-5472.CAN-05-1214

O'Driscoll M (2012) Diseases associated with defective responses to DNA damage. Cold Spring Harb Perspect Biol 4(12):a012773. doi:10.1101/cshperspect.a012773

Ochs K, Sobol RW, Wilson SH, Kaina B (1999) Cells deficient in DNA polymerase β are hypersensitive to alkylating agent-induced apoptosis and chromosomal breakage. Cancer Res 59(7):1544–1551

Ochs K, Lips J, Profittlich S, Kaina B (2002) Deficiency in DNA polymerase β provokes replication-dependent apoptosis via DNA breakage, Bcl-2 decline and caspase-3/9 activation. Cancer Res 62(5):1524–1530

Ogi T, Lehmann AR (2006) The Y-family DNA polymerase kappa (pol kappa) functions in mammalian nucleotide-excision repair. Nat Cell Biol 8(6):640–642. doi:10.1038/ncb1417

Ogi T, Kannouche P, Lehmann AR (2005) Localisation of human Y-family DNA polymerase kappa: relationship to PCNA foci. J Cell Sci 118(Pt 1):129–136. doi:10.1242/jcs.01603

Ogi T, Limsirichaikul S, Overmeer RM, Volker M, Takenaka K, Cloney R, Nakazawa Y, Niimi A, Miki Y, Jaspers NG, Mullenders LH, Yamashita S, Fousteri MI, Lehmann AR (2010) Three DNA polymerases, recruited by different mechanisms, carry out NER repair synthesis in human cells. Mol Cell 37(5):714–727. doi:10.1016/j.molcel.2010.02.009

Ohashi E, Bebenek K, Matsuda T, Feaver WJ, Gerlach VL, Friedberg EC, Ohmori H, Kunkel TA (2000) Fidelity and processivity of DNA synthesis by DNA polymerase kappa, the product of the human DINB1 gene. J Biol Chem 275(50):39678–39684. doi:10.1074/jbc.M005309200

Ohnishi T, Yuba S, Date T, Utsumi H, Matsukage A (1990) Rat DNA polymerase β gene can join in excision repair of Escherichia coli. Nucleic Acids Res 18(19):5673–5676

Ono K, Ohashi A, Tanabe K, Matsukage A, Nishizawa M, Takahashi T (1979) Unique requirements for template primers of DNA polymerase beta from rat ascites hepatoma AH130 cells. Nucleic Acids Res 7(3):715–726

Orlando P, Geremia R, Frusciante C, Tedeschi B, Grippo P (1988) DNA repair synthesis in mouse spermatogenesis involves DNA polymerase beta activity. Cell Differ 23(3):221–230

Otteneder MB, Knutson CG, Daniels JS, Hashim M, Crews BC, Remmel RP, Wang H, Rizzo C, Marnett LJ (2006) In vivo oxidative metabolism of a major peroxidation-derived DNA adduct, M1dG. Proc Natl Acad Sci USA 103(17):6665–6669. doi:10.1073/pnas.0602017103

Overmeer RM, Gourdin AM, Giglia-Mari A, Kool H, Houtsmuller AB, Siegal G, Fousteri MI, Mullenders LH, Vermeulen W (2010) Replication factor C recruits DNA polymerase delta to sites of nucleotide excision repair but is not required for PCNA recruitment. Mol Cell Biol 30 (20):4828–4839. doi:10.1128/MCB.00285-10

Palles C, Cazier JB, Howarth KM, Domingo E, Jones AM, Broderick P, Kemp Z, Spain SL, Guarino E, Salguero I, Sherborne A, Chubb D, Carvajal-Carmona LG, Ma Y, Kaur K, Dobbins S, Barclay E, Gorman M, Martin L, Kovac MB, Humphray S, CORGI Consortium, WGS500 Consortium, Lucassen A, Holmes CC, Bentley D, Donnelly P, Taylor J, Petridis C, Roylance R, Sawyer EJ, Kerr DJ, Clark S, Grimes J, Kearsey SE, Thomas HJ, McVean G, Houlston RS, Tomlinson I (2013) Germline mutations affecting the proofreading domains of POLE and POLD1 predispose to colorectal adenomas and carcinomas. Nat Genet 45(2):136–144. doi:10.1038/ng.2503

Park IS, Park JK, Koh HY, Park SD (1991) DNA single stranded gaps formed during DNA repair synthesis induced by methyl methanesulfonate are filled by sequential action of aphidicolin- and dideoxythymidine sensitive DNA polymerases in HeLa cells. Cell Biol Toxicol 7(1):49–58

Parlanti E, Fortini P, Macpherson P, Laval J, Dogliotti E (2002) Base excision repair of adenine/8-oxoguanine mispairs by an aphidicolin-sensitive DNA polymerase in human cell extracts. Oncogene 21(34):5204–5212. doi:10.1038/sj.onc.1205561

Parsons JL, Tait PS, Finch D, Dianova II, Allinson SL, Dianov GL (2008) CHIP-mediated degradation and DNA damage-dependent stabilization regulate base excision repair proteins. Mol Cell 29(4):477–487

Parsons JL, Tait PS, Finch D, Dianova II, Edelmann MJ, Khoronenkova SV, Kessler BM, Sharma RA, McKenna WG, Dianov GL (2009) Ubiquitin ligase ARF-BP1/Mule modulates base excision repair. EMBO J 28(20):3207–3215. doi:10.1038/emboj.2009.243, emboj2009243 [pii]

Pascucci B, Stucki M, Jonsson ZO, Dogliotti E, Hubscher U (1999) Long patch base excision repair with purified human proteins. DNA ligase I as patch size mediator for DNA polymerases delta and epsilon. J Biol Chem 274(47):33696–33702

Paull TT (2005) Saving the ends for last: the role of pol mu in DNA end joining. Mol Cell 19(3): 294–296. doi:10.1016/j.molcel.2005.07.008

Pawelczak KS, Bennett SM, Turchi JJ (2011) Coordination of DNA-PK activation and nuclease processing of DNA termini in NHEJ. Antioxid Redox Signal 14(12):2531–2543. doi:10.1089/ars.2010.3368

Petta TB, Nakajima S, Zlatanou A, Despras E, Couve-Privat S, Ishchenko A, Sarasin A, Yasui A, Kannouche P (2008) Human DNA polymerase iota protects cells against oxidative stress. EMBO J 27(21):2883–2895. doi:10.1038/emboj.2008.210
Phosphosite (2010) PhosphositePlus. http://www.phosphosite.org
Pinz KG, Bogenhagen DF (2006) The influence of the DNA polymerase gamma accessory subunit on base excision repair by the catalytic subunit. DNA Repair (Amst) 5(1):121–128. doi:10.1016/j.dnarep.2005.08.014
Podlutsky AJ, Dianova II, Podust VN, Bohr VA, Dianov GL (2001) Human DNA polymerase beta initiates DNA synthesis during long-patch repair of reduced AP sites in DNA. EMBO J 20(6):1477–1482. doi:10.1093/emboj/20.6.1477
Poltoratsky V, Horton JK, Prasad R, Wilson SH (2005) REV1 mediated mutagenesis in base excision repair deficient mouse fibroblast. DNA Repair (Amst) 4(10):1182–1188
Poltoratsky V, Horton JK, Prasad R, Beard WA, Woodgate R, Wilson SH (2008) Negligible impact of pol iota expression on the alkylation sensitivity of pol beta-deficient mouse fibroblast cells. DNA Repair (Amst) 7(6):830–833
Prasad R, Dianov GL, Bohr VA, Wilson SH (2000) FEN1 stimulation of DNA polymerase β mediates an excision step in mammalian long patch base excision repair. J Biol Chem 275(6):4460–4466
Prasad R, Lavrik OI, Kim SJ, Kedar P, Yang XP, Vande Berg BJ, Wilson SH (2001) DNA polymerase β-mediated long patch base excision repair. Poly(ADP-ribose)polymerase-1 stimulates strand displacement DNA synthesis. J Biol Chem 276(35):32411–32414
Prasad R, Bebenek K, Hou E, Shock DD, Beard WA, Woodgate R, Kunkel TA, Wilson SH (2003) Localization of the deoxyribose phosphate lyase active site in human DNA polymerase iota by controlled proteolysis. J Biol Chem 278(32):29649–29654
Prasad R, Longley MJ, Sharief FS, Hou EW, Copeland WC, Wilson SH (2009) Human DNA polymerase theta possesses 5′-dRP lyase activity and functions in single-nucleotide base excision repair in vitro. Nucleic Acids Res 37(6):1868–1877. doi:10.1093/nar/gkp035
Preston BD, Albertson TM, Herr AJ (2010) DNA replication fidelity and cancer. Semin Cancer Biol 20(5):281–293. doi:10.1016/j.semcancer.2010.10.009
Price A (1993) The repair of ionising radiation-induced damage to DNA. Semin Cancer Biol 4(2):61–71
Prindle MJ, Loeb LA (2012) DNA polymerase delta in DNA replication and genome maintenance. Environ Mol Mutagen 53(9):666–682. doi:10.1002/em.21745
Ramadan K, Maga G, Shevelev IV, Villani G, Blanco L, Hubscher U (2003) Human DNA polymerase lambda possesses terminal deoxyribonucleotidyl transferase activity and can elongate RNA primers: implications for novel functions. J Mol Biol 328(1):63–72
Ramsden DA (2011) Polymerases in nonhomologous end joining: building a bridge over broken chromosomes. Antioxid Redox Signal 14(12):2509–2519. doi:10.1089/ars.2010.3429
Ramsden DA, Asagoshi K (2012) DNA polymerases in nonhomologous end joining: are there any benefits to standing out from the crowd? Environ Mol Mutagen 53(9):741–751. doi:10.1002/em.21725
Raschle M, Knipscheer P, Enoiu M, Angelov T, Sun J, Griffith JD, Ellenberger TE, Scharer OD, Walter JC (2008) Mechanism of replication-coupled DNA interstrand crosslink repair. Cell 134(6):969–980. doi:10.1016/j.cell.2008.08.030
Reha-Krantz LJ (2010) DNA polymerase proofreading: multiple roles maintain genome stability. Biochim Biophys Acta 1804(5):1049–1063. doi:10.1016/j.bbapap.2009.06.012
Repasky JA, Corbett E, Boboila C, Schatz DG (2004) Mutational analysis of terminal deoxynucleotidyltransferase-mediated N-nucleotide addition in V(D)J recombination. J Immunol 172(9):5478–5488
Revy P, Malivert L, de Villartay JP (2006) Cernunnos-XLF, a recently identified non-homologous end-joining factor required for the development of the immune system. Curr Opin Allergy Clin Immunol 6(6):416–420. doi:10.1097/01.all.0000246623.72365.43

Rivera-Calzada A, Spagnolo L, Pearl LH, Llorca O (2007) Structural model of full-length human Ku70-Ku80 heterodimer and its recognition of DNA and DNA-PKcs. EMBO Rep 8(1):56–62. doi:10.1038/sj.embor.7400847

Ropars V, Drevet P, Legrand P, Baconnais S, Amram J, Faure G, Marquez JA, Pietrement O, Guerois R, Callebaut I, Le Cam E, Revy P, de Villartay JP, Charbonnier JB (2011) Structural characterization of filaments formed by human Xrcc4-Cernunnos/XLF complex involved in nonhomologous DNA end-joining. Proc Natl Acad Sci USA 108(31):12663–12668. doi:10.1073/pnas.1100758108

Ruiz JF, Dominguez O, Lain de Lera T, Garcia-Diaz M, Bernad A, Blanco L (2001) DNA polymerase mu, a candidate hypermutase? Philos Trans R Soc Lond Ser B Biol Sci 356(1405): 99–109. doi:10.1098/rstb.2000.0754

Ruiz JF, Lucas D, Garcia-Palomero E, Saez AI, Gonzalez MA, Piris MA, Bernad A, Blanco L (2004) Overexpression of human DNA polymerase mu in a Burkitt's lymphoma cell line affects the somatic hypermutation rate. Nucleic Acids Res 32(19):5861–5873

Sebesta M, Burkovics P, Haracska L, Krejci L (2011) Reconstitution of DNA repair synthesis in vitro and the role of polymerase and helicase activities. DNA Repair (Amst) 10(6):567–576. doi:10.1016/j.dnarep.2011.03.003

Seki S, Oda T (1986) DNA repair synthesis in bleomycin-pretreated permeable HeLa cells. Carcinogenesis 7(1):77–82

Seki M, Gearhart PJ, Wood RD (2005) DNA polymerases and somatic hypermutation of immunoglobulin genes. EMBO Rep 6(12):1143–1148. doi:10.1038/sj.embor.7400582

Senejani AG, Dalal S, Liu Y, Nottoli TP, McGrath JM, Clairmont CS, Sweasy JB (2012) Y265C DNA polymerase beta knockin mice survive past birth and accumulate base excision repair intermediate substrates. Proc Natl Acad Sci USA 109(17):6632–6637. doi:10.1073/pnas.1200800109

Setlow RB, Setlow JK (1962) Evidence that ultraviolet-induced thymine dimers in DNA cause biological damage. Proc Natl Acad Sci USA 48:1250–1257

Sharma S, Canman CE (2012) REV1 and DNA polymerase zeta in DNA interstrand crosslink repair. Environ Mol Mutagen 53(9):725–740. doi:10.1002/em.21736

Shen X, Jun S, O'Neal LE, Sonoda E, Bemark M, Sale JE, Li L (2006) REV3 and REV1 play major roles in recombination-independent repair of DNA interstrand cross-links mediated by monoubiquitinated proliferating cell nuclear antigen (PCNA). J Biol Chem 281(20): 13869–13872. doi:10.1074/jbc.C600071200

Shivji KK, Kenny MK, Wood RD (1992) Proliferating cell nuclear antigen is required for DNA excision repair. Cell 69(2):367–374

Shivji MK, Podust VN, Hubscher U, Wood RD (1995) Nucleotide excision repair DNA synthesis by DNA polymerase epsilon in the presence of PCNA, RFC, and RPA. Biochemistry 34(15): 5011–5017

Shuck SC, Short EA, Turchi JJ (2008) Eukaryotic nucleotide excision repair: from understanding mechanisms to influencing biology. Cell Res 18(1):64–72. doi:10.1038/cr.2008.2

Siedlecki JA, Szyszko J, Pietrzykowska I, Zmudzka B (1980) Evidence implying DNA polymerase beta function in excision repair. Nucleic Acids Res 8(2):361–375

Simsek D, Furda A, Gao Y, Artus J, Brunet E, Hadjantonakis AK, Van Houten B, Shuman S, McKinnon PJ, Jasin M (2011) Crucial role for DNA ligase III in mitochondria but not in Xrcc1-dependent repair. Nature 471(7337):245–248. doi:10.1038/nature09794

Singhal RK, Wilson SH (1993) Short gap-filling synthesis by DNA polymerase β is processive. J Biol Chem 268(21):15906–15911

Singhal RK, Prasad R, Wilson SH (1995) DNA polymerase β conducts the gap-filling step in uracil-initiated base excision repair in a bovine testis nuclear extract. J Biol Chem 270(2): 949–957

Slupphaug G, Markussen FH, Olsen LC, Aasland R, Aarsaether N, Bakke O, Krokan HE, Helland DE (1993) Nuclear and mitochondrial forms of human uracil-DNA glycosylase are encoded by the same gene. Nucleic Acids Res 21(11):2579–2584

Sobol RW (2007) DNA polymerase β null mouse embryonic fibroblasts harbor a homozygous null mutation in DNA polymerase iota. DNA Repair (Amst) 6(1):3–7
Sobol RW (2008) CHIPping away at base excision repair. Mol Cell 29(4):413–415
Sobol RW (2012a) For MutY, it's all about the OG. Chem Biol 19(3):313–314. doi:10.1016/j.chembiol.2012.03.002
Sobol RW (2012b) Genome instability caused by a germline mutation in the human DNA repair gene POLB. PLoS Genet 8(11):e1003086. doi:10.1371/journal.pgen.1003086
Sobol RW, Horton JK, Kuhn R, Gu H, Singhal RK, Prasad R, Rajewsky K, Wilson SH (1996) Requirement of mammalian DNA polymerase-β in base-excision repair. Nature 379(6561):183–186
Sobol RW, Prasad R, Evenski A, Baker A, Yang XP, Horton JK, Wilson SH (2000) The lyase activity of the DNA repair protein β-polymerase protects from DNA-damage-induced cytotoxicity. Nature 405(6788):807–810
Sobol RW, Watson DE, Nakamura J, Yakes FM, Hou E, Horton JK, Ladapo J, Van Houten B, Swenberg JA, Tindall KR, Samson LD, Wilson SH (2002) Mutations associated with base excision repair deficiency and methylation-induced genotoxic stress. Proc Natl Acad Sci USA 99(10):6860–6865
Sobol RW, Kartalou M, Almeida KH, Joyce DF, Engelward BP, Horton JK, Prasad R, Samson LD, Wilson SH (2003) Base excision repair intermediates induce p53-independent cytotoxic and genotoxic responses. J Biol Chem 278(41):39951–39959
Stachelek GC, Dalal S, Donigan KA, Campisi Hegan D, Sweasy JB, Glazer PM (2010) Potentiation of temozolomide cytotoxicity by inhibition of DNA polymerase beta is accentuated by BRCA2 mutation. Cancer Res 70(1):409–417. doi:10.1158/0008-5472.CAN-09-1353, 0008-5472.CAN-09-1353 [pii]
Starcevic D, Dalal S, Sweasy JB (2004) Is there a link between DNA polymerase beta and cancer? Cell Cycle 3(8):998–1001
Staresincic L, Fagbemi AF, Enzlin JH, Gourdin AM, Wijgers N, Dunand-Sauthier I, Giglia-Mari G, Clarkson SG, Vermeulen W, Scharer OD (2009) Coordination of dual incision and repair synthesis in human nucleotide excision repair. EMBO J 28(8):1111–1120. doi:10.1038/emboj.2009.49
Strittmatter T, Bareth B, Immel TA, Huhn T, Mayer TU, Marx A (2011) Small molecule inhibitors of human DNA polymerase lambda. ACS Chem Biol 6(4):314–319. doi:10.1021/cb100382m
Stuart JA, Mayard S, Hashiguchi K, Souza-Pinto NC, Bohr VA (2005) Localization of mitochondrial DNA base excision repair to an inner membrane-associated particulate fraction. Nucleic Acids Res 33(12):3722–3732. doi:10.1093/nar/gki683
Stucki M, Pascucci B, Parlanti E, Fortini P, Wilson SH, Hubscher U, Dogliotti E (1998) Mammalian base excision repair by DNA polymerases delta and epsilon. Oncogene 17(7):835–843
Sugo N, Aratani Y, Nagashima Y, Kubota Y, Koyama H (2000) Neonatal lethality with abnormal neurogenesis in mice deficient in DNA polymerase β. EMBO J 19(6):1397–1404
Sung P, Klein H (2006) Mechanism of homologous recombination: mediators and helicases take on regulatory functions. Nat Rev Mol Cell Biol 7(10):739–750. doi:10.1038/nrm2008
Svilar D, Goellner EM, Almeida KH, Sobol RW (2011) Base excision repair and lesion-dependent sub-pathways for repair of oxidative DNA damage. Antioxid Redox Signal 14(12):2491–2507. doi:10.1089/ars.2010.3466
Sweasy JB, Lang T, DiMaio D (2006) Is base excision repair a tumor suppressor mechanism? Cell Cycle 5(3):250–259
Symington LS, Gautier J (2011) Double-strand break end resection and repair pathway choice. Annu Rev Genet 45:247–271. doi:10.1146/annurev-genet-110410-132435
Szczesny B, Tann AW, Longley MJ, Copeland WC, Mitra S (2008) Long patch base excision repair in mammalian mitochondrial genomes. J Biol Chem 283(39):26349–26356. doi:10.1074/jbc.M803491200, M803491200 [pii]
Tanabe K, Bohn EW, Wilson SH (1979) Steady-state kinetics of mouse DNA polymerase beta. Biochemistry 18(15):3401–3406

Tang J, Goellner EM, Wang XW, Trivedi RN, St. Croix CM, Jelezcova E, Svilar D, Brown AR, Sobol RW (2010) Bioenergetic metabolites regulate base excision repair-dependent cell death in response to DNA damage. Mol Cancer Res 8(1):67–79. doi:10.1158/1541-7786.MCR-09-0411, 1541-7786.MCR-09-0411 [pii]

Tang JB, Svilar D, Trivedi RN, Wang XH, Goellner EM, Moore B, Hamilton RL, Banze LA, Brown AR, Sobol RW (2011) N-methylpurine DNA glycosylase and DNA polymerase beta modulate BER inhibitor potentiation of glioma cells to temozolomide. Neuro-oncol 13(5): 471–486. doi:10.1093/neuonc/nor011

Tann AW, Boldogh I, Meiss G, Qian W, Van Houten B, Mitra S, Szczesny B (2011) Apoptosis induced by persistent single-strand breaks in mitochondrial genome: critical role of EXOG (5′-EXO/endonuclease) in their repair. J Biol Chem 286(37):31975–31983. doi:10.1074/jbc.M110.215715

Tano K, Nakamura J, Asagoshi K, Arakawa H, Sonoda E, Braithwaite EK, Prasad R, Buerstedde JM, Takeda S, Watanabe M, Wilson SH (2007) Interplay between DNA polymerases beta and lambda in repair of oxidation DNA damage in chicken DT40 cells. DNA Repair (Amst) 6(6):869–875

Terrados G, Capp JP, Canitrot Y, Garcia-Diaz M, Bebenek K, Kirchhoff T, Villanueva A, Boudsocq F, Bergoglio V, Cazaux C, Kunkel TA, Hoffmann JS, Blanco L (2009) Characterization of a natural mutator variant of human DNA polymerase lambda which promotes chromosomal instability by compromising NHEJ. PLoS One 4(10):e7290. doi:10.1371/journal.pone.0007290

Thomas DC, Roberts JD, Kunkel TA (1991) Heteroduplex repair in extracts of human HeLa cells. J Biol Chem 266(6):3744–3751

Tokui T, Inagaki M, Nishizawa K, Yatani R, Kusagawa M, Ajiro K, Nishimoto Y, Date T, Matsukage A (1991) Inactivation of DNA polymerase beta by in vitro phosphorylation with protein kinase C. J Biol Chem 266(17):10820–10824

Tomicic MT, Thust R, Sobol RW, Kaina B (2001) DNA polymerase β mediates protection of mammalian cells against ganciclovir-induced cytotoxicity and DNA breakage. Cancer Res 61 (20):7399–7403

Trivedi RN, Almeida KH, Fornsaglio JL, Schamus S, Sobol RW (2005) The role of base excision repair in the sensitivity and resistance to temozolomide mediated cell death. Cancer Res 65(14): 6394–6400

Trivedi RN, Wang XH, Jelezcova E, Goellner EM, Tang J, Sobol RW (2008) Human methyl purine DNA glycosylase and DNA polymerase β expression collectively predict sensitivity to temozolomide. Mol Pharmacol 74(2):505–516

Ukai A, Maruyama T, Mochizuki S, Ouchida R, Masuda K, Kawamura K, Tagawa M, Kinoshita K, Sakamoto A, Tokuhisa T, O-Wang J (2006) Role of DNA polymerase theta in tolerance of endogenous and exogenous DNA damage in mouse B cells. Genes Cells 11(2): 111–121. doi:10.1111/j.1365-2443.2006.00922.x

van Loon B, Hubscher U (2009) An 8-oxo-guanine repair pathway coordinated by MUTYH glycosylase and DNA polymerase lambda. Proc Natl Acad Sci USA 106(43):18201–18206. doi:10.1073/pnas.0907280106, 0907280106 [pii]

Vens C, Begg AC (2010) Targeting base excision repair as a sensitization strategy in radiotherapy. Semin Radiat Oncol 20(4):241–249. doi:10.1016/j.semradonc.2010.05.005, S1053-4296(10) 00039-1 [pii]

Vens C, Sobol RW (2013) Targeting DNA repair pathways for cancer therapy. In: Johnson DE (ed) Cell death signaling in cancer biology and treatment, vol 1. Springer, New York

Vermeulen C, Bertocci B, Begg AC, Vens C (2007a) Ionizing radiation sensitivity of DNA polymerase lambda-deficient cells. Radiat Res 168(6):683–688. doi:10.1667/RR1057R.1

Vermeulen C, Verwijs-Janssen M, Cramers P, Begg AC, Vens C (2007b) Role for DNA polymerase beta in response to ionizing radiation. DNA Repair (Amst) 6(2):202–212

Vidal AE, Woodgate R (2009) Insights into the cellular role of enigmatic DNA polymerase iota. DNA Repair (Amst) 8(3):420–423. doi:10.1016/j.dnarep.2008.12.007

Wang X, Peterson CA, Zheng H, Nairn RS, Legerski RJ, Li L (2001) Involvement of nucleotide excision repair in a recombination-independent and error-prone pathway of DNA interstrand cross-link repair. Mol Cell Biol 21(3):713–720. doi:10.1128/MCB.21.3.713-720.2001
Wang X, Ira G, Tercero JA, Holmes AM, Diffley JF, Haber JE (2004) Role of DNA replication proteins in double-strand break-induced recombination in Saccharomyces cerevisiae. Mol Cell Biol 24(16):6891–6899. doi:10.1128/MCB.24.16.6891-6899.2004
Waser J, Hubscher U, Kuenzle CC, Spadari S (1979) DNA polymerase beta from brain neurons is a repair enzyme. Eur J Biochem 97(2):361–368
Wawra E, Dolejs I (1979) Evidences for the function of DNA polymerase-beta in unscheduled DNA synthesis. Nucleic Acids Res 7(6):1675–1686
Weissbach A (1977) Eukaryotic DNA polymerases. Annu Rev Biochem 46:25–47
Weissbach A, Baltimore D, Bollum F, Gallo R, Korn D (1975a) Nomenclature of eukaryotic DNA polymerases. Eur J Biochem 59(1):1–2
Weissbach A, Baltimore D, Bollum F, Gallo R, Korn D (1975b) Nomenclature of eukaryotic DNA polymerases. Science 190(4212):401–402
Wiebauer K, Jiricny J (1990) Mismatch-specific thymine DNA glycosylase and DNA polymerase β mediate the correction of G.T mispairs in nuclear extracts from human cells. Proc Natl Acad Sci USA 87(15):5842–5845
Williams HL, Gottesman ME, Gautier J (2012) Replication-independent repair of DNA interstrand crosslinks. Mol Cell 47(1):140–147. doi:10.1016/j.molcel.2012.05.001
Wilson SH, Beard WA, Shock DD, Batra VK, Cavanaugh NA, Prasad R, Hou EW, Liu Y, Asagoshi K, Horton JK, Stefanick DF, Kedar PS, Carrozza MJ, Masaoka A, Heacock ML (2010) Base excision repair and design of small molecule inhibitors of human DNA polymerase beta. Cell Mol Life Sci 67(21):3633–3647. doi:10.1007/s00018-010-0489-1
Wimmer U, Ferrari E, Hunziker P, Hubscher U (2008) Control of DNA polymerase lambda stability by phosphorylation and ubiquitination during the cell cycle. EMBO Rep 9(10): 1027–1033. doi:10.1038/embor.2008.148
Wood RD (1996) DNA repair in eukaryotes. Annu Rev Biochem 65:135–167. doi:10.1146/annurev.bi.65.070196.001031
Wood RD (2010) Mammalian nucleotide excision repair proteins and interstrand crosslink repair. Environ Mol Mutagen 51(6):520–526. doi:10.1002/em.20569
Wood RD, Mitchell M, Sgouros J, Lindahl T (2001) Human DNA repair genes. Science 291 (5507):1284–1289
Wood RD, Mitchell M, Lindahl T (2005) Human DNA repair genes. Mutat Res 577 (1–2):275–283. doi:10.1016/j.mrfmmm.2005.03.007, S0027-5107(05)00163-6 [pii]
Woods NT, Mesquita RD, Sweet M, Carvalho MA, Li X, Liu Y, Nguyen H, Thomas CE, Iversen ES Jr, Marsillac S, Karchin R, Koomen J, Monteiro AN (2012) Charting the landscape of tandem BRCT domain-mediated protein interactions. Sci Signal 5(242):rs6. doi:10.1126/scisignal.2002255
Wyatt MD, Pittman DL (2006) Methylating agents and DNA repair responses: methylated bases and sources of strand breaks. Chem Res Toxicol 19(12):1580–1594
Yano K, Morotomi-Yano K, Wang SY, Uematsu N, Lee KJ, Asaithamby A, Weterings E, Chen DJ (2008) Ku recruits XLF to DNA double-strand breaks. EMBO Rep 9(1):91–96. doi:10.1038/sj.embor.7401137
Yano K, Morotomi-Yano K, Lee KJ, Chen DJ (2011) Functional significance of the interaction with Ku in DNA double-strand break recognition of XLF. FEBS Lett 585(6):841–846. doi:10.1016/j.febslet.2011.02.020
Yeh JI, Levine AS, Du S, Chinte U, Ghodke H, Wang H, Shi H, Hsieh CL, Conway JF, Van Houten B, Rapic-Otrin V (2012) Damaged DNA induced UV-damaged DNA-binding protein (UV-DDB) dimerization and its roles in chromatinized DNA repair. Proc Natl Acad Sci USA 109(41):E2737–E2746. doi:10.1073/pnas.1110067109
Yi C, He C (2013) DNA repair by reversal of DNA damage. Cold Spring Harb Perspect Biol 5(1): a012575. doi:10.1101/cshperspect.a012575

Yoshida S, Yamada M, Masaki S (1979) Novel properties of DNA polymerase beta with poly(rA).oligo(dT) template-primer. J Biochem (Tokyo) 85(6):1387–1395

Yousefzadeh MJ, Wood RD (2013) DNA polymerase POLQ and cellular defense against DNA damage. DNA Repair (Amst) 12(1):1–9. doi:10.1016/j.dnarep.2012.10.004

Zheng L, Zhou M, Guo Z, Lu H, Qian L, Dai H, Qiu J, Yakubovskaya E, Bogenhagen DF, Demple B, Shen B (2008) Human DNA2 is a mitochondrial nuclease/helicase for efficient processing of DNA replication and repair intermediates. Mol Cell 32(3):325–336. doi:10.1016/j.molcel.2008.09.024, S1097-2765(08)00698-9 [pii]

Ziv O, Geacintov N, Nakajima S, Yasui A, Livneh Z (2009) DNA polymerase zeta cooperates with polymerases kappa and iota in translesion DNA synthesis across pyrimidine photodimers in cells from XPV patients. Proc Natl Acad Sci USA 106(28):11552–11557. doi:10.1073/pnas.0812548106

Chapter 4
Eukaryotic Y-Family Polymerases: A Biochemical and Structural Perspective

John M. Pryor, Lynne M. Dieckman, Elizabeth M. Boehm, and M. Todd Washington

Abstract Classical DNA polymerases, which replicate DNA rapidly and with high fidelity, stall upon encountering DNA damage. Thus nonclassical polymerases, which have evolved to accommodate DNA damage, are necessary to overcome these replication blocks. These nonclassical polymerases mainly belong to the Y-family and replicate DNA slower and with lower fidelity than their classical counterparts. Y-family polymerases employ surprising strategies to incorporate nucleotides opposite DNA damage. These include the use of larger and less constrained active sites, the use of Hoogsteen base pairing, and the use of amino acid side chains as templates. Y-family polymerases also engage in protein–protein interactions that are important for their recruitment to stalled replication forks and the coordination of their activities on the DNA. These polymerases function within a dynamic network of protein–protein interactions that are mediated by intrinsically disordered regions of these enzymes. This review focuses on the biochemical and structural studies of the Y-family polymerases, which have provided clear insights into their function.

Keywords DNA replication • DNA repair • DNA polymerase • Protein–protein interactions • PCNA • Rev1 • Translesion synthesis • Intrinsically disordered proteins

Abbreviations

BRCT	BRCA1 C-terminal
CTD	C-terminal domain
NMR	Nuclear magnetic resonance

J.M. Pryor • L.M. Dieckman • E.M. Boehm • M.T. Washington (✉)
Department of Biochemistry, University of Iowa College of Medicine, 4-403 Bowen Science Building, Iowa City, IA 52242-1109, USA
e-mail: todd-washington@uiowa.edu

K.S. Murakami and M.A. Trakselis (eds.), *Nucleic Acid Polymerases*, Nucleic Acids and Molecular Biology 30, DOI 10.1007/978-3-642-39796-7_4,

PAD	Polymerase-associated domain
PCNA	Proliferating cell nuclear antigen
PIP	PCNA-interacting protein
Pol	Polymerase
RIR	Rev1-interacting region
SAXS	Small-angle X-ray scattering
TT	Thymine–thymine
UBM	Ubiquitin-binding motif
UBZ	Ubiquitin-binding zinc finger
UV	Ultraviolet
XPV	Xeroderma pigmentosum variant form

4.1 Introduction

Classical DNA polymerases, despite their remarkable rate of DNA synthesis and their high fidelity, are unable to efficiently incorporate nucleotides opposite DNA damage. The structural distortions in the DNA caused by the damage are generally not tolerated by their highly constrained active sites. Thus, when one of these enzymes encounters a DNA lesion, it is blocked and the replication fork stalls. Translesion synthesis (TLS) is one strategy for overcoming such replication blocks. In TLS, one or more nonclassical DNA polymerases are recruited to the stalled replication fork. These nonclassical polymerases, which have evolved to accommodate DNA damage, replace the stalled classical polymerase and catalyze DNA synthesis past the lesion. The most studied of these nonclassical polymerases are the Y-family polymerases (Burgers et al. 2001; Ohmori et al. 2001). In prokaryotes, these include the SOS-inducible DNA polymerase IV and DNA polymerase V. In eukaryotes, these include DNA polymerase eta (Pol η), DNA polymerase iota (Pol ι), DNA polymerase kappa (Pol κ), and Rev1. This chapter focuses on the biochemical activities and structures of the eukaryotic Y-family polymerases.

Although most Y-family polymerases can inefficiently incorporate nucleotides opposite a wide range of DNA lesions, each polymerase appears to be specific for efficiently incorporating nucleotides opposite a small number of lesions or families of closely related lesions. These are referred to as the cognate lesions for a given polymerase (Friedberg et al. 2002; Lehmann et al. 2007; Waters et al. 2009). The clearest example of this concept is Pol η and the ultraviolet (UV) radiation-induced *cis–syn* thymine–thymine (TT) dimer (Johnson et al. 1999b). Another cognate lesion of Pol η is 8-oxoguanine (Haracska et al. 2000). The cognate lesions of Pol ι are some minor groove purine adducts and some exocyclic purine adducts (Washington et al. 2004c; Wolfle et al. 2005; Pence et al. 2009; Nair et al. 2006). The cognate lesions of Pol κ are a few minor groove guanine adducts (Choi et al. 2006). The cognate lesions of Rev1 are abasic sites (Pryor and Washington 2011), several minor groove guanine adducts, and some exocyclic guanine adducts

(Washington et al. 2004b; Nair et al. 2008; Choi and Guengerich 2008; Zhang et al. 2002b).

DNA lesions present two kinetic barriers to nucleotide incorporation by DNA polymerases. The first occurs during the incorporation step, when the lesion is in the nascent base pair itself. The second occurs during the subsequent nucleotide-incorporation step (i.e., the extension step), when the lesion is in the primer-terminal base pair. In some cases, Y-family polymerases have evolved to overcome both of these barriers. Pol η, for example, readily inserts nucleotides opposite TT dimers and extends beyond the lesion (Johnson et al. 1999b). In many cases, however, Y-family polymerases have evolved to overcome only one of these two barriers. Thus to bypass these lesions, multiple Y-family polymerases have to work in tandem with one polymerase incorporating a nucleotide opposite the lesion and another catalyzing subsequent extension (Prakash and Prakash 2002; Livneh et al. 2010). For example, Pol ι can efficiently incorporate a nucleotide opposite the acrolein-derived γ-hydroxy-1,N^2-propanoguanine, and Pol κ can efficiency extend beyond this lesion (Washington et al. 2004c). It should be noted that in most cases, this extension step is catalyzed by DNA polymerase zeta (Pol ξ), another nonclassical polymerase. This enzyme is a B-family polymerase, and many excellent review articles describing the biological function and the biochemical properties of Pol ξ are available (Prakash and Prakash 2002; Lawrence 2002, 2004; Waters et al. 2009; Livneh et al. 2010).

4.2 Overview of the Structures of Y-Family Polymerases

The overall structures of eukaryotic Y-family polymerases are similar to each other. These enzymes have a conserved catalytic core region comprised of approximately 400–500 amino acid residues that constitute a polymerase domain and a polymerase-associated domain (PAD). X-ray crystal structures of the catalytic core regions of all four human Y-family polymerases have been determined (Ummat et al. 2012; Nair et al. 2004; Lone et al. 2007; Swan et al. 2009). These structures show that the polymerase domains all contain fingers, thumb, and palm sub-domains analogous to those found in classical DNA polymerases. The polymerase domain binds to the primer-template DNA and an incoming nucleotide and catalyzes the incorporation reaction. The structures and mechanisms of the catalytic core regions of these polymerases are discussed in more detail below (see Sect. 4.3).

In all four enzymes, the catalytic core region is followed by a C-terminal region comprised of 300–400 amino acid residues. An analysis of the C-terminal regions of these polymerases shows that they are mostly intrinsically disordered (Fig. 4.1) (Ohmori et al. 2009). It has been estimated that as many as a third of eukaryotic proteins and a half of mammalian proteins are partially or fully disordered (Fink 2005; Dunker et al. 2008). In general, the disordered regions of proteins are often involved in interactions with multiple protein partners (Fink 2005; Dunker et al. 2008; Cortese et al. 2008). This clearly is the case with the C-terminal

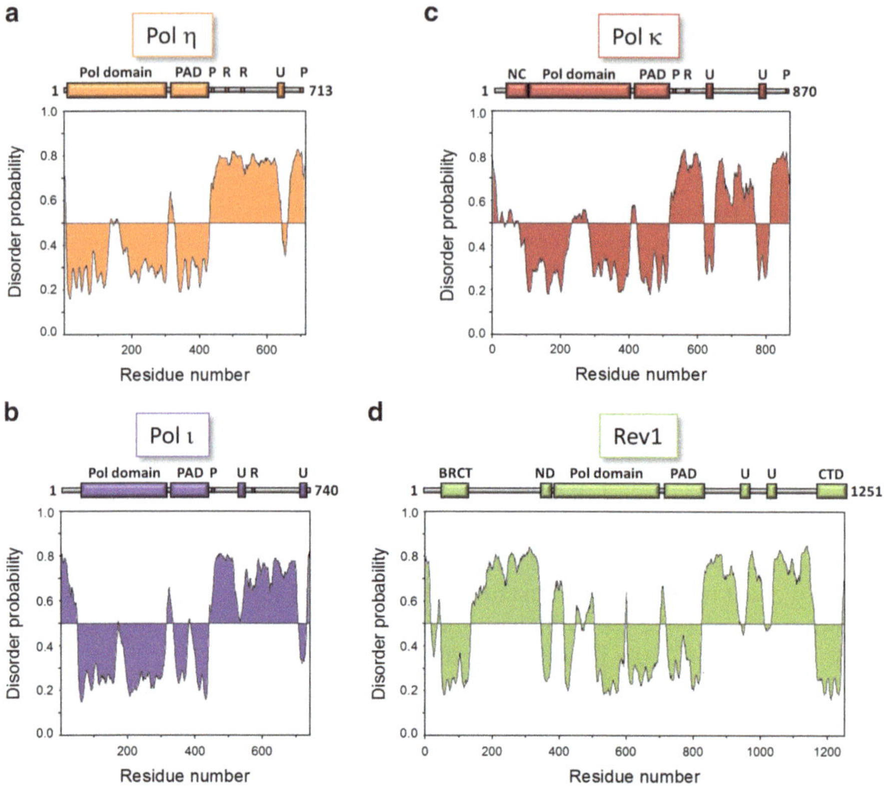

Fig. 4.1 The structured and unstructured regions of Y-family polymerases. The graphs of disorder probability for (**a**) Pol η, (**b**) Pol ι, (**c**) Pol κ, and (**d**) Rev1 were obtained using the meta-approach for predicting disordered regions of proteins (Ishida and Kinoshita 2008). In the diagrams of each polymerase, the structured regions are shown as *thick rectangles*, and the disordered regions are shown as *thin rectangles*. The polymerase (Pol) domain and PAD of each protein are indicated. The N-clasp (NC) of Pol κ as well as the N-digit (ND), the BRCT domain, and the CTD of Rev1 are indicated. PCNA-binding, ubiquitin-binding, and Rev1-binding motifs are indicated by P, U, and R, respectively

disordered regions of the Y-family polymerases. The C-terminal regions of Pol η, Pol ι, and Pol κ all contain small structured and unstructured motifs involved in binding several proteins including the key replication accessory factor proliferating cell nuclear antigen (PCNA), ubiquitin, and Rev1. The C-terminal region of Rev1 contains two small ubiquitin-binding motifs as well as a small, structured C-terminal domain (CTD) that binds Pol η, Pol ι, and Pol κ. In addition, Rev1 is the only eukaryotic Y-family polymerase possessing an N-terminal region comprised of approximately 350 amino acid residues that is also mostly disordered. This region contains a small, structured BRCA1 C-terminal (BRCT) domain. The structure and function of the motifs and small domains in the disordered regions of Y-family polymerases are discussed in more detail below (see Sects. 4.4 and 4.5).

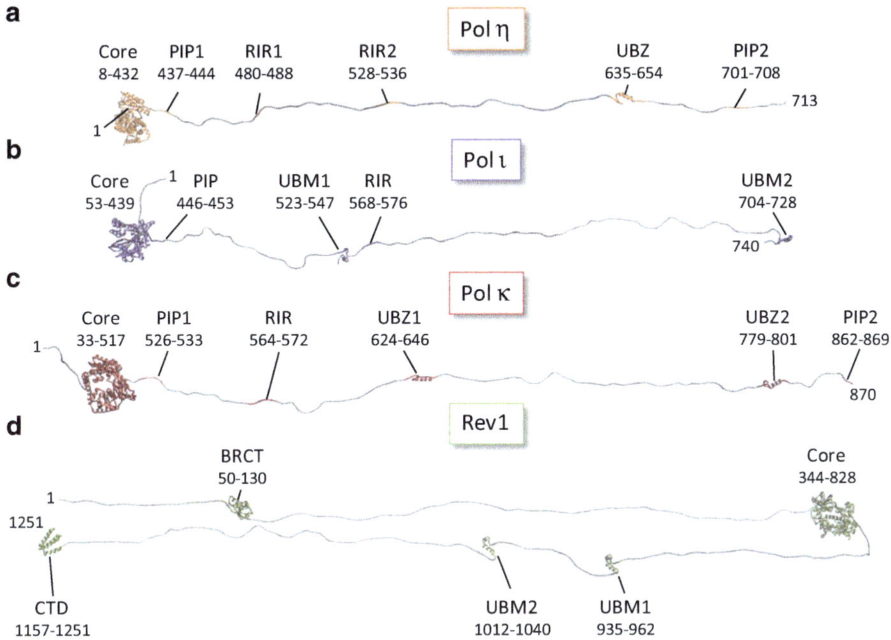

Fig. 4.2 Structural models of the full-length Y-family polymerases. The models of full-length (**a**) Pol η, (**b**) Pol ι, (**c**) Pol κ, and (**d**) Rev1 were built using Coot (Emsley and Cowtan 2004) starting with the X-ray crystal structures of the catalytic core regions of these polymerases (PDB ID: 3TQ1, 1T3N, 2OH2, and 3GQC, respectively) (Ummat et al. 2012; Nair et al. 2004; Lone et al. 2007; Swan et al. 2009) and the NMR structures of the UBZ of Pol η (PDB ID: 2I5O) (Bomar et al. 2007), the UBM of Pol ι (PDB ID: 2KHU) (Bomar et al. 2010), the Rev1 CTD (PDB ID: 2LSY) (Pozhidaeva et al. 2012), and the Rev1 BRCT domain (PDB ID: 2EBW). The UBZ of Pol κ was modeled based on the UBZ of Pol η, and the UBM of Rev1 was modeled based on the UBM of Pol ι. The disordered regions were then built as random coils. The various PIP motifs, UBZs, UBMs, and RIR motifs are indicated

We have built structural models of the full-length versions of all four human Y-family polymerases (Fig. 4.2). We started with the X-ray crystal structures of the catalytic core regions of these polymerases (Ummat et al. 2012; Nair et al. 2004; Lone et al. 2007; Swan et al. 2009) as well as the NMR structure of the ubiquitin-binding zinc finger (UBZ) of Pol η (Bomar et al. 2007), the ubiquitin-binding motif (UBM) of Pol ι (Bomar et al. 2010), the Rev1 CTD (Pozhidaeva et al. 2012), and the Rev1 BRCT domain (Riken Structural Genomics/Proteomics Initiative). We then built the disordered regions as random coils. Given the accuracy of disorder predictions, these extremely long regions are unlikely to contain any folded domains other than the few small domains (the ubiquitin-binding elements, the Rev1 BRCT domain, and the Rev1 CTD) that have already been identified. These disordered regions, however, may still contain some small secondary structural elements that are not shown in the models. Nevertheless, these models provide an excellent global view of the size and extended nature of the disordered regions of these polymerases relative to the structured catalytic core regions.

4.3 The Catalytic Activity of Y-Family Polymerases

Classical polymerases have evolved to synthesize DNA with high fidelity. Consequently, the active sites of classical polymerases are tightly constrained so that only the correct Watson–Crick base pairs can fit. The structural distortions in the DNA caused by lesions generally do not fit properly in the active sites of classical polymerases, and as a result, classical polymerases cannot efficiently incorporate nucleotides opposite DNA lesions. By contrast, nonclassical polymerases such as the Y-family polymerases have evolved to efficiently incorporate nucleotides opposite DNA damage using a variety of strategies. However, their active sites are not as tightly constrained, and as a result, they synthesize DNA with considerably lower fidelity. Because of their intrinsic low fidelity, replication by Y-family polymerases is highly error prone. Thus to reduce the likelihood of errors, these enzymes all synthesize DNA with low processivity. In this section, we discuss the biochemical activities and structures of the catalytic core regions of the four eukaryotic Y-family polymerases.

4.3.1 DNA Polymerase η

Pol η, the best studied of the Y-family polymerases, is found in all eukaryotes. In yeast, the lack of Pol η leads to an increase in UV-induced mutagenesis (McDonald et al. 1997; Roush et al. 1998). In humans, the lack of Pol η results in the genetic disorder xeroderma pigmentosum variant form (XPV), which is characterized by an extreme sensitivity to sunlight and a predisposition to skin cancer (Johnson et al. 1999a; Masutani et al. 1999). Steady-state and pre-steady-state kinetic studies showed that Pol η synthesizes DNA with low fidelity with error frequencies ranging from 10^{-2} to 10^{-3} (Washington et al. 1999, 2001; Matsuda et al. 2000; Johnson et al. 2000c). This error frequency is approximately 1,000-fold greater than those of classical DNA polymerases. While Pol η can incorporate nucleotides opposite all four non-damaged template residues, it synthesizes DNA with low processivity only incorporating about four or five nucleotides per DNA-binding event (Washington et al. 1999).

The cognate lesions of Pol η include the *cis–syn* TT dimer, which is caused by exposure to UV radiation, and 8-oxoguanine, which is a common form of oxidative base damage. In the case of the TT dimer, purified Pol η incorporates adenines opposite both bases of the dimer with the same kinetics as it incorporates opposite non-damaged thymine (Washington et al. 2000, 2003; Johnson et al. 2000c). In the case of 8-oxoguanine, it incorporates cytosine opposite the lesion with the same kinetics as it incorporates opposite non-damaged guanine (Haracska et al. 2000; Carlson and Washington 2005). Moreover, it incorporates cytosine 20–60-fold more efficiently than it incorporates adenine. By contrast, classical polymerases often preferentially incorporate adenine opposite 8-oxoguanine because the 8-oxoguanine–adenine base pair is less distorting to the DNA geometry than is the 8-oxoguanine–cytosine base pair.

Insight into the structural basis of the ability of Pol η to replicate through TT dimers was provided by X-ray crystal structures of the catalytic core region of Pol η bound to DNA and incoming nucleotide substrates (Silverstein et al. 2010; Ummat et al. 2012; Biertuempfel et al. 2010). The DNA is contacted by the finger, thumb, and palm sub-domains of the polymerase domain and by the PAD (Fig. 4.3a). The active site of Pol η is larger than those of classical polymerases due in part to a truncation of the fingers sub-domain. This larger active site allows Pol η to accommodate the two cross-linked bases in its active site without steric clashes. The structure of Pol η bound to DNA containing a TT dimer shows that the incoming dATP forms Watson–Crick base pairs with the damaged thymines of the dimer in exactly the same way that it forms base pairs with non-damaged thymines (Fig. 4.3a) (Silverstein et al. 2010). In addition, Pol η also makes several direct contacts with the template strand, and these contacts ensure that the DNA bound in the active site is in the same stable, nearly B-form conformation whether it is damaged or non-damaged. Together, these properties allow Pol η to catalyze nucleotide incorporation opposite the TT dimer with nearly identical kinetics and fidelity as it does opposite non-damaged DNA.

4.3.2 DNA Polymerase ι

Pol ι is found in many higher eukaryotes including insects and mammals; it is not found in yeast and nematodes. The biological role of Pol ι is not well understood, and the lack of Pol ι is not associated with any disease. However, mice lacking Pol ι are at a higher risk for urethane-induced lung cancers (Lee and Matsushita 2005), and human fibroblasts lacking Pol ι were shown to be sensitive to oxidative stress (Petta et al. 2008). Steady-state and pre-steady-state kinetic studies showed that the fidelity and efficiency of nucleotide incorporation by Pol ι varies greatly depending on the template base (Johnson et al. 2000b; Tissier et al. 2000; Zhang et al. 2000; Washington et al. 2004a). Opposite a template adenine, Pol ι carries out nucleotide incorporation with a moderate-to-high efficiency and fidelity with error frequencies ranging from 10^{-4} to 10^{-5}. Opposite both a template guanine and cytosine, the efficiency of incorporation is reduced about tenfold and the fidelity is significantly lower with error frequencies ranging from 10^{-1} to 10^{-2}. Opposite a template thymine, incorporation is very inefficient and the fidelity is extraordinarily low with error frequencies ranging from 10^{-1} to 10^{1}. In fact, Pol ι incorporates the incorrect guanine opposite a template thymine about tenfold more efficiently than it incorporates the correct adenine.

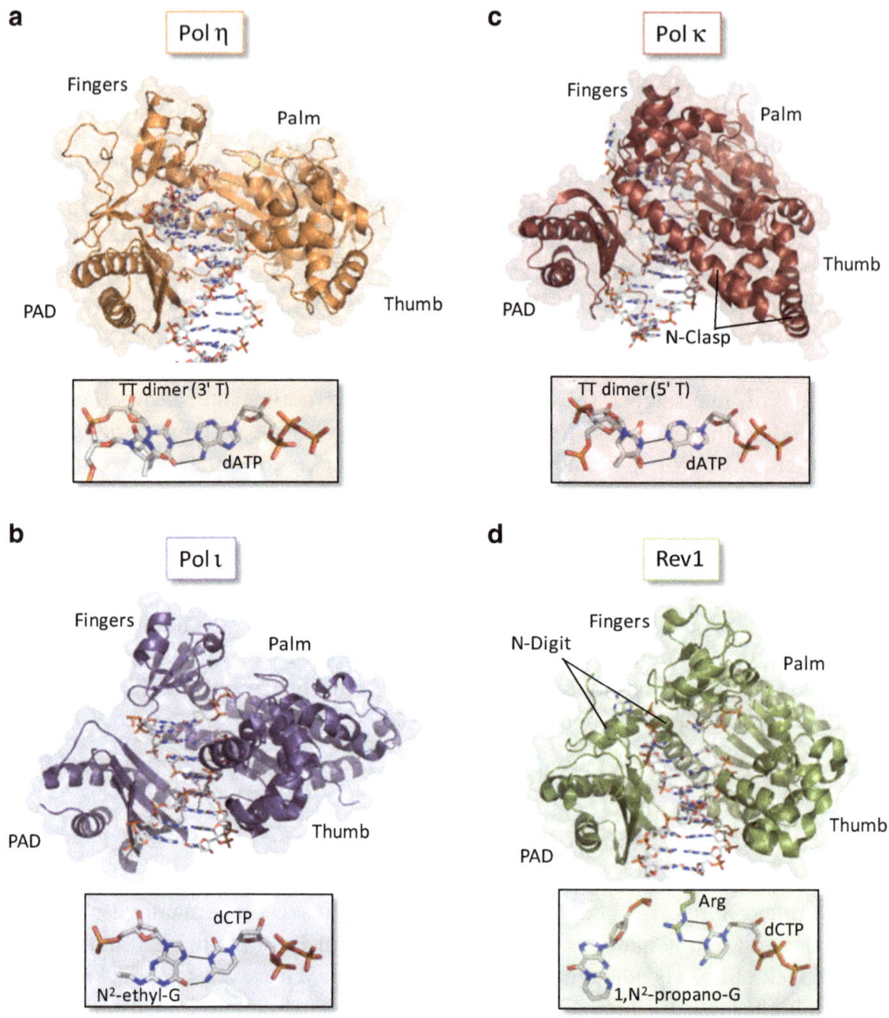

Fig. 4.3 Structure of the catalytic core regions and active sites of Y-family polymerases. Ribbon diagrams of the catalytic core regions of (**a**) Pol η, (**b**) Pol ι, (**c**) Pol κ, and (**d**) Rev1 are shown bound to DNA and an incoming nucleotide. In the case of Pol η, the template base is the 3′ thymine of a TT dimer (PDB ID: 3MFI) (Silverstein et al. 2010). In the case of Pol ι, the template base is N^2-ethylguanine (PDB ID: 3EPI) (Pence et al. 2009). In the case of Pol κ, the template base is the 5′ thymine of a TT dimer (PDB ID: 3PZP) (Vasquez-Del Carpio et al. 2011). In the case of Rev1, the template base is γ-hydroxy-1,N^2-propanoguanine (PDB ID: 3BJY) (Nair et al. 2008). The fingers, palm, and thumb sub-domains of the polymerase domain and the PAD are indicated. The N-clasp and the N-digit of Pol κ and Rev1, respectively, are indicated. The *insets* below the ribbon diagrams show the nascent base pair in the polymerase active site

Kinetic studies of purified Pol ι suggest that minor groove purine adducts and some exocyclic purine adducts are cognate lesions for this polymerase. For example, Pol ι incorporates cytosine opposite N^2-ethylguanine with the same efficiency with which it incorporates opposite a non-damaged guanine (Pence et al. 2009). This is remarkable, because minor groove adducts like this are a strong block to most polymerases because of steric clashes in their active sites. Even more striking is the ability of Pol ι to efficiently incorporate the correct nucleotide opposite lesions such as γ-hydroxy-1,N^2-propanoguanine (Washington et al. 2004c; Wolfle et al. 2005), which has an exocyclic ring connecting N1 and N^2 of the guanine, and 1,N^6-ethenoadenine (Nair et al. 2006), which has an exocyclic ring connecting N1 and N^6 of the adenine. In both of these lesions, the exocyclic ring disrupts their Watson–Crick base-pairing edge preventing them from forming Watson–Crick base pairs.

Insight into the structural basis of the ability of Pol ι to incorporate nucleotides opposite both minor groove purine adducts as well as exocyclic purine adducts was provided by X-ray crystal structures of the catalytic core region of Pol ι bound to DNA and incoming nucleotide substrates (Nair et al. 2004, 2005a, 2006; Pence et al. 2009). When Pol ι forms a ternary complex with the incoming nucleotide and DNA, the template backbone is held in a position that reduces the distance between the two C1′ atoms of the nascent base pair by approximately 2 Å relative to other DNA polymerases. As a result of this, when the template is a purine, Pol ι rotates the template base from the normal *anti* configuration around the N-glycosidic bond to the *syn* configuration in order to accommodate this reduced distance. This forces the incoming nucleotide to bind the template base using Hoogsteen base pairs, not Watson–Crick base pairs (Nair et al. 2004). In the case of the minor groove purine adducts, such as N^2-ethylguanine, Hoogsteen base pairing allows for efficient incorporation because the rotation of the damaged bases from the *anti* to the *syn* configuration places these adducts in the major groove where there are no steric clashes (Fig. 4.3b) (Pence et al. 2009). In the case of the exocyclic purine adducts, the rotation of the damaged base to the *syn* configuration allows for efficient incorporation because these exocyclic rings are easily accommodated in the active site and do not interfere with Hoogsteen base pairing.

4.3.3 DNA Polymerase κ

Pol κ is found in many higher eukaryotes including mammals, but it is not found in yeast and insects. The biological role of Pol κ is also poorly understood. Mouse cells lacking Pol κ have an enhanced spontaneous mutation rate (Stancel et al. 2009) and are sensitive to benzo(*a*)pyrene diol epoxide (Ogi et al. 2002). Unlike other Y-family polymerases, Pol κ is capable of synthesizing DNA with moderate fidelity with error frequencies ranging from 10^{-3} to 10^{-4} (Ohashi et al. 2000; Johnson et al. 2000a). Moreover, Pol κ synthesizes DNA with a higher processivity than other Y-family polymerases; it can incorporate up to

25 nucleotides per DNA-binding event (Ohashi et al. 2000). Thus among the Y-family polymerases, Pol κ is most similar to classical DNA polymerases.

Kinetic studies of Pol κ have led to two proposed roles for this enzyme in TLS. The first is that Pol κ catalyzes nucleotide incorporation opposite cognate lesions as do the other Y-family polymerases. The possible cognate lesions for Pol κ include several minor groove guanine adducts. Kinetic studies have shown that Pol κ incorporates nucleotides opposite N^2-methylguanine, N^2-ethylguanine, N^2-isobutylguanine, N^2-benzylguanine, and N^2-naphthylguanine with nearly the same efficiency as opposite non-damaged guanine (Choi et al. 2006). The second proposed role is that Pol κ functions during the extension step of TLS when two Y-family polymerases work in tandem to bypass a lesion. This comes from the observation that Pol κ efficiently extends from aberrant primer-terminal pairs containing mismatches (Washington et al. 2002). Pol κ also efficiently extends from primer-terminal pairs containing O^6-methylguanine and 8-oxoguanine (Haracska et al. 2002a), N^2-benzo(*a*)pyrenyl guanine adducts (Zhang et al. 2002a), and the TT dimer (Washington et al. 2002). In this latter case, Pol κ extends from the nucleotide already inserted opposite the first base of the TT dimer (the 3′ thymine) by directly incorporating an adenine opposite the second base of the dimer (the 5′ thymine). Given the strong biochemical support for both of these proposed roles, it is likely that Pol κ functions in some contexts to insert nucleotides opposite DNA lesions and in other contexts to extend from nucleotides inserted opposite lesions by other polymerases.

X-ray crystal structures of the catalytic core region of Pol κ in ternary complexes with DNA and incoming nucleotides have shown that like Pol η, it utilizes Watson–Crick base pairing to form the nascent base pair. Its active site, however, is more constrained than those of other Y-family polymerases (Uijon et al. 2004; Lone et al. 2007), which is presumably why this enzyme is capable of higher fidelity DNA synthesis. Pol κ possesses a structural feature called an N-clasp comprised of about 75 amino acid residues that immediately precedes the polymerase domain (Fig. 4.3c). The N-clasp allows Pol κ to completely encircle the DNA substrate when it binds, and deletion of the N-clasp significantly reduces DNA binding by Pol κ (Lone et al. 2007). The N-clasp may provide additional stability to the polymerase–DNA complex and in so doing may contribute to the ability of Pol κ to extend from aberrant primer-terminal pairs. An X-ray crystal structure of Pol κ shows this enzyme in the act of extending from an aberrant base pair, in this case a TT dimer (Fig. 4.3c) (Vasquez-Del Carpio et al. 2011). While the highly restrictive active site does not allow Pol κ to incorporate a nucleotide opposite the first base of this lesion (the 3′ thymine), it does readily accommodate the TT dimer in the active site when extending from this aberrant base pair (i.e., when incorporating opposite the second base of the lesion, the 5′ thymine).

4.3.4 Rev1

Rev1, which is found in all eukaryotes, has highly unusual substrate specificity. It preferentially incorporates cytosine opposite every template base, albeit with different efficiencies (Haracska et al. 2002b; Masuda et al. 2002; Masuda and Kamiya 2002). Opposite a template guanine, Rev1 incorporates cytosine with high efficiency. Although Rev1 is capable of misincorporating other incoming nucleotides, it does so at relatively low frequencies ranging from 10^{-3} to 10^{-4}. Thus opposite a template guanine, Rev1 has a moderate fidelity. Opposite the other three non-damaged templates, Rev1 still preferentially incorporates cytosine, although the efficiency of incorporation is reduced by as much as 500-fold. Because of this unique substrate specificity, Rev1 has not been officially designated as a DNA polymerase and assigned a Greek letter. Nevertheless, it has the same general structure and chemical mechanism as other DNA polymerases. It can even processively synthesize DNA if the template contains a series of guanines (Haracska et al. 2002b). Thus Rev1 is a DNA polymerase in everything but name.

Kinetic studies of purified Rev1 suggest that abasic sites as well as several guanine adducts are cognate lesions for this polymerase. Pre-steady-state kinetic studies have shown that Rev1 is capable of highly efficient incorporation opposite an abasic site (Pryor and Washington 2011). Similarly, Rev1 incorporates nucleotides opposite minor groove N^2-guanine adducts including N^2-methylguanine, N^2-ethylguanine, N^2-isobutylguanine, N^2-benzylguanine, N^2-naphthylguanine, N^2-anthracenylguanine, and N^2-benzo(*a*)pyrenyl guanine with nearly the same efficiency as opposite non-damaged guanine (Choi and Guengerich 2008). Incorporation opposite similar major groove guanine adducts is much less efficient, except for O^2-benzylguanine, opposite which Rev1 incorporates nucleotides efficiently. In addition, efficient incorporation by Rev1 was also observed opposite the exocyclic γ-hydroxy-1,N^2-propanoguanine adduct (Washington et al. 2004b). Rev1 also incorporates opposite 8-oxoguanine but does so with low efficiency (Haracska et al. 2002b; Pryor and Washington 2011).

Insight into the structural basis of this unique preference for cytosine incorporation was provided by X-ray crystal structures of Rev1 in a ternary complex with an incoming dCTP and a DNA substrate containing a template guanine (Nair et al. 2005b; Swan et al. 2009). Surprisingly, this structure showed that the template guanine and the incoming dCTP are not base-paired. Instead, the template base is flipped out of the DNA double helix and into a binding pocket in the enzyme comprised in part from amino acid residues in the PAD. Situated between the polymerase domain and the PAD is the N-digit, a structural element comprised of about 50 amino acid residues (Fig. 4.3d). The N-digit contains a highly conserved leucine and a highly conserved arginine that are both critical for catalysis. The leucine is responsible for flipping out the template base into its binding pocket, and the arginine is responsible for hydrogen bonding with the incoming dCTP. Thus Rev1 utilizes an amino acid side chain as a template to direct the incorporation of cytosine irrespective of the identity of the template base, provided that the flipped out template base fits nicely within its binding pocket. An X-ray crystal structure of Rev1 bound to a DNA substrate containing an exocyclic γ-hydroxy-1,N^2-propanoguanine shows that this lesion is readily accommodated by the template base-binding pocket (Fig. 4.3d) (Nair et al. 2008). Another structure of Rev1 bound to DNA containing an abasic site shows that this lesion does not distort the template base-binding pocket because the pocket is occupied by a series of ordered water molecules (Nair et al. 2011).

4.4 Interactions of Y-Family Polymerases with PCNA

The Y-family polymerases are recruited to stalled replication forks and regulated in part by their interactions with the key replication accessory factor PCNA. PCNA is a ring-shaped homo-trimer that encircles the DNA and greatly enhances the processivity of classical DNA polymerases (Krishna et al. 1994). It also interacts with a wide range of other proteins and in so doing recruits them to sites of DNA replication (Maga and Hubscher 2003; Moldovan et al. 2007; Naryzhny 2008; Zhuang and Ai 2010; Dieckman et al. 2012). When cells are exposed to DNA damaging agents, PCNA is ubiquitylated on lysine-164 by the Rad6–Rad18 ubiquitin-conjugating complex (Hoege et al. 2002; Stelter and Ulrich 2003; Kannouche et al. 2004), and ubiquitin-modified PCNA recruits Y-family polymerases to replication forks. In the structure of ubiquitin-modified PCNA, the ubiquitin moiety sits on the back face of the PCNA ring (Freudenthal et al. 2010). In this section, we will discuss the interactions of Y-family polymerases with unmodified and ubiquitin-modified PCNA.

4.4.1 Interactions with Unmodified PCNA

Pol η, Pol ι, and Pol κ all possess one or more PCNA-interacting protein (PIP) motifs in their disordered C-terminal regions (Figs. 4.1 and 4.2). These motifs contain eight amino acid residues and are found in a large number of proteins that interact with PCNA (Tsurimoto 1999; Hingorani and O'Donnell 2000; Maga and Hubscher 2003). The fourth residue of the motif is hydrophobic (usually a leucine, an isoleucine, or a methionine) and the seventh and eighth residues are aromatic (usually phenylalanine or tyrosine). The X-ray crystal structures of PCNA bound to a variety of PIP motifs have been determined, and these motifs all bind on the front face of the PCNA ring near the inter-domain connector loop. The conserved hydrophobic and aromatic residues bind in a pocket at the interface of the two domains of PCNA. Structures of PCNA bound to the PIP motifs of Pol η and Pol ι have been determined (Hishiki et al. 2009). While the Pol η PIP motif binds to PCNA by forming the same 3_{10} helix that other PIP motifs form (Fig. 4.4a, b), the Pol ι PIP binds to PCNA by forming a novel β-bend structure. The significance of

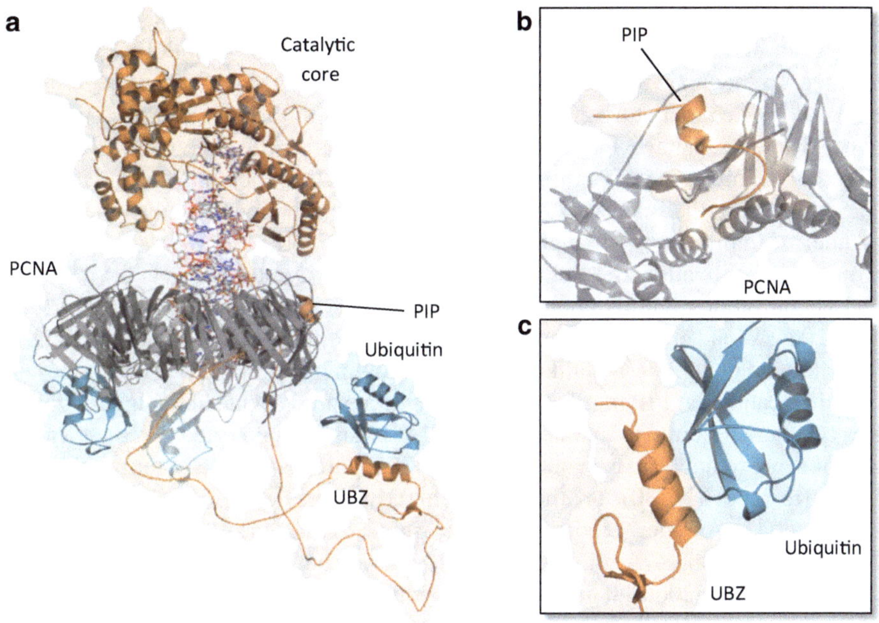

Fig. 4.4 Structural model of full-length Pol η bound to ubiquitin-modified PCNA. (**a**) The model of Pol η bound to ubiquitin-modified PCNA was built using Coot (Emsley and Cowtan 2004) starting with the X-ray crystal structures of the catalytic core region of Pol η (PDB ID: 3TQ1) (Ummat et al. 2012), ubiquitin-modified PCNA (PDB ID: 3L10) (Freudenthal et al. 2010), and PCNA bound to the Pol η-PIP motif (PDB ID: 2ZVK) (Hishiki et al. 2009) and with the NMR structure of the Pol η UBZ (PDB ID: 2I5O) (Bomar et al. 2007). (**b**) A close-up of the Pol η PIP motif bound to the PCNA portion of ubiquitin-modified PCNA is shown. (**c**) A close-up of the Pol η UBZ bound to the ubiquitin portion of ubiquitin-modified PCNA is shown

this unusual PIP conformation is unclear. It should be noted that a structure of PCNA bound to the Pol κ PIP has also been determined (Hishiki et al. 2009), but in this case, additional amino acid residues not found in Pol κ were added to the PIP construct to allow PCNA binding. The native Pol κ PIP does not seem to bind PCNA, so this particular structure is of limited value.

Purified Pol η, Pol ι, and Pol κ physically interact with unmodified PCNA (Haracska et al. 2001a–c, 2002c). Unlike the interactions between PCNA and classical polymerases, the interactions between PCNA and the Y-family polymerases do not substantially increase the processivity of DNA synthesis. Nevertheless, steady-state kinetics shows that interacting with PCNA significantly increases the catalytic efficiency of nucleotide incorporation by all three of these enzymes on both non-damaged and damaged templates. For example, in the case of Pol η, the increase in efficiency of incorporation opposite non-damaged DNA ranges from three- to tenfold and the increase in efficiency on a template abasic site, a non-cognate lesion, ranges from 3-fold to as much as 300-fold depending on experimental conditions (Haracska et al. 2001a, c; Freudenthal et al. 2008). These physical and functional interactions with PCNA are dependent on intact PIP motifs. Moreover, in human cells, intact PIP motifs are required for both Pol η and Pol ι to localize to nuclear foci containing PCNA following DNA damage (Bienko et al. 2005; Acharya et al. 2008; Vidal et al. 2004). In the case of Pol η, there are two PIP motifs, named PIP1 and PIP2. Disruptions of the individual PIP motifs have only a moderate affect on localization to nuclear foci and Pol η-dependent TLS suggesting that the two PIP motifs are able to functionally substitute for one another. Simultaneous disruption of both PIP motifs, however, completely eliminates localization and Pol η-dependent TLS in vivo (Acharya et al. 2008).

Like the other Y-family polymerases, Rev1 physically interacts with PCNA (Guo et al. 2006; Wood et al. 2007), and this interaction stimulates the catalytic activity of Rev1 (Wood et al. 2007). Unlike these other polymerases, however, Rev1 does not contain a canonical PIP motif, and there has been some debate about the regions of Rev1 that are required to interact with PCNA. It has been reported that the localization of Rev1 to nuclear foci containing PCNA requires either the N-terminal half of Rev1 (residues 1–730) or the C-terminal half (residues 730–1251) (Tissier et al. 2004). Another report, however, showed that localization requires the C-terminal region of Rev1 (residues 826–1251), but not the N-terminal region (Murakumo et al. 2006). It has also been reported that the N-terminal BRCT domain of Rev1 is required for localization to foci in non-damaged cells but is not required in UV-treated cells (Guo et al. 2006). This too is controversial as another study failed to detect a direct interaction between PCNA and the Rev1 BRCT domain (de Groote et al. 2011). Moreover, the stimulation of Rev1's catalytic activity by PCNA does not require an intact BRCT domain (Wood et al. 2007). Thus questions remain regarding the structural basis of the PCNA–Rev1 interaction.

4.4.2 Interactions with Ubiquitin-Modified PCNA

All four Y-family polymerases possess one or more small ubiquitin-binding domains in their disordered C-terminal regions (Figs. 4.1 and 4.2). In the case of Pol η and Pol κ, these small domains are UBZs, which contain about 20 amino acid residues and form a short, two-stranded antiparallel β-sheet followed by an α-helix (Fig. 4.4a, c) (Bomar et al. 2007). Two conserved cysteine residues and two conserved histidine residues coordinate a zinc ion, which likely provides structural stability to this small domain. In the case of Pol ι and Rev1, these small domains are UBMs, which contain about 30 amino acid residues and form a helix-turn-helix motif (Bomar et al. 2010; Burschowsky et al. 2011). NMR titrations have shown that the UBZs and the UBMs interact in slightly different ways with the canonical protein–protein interaction surface of ubiquitin, which is made up of a conserved hydrophobic patch containing leucine-8, isoleucine-44, and valine-70. Neither the conformation of the ubiquitin nor the conformation of the ubiquitin-binding domains seems to change upon complex formation.

In vitro pull-downs using purified Rev1 have shown that this polymerase interacts with ubiquitin-modified PCNA with qualitatively higher affinity than it interacts with unmodified PCNA (Wood et al. 2007). We have quantitatively analyzed the interaction of the disordered C-terminal region of Pol η, which contains the PIP and UBZ motifs, with unmodified PCNA and ubiquitin-modified PCNA, and we find that the attachment of ubiquitin to PCNA enhances the binding affinity of Pol η by approximately 20-fold (unpublished data). In human cells, Pol η specifically interacts with ubiquitin-modified PCNA, but not unmodified PCNA. Immunoprecipitation of PCNA from normal cells pulled down only unmodified PCNA, whereas immunoprecipitation of PCNA from UV-irradiated cells pulled down ubiquitin-modified PCNA and Pol η. Moreover, localization of Pol η to nuclear foci and Pol η-dependent TLS require that the UBZ be intact (Bienko et al. 2005). Localization of Pol ι to foci requires that both UBMs be intact (Bienko et al. 2005; Bomar et al. 2010). Thus ubiquitin-binding domains are important for localization to nuclear foci.

The complex of Y-family polymerases and ubiquitin-modified PCNA is likely flexible. First, the PIP and ubiquitin-binding domains are located within large regions of the Y-family polymerases that are intrinsically disordered. Second, experimental evidence obtained using small-angle X-ray scattering (SAXS) and computational studies using Brownian dynamics simulations show that the ubiquitin moieties of ubiquitin-modified PCNA are dynamic (Tsutakawa et al. 2011). Nevertheless, while the ubiquitin moieties (still attached to lysine-164 of PCNA) are capable of moving around, they have preferred positions on the back face and the side of the PCNA ring. This is important as nearly all PCNA-binding proteins interact with the front face of PCNA. This suggests that Y-family polymerases can bind to the back or side of the PCNA ring without affecting ongoing activity of other proteins bound to the front face of PCNA. Thus the Y-family polymerases can be held in reserve on the back or side of PCNA until their activities are required. Then because of the flexible nature of this complex, they can move to the front face of PCNA and engage the primer-terminus of the DNA substrate. A model of Pol η bound to the DNA substrate on the front face of ubiquitin-modified PCNA is shown in Fig. 4.4a.

4.5 Other Interactions of Y-Family Polymerases

Y-family polymerases function within a dynamic network of protein–protein interactions. The interactions in this network, which are mediated by intrinsically disordered regions of the polymerases, govern their recruitment and regulation. In this network, ubiquitin-modified PCNA, Rev1, and (to a lesser extent) Pol η function as hub proteins, which interact with multiple binding partners (Fig. 4.5a). We discuss the interactions of Y-family polymerases with ubiquitin-modified PCNA above (see Sect. 4.4). In this section, we discuss the other interactions of Pol η and Rev1. We also discuss speculations that Pol η functions as a "first responder" during TLS and that Rev1 functions as a scaffold to recruit other nonclassical polymerases.

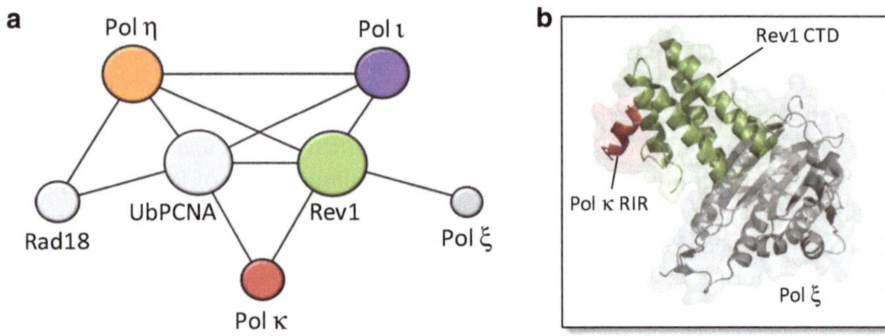

Fig. 4.5 The network of protein–protein interactions of Y-family polymerases. (**a**) The network diagram shows the known protein–protein interactions among Pol η, Pol ι, Pol κ, Rev1, ubiquitin-modified PCNA (Ub-PCNA), Rad18, and the B-family Pol ξ. The size of the *circles* is proportional to the number of contacts in the network. Ubiquitin-modified PCNA and Rev1 are hub proteins each contacting five other proteins in the network. (**b**) A structure of the Rev1 CTD bounds simultaneously to both the Pol κ RIR motif and the Rev7 non-catalytic subunit of Pol ξ (PDB ID: 4FJO) (Wojtaszek et al. 2012a)

4.5.1 Interactions of DNA Polymerase η

Pol η directly interacts with Rad18 (Watanabe et al. 2004), the E3 ubiquitin ligase that, along with Rad6 (the E2 ubiquitin-conjugating enzyme), catalyzes the ubiquitylation of PCNA in the presence of DNA damage. Rad18 binds to the intrinsically disordered C-terminal region of Pol η, and Rad18 and Pol η co-localize to nuclear foci following DNA damage (Watanabe et al. 2004). Moreover, cells expressing a mutant form of Rad18 that is unable to bind Pol η are defective in forming Pol η-containing nuclear foci and are sensitive to UV radiation (suggesting a defect in Pol η-dependent TLS). Moreover, Rad18 is phosphorylated in its Pol η-binding site in response to DNA damage, and this modification enhances the association of Pol η and Rad18 (Day et al. 2010; Barkley et al. 2012). These studies have led to a model in which Rad18 chaperones Pol η to PCNA. Then Rad18 (along with Rad6) catalyzes the ubiquitylation of PCNA and subsequently hands off Pol η to the ubiquitin-modified PCNA. Such a model would ensure that Pol η is the first Y-family polymerase to arrive at the stalled replication fork. Once there, Pol η could either catalyze nucleotide incorporation opposite the lesion or recruit other Y-family polymerases to the fork. While the idea that Pol η functions as a "first responder" during TLS is speculative, it does make sense given that the replication of TT dimers and 8-oxoguanines by Pol η actually reduces the mutagenic potential of these lesions.

Pol η also directly interacts with both Rev1 and Pol ι. Rev1 binds to a short motif containing approximately 10 amino acids located in the disordered C-terminal region of Pol η. This motif on Pol η is called the Rev1-interacting region (RIR), and it binds to the Rev1 CTD (Ohashi et al. 2004, 2009). The RIR motif of Pol η forms an α-helix upon binding to the CTD of Rev1 (see Sect 4.5.2 below)

(Pozhidaeva et al. 2012). Cells producing Pol η mutant proteins that are unable to interact with Rev1 showed defects in forming Rev1-containing foci (Akagi et al. 2009). Pol ι also interacts with Pol η, and the intrinsically unstructured C-terminal regions of both polymerases are involved in this interaction (Kannouche et al. 2002). In cells lacking Pol η, there is a significant reduction in the number of Pol ι-containing nuclear foci. Taken together, these studies suggest that interactions with Pol η are important for recruiting Rev1 and Pol ι to stalled replication forks.

4.5.2 Interactions of Rev1

In addition to its role as a polymerase, Rev1 also plays an essential non-catalytic role in TLS. Yeast cells producing a mutant form of Rev1 with an amino acid substitution in its BRCT domain have very low frequencies of UV-induced mutagenesis suggesting that error-prone TLS is not occurring in these cells (Lemontt 1971). This defect in TLS cannot be attributed to a decrease in the catalytic activity of Rev1 for two reasons. First, the mutant protein with a substitution in the BRCT domain retains a substantial amount of enzymatic activity (Nelson et al. 2000). Second, cells producing a catalytically inactive Rev1 mutant protein have nearly normal frequencies of UV-induced mutagenesis suggesting that error-prone TLS is occurring normally in these cells (Haracska et al. 2001d; Zhou et al. 2010; Otsuka et al. 2005). These results have led to the suggestion that Rev1 functions as a scaffold protein that mediates protein–protein interactions that are required for error-prone TLS.

The CTD of Rev1 has been shown to be essential for its non-catalytic role in TLS. This small domain is comprised of approximately 100 amino acid residues that bind Pol η, Pol ι, and Pol κ (Tissier et al. 2004; Ohashi et al. 2004; Guo et al. 2003). This domain interacts with short RIR motifs found in Pol η, Pol ι, and Pol κ that contain two conserved phenylalanine residues (Ohashi et al. 2009). Recent structures of the RIR motifs of Pol η and Pol κ bound to the Rev1 CTD showed that this domain forms a four-helix bundle, and when the RIR from another polymerase binds, it forms an α-helix that contacts two of the helices in the CTD (Wojtaszek et al. 2012a, b; Pozhidaeva et al. 2012). Expression of a Pol κ mutant protein that had substitutions in its RIR that blocked interactions with Rev1 was not able to rescue the decreased viability of a Pol κ-deficient cell line after induction of DNA damage (Ohashi et al. 2009). This suggests that the interaction with Rev1 is necessary for the function of Pol κ in vivo.

Rev1 also interacts with the non-catalytic Rev7 subunit of Pol ζ (Murakumo et al. 2001), a nonclassical B-family polymerase that functions in the extension step during the bypass of a wide range of DNA lesions. Interactions with Rev1 are necessary for the proper localization of Pol ζ to nuclear foci and for the enhancement of its catalytic activity in yeast (Acharya et al. 2006). Recent structural studies have shown that the Rev7 subunit of Pol ζ is able to interact with the Rev1 CTD via a linker region between two CTD α-helices (Wojtaszek et al. 2012a; Kikuchi et al. 2012). Moreover, an X-ray crystal structure was determined for the complex containing the Rev7 subunit of Pol ζ, the Rev1 CTD, and the Pol κ RIR, demonstrating that the Rev1 CTD can bind multiple DNA polymerases simultaneously (Fig. 4.5b) (Wojtaszek et al. 2012a). Together, these data suggest that Rev1 plays a critical role in recruiting other TLS polymerases including Pol ξ and perhaps even Pol κ and Pol ι to stalled replication forks.

4.6 Concluding Remarks

The biochemical and structural studies reviewed here have provided tremendous insight into how Y-family polymerases differ from their classical counterparts and can efficiently incorporate nucleotides opposite DNA lesions. They have revealed some surprising strategies employed by Y-family polymerases to accommodate these lesions in their active sites. These strategies include having larger and less constrained active sites, forming Hoogsteen base pairs, and using amino acid side chains as templates. Unfortunately, far less is known about how Y-family polymerases are recruited to stalled replication forks and how their activities are coordinated on the DNA. Biochemical studies of the protein–protein interactions of these polymerases have begun to provide some insight into the recruitment and regulation of Y-family polymerases. It is becoming clear that these enzymes function within a dynamic network of protein–protein interactions in which ubiquitin-modified PCNA and Rev1 play critical roles. Moreover, these interactions are mediated largely via the intrinsically disordered regions of these polymerases. Exciting new breakthroughs will almost certainly emerge from further biochemical and structural studies of this network of protein–protein interactions.

Acknowledgment This article was supported by Award Number GM081433 from the National Institute of General Medical Sciences to M.T.W. The content is solely the responsibility of the authors and does not necessarily represent the official views of the National Institute of General Medical Sciences or the National Institutes of Health.

We thank Christine Kondratick, Maria Spies, Adrian Elcock, and Marc Wold for valuable discussions.

References

Acharya N, Johnson RE, Prakash S, Prakash L (2006) Complex formation with Rev1 enhances the proficiency of Saccharomyces cerevisiae DNA polymerase for mismatch extension and for extension opposite from DNA lesions. Mol Cell Biol 26(24):9555–9563

Acharya N, Yoon JH, Gali H, Unk I, Haracska L, Johnson RE, Hurvvitz J, Prakash L, Prakash S (2008) Roles of PCNA-binding and ubiquitin-binding domains in human DNA polymerase eta in translesion DNA synthesis. Proc Natl Acad Sci USA 105(46):17724–17729

Akagi J, Masutani C, Kataoka Y, Kan T, Ohashi E, Mori T, Ohmori H, Hanaoka F (2009) Interaction with DNA polymerase eta is required for nuclear accumulation of REV1 and suppression of spontaneous mutations in human cells. DNA Repair 8(5):585–599. doi:10.1016/j.dnarep.2008.12.006

Barkley LR, Palle K, Durando M, Day TA, Gurkar A, Kakusho N, Li J, Masai H, Vaziri C (2012) c-Jun N-terminal kinase-mediated Rad18 phosphorylation facilitates Pol eta recruitment to stalled replication forks. Mol Biol Cell 23(10):1943–1954. doi:10.1091/mbc.E11-10-0829

Bienko M, Green CM, Crosetto N, Rudolf F, Zapart G, Coull B, Kannouche P, Wider G, Peter M, Lehmann AR, Hofmann K, Dikic I (2005) Ubiquitin-binding domains in Y-family polymerases regulate translesion synthesis. Science 310(5755):1821–1824

Biertuempfel C, Zhao Y, Kondo Y, Ramon-Maiques S, Gregory M, Lee JY, Masutani C, Lehmann AR, Hanaoka F, Yang W (2010) Structure and mechanism of human DNA polymerase eta. Nature 465(7301):1044–1048. doi:10.1038/nature09196

Bomar MG, Pai MT, Tzeng SR, Li SSC, Zhou P (2007) Structure of the ubiquitin-binding zinc finger domain of human DNA Y-polymerase eta. EMBO Rep 8(3):247–251

Bomar MG, D'Souza S, Bienko M, Dikic I, Walker GC, Zhou P (2010) Unconventional ubiquitin recognition by the ubiquitin-binding motif within the Y Family DNA polymerases iota and Rev1. Mol Cell 37(3):408–417. doi:10.1016/j.molcel.2009.12.038

Burgers PMJ, Koonin EV, Bruford E, Blanco L, Burtis KC, Christman MF, Copeland WC, Friedberg EC, Hanaoka F, Hinkle DC, Lawrence CW, Nakanishi M, Ohmori H, Prakash L, Prakash S, Reynaud CA, Sugino A, Todo T, Wang ZG, Weill JC, Woodgate R (2001) Eukaryotic DNA polymerases: proposal for a revised nomenclature. J Biol Chem 276(47):43487–43490

Burschowsky D, Rudolf F, Rabut G, Herrmann T, Matthias P, Wider G (2011) Structural analysis of the conserved ubiquitin-binding motifs (UBMs) of the translesion polymerase iota in complex with ubiquitin. J Biol Chem 286(2):1364–1373. doi:10.1074/jbc.M110.135038

Carlson KD, Washington AT (2005) Mechanism of efficient and accurate nucleotide incorporation opposite 7,8-dihydro-8-oxoguanine by Saccharomyces cerevisiae DNA polymerase eta. Mol Cell Biol 25(6):2169–2176

Choi JY, Guengerich FP (2008) Kinetic analysis of translesion synthesis opposite bulky N-2- and O-6-alkylguanine DNA adducts by human DNA polymerase REV1. J Biol Chem 283(35):23645–23655. doi:10.1074/jbc.M801686200

Choi JY, Angel KC, Guengerich FP (2006) Translesion synthesis across bulky N-2-alkyl guanine DNA adducts by human DNA polymerase kappa. J Biol Chem 281(30):21062–21072

Cortese MS, Uversky VN, Keith Dunker A (2008) Intrinsic disorder in scaffold proteins: getting more from less. Prog Biophys Mol Biol 98(1):85–106. doi:10.1016/j.pbiomolbio.2008.05.007

Day TA, Palle K, Barkley LR, Kakusho N, Zou Y, Tateishi S, Verreault A, Masai H, Vaziri C (2010) Phosphorylated Rad18 directs DNA polymerase eta to sites of stalled replication. J Cell Biol 191(5):953–966. doi:10.1083/jcb.201006043

de Groote FH, Jansen JG, Masuda Y, Shah DM, Kamiya K, de Wind N, Siegal G (2011) The Rev1 translesion synthesis polymerase has multiple distinct DNA binding modes. DNA Repair 10(9):915–925. doi:10.1016/j.dnarep.2011.04.033

Dieckman LM, Freudenthal BD, Washington MT (2012) PCNA structure and function: insights from structures of PCNA complexes and post-translationally modified PCNA. Subcell Biochem 62:281–299

Dunker AK, Silman I, Uversky VN, Sussman JL (2008) Function and structure of inherently disordered proteins. Curr Opin Struct Biol 18(6):756–764. doi:10.1016/j.sbi.2008.10.002

Emsley P, Cowtan K (2004) Coot: model-building tools for molecular graphics. Acta Crystallogr D Biol Crystallogr 60:2126–2132

Fink AL (2005) Natively unfolded proteins. Curr Opin Struct Biol 15(1):35–41. doi:10.1016/j.sbi.2005.01.002

Freudenthal BD, Ramaswamy S, Hingorani MM, Washington MT (2008) Structure of a mutant form of proliferating cell nuclear antigen that blocks translesion DNA synthesis. Biochemistry 47(50):13354–13361

Freudenthal BD, Gakhar L, Ramaswamy S, Washington MT (2010) Structure of monoubiquitinated PCNA and implications for translesion synthesis and DNA polymerase exchange. Nat Struct Mol Biol 17(4):479–484. doi:10.1038/nsmb.1776

Friedberg EC, Wagner R, Radman M (2002) Molecular biology – specialized DNA polymerases, cellular survival, and the genesis of mutations. Science 296(5573):1627–1630

Guo CX, Fischhaber PL, Luk-Paszyc MJ, Masuda Y, Zhou J, Kamiya K, Kisker C, Friedberg EC (2003) Mouse Rev1 protein interacts with multiple DNA polymerases involved in translesion DNA synthesis. EMBO J 22(24):6621–6630

Guo CX, Sonoda E, Tang TS, Parker JL, Bielen AB, Takeda S, Ulrich HD, Friedberg EC (2006) REV1 protein interacts with PCNA: significance of the REV1 BRCT domain in vitro and in vivo. Mol Cell 23(2):265–271

Haracska L, Yu SL, Johnson RE, Prakash L, Prakash S (2000) Efficient and accurate replication in the presence of 7,8-dihydro-8-oxoguanine by DNA polymerase eta. Nat Genet 25(4):458–461

Haracska L, Johnson RE, Unk I, Phillips B, Hurwitz J, Prakash L, Prakash S (2001a) Physical and functional interactions of human DNA polymerase eta with PCNA. Mol Cell Biol 21(21):7199–7206

Haracska L, Johnson RE, Unk I, Phillips BB, Hurwitz J, Prakash L, Prakash S (2001b) Targeting of human DNA polymerase iota to the replication machinery via interaction with PCNA. Proc Natl Acad Sci USA 98(25):14256–14261

Haracska L, Kondratick CM, Unk I, Prakash S, Prakash L (2001c) Interaction with PCNA is essential for yeast DNA polymerase eta function. Mol Cell 8(2):407–415

Haracska L, Unk I, Johnson RE, Johansson E, Burgers PMJ, Prakash S, Prakash L (2001d) Roles of yeast DNA polymerases delta and zeta and of Rev1 in the bypass of abasic sites. Genes Dev 15(8):945–954

Haracska L, Prakash L, Prakash S (2002a) Role of human DNA polymerase kappa as an extender in translesion synthesis. Proc Natl Acad Sci USA 99(25):16000–16005

Haracska L, Prakash S, Prakash L (2002b) Yeast Rev1 protein is a G template-specific DNA polymerase. J Biol Chem 277(18):15546–15551

Haracska L, Unk I, Johnson RE, Phillips BB, Hurwitz J, Prakash L, Prakash S (2002c) Stimulation of DNA synthesis activity of human DNA polymerase kappa by PCNA. Mol Cell Biol 22(3):784–791

Hingorani MM, O'Donnell M (2000) Sliding clamps: a (tail)ored fit. Curr Biol 10(1):R25–R29

Hishiki A, Hashimoto H, Hanafusa T, Kamei K, Ohashi E, Shimizu T, Ohmori H, Sato M (2009) Structural basis for novel interactions between human translesion synthesis polymerases and proliferating cell nuclear antigen. J Biol Chem 284(16):10552–10560

Hoege C, Pfander B, Moldovan GL, Pyrowolakis G, Jentsch S (2002) RAD6-dependent DNA repair is linked to modification of PCNA by ubiquitin and SUMO. Nature 419(6903):135–141

Ishida T, Kinoshita K (2008) Prediction of disordered regions in proteins based on the meta approach. Bioinformatics 24(11):1344–1348. doi:10.1093/bioinformatics/btn195

Johnson RE, Kondratick CM, Prakash S, Prakash L (1999a) hRAD30 mutations in the variant form of xeroderma pigmentosum. Science 285(5425):263–265

Johnson RE, Prakash S, Prakash L (1999b) Efficient bypass of a thymine-thymine dimer by yeast DNA polymerase, Pol eta. Science 283(5404):1001–1004

Johnson RE, Prakash S, Prakash L (2000a) The human DINB1 gene encodes the DNA polymerase Pol theta. Proc Natl Acad Sci USA 97(8):3838–3843

Johnson RE, Washington MT, Haracska L, Prakash S, Prakash L (2000b) Eukaryotic polymerases iota and zeta act sequentially to bypass DNA lesions. Nature 406(6799):1015–1019

Johnson RE, Washington MT, Prakash S, Prakash L (2000c) Fidelity of human DNA polymerase eta. J Biol Chem 275(11):7447–7450

Kannouche P, de Henestrosa ARF, Coull B, Vidal AE, Gray C, Zicha D, Woodgate R, Lehmann AR (2002) Localization of DNA polymerases eta and iota to the replication machinery is tightly co-ordinated in human cells. EMBO J 21(22):6246–6256

Kannouche PL, Wing J, Lehmann AR (2004) Interaction of human DNA polymerase eta with monoubiquitinated PCNA: a possible mechanism for the polymerase switch in response to DNA damage. Mol Cell 14(4):491–500
Kikuchi S, Hara K, Shimizu T, Sato M, Hashimoto H (2012) Structural basis of recruitment of DNA polymerase zeta by interaction between REV1 and REV7. J Biol Chem 287:33847–33852. doi:10.1074/jbc.M112.396838, M112.396838 [pii]
Krishna TSR, Kong X-P, Gary S, Burgers PM, Kuriyan J (1994) Crystal structure of the eukaryotic DNA polymerase processivity factor PCNA. Cell 79(7):1233–1243
Lawrence CW (2002) Cellular roles of DNA polymerase zeta and Rev1 protein. DNA Repair (Amst) 1(6):425–435
Lawrence CW (2004) Cellular functions of DNA polymerase zeta and Rev1 protein. Adv Protein Chem 69:167–203
Lee GH, Matsushita H (2005) Genetic linkage between Polt deficiency and increased susceptibility to lung tumors in mice. Cancer Sci 96(5):256–259. doi:10.1111/j.1349-7006.2005.00042.x
Lehmann AR, Niimi A, Ogi T, Brown S, Sabbioneda S, Wing JF, Kannouche PL, Green CM (2007) Translesion synthesis: Y-family polymerases and the polymerase switch. DNA Repair 6(7):891–899. doi:10.1016/j.dnarep.2007.02.003
Lemontt JF (1971) Mutants of yeast defective in mutation induced by ultraviolet light. Genetics 68 (1):21–33
Livneh Z, Ziv O, Shachar S (2010) Multiple two-polymerase mechanisms in mammalian translesion DNA synthesis. Cell Cycle 9(4):729–735. doi:10.4161/cc.9.4.10727
Lone S, Townson SA, Uljon SN, Johnson RE, Brahma A, Nair DT, Prakash S, Prakash L, Aggarwall AK (2007) Human DNA polymerase kappa encircles DNA: implications for mismatch extension and lesion bypass. Mol Cell 25(4):601–614
Maga G, Hubscher U (2003) Proliferating cell nuclear antigen (PCNA): a dancer with many partners. J Cell Sci 116(15):3051–3060
Masuda Y, Kamiya K (2002) Biochemical properties of the human REV1 protein. FEBS Lett 520(1–3):88–92
Masuda Y, Takahashi M, Fukuda S, Sumii M, Kamiya K (2002) Mechanisms of dCMP transferase reactions catalyzed by mouse Rev1 protein. J Biol Chem 277(4):3040–3046
Masutani C, Kusumoto R, Yamada A, Dohmae N, Yokoi M, Yuasa M, Araki M, Iwai S, Takio K, Hanaoka F (1999) The XPV (xeroderma pigmentosum variant) gene encodes human DNA polymerase eta. Nature 399(6737):700–704
Matsuda T, Bebenek K, Masutani C, Hanaoka F, Kunkel TA (2000) Low fidelity DNA synthesis by human DNA polymerase-eta. Nature 404(6781):1011–1013
McDonald JP, Levine AS, Woodgate R (1997) The Saccharomyces cerevisiae RAD30 gene, a homologue of Escherichia coli dinB and umuC, is DNA damage inducible and functions in a novel error-free postreplication repair mechanism. Genetics 147(4):1557–1568
Moldovan GL, Pfander B, Jentsch S (2007) PCNA, the maestro of the replication fork. Cell 129:665–679. doi:10.1016/j.cell.2007.05.003
Murakumo Y, Ogura Y, Ishii H, Numata S, Ichihara M, Croce CM, Fishel R, Takahashi M (2001) Interactions in the error-prone postreplication repair proteins hREV1, hREV3, and hREV7. J Biol Chem 276(38):35644–35651. doi:10.1074/jbc.M102051200
Murakumo Y, Mizutani S, Yamaguchi M, Ichihara M, Takahashi M (2006) Analyses of ultraviolet-induced focus formation of hREV1 protein. Genes Cells 11(3):193–205
Nair DT, Johnson RE, Prakash S, Prakash L, Aggarwal AK (2004) Replication by human DNA polymerase-iota occurs by Hoogsteen base-pairing. Nature 430(6997):377–380
Nair DT, Johnson RE, Prakash L, Prakash S, Aggarwal AK (2005a) Human DNA polymerase iota incorporates dCTP opposite template G via a G.C plus Hoogsteen base pair. Structure 13(10):1569–1577
Nair DT, Johnson RE, Prakash L, Prakash S, Aggarwal AK (2005b) Rev1 employs a novel mechanism of DNA synthesis using a protein template. Science 309(5744):2219–2222

Nair DT, Johnson RE, Prakash L, Prakash S, Aggarwal AK (2006) Hoogsteen base pair formation promotes synthesis opposite the 1, N6-ethenodeoxyadenosine lesion by human DNA polymerase iota. Nat Struct Mol Biol 13(7):619–625

Nair DT, Johnson RE, Prakash L, Prakash S, Aggarwal AK (2008) Protein-template-directed synthesis across an acrolein-derived DNA adduct by yeast rev1 DNA polymerase. Structure 16(2):239–245

Nair DT, Johnson RE, Prakash L, Prakash S, Aggarwal AK (2011) DNA synthesis across an abasic lesion by yeast Rev1 DNA polymerase. J Mol Biol 406(1):18–28. doi:10.1016/j.jmb.2010.12.016

Naryzhny SN (2008) Proliferating cell nuclear antigen: a proteomics view. Cell Mol Life Sci 65 (23):3789–3808. doi:10.1007/s00018-008-8305-x

Nelson JR, Gibbs PE, Nowicka AM, Hinkle DC, Lawrence CW (2000) Evidence for a second function for Saccharomyces cerevisiae Rev1p. Mol Microbiol 37(3):549–554, mmi1997 [pii]

Ogi T, Shinkai Y, Tanaka K, Ohmori H (2002) Pol kappa protects mammalian cells against the lethal and mutagenic effects of benzo a pyrene. Proc Natl Acad Sci USA 99(24):15548–15553. doi:10.1073/pnas.222377899

Ohashi E, Bebenek K, Matsuda T, Feaver WJ, Gerlach VL, Friedberg EC, Ohmori H, Kunkel TA (2000) Fidelity and processivity of DNA synthesis by DNA polymerase kappa, the product of the human DINB1 gene. J Biol Chem 275(50):39678–39684

Ohashi E, Murakumo Y, Kanjo N, Akagi J, Masutani C, Hanaoka F, Ohmori H (2004) Interaction of hREV1 with three human Y-family DNA polymerases. Genes Cells 9(6):523–531

Ohashi E, Hanafusa T, Kamei K, Song I, Tomida J, Hashimoto H, Vaziri C, Ohmori H (2009) Identification of a novel REV1-interacting motif necessary for DNA polymerase kappa function. Genes Cells 14(2):101–111

Ohmori H, Friedberg EC, Fuchs RPP, Goodman MF, Hanaoka F, Hinkle D, Kunkel TA, Lawrence CW, Livneh Z, Nohmi T, Prakash L, Prakash S, Todo T, Walker GC, Wang ZG, Woodgate R (2001) The Y-family of DNA polymerases. Mol Cell 8(1):7–8. doi:10.1016/s1097-2765(01)00278-7

Ohmori H, Hanafusa T, Ohashi E, Vaziri C (2009) Separate roles of structured and unstructured regions of Y-family DNA polymerases. In: McPherson A (ed) Advances in protein chemistry and structural biology, vol 78. pp 99–146. doi:10.1016/s1876-1623(09)78004-0

Otsuka C, Kunitomi N, Iwai S, Loakes D, Negishi K (2005) Roles of the polymerase and BRCT domains of Rev1 protein in translesion DNA synthesis in yeast in vivo. Mutat Res 578(1–2):79–87

Pence MG, Blans P, Zink CN, Hollis T, Fishbein JC, Perrino FW (2009) Lesion bypass of N-2-ethylguanine by human DNA polymerase i. J Biol Chem 284(3):1732–1740

Petta TB, Nakajima S, Zlatanou A, Despras E, Couve-Privat S, Ishchenko A, Sarasin A, Yasui A, Kannouche P (2008) Human DNA polymerase iota protects cells against oxidative stress. EMBO J 27(21):2883–2895. doi:10.1038/emboj.2008.210

Pozhidaeva A, Pustovalova Y, D'Souza S, Bezsonova I, Walker GC, Korzhnev DM (2012) NMR structure and dynamics of the C-terminal domain from human Rev1 and its complex with Rev1 interacting region of DNA polymerase eta. Biochemistry 51(27):5506–5520. doi:10.1021/bi300566z

Prakash S, Prakash L (2002) Translesion DNA synthesis in eukaryotes: a one- or two-polymerase affair. Genes Dev 16(15):1872–1883

Pryor JM, Washington MT (2011) Pre-steady state kinetic studies show that an abasic site is a cognate lesion for the yeast Rev1 protein. DNA Repair 10(11):1138–1144. doi:10.1016/j.dnarep.2011.08.011

Roush AA, Suarez M, Friedberg EC, Radman M, Siede W (1998) Deletion of the Saccharomyces cerevisiae gene RAD30 encoding an Escherichia coli DinB homolog confers UV radiation sensitivity and altered mutability. Mol Gen Genet 257(6):686–692

Silverstein TD, Johnson RE, Jain R, Prakash L, Prakash S, Aggarwal AK (2010) Structural basis for the suppression of skin cancers by DNA polymerase eta. Nature 465(7301):1039–1043. doi:10.1038/nature09104

Stancel JNK, McDaniel LD, Velasco S, Richardson J, Guo C, Friedberg EC (2009) Polk mutant mice have a spontaneous mutator phenotype. DNA Repair 8(12):1355–1362. doi:10.1016/j.dnarep.2009.09.003

Stelter P, Ulrich HD (2003) Control of spontaneous and damage-induced mutagenesis by SUMO and ubiquitin conjugation. Nature 425(6954):188–191

Swan MK, Johnson RE, Prakash L, Prakash S, Aggarwal AK (2009) Structure of the human Rev1-DNA-dNTP ternary complex. J Mol Biol 390(4):699–709. doi:10.1016/j.jmb.2009.05.026

Tissier A, McDonald JP, Frank EG, Woodgate R (2000) Pol iota, a remarkably error-prone human DNA polymerase. Genes Dev 14(13):1642–1650

Tissier A, Kannouche P, Reck MP, Lehmann AR, Fuchs RPP, Cordonnier A (2004) Co-localization in replication foci and interaction of human Y-family members, DNA polymerase pol eta and REV1 protein. DNA Repair 3(11):1503–1514

Tsurimoto T (1999) PCNA binding proteins. Front Biosci 4:d849–d858

Tsutakawa SE, Van Wynsberghe AW, Freudenthal BD, Weinacht CP, Gakhar L, Washington MT, Zhuang Z, Tainer JA, Ivanov I (2011) Solution X-ray scattering combined with computational modeling reveals multiple conformations of covalently bound ubiquitin on PCNA. Proc Natl Acad Sci USA 108(43):17672–17677. doi:10.1073/pnas.1110480108

Uijon SN, Johnson RE, Edwards TA, Prakash S, Prakash L, Aggarwal AK (2004) Crystal structure of the catalytic core of human DNA polymerase kappa. Structure 12(8):1395–1404

Ummat A, Silverstein TD, Jain R, Buku A, Johnson RE, Prakash L, Prakash S, Aggarwal AK (2012) Human DNA polymerase eta is pre-aligned for dNTP binding and catalysis. J Mol Biol 415(4):627–634. doi:10.1016/j.jmb.2011.11.038

Vasquez-Del Carpio R, Silverstein TD, Lone S, Johnson RE, Prakash L, Prakash S, Aggarwal AK (2011) Role of human DNA polymerase kappa in extension opposite from a cis-syn thymine dimer. J Mol Biol 408(2):252–261. doi:10.1016/j.jmb.2011.02.042

Vidal AE, Kannouche P, Podust VN, Yang W, Lehmann AR, Woodgate R (2004) Proliferating cell nuclear antigen-dependent coordination of the biological functions of human DNA polymerase iota. J Biol Chem 279(46):48360–48368

Washington MT, Johnson RE, Prakash S, Prakash L (1999) Fidelity and processivity of Saccharomyces cerevisiae DNA polymerase eta. J Biol Chem 274(52):36835–36838

Washington MT, Johnson RE, Prakash S, Prakash L (2000) Accuracy of thymine-thymine dimer bypass by Saccharomyces cerevisiae DNA polymerase eta. Proc Natl Acad Sci USA 97(7):3094–3099

Washington MT, Prakash L, Prakash S (2001) Yeast DNA polymerase eta utilizes an induced-fit mechanism of nucleotide incorporation. Cell 107(7):917–927

Washington MT, Johnson RE, Prakash L, Prakash S (2002) Human DINB1-encoded DNA polymerase kappa is a promiscuous extender of mispaired primer termini. Proc Natl Acad Sci USA 99(4):1910–1914

Washington MT, Prakash L, Prakash S (2003) Mechanism of nucleotide incorporation opposite a thymine-thymine dimer by yeast DNA polymerase eta. Proc Natl Acad Sci USA 100(21):12093–12098

Washington MT, Johnson RE, Prakash L, Prakash S (2004a) Human DNA polymerase iota utilizes different nucleotide incorporation mechanisms dependent upon the template base. Mol Cell Biol 24(2):936–943

Washington MT, Minko IG, Johnson RE, Haracska L, Harris TM, Lloyd RS, Prakash S, Prakash L (2004b) Efficient and error-free replication past a minor-groove N-2-guanine adduct by the sequential action of yeast Rev1 and DNA polymerase zeta. Mol Cell Biol 24(16):6900–6906

Washington MT, Minko IG, Johnson RE, Wolfle WT, Harris TM, Lloyd RS, Prakash S, Prakash L (2004c) Efficient and error-free replication past a minor-groove DNA adduct by the sequential action of human DNA polymerases iota and kappa. Mol Cell Biol 24(13):5687–5693

Watanabe K, Tateishi S, Kawasuji M, Tsurimoto T, Inoue H, Yamaizumi M (2004) Rad18 guides pol eta to replication stalling sites through physical interaction and PCNA monoubiquitination. EMBO J 23(19):3886–3896

Waters LS, Minesinger BK, Wiltrout ME, D'Souza S, Woodruff RV, Walker GC (2009) Eukaryotic translesion polymerases and their roles and regulation in DNA damage tolerance. Microbiol Mol Biol Rev 73(1):134–154. doi:10.1128/mmbr.00034-08

Wojtaszek J, Lee CJ, D'Souza S, Minesinger B, Kim H, D'Andrea AD, Walker GC, Zhou P (2012a) Structural basis of Rev1-mediated assembly of a quaternary vertebrate translesion polymerase complex consisting of Rev1, heterodimeric Pol zeta and Pol kappa. J Biol Chem 287(40):33836–33846. doi:10.1074/jbc.M112.394841, M112.394841 [pii]

Wojtaszek J, Liu J, D'Souza S, Wang S, Xue Y, Walker GC, Zhou P (2012b) Multifaceted recognition of vertebrate Rev1 by translesion polymerases zeta and kappa. J Biol Chem 287(31):26400–26408. doi:10.1074/jbc.M112.380998

Wolfle WT, Johnson RE, Minko IG, Lloyd RS, Prakash S, Prakash L (2005) Human DNA polymerase iota promotes replication through a ring-closed minor-groove adduct that adopts a syn conformation in DNA. Mol Cell Biol 25(19):8748–8754

Wood A, Garg P, Burgers PMJ (2007) A ubiquitin-binding motif in the translesion DNA polymerase Rev1 mediates its essential functional interaction with ubiquitinated proliferating cell nuclear antigen in response to DNA damage. J Biol Chem 282(28):20256–20263. doi:10.1074/jbc.M702366200

Zhang YB, Yuan FH, Wu XH, Wang ZG (2000) Preferential incorporation of G opposite template T by the low-fidelity human DNA polymerase iota. Mol Cell Biol 20(19):7099–7108

Zhang YB, Wu XH, Guo DY, Rechkoblit O, Wang ZG (2002a) Activities of human DNA polymerase kappa in response to the major benzo[a]pyrene DNA adduct: error-free lesion bypass and extension synthesis from opposite the lesion. DNA Repair 1(7):559–569

Zhang YB, Wu XH, Rechkoblit O, Geacintov NE, Taylor JS, Wang ZG (2002b) Response of human REV1 to different DNA damage: preferential dCMP insertion opposite the lesion. Nucleic Acids Res 30(7):1630–1638

Zhou Y, Wang J, Zhang Y, Wang Z (2010) The catalytic function of the Rev1 dCMP transferase is required in a lesion-specific manner for translesion synthesis and base damage-induced mutagenesis. Nucleic Acids Res 38(15):5036–5046. doi:10.1093/nar/gkq225

Zhuang ZH, Ai YX (2010) Processivity factor of DNA polymerase and its expanding role in normal and translesion DNA synthesis. Biochim Biophys Acta 1804(5):1081–1093. doi:10.1016/j.bbapap.2009.06.018

Chapter 5
DNA Polymerases That Perform Template-Independent DNA Synthesis

Anthony J. Berdis

Abstract DNA polymerases typically catalyze the incorporation of mononucleotides into a growing primer using a DNA or RNA template to properly guide each incorporation event. However, several members of the X-family of DNA polymerases are capable of replicating DNA in the complete absence of a templating strand. This form of template-independent DNA synthesis typically occurs during the repair of double-strand DNA breaks. In addition, there is one exceptional polymerase, denoted as terminal deoxynucleotidyl transferase, that exclusively replicates single-strand DNA rather than duplex DNA. This chapter describes the biological roles for the ability of these DNA polymerases to perform template-independent DNA synthesis during nonhomologous end joining. The biochemical mechanisms for how members of the X-family of DNA polymerases perform this type of replication are compared with other "conventional" DNA polymerases that perform template-dependent synthesis. Enzymatic steps encompassing the binding of DNA and dNTP substrates, the involvement of conformational changes that precede chemistry, and kinetic steps associated with product release are described. The influence of other cellular proteins on the activity of these DNA polymerases during nonhomologous end joining is discussed. Finally, the roles of these specialized DNA polymerases in pathological conditions such as cancer are described with a special emphasis on several new nucleoside analogs that function as therapeutic agents against these DNA polymerases.

Keywords DNA polymerases • Template-independent DNA synthesis • Immunology • Cancer • Chemotherapy

---

A.J. Berdis (✉)
Department of Chemistry, Cleveland State University, 2121 Euclid Avenue, Cleveland, OH 44106, USA
e-mail: a.berdis@csuohio.edu

K.S. Murakami and M.A. Trakselis (eds.), *Nucleic Acid Polymerases*, Nucleic Acids and Molecular Biology 30, DOI 10.1007/978-3-642-39796-7_5,

Abbreviations

3-Eth-5-NIdR	3-Ethynyl-5-nitoindolyl-deoxyribose
3-Eth-5-NITP	3-Ethynyl-5-nitoindolyl-deoxyribose triphosphate
5-NITP	5-Nitroindoyl-deoxyribose triphosphate
AID	Activation-induced deaminase
BRCT	BRCA1 c-terminal domain
DNA-PKcs	Serine/threonine protein kinase that is stimulated by free DNA ends
HR	Homologous recombination
Ig	Immunoglobulin
NHEJ	Nonhomologous DNA end joining
PCNA	Proliferating cell nuclear antigen
PK	Protein kinase
Pol λ	DNA polymerase lambda
Pol μ	DNA polymerase mu
PP_i	Inorganic pyrophosphate
RAG-1	Recombination-activating gene 1
RAG-2	Recombination-activating gene 2
RSS	Recombination signal sequences
SHM	Somatic hypermutation
TCR	T-cell receptor
TdT	Terminal deoxynucleotidyl transferase
V(D)J	Variable diversity and joining
XLF	XRCC4-like factor (also called Cernunnos)

5.1 Introduction

DNA polymerases extend nucleic acid primers using a DNA template (or RNA during reverse transcription) to guide each nucleotide incorporation event (Fig. 5.1a). However, using a template is not a universal requirement for all DNA polymerases as there are several polymerases that can perform this process in the absence of templating information (Garcia-Diaz et al. 2005; Paull 2005; Bollum 1960). These include specialized DNA polymerases such as polymerase lambda (pol λ) and polymerase mu (pol μ) which can incorporate nucleotides at the blunt end of duplex DNA (Fig. 5.1b). In addition, there is another unique polymerase denoted as *t*erminal *d*eoxynucleotidyl *t*ransferase (TdT) that has the unusual ability to perform polymerization in a completely template-independent manner using only single-strand DNA as the substrate (Fig. 5.1c). At face value, performing DNA synthesis in the absence of any templating information appears to be a highly risky endeavor for the cell because of the implied risk of introducing genetic mutations into chromosomal DNA. Indeed, this is typically true for cells that respond to various DNA damaging agents. However, template-independent DNA

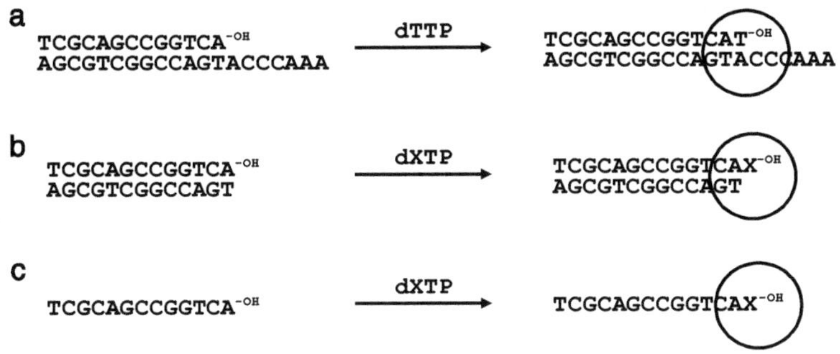

Fig. 5.1 Models for template-dependent and template-independent DNA synthesis. (**a**) Conventional DNA polymerases involved in replicating genomic DNA require double-stranded DNA as a substrate in which the 5′ → 3′ strand is used as a primer and the complementary strand 3′ → 5′ is used as a template. (**b**) Members of the X-family of DNA polymerases including pol λ and pol μ can perform template-independent DNA synthesis using blunt-end DNA as the substrate. (**c**) Terminal deoxynucleotidyl transferase is unique in its ability to catalyze phosphoryl transfer in the absence of a template that cannot be accommodated in its active site

synthesis catalyzed by TdT, for example, is very important for generating random mutations in specific regions of DNA during V(D)J recombination (Kepler et al. 1996; Kunkel et al. 1986). By subtly randomizing genetic material, TdT plays a key role in the maturation of the vertebrate immune system (Komori et al. 1993; Bertocci et al. 2006a; Kallenbach et al. 1990; Landau et al. 1987a). In fact, randomly incorporating nucleotides during V(D)J recombination serves as a viable way to increase antigen receptor diversity to produce approximately 10^{14} different immunoglobulins and 10^{18} unique T-cell antigen receptors that can neutralize potential antigens (Sadofsky 2001; Janeway 1999).

This chapter describes the cellular roles for template-independent DNA synthesis catalyzed by three distinct DNA polymerases including pol λ, pol μ, and TdT. Particular emphasis is placed on TdT as this enzyme plays an important role during V(D)J recombination. The molecular mechanisms responsible for the unique activities of these specialized DNA polymerases are also described. These discussions include the use of the reported structure of TdT which provides important insights into several properties of the polymerase such as recognition of nucleic acid and nucleotide substrates as well as the utilization of various metal ion cofactors. Finally, the biomedical importance of these specialized DNA polymerases are discussed in the context of pathological conditions including leukemia and Merkel cell carcinoma.

5.2 Why Is Template-Independent DNA Synthesis Necessary?

Double-strand breaks (DSBs) arise in dividing cells about ten times per cell per day (reviewed in Ohnishi et al. 2009). DSBs are formed by several mutually exclusive mechanisms including replication across nicked DNA, damage from free radicals caused by oxidants or ionizing radiation, and by the inadvertent action of enzymes involved in DNA metabolism. The two major pathways involved in repairing DSBs are homologous recombination (HR) and nonhomologous end joining (NHEJ) (Kass and Jasin 2010). Of the two pathways, HR is considered to be more faithful since it requires that there be sequence identity in regions that undergo recombination. However, the repair capacity of HR is limited as it occurs only during S-phase of the cell cycle and typically requires that there be hundreds of base pairs of homology in order to be fully efficient. The more commonly used pathway is NHEJ which can repair a DSB at any time during the cell cycle. In addition, when two DSBs occur on different chromosomes, their rejoining is almost always performed by NHEJ. This ability reflects the fact that NHEJ does not require significant amounts of sequence homology, although a few nucleotides of terminal microhomology are often utilized by the NHEJ enzymes. The proteins involved in NHEJ include Ku, DNA-PKcs, Artemis, pol μ, pol λ, XLF (aka Cernunnos), XRCC4, and DNA ligase IV.

5.2.1 *The X-Family of DNA Polymerases*

Mammals possess four DNA polymerase X-family members that include pol β, pol μ, pol λ, and TdT (Yamtich and Sweasy 2010) and Chap. 2. Pol β is a single polypeptide with an apparent molecular weight of 39 kDa that primarily functions during base excision repair. Pol β possesses both template-dependent polymerase activity and deoxyribose phosphate (dRP) lyase activity (Matsumoto and Kim 1995). Pol λ is most closely related to pol β as both possess dRP lyase activity (Aoufouchi et al. 2000; Nagasawa et al. 2000). In addition, pol λ can substitute for pol β in single-nucleotide base excision repair reactions, at least under in vitro conditions (Braithwaite et al. 2005). Pol μ participates in the resynthesis of missing nucleotides during NHEJ repair of DNA breaks (Ruiz et al. 2001). While pol μ can replicate normal DNA, it also has the ability to incorporate nucleotides opposite damaged DNA. With respect to primary amino acid sequence, pol μ and TdT are approximately 40 % identical to each other and approximately 20 % identical to other Pol X-family members including pol β (Domínguez et al. 2000; Yang 2003). On the other hand, TdT is a very specialized DNA polymerase that adds random nucleotides to DNA ends during V(D)J recombination. TdT activity is generally confined to B- and T-cells and is used to generate diversity during an immune response. During V(D)J recombination, TdT and pol μ participate during heavy chain rearrangements, while pol λ is believed to function only during light chain rearrangements of antibodies.

5.2.2 Polymerase Activity During Nonhomologous End Joining

The relatively high degree of similarity between TdT and pol μ suggests that both polymerases might interact with other core end-joining proteins when processing DSBs (Nick McElhinny et al. 2005). However, their contrasting activities would produce opposite effects on genomic fidelity. For example, the function of TdT is to produce diversity in genomic material during V(D)J recombination, while the template-dependent activity of pol μ should maintain genomic fidelity by allowing accurate gap-filling DNA synthesis during DSB repair. Although pol μ is efficient during alignment-based gap fill-in synthesis in vitro (García-Díaz et al. 2000a; Lee et al. 2004), the polymerase is also prone to frameshift synthesis (Zhang et al. 2001) and is unable to perform strand displacement DNA synthesis (Lee et al. 2004). Ex vivo data also suggests that pol μ can participate in general DSB repair. This is evident as cells exposed to DNA damaging agents show higher levels of pol μ expression and co-localization with γH2AX, a biochemical marker associated with DSB formation (Lee et al. 2004). This is in contrast with the activity of pol β, which does not associate with end-joining factors nor performs alignment-based gap fill-in synthesis (García-Díaz et al. 2000a; Lee et al. 2004).

Pol λ, another Pol X-family member, is similar to TdT and pol μ as all three have an amino-terminal BRCT domain (conserved C-terminal domain in BRCA1) and a carboxy-terminal catalytic domain (Fig. 5.2). The BRCT domain is an important structural feature as it mediates protein/protein and protein/DNA interactions during DNA repair pathways and cell cycle checkpoint regulation. However, the role of this domain in coordinating polymerase activity during DNA repair is still poorly understood. For example, while pol λ can substitute for pol β during in vitro base excision repair reactions (García-Díaz et al. 2001), it is not clear if pol λ performs this function in vivo. Despite this uncertainty, the dRP lyase activity of pol λ has been proposed to participate in end-joining reactions as a way to deal with "abortive" DSB intermediates that form during base excision repair of radiation damage (Bebenek et al. 2003). Mechanistically, pol λ is similar to pol μ as both are prone to frameshift DNA synthesis (Bebenek et al. 2003) and neither performs strand displacement DNA synthesis (Capp et al. 2006).

Depletion of pol λ can block the end joining of substrates that require gap-filling synthesis in cell extracts (Bertocci et al. 2002). However, a deficiency in pol λ activity does not influence the efficiency of end-joining during V(D)J recombination (Maga et al. 2005). In contrast, the efficiency of V(D)J recombination is affected by deficiencies in pol μ activity. Collectively, these results suggest that the cellular role of pol μ is limited to V(D)J recombination while pol λ functions

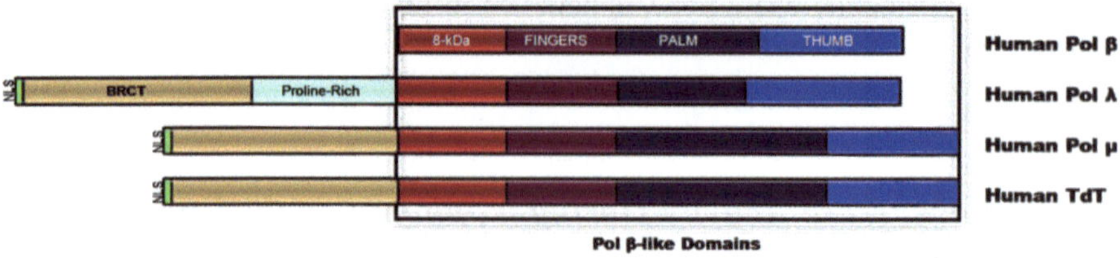

Fig. 5.2 Schematic representations of the different domains found in the four X-family DNA polymerases. Each domain is labeled and colored for clarity. *NLS* represents the nuclear localization signal motif, and *BRCT* indicates the BRCA1 carboxy terminus domain

almost exclusively during general DSB repair. The exact details of how various X-family members interact with end-joining factors to coordinate their activities remain unclear. Current models suggest that TdT, pol μ, and pol λ are recruited to DSBs by interactions of their BRCT domains with ligase IV. Competition amongst these polymerases has only been observed between pol μ and TdT (Nick McElhinny et al. 2005). The details of these studies are described later as they occur only during V(D)J recombination.

Another important consideration for regulating polymerase activity during DSB repair is the availability of dNTP substrates. Since dNTPs are at their highest concentration during the S-phase of the cell cycle, most template-dependent polymerases are highly active during this stage (Bjursell and Skoog 1980). However, most DSB repair occurs by NHEJ outside of S-phase. In fact, most repair takes place during the G1 phase of the cell cycle (Takata et al. 1998) when dNTP pools are significantly lower (McCormick et al. 1983). Thus, polymerases that participate in NHEJ face a significant challenge in performing DNA synthesis since the availability of nucleotide substrates may be a limiting factor. However, the polymerases involved in NHEJ appear to use extraordinary means to overcome low dNTP levels. For example, pol λ has an unusually high affinity for dNTPs as the K_d for dATP is ~30-fold lower than that for any other X-family member (García-Díaz et al. 2000b; Ruiz et al. 2003). This higher binding affinity means that pol λ can remain active even when dNTP levels are very low. Pol μ and TdT appear to employ an alternative strategy by using rNTPs rather than dNTPs. While most template-dependent polymerases display a 1,000-fold preference for dNTPs over rNTPs, both pol μ (Roychoudhury and Kössel 1971) and TdT (Boulé et al. 2001) show little discrimination between either substrate. This lack of discrimination is thought to occur due to the absence of amino acids that can function as a "steric gate" to prevent the utilization of rNTPs. In any event, it may be advantageous for these polymerases to use either ribo- or deoxyribose nucleotides since rNTP pools typically remain high throughout the cell cycle (Traut 1994). As such, the ability to incorporate rNTPs could allow pol μ and TdT to temporarily repair DSBs formed outside of the S-phase of the cell cycle where dNTP levels are precariously low. This "temporary fix" would allow the cell to survive rather than undergo cell death via apoptosis.

5.2.3 The Involvement of DNA Polymerases During V(D)J Recombination

The T- and B-cells of the adaptive immune system function to mount a rapid and robust protective response against foreign entities present in the systemic circulation. This is achieved by expanding the number of pathogen-specific T-cells after an antibody receptor binds an antigen (Papermaster et al. 1964). Once binding of an antigen and activation occurs, thousands of clones are generated over the course of a week to produce effector functions (Landau et al. 1987b; Lieber et al. 1988; Bertocci et al. 2006b). While the vast majority (90–95 %) of these activated T-cells undergo apoptosis (Haeryfar et al. 2008), a small population of T-cells persist in lymphoid and non-lymphoid tissues (Benedict et al. 2000; Baltimore 1974), and this population of cells scans the body for previously encountered pathogens. The impressive immunological memory displayed by the adaptive immune system provides long-term protection against subsequent infection, and this effect can last for several decades.

To increase acquired immunity against biological and chemical antigens, B- and T-cells use a genetic strategy known as V(D)J recombination (Beutler 2003). V(D)J recombination, also known as somatic recombination, is a mechanism for recombining genetic fragments during the initial stages of immunoglobulin and T-cell receptor production of the vertebrate immune system. This process occurs only in primary lymphoid tissue (bone marrow for B-cells and the thymus for T-cells). During this process, the variable (V), diversity (D), and joining (J) gene segments are randomly rearranged to increase the number and diversity of antigen receptors. By creating unique antibodies with distinct antigen specificity, this process creates a highly versatile and competent immune system (Schatz et al. 1992). The ability to cleave, rearrange, and rejoin the V, D, and J regions of germline immunoglobulin genes requires the concerted efforts of the three distinct enzyme activities that include nucleases, polymerases, and ligases (Fig. 5.3).

The early steps of V(D)J recombination occur during the RAG (recombination-activating gene) cleavage phase. This process is initiated by the introduction of a DSB at the edge of the selected gene segment and is catalyzed by the *RAG-1* and *RAG-2* proteins which selectively bind to specific recombination signal sequences (RSS) containing heptamer and nonamer elements with 12 or 23 bp spacer regions in between (Schatz 2004; Early et al. 1980). Recognition of complementary RSS allows the *RAG* complex to introduce a nick between the D and J coding segment as well as with the adjoining recombination signal sequence. The RAG complex also catalyzes the formation of hairpins at each coding end. This occurs using the 3′-OH group at each nick as the nucleophile to catalyze the reaction.

DSBs introduced during the RAG cleavage phase are subsequently repaired during NHEJ (Fig. 5.3). This second stage depends upon the activity of the Artemis:DNA-PK_{cs} complex which functions as a nuclease to trim 5′ and 3′ overhangs (Steen et al. 1996). This activity opens the hairpins present at the coding ends of the gene segment to produce palindromic nucleotides (P-nucleotides). TdT

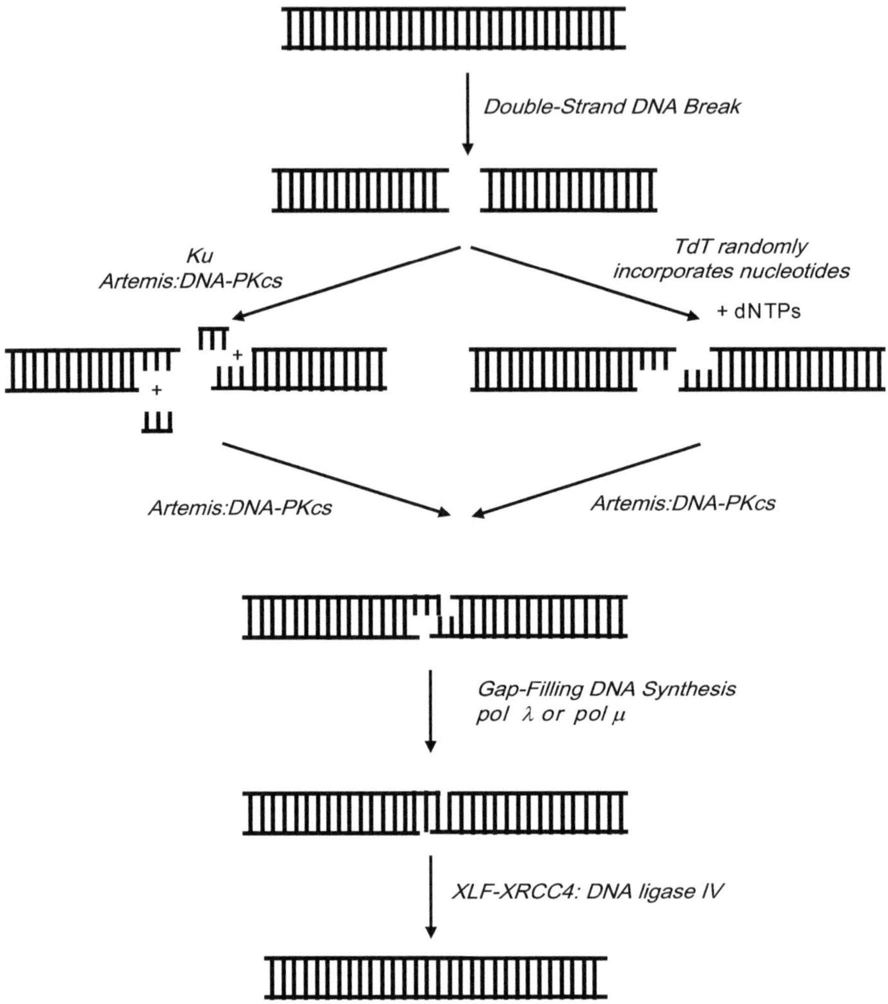

Fig. 5.3 Overview of V(D)J recombination. A simplified model for the RAG cleavage phase generating double-strand breaks and DNA repair through nonhomologous end-joining pathway during DJ gene segment assembly of the V(D)J recombination mechanism. Simplified overview of the enzymatic steps and the role of terminal deoxynucleotidyl transferase in lymphocyte gene rearrangement. The variability of the recombined gene segments is increased through the random addition of non-templated (N) nucleotides catalyzed by the terminal deoxynucleotidyl transferase prior to complementary pairing and extension by template-dependent DNA polymerases

participates at this stage by randomly incorporating nucleotides to an available 3′-OH. TdT is a template-independent DNA polymerase that is expressed only in T- and B-cells that are cells active in V(D)J recombination. This biological activity is designed to increase the diversity of the antigen receptor repertoire by mediating N-addition (nongermline encoded sequence addition) in receptor junctions. It is likely that TdT activity is solely linked to V(D)J recombination as its expression pattern is restricted to only cells that are active in V(D)J recombination. Highly restricted expression of TdT is necessary since its template-independent activity would undoubtedly produces mutagenic consequences during general DSB repair in other cell types. As described later, certain forms of cancer show altered expression levels of TdT that correlate with pathophysiological cellular activity.

In vitro studies have shown that TdT can utilize all four natural dNTPs (Bollum 1960; Boulé et al. 2001). However, in vivo studies show a distinctive bias for preferential incorporation of dGTP and dCTP versus dATP or dTTP (Mickelsen et al. 1999; Feeney 1990; Bangs et al. 1991; Basu et al. 1983; Coleman et al. 1974a; Bollum 1979a; Cabaniols et al. 2001). This preference offers a molecular mechanism to explain the high G/C content that is present in immunoglobulin (Ig) and T-cell receptor (TCR) N-regions (Mickelsen et al. 1999). In addition, the average length of the N-nucleotide segment created by TdT is between 2 and 5 base pairs per coding joint (Bollum 1979a). While this length appears short, it is sufficiently long enough to allow the extended DNA strands to anneal using microhomology alignment via Watson–Crick base-pairing. Again, the role of TdT to increase immunological diversity has been also been validated by the near exclusive expression and localization of this polymerase in primary lymphoid tissues including thymus and bone marrow (Coleman et al. 1974a; Bollum 1979a). Furthermore, TdT-knockout mice display tenfold reductions in TCR diversity compared to wild-type mice (Cabaniols et al. 2001), further validating the role of this specialized polymerase in V(D)J recombination.

The last stage of V(D)J recombination requires the action of the Artemis:DNA-PK_{cs} complex which excises any unpaired bases that form during the annealing process (Ma et al. 2002). Gaps that exist in the recombined section of DNA are filled in by template-dependent DNA polymerases, including X-family members such as pol λ, pol μ, or replicative polymerases such as pol δ or ε. Ligation of the coding ends is performed by the XRCC4-DNA ligase IV complex (Sibanda et al. 2001).

Recently, it was shown that competition between TdT and pol μ is a biologically relevant reaction with important ramifications on efficient immune function. In this case, overexpression of pol μ leads to a reduction in the frequency of TdT-dependent N-addition in a plasmid-based assay for V(D)J recombination (Bertocci et al. 2003). In addition, mice lacking pol μ have a mild B-cell deficiency apparently arising as a result of deletions in V(D)J recombination sites at immunoglobulin light chain loci (Bertocci et al. 2003).

5.3 Terminal Deoxynucleotidyl Transferase

Chapter 2 in this book discusses the mechanism of other X-family polymerase members including pol λ and pol μ that can also perform template-independent DNA synthesis. As such, the remaining focus of this chapter will be on the mechanism and regulation of TdT activity.

5.3.1 Sources and Purification of TdT

There are numerous complexities in the number of different TdT isoforms that are present in mammalian systems. Discussions here are limited to human (*h*) TdT isoforms. Humans possess three alternative splice variants designated as *h*TdTS (short), *h*TdTL1 (long), and *h*TdTL2 (long) (Takahara et al. 1994; Thai and Kearney 2004, 2005). Although *h*TdTL1 and *h*TdTL2 both localize in the nucleus, *h*TdTL2 is expressed more abundantly in normal lymphocytes compared to *h*TdTL1 (Thai and Kearney 2005). The long isoforms of human TdT are both reported to possess $3' \rightarrow 5'$ exonuclease activity for nucleotide removal whereas the short isoform performs nucleotide elongation of the coding ends during V(D)J recombination (Takahara et al. 1994; Thai and Kearney 2004, 2005).

Overexpression of either *h*TdTS or *h*TdTL2 reduces the efficiency of V(D)J recombination greatly (Thai and Kearney 2005) while simultaneous overexpression of *h*TdTS and *h*TdTL2 results in normal recombination frequencies. In addition, the presence of all three human TdT variants during antigen receptor gene rearrangement drastically diminishes recombination frequencies. These results collectively suggest that *h*TdTL1 serves to modulate the activities of either *h*TdTL2 or *h*TdTS to maintain proper recombination.

TdT has been purified from a variety of different sources including calf thymus glands and cultured cell lines propagated from patients with acute lymphoblastic leukemia (Chang and Bollum 1971; Deibel and Coleman 1980a, b; Chang et al. 1982; Bollum and Chang 1981; Nakamura et al. 1981). In its purified form, hTdTL2 (designated from this point on simply as TdT) is a monomeric protein with a molecular weight of approximately 58 kDa (Nakamura et al. 1981). In general, large quantities of human TdT can be purified from cultured cells. However, this approach is generally cost prohibitive for most laboratories. To circumvent this complication, several attempts have been made to overexpress TdT in bacterial systems (Peterson et al. 1985; Boule et al. 1998). These efforts have generally been met with limited success, and the associated failures are attributed to differences in codon frequencies and tRNA pools in *Escherichia coli* versus eukaryotes as well as the low solubility of expressed TdT in these bacterial systems (Peterson et al. 1985; Boule et al. 1998). The production of soluble and active forms of TdT have been achieved by overexpressing a rare *argU* tRNA in the *E. coli* system coupled with growing cultures at a lower temperature (15 °C) to optimize protein folding (Boule et al. 1998). Recombinant human TdT has also been overexpressed using a baculovirus expression system (Chang et al. 1988).

5.3.2 Enzymatic Properties of TdT

The ability of TdT to incorporate nucleotides in a template-independent manner has been investigated by several different research groups (Karkow and Kamen 1966; Kato et al. 1967; Chang et al. 1972; Roychoudhury 1972; Chang and Bollum 1980). In most reports, the template-independent activity of TdT was distinguished from that of template-dependent DNA polymerases by measuring polymerization activity with single-strand versus double-strand DNA. Under in vitro conditions, the replication of homopolymers by TdT requires an initiator chain of six or more nucleotides for poly(dA) and more than five nucleotides for poly(dT) (Kato et al. 1967). Interestingly, inactivating the exonuclease domain of the archaeal B-family DNA polymerase from *Sulfolobus solfataricus* results in the ability of this polymerase to robustly extend short single-strand DNA into several thousand bases using a hybrid mechanism of template-dependent and independent activities (Zuo et al. 2011). TdT, however, proves to be a unique DNA polymerase as it can catalyze the de novo synthesis of polynucleotides ranging in size from 2- to 15-mers when provided with dNTPs in absence of a primer (Chang et al. 1972). As indicated earlier, TdT can also utilize rNTPs, under in vitro conditions. In these instances, TdT can elongate a primer containing a 3′-terminal ribonucleotide 5′-monophosphate (rNMP) (Roychoudhury 1972). However, the addition of rNMPs to a DNA template can significantly impair the kinetics of elongation. For example, the addition of more than two rNMPs does not occur, and this lack of activity may reflect the ability of single-strand RNA to form secondary structures that resemble duplex DNA. As discussed earlier, most template-dependent polymerases utilize dNTPs 1,000-fold more efficiently than rNTPs. The ability of TdT to show little discrimination against rNTPs as potential substrates may have important ramifications for temporarily repairing DSBs under in vivo conditions.

Like all other DNA polymerases, TdT requires divalent metal ions as cofactors to catalyze nucleotide incorporation. However, TdT is again unique amongst all polymerases as it can perform polymerization with a wide number of different divalent cations including Co^{2+}, Mn^{2+}, Zn^{2+}, and Mg^{2+} (Chang and Bollum 1980). Under in vitro conditions, each metal ion produces a different effect on the efficiency and kinetics of nucleotide incorporation. For example, the presence of Mg^{2+} leads to the preferential utilization of dGTP and dATP whereas Co^{2+} leads to the preferential utilization of the pyrimidines, dCTP, and dTTP (Chang and Bollum 1980). Zn^{2+} behaves as a unique cofactor as it functions as a positive allosteric effector for TdT. This is based on experiments demonstrating that polymerization rates by TdT in the presence of Mg^{2+} are stimulated by the addition of micromolar quantities of Zn^{2+} (Chang and Bollum 1980). This rate enhancement could be caused by Zn^{2+} producing different conformational changes in TdT, each of which displays higher intrinsic polymerization efficiencies (Chang and Bollum 1980). However, further work is needed to verify this proposal. Surprisingly, polymerization rates are lower in the presence of Mn^{2+} compared to Mg^{2+} (Deibel and Coleman 1980b). This is intriguing since opposite effects are observed with most template-dependent DNA polymerases. Indeed, Mn^{2+} has been shown by several groups to be a highly pro-mutagenic metal ion as its presence has a negative effect on polymerization fidelity, increasing both the frequency and rate of misincorporation events (Beckman et al. 1985).

5.3.3 Kinetic Mechanism of TdT

5.3.3.1 Order of Substrate Binding

Template-dependent DNA synthesis is achieved through the strictly ordered binding of DNA substrate to the polymerase prior to the binding of dNTP. From a biological perspective, the binding of DNA before dNTP is intuitively obvious as this order allows the polymerase to "sample" each dNTP prior to making a commitment to incorporate the correct nucleotide. The alternative scenario, a prerequisite binding of dNTP prior to DNA, is unattractive since this would allow the polymerase to bind the "correct" complementary nucleotide only 25 % of this time. This mechanism would be highly inefficient for cellular replication. In fact, forcing a polymerase to bind a dNTP substrate first often causes the formation of a dead-end complex that reduces polymerization efficiency. For example, the use of exceedingly high concentrations of dNTPs (>1 mM) can produce substrate-induced inhibition that is likely caused by the inappropriate binding of dNTP prior to DNA substrate (Zhang et al. 2011). However, TdT again appears to be the lone exception to this general rule as the order for substrate binding follows a rapid-equilibrium random mechanism as determined by initial velocity studies in the absence and presence of product inhibitors (Deibel and Coleman 1980b). There are several important ramifications for the ability of TdT to randomly bind substrates. Since the relative "lifetime" of blunt-end DNA formed during V(D)J recombination is likely to be very short, "preloading" TdT with a dNTP could increase the efficiency of template-independent synthesis and aid to produce random mutations. Alternatively, potential interactions with PCNA and Ku70/80, both of which participate in recombination, could also affect the kinetic mechanism of TdT (Hoek et al. 2011). For example, interactions with the clamp protein, PCNA, could force TdT to be more stably associated with DNA substrate and thus mandate an ordered mechanism for DNA binding prior to dNTP selection.

5.3.3.2 Conformational Changes During Catalysis

During template-dependent DNA synthesis, most DNA polymerases undergo at least one enzymatic conformational change after the binding of nucleotide to the polymerase-DNA complex. The existence of these conformational changes has been demonstrated using various kinetic, structural, and spectroscopic techniques (Echols and Goodman 1991; Kiefer et al. 1998; Hogg et al. 2004; Bloom et al. 1993; Hariharan and Reha-Krantz 2005; Johnson 1993; Dunlap and Tsai 2002). While it is clear that conformational changes occur during catalysis, there is still much debate regarding their role and importance during DNA polymerization. For example, conformational changes have been proposed to be a key step in maintaining replication fidelity by aligning the incoming dNTP into a precise geometrical shape with the templating nucleobase. In this model, proper alignment allows for efficient phosphoryl transfer. In addition, this step imposes discrimination against the misinsertion of an incorrect nucleotide as the geometry of the polymerase's active site perturbed to inhibit efficient phosphoryl transfer (Johnson 1993).

Since TdT uses only single-strand DNA, it is unclear if this template-independent DNA polymerase undergoes a rate-limiting conformational change to achieve efficient polymerization. This represents an important question since, as outlined above, this conformational step is often associated with a kinetic control point that plays a pivotal role in maintaining polymerization fidelity. Since TdT is designed to randomly incorporate nucleotides, it should not be limited by constraints imposed by fidelity. Thus, the lack of "faithful" polymerization displayed by TdT should negate any need for a conformational change step. As such, the phosphoryl transfer should be the rate-limiting step for nucleotide incorporation. Initial velocity studies performed by Coleman's group (Deibel and Coleman 1980b) provided evidence that TdT possesses a rapid-equilibrium random kinetic mechanism, a result consistent with phosphoryl transfer being the rate-limiting step for enzyme turnover. Indeed, similarly mechanistic deductions have been reported for "error-prone" polymerases that have lower constraints in fidelity (Dunlap and Tsai 2002).

5.3.3.3 Product Release

After nucleotide incorporation, template-dependent polymerases show an obligatory release in products in which pyrophosphate is first product to be released. Following pyrophosphate release, the polymerase can either remain bound to DNA and continue primer elongation (processive DNA synthesis) or dissociate from the extended primer to reinitiate DNA synthesis on another usable primer (distributive DNA synthesis). Again, since TdT follows a rapid-equilibrium random kinetic mechanism, it is likely that it uses a distributive mode for replication rather that a processive mechanism. While this mechanism is reasonable, it has yet to be conclusively established especially within the context of other interacting recombination proteins such as PCNA and Ku70/80.

5.3.4 Mechanism of Nucleotide Selection

With most DNA polymerases, the binding of a dNTP is highly influenced by steric constraints and hydrogen-bonding interactions imposed by the presence of a templating strand (Beckman and Loeb 1993; Kool 2002; Lee and Berdis 2010). However, the molecular details regarding nucleotide binding and selection by TdT remain elusive since this polymerase does not rely on a templating strand for polymerization activity. At face value, the non-reliance on coding information predicts that TdT would utilize all four dNTPs with equal efficiency. This mechanism appears unlikely as work from the Coleman and Modak laboratories have independently demonstrated that TdT shows an unequal bias for incorporating nucleotides (Deibel and Coleman 1980b; Modak 1978). For example, TdT binds dGTP with a ~4-fold higher affinity than dATP (compare K_m values of 120 μM versus 540 μM, respectively) (Deibel and Coleman 1980b; Modak 1978). In addition, it was more recently demonstrated that recombinant TdT utilizes dGTP, dCTP, and dTTP much more efficiently than dATP (Berdis and McCutcheon 2007). Collectively, these studies indicate that TdT actively discriminates against utilizing dATP. At the molecular level, this could be achieved by the use of hydrogen-bonding information present on amino acids in the active site of TdT. This type of "negative selection" mechanism has been documented by various DNA polymerases (Urban et al. 2010; Beckman et al. 2007) and would be similar to that displayed by Rev1, an error-prone DNA polymerase that preferentially incorporates dCTP via direct interactions with an active site arginine (Nair et al. 2005). At the cellular level, "negative" selection against dATP could be necessary since the intracellular concentration of dATP is higher than that for the other three natural dNTPs. As a result, active discrimination against dATP utilization would compensate against a higher concentration of this nucleotide to ensure randomization. An alternative mechanism is that TdT may preferentially incorporate nucleotides that form complementary base pairs (G/C or A/T) which are predicted to facilitate annealing during recombination. Complementary pairing combinations of G and C are predicted since they would form more thermodynamically stable base pairs.

Several groups have used various nonnatural nucleotides lacking hydrogen-bonding functional groups to probe their importance for nucleotide recognition by TdT (Berdis and McCutcheon 2007; Arzumanov et al. 2000; Sosunov et al. 2000; Krayevsky et al. 2000a; Horáková et al. 2011; Jarchow-Choy et al. 2011; Motea et al. 2012). One recent example is the recent demonstration that TdT incorporates nonnatural analogs such as 5-nitroindolyl-2′-deoxyriboside triphosphate (5-NITP) (Motea et al. 2012). In this case, 5-NITP is incorporated by TdT with an overall catalytic efficiency equal to dGTP. These data suggest that the primary molecular determinant for nucleotide binding is the triphosphate moiety of the incoming nucleotide which can favorably interact with positively charged amino acids lining the active site of TdT. However, this mechanism cannot fully account for productive binding interactions since it cannot adequately explain the significant differences observed in kinetic parameters for the utilization of various natural and nonnatural nucleotides (Berdis and McCutcheon 2007; Motea et al. 2012). In addition, replacement of an active site arginine residue (R336) that interacts with the triphosphate moiety of a dNTP with either glutamine (R336Q) or alanine (R336A) reduces the binding affinity for dGTP and dATP by only tenfold (Yang et al. 1994). This reduction indicates that ionic interactions between active site residues and the triphosphate of an incoming dNTP are important for binding. However, they are not essential for catalysis.

5.4 Structural Insights into Template-Independent DNA Synthesis

5.4.1 Primary Amino Acid Sequence Information

Sequence alignment of the C-terminus of the X-family DNA polymerases shows that they all possess fingers, palm, and thumb subdomains that are universally associated with members of the A-, B-, Y-, and RT-families of DNA polymerases (reviewed in Brautigam and Steitz 1998). TdT, pol λ, and pol μ are distinct from pol β as their N-terminal domain contains nuclear localization signal motifs as well as the breast cancer susceptibility protein BRCA1 C-terminal (BRCT) domains (Rodriguez and Songyang 2008) (Fig. 5.2). The BRCT domain typically functions to mediate protein/protein and protein/DNA interactions during DNA repair pathways and cell cycle checkpoint regulation. In addition, the BRCT domain of TdT, pol λ, and pol μ may interact with Ku70/86, a heterodimeric protein involved in recognizing and binding the ends of DSBs formed during V(D)J recombination (Morozov and Wawrousek 2008).

A more thorough comparison of the primary amino acid sequences reveals that these polymerases have very little overall identity with each other. In fact, while TdT and pol μ are the most closely related members of this family, they only share 42 % amino acid identity. Despite this low identity, TdT and pol μ appear to share a common ancestry as their template-independent activity coincides with the development of V(D)J recombination in mammals (Bartl et al. 2003).

5.4.2 Tertiary Structure of TdT

The structures of all template-dependent DNA polymerases characterized to date reveal a common molecular architecture resembling a right hand containing thumb, fingers, and palm subdomains (Fig. 5.4a) (reviewed in Morozov and Wawrousek 2008). The palm subdomain is viewed as the catalytic core since this is where the phosphoryl transfer reaction occurs. In the case of TdT, this subdomain contains three carboxylic acid amino acids, two of which are highly conserved amongst all DNA polymerases (Steitz 1999). These carboxylic amino acids coordinate metal ions that act as Lewis acids to lower the activation energy barrier needed for efficient phosphoryl transfer (Steitz and Steitz 1993). With template-dependent DNA polymerases, the fingers subdomain plays a large role in coordinating interactions between the templating base and the incoming dNTP. This subdomain mediates the conformational change step that allows for proper alignment of the nucleobases prior to phosphoryl transfer. Finally, the thumb subdomain of most template-dependent DNA polymerases serves a dual role by positioning duplex DNA for accepting the incoming dNTP as well as for polymerase translocation to the next templating base position after phosphoryl transfer.

The structure of TdT (Fig. 5.4a) shows remarkable similarities to template-dependent DNA polymerase as it also contains thumb, fingers, and palm subdomains. This model shows the presence of two Co^{2+} ions in the palm domain, and this stoichiometry of 2 is consistent with the "two divalent metal ion" mechanism proposed for phosphoryl transfer (Delarue et al. 2002). Finally, these metal ions are coordinated by the oxygens of the triphosphate moiety of the incoming dNTP as well as by three conserved aspartate residues.

Despite these similarities, however, there are several structural variations in TdT that distinguish it from conventional template-dependent polymerases. For example, TdT possesses a "lariat-like loop" that is not present in the structures of conventional template-dependent DNA polymerases (Fig. 5.4a) (Delarue et al. 2002). Superimposing the structure of TdT with the closed form of pol λ shows that the 16 amino acids forming the "lariat-like loop" in TdT would hinder the polymerase from interacting with duplex DNA. This unique structural element provides the physical basis for why TdT only replicates single-strand DNA. In addition, TdT contains an 8 kDa domain, referred to as the "index finger" domain that contacts the thumb subdomain to form a channel allowing dNTPs to diffuse into the enzyme's active site. Based on this structural model, Delarue and colleagues proposed that TdT bound with either DNA or ddATP resembles the "closed conformation" of pol β complexed with DNA and nucleoside (Delarue et al. 2002). This type of "closed" complex suggests that TdT can incorporate nucleotides without requiring a conformational change step. However, further structural and kinetic studies are needed to truly validate this proposal.

Despite having binary complexes of TdT bound with ddATP (PDB ID code 1KEJ or DNA [PDB ID code 1KDH (Delarue et al. 2002)], the molecular details accounting for nucleotide selectivity and specificity remain elusive. Inspection of

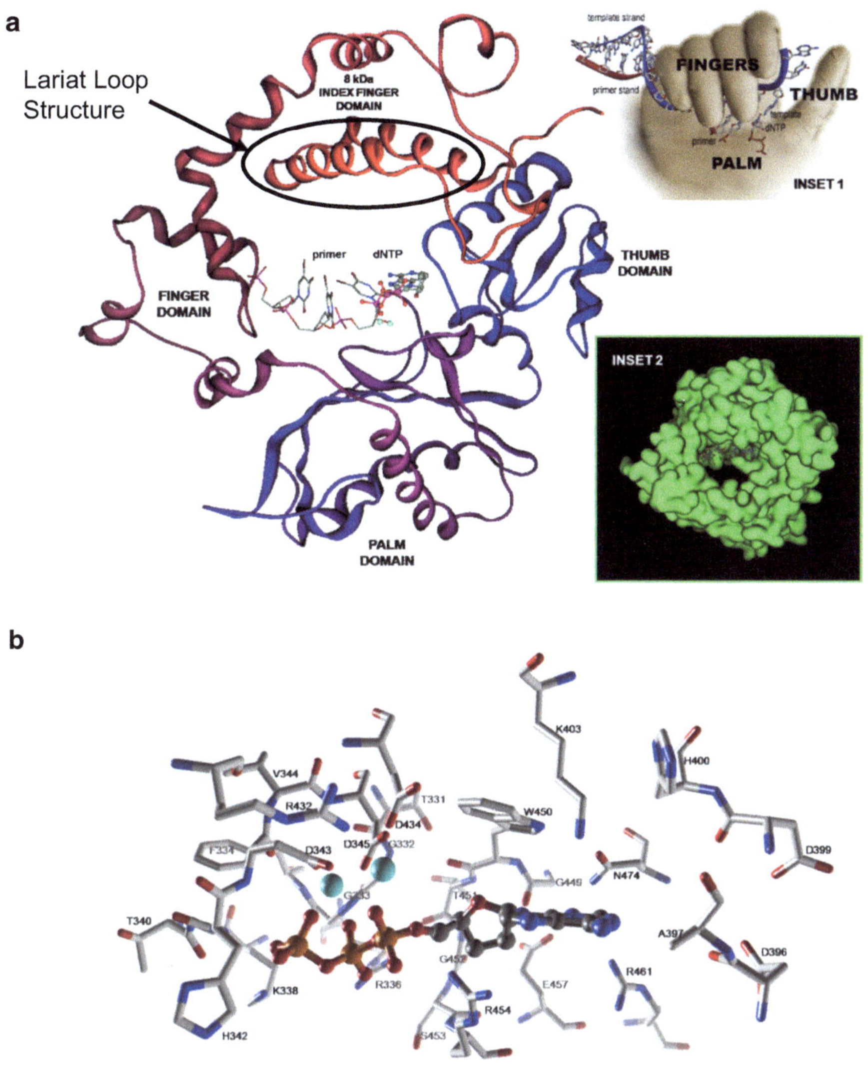

Fig. 5.4 (**a**) Crystal structure of terminal deoxynucleotidyl transferase. *Highlighted* are the fingers, thumb, palm, and index finger (8 kDa) subdomains that work synergistically to catalyze nucleotide incorporation. The ternary complex structure was prepared using the available binary crystal structures of murine TdT (Delarue et al. 2002) and the PDB ID codes 1KEJ (TdT•ddATP) and 1KDH (TdT•ssDNA). MOE (http://www.chemcomp.com) was used for all structural modeling. (**b**) The active site of TdT (PDB ID code: 1KEJ) as defined by amino acids that exist within 6 Å of the bound nucleotide substrate, ddATP. The incoming nucleotide is shown in *ball-stick* representation in CPK color scheme. The two cobalt ions are colored as *cyan*. This figure was prepared using the UCSF Chimera package (http://www.cgl.ucsf.edu/chimera). This figure was adapted from Motea and Berdis (2010)

amino acids that exist within a 6 Å radius of the bound ddATP (Fig. 5.4b) shows that three positively charged residues, K338, R336, and R454, point toward the triphosphate moiety. Based on their proximity and orientation in the active site, these amino acids likely function to neutralize the negatively charged phosphates of the incoming dNTP. Other interactions include the aromatic ring of W450 that is located only 3.6 Å away from the adenine and is positioned parallel to this nucleobase. The location and orientation of this amino acid could provide favorable pi–pi stacking interactions with the adenine base. The positively charged ε amino group of K403 lies only 4 Å away from the adenine base and thus could also provide favorable pi–cation interactions. Likewise, R454 may also participate in pi–cation interactions with the aromatic adenine to influence its binding. There are no amino acid side chains in close proximity with either the 2′ or the 3′ position of the ddATP sugar to provide any electrostatic or hydrogen-bonding interactions. The conspicuous absence of amino acids that could function as a "steric gate" to select against a 2′-OH could explain the rather promiscuous activity of TdT to utilize both ribo- and deoxyribonucleotides (Roychoudhury 1972). Finally, three conserved aspartates, D343, D345, and D434, exist in the catalytic palm subdomain that likely function to position the two metal ions for catalysis.

5.5 Regulation of TdT Activity

The activities of the enzymes involved in V(D)J recombination are tightly regulated, primarily since the formation of DSBs can produce adverse cellular effects such as apoptosis (Roos and Kaina 2012) or an increase in the frequency of genetic mutations (Nickoloff et al. 2008). TdT is regulated at multiple levels including transcriptional control, posttranslational modifications, and through protein–protein interactions. TdT expression is primarily confined to lymphoid tissues including bone marrow and the thymus (Coleman et al. 1974b; Bollum 1979b) in which its transcription is regulated by several factors including AP-1 (Peralta-Zaragoza et al. 2004). Posttranslational regulation of TdT comes primarily through phosphorylation. Ex vivo experiments proved that TdT can be phosphorylated in lymphoblastoid cells (Elias et al. 1982). In addition, recombinant human TdT can be phosphorylated in vitro by protein kinase C (Trubiani et al. 1995). Calf thymus TdT can be phosphorylated by beef heart cAMP-dependent protein kinase at multiple sites corresponding to S7 and T19 in human TdT (Chang and Bollum 1982). While TdT can be phosphorylated at several different sites, it is still unknown what the exact role of this posttranslational modification plays in regulating TdT activity.

Protein–protein interactions between TdT and other DNA-binding proteins can produce both positive and negative effects on TdT activity (Yamashita et al. 2001). For example, TdT interacting factors (TdiFs) can increase the polymerase activity of TdT by ~4-fold through binding interactions with the C-terminus of TdT (Yamashita et al. 2001). In contrast, the binding of TdiF2 to TdT via its C-terminus causes a twofold decrease in polymerase activity.

PCNA can also physically interact with TdT (Ibe et al. 2001). The cellular role of PCNA is to coordinate several aspects of DNA metabolism including chromosomal DNA replication, DNA repair, translesion DNA synthesis, and recombination (Moldovan et al. 2007). PCNA appears to function similarly to TdiF2 as it reduces the polymerase activity of TdT by ~2-fold (Ibe et al. 2001). It is proposed that PCNA and TdiF1 compete for binding to the C-terminal region of TdT, and this competition for a single binding site can produce positive or negative effects against TdT. Other proteins including the Ku proteins and various DNA polymerases may also regulate TdT activity. For example, "N" regions generated by TdT are unusually longer when Ku80 is knocked-out compared to when Ku80 is present (Sandor et al. 2004). As described earlier, TdT activity can also be indirectly regulated by other X-family polymerase family members. For example, the ability of pol μ to compete with TdT for the same binding site on DNA can inhibit "N" region synthesis catalyzed by TdT (Bertocci et al. 2003).

5.6 TdT in Disease

5.6.1 The Involvement of TdT in Cancer

Alterations in TdT activity and/or its expression level can play a significant role in the initiation and progression of various cancers as well as in the response of these cancers to chemotherapy. For example, TdT overexpression is a common feature in both *a*cute *l*ymphocytic *l*eukemia (ALL) and in *a*cute *m*yelocytic *l*eukemia (AML) (Greaves et al. 1980; Hoffbrand et al. 1977; Kung et al. 1978; Venditti et al. 1997; McCaffrey et al. 1983). Greater than 80 % of ALL patients present with significant higher levels of TdT expression in addition to displaying multiple TdT isoforms in their blast cells (Greaves et al. 1980; Hoffbrand et al. 1977; Kung et al. 1978; Venditti et al. 1997). Although TdT overexpression is less frequent in patients with AML (~20 %), levels of TdT are still significantly higher than those found in other lymphoid malignancies such as *c*hronic *l*ymphocytic *l*eukemia (CLL) (Venditti et al. 1997). In addition, higher levels of TdT activity correlate with a poor prognosis due to suboptimal responses to chemotherapy. Prognosis and survival studies of patients with ALL show that remission rates are nearly twofold higher in TdT-negative (61 %) versus TdT-positive (36 %) patients.

Based on these epidemiological data, several attempts have been made to develop anticancer agents that target TdT activity (McCaffrey et al. 1983; Spigelman et al. 1988; Kodama et al. 2000). One important example is the nucleoside analogue, cordycepin (3′-deoxyadenosine) (Fig. 5.5). Since this analogue lacks

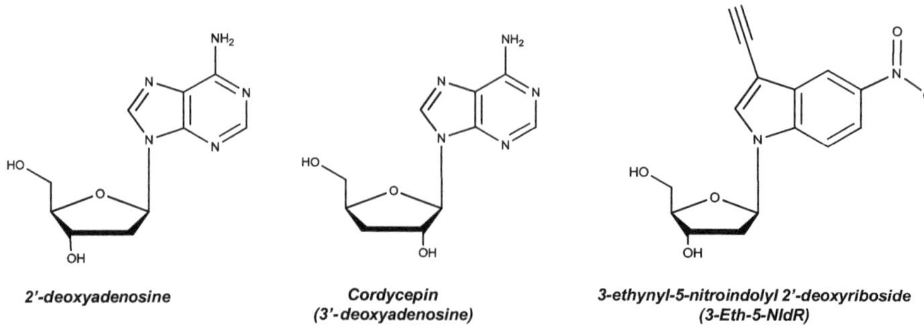

Fig. 5.5 Comparison of the chemical structures for natural nucleotide substrate, dATP, with TdT inhibitors including cordycepin and 3-Eth-5-NITP

a 3′-OH, it can function as a chain terminator after the corresponding nucleoside triphosphate is incorporated into DNA. Unfortunately, cordycepin is not a truly selective inhibitor of TdT as it can be utilized by several template-dependent DNA polymerases involved in chromosomal DNA synthesis (Plunkett and Gandhi 2001). This off-target activity can produce adverse side effects including immunosuppression, fatigue, nausea, and vomiting (Plunkett and Gandhi 2001).

To overcome this problem, Motea et al. (2012) recently developed a nonnatural nucleotide, designated 5-nitroindolyl-2′-deoxynucleoside triphosphate (5-NITP) (Fig. 5.4). The advantage of this nonnatural nucleotide is that it is efficiently utilized by TdT while poorly utilized by conventional template-dependent DNA polymerases. The addition of an ethynyl moiety at the 3-position of the 5-NITP produced a chemical probe that could be used to visualize and quantify replication catalyzed by TdT (Motea et al. 2012). The corresponding "theranostic nucleotide," designated 3-ethynyl-5-nitroindolyl-2′-deoxynucleoside triphosphate (3-Eth-5-NITP), functions as an efficient and potent chain-terminating nucleotide for TdT. The K_m for 3-Eth-5-NITP is ~200 nM while the inhibition constant for terminating in vitro replication catalyzed by TdT is ~350 nM. This group also demonstrated that the corresponding nucleoside, 3-ethynyl-5-nitroindolyl-2′-deoxynucleoside (3-Eth-5-NIdR), produces both cytostatic and cytotoxic effects against leukemia cells that overexpress TdT. Finally, the strategic placement of the ethynyl moiety allows the incorporated nonnatural nucleotide to be efficiently and selectively tagged with an azide-containing fluorophore via "click" chemistry. The application of Cu^{2+}-catalyzed "click" chemistry allows one to quantify the amount of nucleotide incorporation. Using this technique, it was shown that the anticancer effects of the corresponding nonnatural nucleoside could be quantified, and a dose-dependent relationship on the cytostatic effects of the analog was established. Furthermore, a distinct correlation between therapeutic activity and cellular levels of TdT were demonstrated. Collectively, these studies highlight the development of a first-in-class "theranostic" agent against TdT-positive ALL. The use of this agent can hopefully improve the accuracy of dosing regimens and thus accelerate clinical decisions regarding therapeutic intervention in leukemia patients.

5.6.1.1 The Involvement of TdT in Other Cancers

Merkel cell carcinoma is a rare form of skin cancer that primarily affects patients over the age of 60 (Pectasides et al. 2006). While this form of skin cancer accounts for less than 1 % of all cutaneous malignancies (Albores-Saavedra et al. 2010), Merkel cell carcinoma can display aggressive biological behavior and rapidly metastasize to regional lymph nodes and other organs including liver, bones, lungs, and brain. The most frequent change is the loss of heterozygosity caused by deletions or chromosomal translocations (Ronan et al. 1993).

Differential diagnosis of Merkel cell carcinoma historically involves histochemical validation to distinguish it from small cell carcinomas that originate in other organs and hematological malignancies. These include acute lymphoblastic lymphoma, acute myeloid leukemia, and cutaneous natural killer (NK)/T-cell lymphoma and hematodermic CD56/CD4 neoplasms. Several recent studies have shown that TdT is expressed in Merkel cell carcinoma. This unusual feature provides a new biomarker to further diagnose Merkel cell carcinoma and avoid diagnostic pitfalls associated with identifying other small round cell tumors presenting in cutaneous and soft tissues. An early study performed by Sur et al. (2007) reported that 8 of 15 cases (53 %) of Merkel cell carcinomas displayed strong yet diffuse TdT nuclear staining. A subsequent study by Buresh et al. (2008) expanded upon these results to shown that nuclear TdT immunoreactivity was present in 19 of 26 (73 %) cases of Merkel cell carcinoma. Furthermore, TdT expression was not observed in the epidermis, the epithelium of cutaneous appendages, endothelial cells, or stromal cells, and thus difference provides a biomarker for selectivity. Finally, Sidiropoulos et al. (2011) showed that 28 of 40 (70 %) cases of Merkel cell carcinoma were positive for nuclear staining of TdT. Equally important, they showed that only 2 of 30 (7 %) cases of small cell lung carcinoma (SCLC) were positive for TdT. In addition, positive staining for TdT was not detected in pulmonary carcinoid tumors (0/6). Collectively, these findings provide an important way to differentiate Merkel cell carcinomas from SCLC carcinomas, both of which are very aggressive neuroendocrine cancers and that display high mortality rates. While TdT can be used in the differential diagnosis of Merkel cell carcinoma, it has yet to be demonstrated if TdT can be targeted for therapeutic intervention against this malignancy.

5.6.2 Biochemical Applications of TdT

The ability of TdT to incorporate a wide variety of nucleotide analogs (Arzumanov et al. 2000; Sosunov et al. 2000; Krayevsky et al. 2000a; Horáková et al. 2011; Jarchow-Choy et al. 2011; Motea et al. 2012) has led to the development of a very effective method for in vivo labeling of DNA broken ends. Perhaps the most widely used technique is the *T*dT-mediated d*U*TP-biotin *n*ick *e*nd-*l*abeling (TUNEL) assay

(Gorczyca et al. 1993). This assay is based on the ability of TdT to efficiently incorporate biotinylated dUTP into single-strand DNA at sites of DNA breaks. Biotin dUTP that is incorporated at the ends of broken DNA can be visualized using fluorescently labeled avidin or streptavidin. The application of this technique allows the number and location of DNA breaks to be directly quantified. This technology is widely used to detect apoptosis, a form of programmed cell death, in eukaryotic cells.

TdT is also used ubiquitously to label the 3′-termini of synthetic oligonucleotides with a radioactive nucleotide or with various fluorescent probes (Krayevsky et al. 2000b). The labeled primers can then be annealed to a complementary strand as used as substrates for various enzymes involved in nucleic acid metabolism including restriction endonucleases, DNA glycosylases, and, of course, template-dependent DNA polymerases.

5.7 Conclusion

Several DNA polymerases including polymerase μ, polymerase λ, and TdT display unique catalytic activity as they can perform DNA synthesis in the absence of a DNA template. This unusual activity defies that displayed by conventional DNA polymerases that are charged with maintaining genomic fidelity by replicating undamaged DNA or that properly replicate various forms of damaged DNA. While this template-independent polymerase activity can produce mutagenic consequences, it also plays an indispensible role in the efficient and proper repair of double-strand DNA breaks. In fact, the activity of TdT during V(D)J recombination highlights the importance of template-independent DNA synthesis in generating immunological diversity that is needed for an efficient immune system. While kinetic and structural studies have provided significant insights into the catalytic mechanisms of these DNA polymerases, there is clearly much more work needed to completely understand their biological roles, especially during the repair of damaged DNA. Again, TdT provides an important example of this feature as its unregulated activity is linked with various cancers including acute lymphoblastic leukemia and Merkel cell carcinoma. It is highly likely that dysfunctional or unregulated activity of either polymerase μ and polymerase λ also play key roles in genetic diseases such as cancer.

References

Albores-Saavedra J, Batich K, Chable-Montero F, Sagy N, Schwartz AM, Henson DE (2010) Merkel cell carcinoma demographics, morphology, and survival based on 3870 cases: a population based study. J Cutan Pathol 37:20–27

Aoufouchi S, Flatter E, Dahan A, Faili A, Bertocci B, Storck S, Delbos F, Cocea L, Gupta N, Weill JC, Reynaud CA (2000) Two novel human and mouse DNA polymerases of the pol X family. Nucleic Acids Res 28:3684–3693

Arzumanov AA, Victorova LS, Jasko MV, Yesipov DS, Krayevsky AA (2000) Terminal deoxynucleotidyl transferase catalyzes the reaction of DNA phosphorylation. Nucleic Acids Res 28:1276–1281

Baltimore D (1974) Is terminal deoxynucleotidyl transferase a somatic mutagen in lymphocytes? Nature 248:409–411

Bangs LA, Sanz IE, Teale JM (1991) Comparison of D, JH, and junctional diversity in the fetal, adult, and aged B cell repertoires. J Immunol 146:1996–2004

Bartl S, Miracle AL, Rumfelt LL, Kepler TB, Mochon E, Litman GW, Flajnik MF (2003) Terminal deoxynucleotidyl transferases from elasmobranchs reveal structural conservation within vertebrates. Immunogenetics 55:594–604

Basu M, Hegde MV, Modak MJ (1983) Synthesis of compositionally unique DNA by terminal deoxynucleotidyl transferase. Biochem Biophys Res Commun 111:1105–1112

Bebenek K, Garcia-Diaz M, Blanco L, Kunkel TA (2003) The frameshift infidelity of human DNA polymerase lambda. Implications for function. J Biol Chem 278:34685–34690

Beckman RA, Loeb LA (1993) Multi-stage proofreading in DNA replication. Q Rev Biophys 26:225–331

Beckman RA, Mildvan AS, Loeb LA (1985) On the fidelity of DNA replication: manganese mutagenesis in vitro. Biochemistry 24:5810–5817

Beckman J, Kincaid K, Hocek M, Spratt T, Engels J, Cosstick R, Kuchta RD (2007) Human DNA polymerase alpha uses a combination of positive and negative selectivity to polymerize purine dNTPs with high fidelity. Biochemistry 46:448–460

Benedict CL, Gilfillan S, Thai TH, Kearney JF (2000) Terminal deoxynucleotidyl transferase and repertoire development. Immunol Rev 175:150–157

Berdis AJ, McCutcheon D (2007) The use of non-natural nucleotides to probe template-independent DNA synthesis. Chembiochem 8:1399–1408

Bertocci B, De Smet A, Flatter E, Dahan A, Bories JC, Landreau C, Weill JC, Reynaud CA (2002) Cutting edge: DNA polymerases mu and lambda are dispensable for Ig gene hypermutation. J Immunol 168:3702–3706

Bertocci B, De Smet A, Berek C, Weill JC, Reynaud CA (2003) Immunoglobulin kappa light chain rearrangement is impaired in mice deficient for DNA polymerases mu. Immunity 19:203–211

Bertocci B, De Smet A, Weill JC, Reynaud CA (2006) Nonoverlapping functions of DNA polymerases mu, lambda, and terminal deoxynucleotidyltransferase during immunoglobulin V(D)J recombination *in vivo*. Immunity 25:31–41

Beutler B (2003) Not "molecular patterns" but molecules. Immunity 19:155–156

Bjursell G, Skoog L (1980) Control of nucleotide pools in mammalian cells. Antibiot Chemother 28:78–85

Bloom LB, Otto MR, Beechem JM, Goodman MF (1993) Influence of 5′-nearest neighbors on the insertion kinetics of the fluorescent nucleotide analog 2-aminopurine by Klenow fragment. Biochemistry 32:1247–11258

Bollum FJ (1960) Calf thymus polymerase. J Biol Chem 235:2399–2403

Bollum FJ (1979) Terminal deoxynucleotidyl transferase as a hematopoietic cell marker. Blood 54:1203–1215

Bollum FJ, Chang LM (1981) Immunological detection of a conserved structure for terminal deoxynucleotidyltransferase. J Biol Chem 256:8767–8770

Boule JB, Johnson E, Rougeon F, Papanicolaou C (1998) High-level expression of murine terminal deoxynucleotidyl transferase in *Escherichia coli* grown at low temperature and overexpressing argU tRNA. Mol Biotechnol 10:199–208

Boulé JB, Rougeon F, Papanicolaou C (2001) Terminal deoxynucleotidyl transferase indiscriminately incorporates ribonucleotides and deoxyribonucleotides. J Biol Chem 276:31388–31393

Braithwaite EK, Prasad R, Shock DD, Hou EW, Beard WA, Wilson SH (2005) DNA polymerase lambda mediates a back-up base excision repair activity in extracts of mouse embryonic fibroblasts. J Biol Chem 280:18469–18475

Brautigam CA, Steitz TA (1998) Structural and functional insights provided by crystal structures of DNA polymerases and their substrate complexes. Curr Opin Struct Biol 8:54–63

Buresh CJ, Oliai BR, Miller RT (2008) Reactivity with TdT in Merkel cell carcinoma: a potential diagnostic pitfall. Am J Clin Pathol 129:894–898

Cabaniols JP, Fazilleau N, Casrouge A, Kourilsky P, Kanellopoulos JM (2001) Most alpha/beta T cell receptor diversity is due to terminal deoxynucleotidyl transferase. J Exp Med 194:1385–1390

Capp JP, Boudsocq F, Bertrand P, Laroche-Clary A, Pourquier P, Lopez BS, Cazaux C, Hoffmann JS, Canitrot Y (2006) The DNA polymerase lambda is required for the repair of non-compatible DNA double strand breaks by NHEJ in mammalian cells. Nucleic Acids Res 34:2998–3007

Chang LM, Bollum FJ (1971) Deoxynucleotide-polymerizing enzymes of calf thymus gland. V. Homogeneous terminal deoxynucleotidyl transferase. J Biol Chem 246:909–916

Chang LM, Bollum FJ (1980) Multiple roles of divalent cation in the terminal deoxynucleotidyl-transferase reaction. J Biol Chem 265:17436–17440

Chang LM, Bollum FJ (1982) Cyclic AMP-dependent phosphorylation of terminal deoxynucleotidyl transferase. J Biol Chem 257:9588–9592

Chang LM, Cassani GR, Bollum FJ (1972) Deoxynucleotide-polymerizing enzymes of calf thymus gland. VII. Replication of homopolymers. J Biol Chem 247:7718–7723

Chang LM, Plevani P, Bollum FJ (1982) Proteolytic degradation of calf thymus terminal deoxynucleotidyl transferase. J Biol Chem 257:5700–5706

Chang LM, Rafter E, Rusquet-Valerius R, Peterson RC, White ST, Bollum FJ (1988) Expression and processing of recombinant human terminal transferase in the baculovirus system. J Biol Chem 263:12509–12513

Coleman MS, Hutton JJ, De Simone P, Bollum FJ (1974) Terminal deoxyribonucleotidyl transferase in human leukemia. Proc Natl Acad Sci USA 71:4404–4408

Deibel MR Jr, Coleman MS (1980a) Limited proteolysis of calf thymus terminal deoxynucleotidyl transferase. Arch Biochem Biophys 202:414–419

Deibel MR Jr, Coleman MS (1980b) Biochemical properties of purified human terminal deoxynucleotidyltransferase. J Biol Chem 255:4206–4212

Delarue M, Boule JB, Lescar J, Expert-Bezancon N, Jourdan N, Sukumar N, Rougeon F, Papanicolaou C (2002) Crystal structures of a template-independent DNA polymerase: murine terminal deoxynucleotidyltransferase. EMBO J 21:427–439

Domínguez O, Ruiz JF, Laín de Lera T, García-Díaz M, González MA, Kirchhoff T, Martínez-A C, Bernad A, Blanco L (2000) DNA polymerase mu (Pol mu), homologous to TdT, could act as a DNA mutator in eukaryotic cells. EMBO J 19:1731–1742

Dunlap CA, Tsai MD (2002) Use of 2-aminopurine and tryptophan fluorescence as probes in kinetic analyses of DNA polymerase beta. Biochemistry 41:11226–11235

Early P, Huang H, Davis M, Calame K, Hood L (1980) An immunoglobulin heavy chain variable region gene is generated from three segments of DNA: VH, D and JH. Cell 19:981–992

Echols H, Goodman MF (1991) Fidelity mechanisms in DNA replication. Annu Rev Biochem 60:477–511

Elias L, Longmire J, Wood A, Ratliff R (1982) Phosphorylation of terminal deoxynucleotidyl transferase in leukemic cells. Biochem Biophys Res Commun 106:458–465

Feeney AJ (1990) Lack of N regions in fetal and neonatal mouse immunoglobulin V-D-J junctional sequences. J Exp Med 172:1377–1390

García-Díaz M, Domínguez O, López-Fernández LA, de Lera LT, Saníger ML, Ruiz JF, Párraga M, García-Ortiz MJ, Kirchhoff T, del Mazo J, Bernad A, Blanco L (2000) DNA polymerase lambda (Pol lambda), a novel eukaryotic DNA polymerase with a potential role in meiosis. J Mol Biol 301:851–867

García-Díaz M, Bebenek K, Kunkel TA, Blanco L (2001) Identification of an intrinsic 5′-deoxyribose-5-phosphate lyase activity in human DNA polymerase lambda: a possible role in base excision repair. J Biol Chem 276:34659–34663

Garcia-Diaz M, Bebenek K, Gao G, Pedersen LC, London RE, Kunkel TA (2005) Structure-function studies of DNA polymerase lambda. DNA Repair (Amst) 4:1358–1367

Gorczyca W, Gong J, Darzynkiewicz Z (1993) Detection of DNA strand breaks in individual apoptotic cells by the in situ terminal deoxynucleotidyl transferase and nick translation assays. Cancer Res 53:1945–1951

Greaves M, Paxton A, Janossy G, Pain C, Johnson S, Lister TA (1980) Acute lymphoblastic leukaemia associated antigen. III Alterations in expression during treatment and in relapse. Leuk Res 4:1–14

Haeryfar SM, Hickman HD, Irvine KR, Tscharke DC, Bennink JR, Yewdell JW (2008) Terminal deoxynucleotidyl transferase establishes and broadens antiviral CD8+ T cell immunodominance hierarchies. J Immunol 181:649–659

Hariharan C, Reha-Krantz LJ (2005) Using 2-aminopurine fluorescence to detect bacteriophage T4 DNA polymerase-DNA complexes that are important for primer extension and proofreading reactions. Biochemistry 44:15674–15684

Hoek M, Myers MP, Stillman B (2011) An analysis of CAF-1-interacting proteins reveals dynamic and direct interactions with the KU complex and 14-3-3 proteins. J Biol Chem 286:10876–10887

Hoffbrand AV, Ganeshaguru K, Janossy G, Greaves MF, Catovsky D, Woodruff RK (1977) Terminal deoxynucleotidyl-transferase levels and membrane phenotypes in diagnosis of acute leukaemia. Lancet 2:520–523

Hogg M, Wallace SS, Doublie S (2004) Crystallographic snapshots of a replicative DNA polymerase encountering an abasic site. EMBO J 23:1483–1493

Horáková P, Macíčková-Cahová H, Pivoňková H, Spaček J, Havran L, Hocek M, Fojta M (2011) Tail-labelling of DNA probes using modified deoxynucleotide triphosphates and terminal deoxynucleotidyl transferase. Application in electrochemical DNA hybridization and protein-DNA binding assays. Org Biomol Chem 9:1366–1371

Ibe S, Fujita K, Toyomoto T, Shimazaki N, Kaneko R, Tanabe A, Takebe I, Kuroda S, Kobayashi T, Toji S, Tamai K, Yamamoto H, Koiwai O (2001) Terminal deoxynucleotidyl transferase is negatively regulated by direct interaction with proliferating cell nuclear antigen. Genes Cells 6:815–824

Janeway CA (1999) Immunobiology: the immune system in health and disease, vol 4. Current Biology Publications, London

Jarchow-Choy SK, Krueger AT, Liu H, Gao J, Kool ET (2011) Fluorescent xDNA nucleotides as efficient substrates for a template-independent polymerase. Nucleic Acids Res 39:1586–1594

Johnson KA (1993) Conformational coupling in DNA polymerase fidelity. Annu Rev Biochem 62:685–713

Kallenbach S, Goodhardt M, Rougeon F (1990) A rapid test for V(D)J recombinase activity. Nucleic Acids Res 18:6730

Karkow JS, Kamen LHO (1966) DNA synthesis in thymus gland. Fed Proc 2:307

Kass EM, Jasin M (2010) Collaboration and competition between DNA double-strand break repair pathways. FEBS Lett 584:3703–3708

Kato KI, Goncalves JM, Houts GE, Bollum FJ (1967) Deoxynucleotide-polymerizing enzymes of calf thymus gland. II. Properties of the terminal deoxynucleotidyltransferase. J Biol Chem 242:2780–2789

Kepler TB, Borrero M, Rugerio B, McCray SK, Clarke SH (1996) Interdependence of N nucleotide addition and recombination site choice in V(D)J rearrangement. J Immunol 157:4451–4457

Kiefer JR, Mao C, Braman JC, Beese LS (1998) Visualizing DNA replication in a catalytically active Bacillus DNA polymerase crystal. Nature 391:304–307

Kodama EN, McCaffrey RP, Yusa K, Mitsuya H (2000) Antileukemic activity and mechanism of action of cordycepin against terminal deoxynucleotidyl transferase-positive (TdT+) leukemic cells. Biochem Pharmacol 59:273–281

Komori T, Okada A, Stewart V, Alt FW (1993) Lack of N regions in antigen receptor variable region genes of TdT-deficient lymphocytes. Science 261:1171–1175

Kool ET (2002) Active site tightness and substrate fit in DNA replication. Annu Rev Biochem 71:191–219

Krayevsky AA, Victorova LS, Arzumanov AA, Jasko MV (2000) Terminal deoxynucleotidyl transferase. catalysis of DNA (oligodeoxynucleotide) phosphorylation. Pharmacol Ther 85:165–173

Kung PC, Long JC, McCaffrey RP, Ratliff RL, Harrison TA, Baltimore D (1978) Terminal deoxynucleotidyl transferase in the diagnosis of leukemia and malignant lymphoma. Am J Med 64:788–794

Kunkel TA, Gopinathan KP, Dube DK, Snow ET, Loeb LA (1986) Rearrangements of DNA mediated by terminal transferase. Proc Natl Acad Sci USA 83:1867–1871

Landau NR, Schatz DG, Rosa M, Baltimore D (1987) Increased frequency of N-region insertion in a murine pre-B-cell line infected with a terminal deoxynucleotidyl transferase retroviral expression vector. Mol Cell Biol 7:3237–3243

Lee I, Berdis AJ (2010) Non-natural nucleotides as probes for the mechanism and fidelity of DNA polymerases. Biochim Biophys Acta 1804:1064–1080

Lee JW, Blanco L, Zhou T, Garcia-Diaz M, Bebenek K, Kunkel TA, Wang Z, Povirk LF (2004) Implication of DNA polymerase lambda in alignment-based gap filling for nonhomologous DNA end joining in human nuclear extracts. J Biol Chem 279:805–811

Lieber MR, Hesse JE, Mizuuchi K, Gellert M (1988) Lymphoid V(D)J recombination: nucleotide insertion at signal joints as well as coding joints. Proc Natl Acad Sci USA 85:8588–8592

Ma Y, Pannicke U, Schwarz K, Lieber MR (2002) Hairpin opening and overhang processing by an Artemis/DNA-dependent protein kinase complex in nonhomologous end joining and V(D)J recombination. Cell 108:781–794

Maga G, Ramadan K, Locatelli GA, Shevelev I, Spadari S, Hübscher U (2005) DNA elongation by the human DNA polymerase lambda polymerase and terminal transferase activities are differentially coordinated by proliferating cell nuclear antigen and replication protein A. J Biol Chem 280:1971–1981

Matsumoto Y, Kim K (1995) Excision of deoxyribose phosphate residues by DNA polymerase beta during DNA repair. Science 269:699–702

McCaffrey R, Bell R, Lillquist A, Wright G, Baril E, Minowada J (1983) Selective killing of leukemia cells by inhibition of TdT. Haematol Blood Transfus 28:24–27

McCormick PJ, Danhauser LL, Rustum YM, Bertram JS (1983) Changes in ribo- and deoxyribonucleoside triphosphate pools within the cell cycle of a synchronized mouse fibroblast cell line. Biochim Biophys Acta 756:36–40

Mickelsen S, Snyder C, Trujillo K, Bogue M, Roth DB, Meek K (1999) Modulation of terminal deoxynucleotidyltransferase activity by the DNA-dependent protein kinase. J Immunol 163:834–843

Modak MJ (1978) Biochemistry of terminal deoxynucleotidyltransferase: mechanism of inhibition by adenosine 5′-triphosphate. Biochemistry 17:3116–3120

Moldovan GL, Pfander B, Jentsch S (2007) PCNA, the maestro of the replication fork. Cell 129:665–679

Morozov V, Wawrousek EF (2008) Single-strand DNA-mediated targeted mutagenesis of genomic DNA in early mouse embryos is stimulated by Rad51/54 and by Ku70/86 inhibition. Gene Ther 15:468–472

Motea EA, Berdis AJ (2010) Terminal deoxynucleotidyl transferase: the story of a misguided DNA polymerase. Biochim Biophys Acta 1804:1151–1166

Motea EA, Lee I, Berdis AJ (2012) A non-natural nucleoside with combined therapeutic and diagnostic activities against leukemia. ACS Chem Biol 7:988–998

Nagasawa K, Kitamura K, Yasui A, Nimura Y, Ikeda K, Hirai M, Matsukage A, Nakanishi M (2000) Identification and characterization of human DNA polymerase beta 2, a DNA polymerase beta -related enzyme. J Biol Chem 275:31233–31238

Nair DT, Johnson RE, Prakash L, Prakash S, Aggarwal AK (2005) Rev1 employs a novel mechanism of DNA synthesis using a protein template. Science 309:2219–2222

Nakamura H, Tanabe K, Yoshida S, Morita T (1981) Terminal deoxynucleotidyltransferase of 60,000 daltons from mouse, rat, and calf thymus. Purification by immunoadsorbent chromatography and comparison of peptide structures. J Biol Chem 256:8745–8751

Nick McElhinny SA, Havener JM, Garcia-Diaz M, Juárez R, Bebenek K, Kee BL, Blanco L, Kunkel TA, Ramsden DA (2005) A gradient of template dependence defines distinct biological roles for family X polymerases in nonhomologous end joining. Mol Cell 19:357–366

Nickoloff JA, De Haro LP, Wray J, Hromas R (2008) Mechanisms of leukemia translocations. Curr Opin Hematol 15:338–345

Ohnishi T, Mori E, Takahashi A (2009) DNA double-strand breaks: their production, recognition, and repair in eukaryotes. Mutat Res 669:8–12

Papermaster BW, Condie RM, Finstad J, Good RA (1964) Evolution of the immune response. I. The phylogenetic development of adaptive immunologic responsiveness in vertebrates. J Exp Med 119:105–130

Paull TT (2005) Saving the ends for last: the role of pol mu in DNA end joining. Mol Cell 19:294–296

Pectasides D, Pectasides M, Economopoulos T (2006) Merkel cell cancer of the skin. Ann Oncol 17:1489–1495

Peralta-Zaragoza O, Recillas-Targa F, Madrid-Marina V (2004) Terminal deoxynucleotidyl transferase is down-regulated by AP-1-like regulatory elements in human lymphoid cells. Immunology 111:195–203

Peterson RC, Cheung LC, Mattaliano RJ, White ST, Chang LM, Bollum FJ (1985) Expression of human terminal deoxynucleotidyl transferase in *Escherichia coli*. J Biol Chem 260:10495–10502

Plunkett W, Gandhi V (2001) Purine and pyrimidine nucleoside analogs. Cancer Chemother Biol Response Modif 19:21–45

Rodriguez MC, Songyang Z (2008) BRCT domains: phosphopeptide binding and signaling modules. Front Biosci 13:5905–5915

Ronan SG, Green AD, Shilkaitis A, Huang TS, Das Gupta TK (1993) Merkel cell carcinoma: in vitro and in vivo characteristics of a new cell line. J Am Acad Dermatol 29:715–722

Roos WP, Kaina B (2012) DNA damage-induced apoptosis: from specific DNA lesions to the DNA damage response and apoptosis. Cancer Lett 332(2):237–248

Roychoudhury R (1972) Enzymic synthesis of polynucleotides. Oligodeoxynucleotides with one 3′-terminal ribonucleotide as primers for polydeoxynucleotide synthesis. J Biol Chem 247:3910–3917

Roychoudhury R, Kössel H (1971) Synthetic polynucleotides. Enzymic synthesis of ribonucleotide terminated oligodeoxynucleotides and their use as primers for the enzymic synthesis of polydeoxynucleotides. Eur J Biochem 22:310–320

Ruiz JF, Domínguez O, Laín de Lera T, Garcia-Díaz M, Bernad A, Blanco L (2001) DNA polymerase mu, a candidate hypermutase? Philos Trans R Soc Lond B Biol Sci 356:99–109

Ruiz JF, Juárez R, García-Díaz M, Terrados G, Picher AJ, González-Barrera S, Fernández de Henestrosa AR, Blanco L (2003) Lack of sugar discrimination by human Pol mu requires a single glycine residue. Nucleic Acids Res 31:4441–4449

Sadofsky MJ (2001) The RAG proteins in V(D)J recombination: more than just a nuclease. Nucleic Acids Res 29:1399–1409

Sandor Z, Calicchio ML, Sargent RG, Roth DB, Wilson JH (2004) Distinct requirements for Ku in N nucleotide addition at V(D)J- and non-V(D)J-generated double-strand breaks. Nucleic Acids Res 32:1866–1873

Schatz DG (2004) V(D)J recombination. Immunol Rev 200:5–11

Schatz DG, Oettinger MA, Schlissel MS (1992) V(D)J recombination: molecular biology and regulation. Annu Rev Immunol 10:359–383

Sibanda BL, Critchlow SE, Begun J, Pei XY, Jackson SP, Blundell TL, Pellegrini L (2001) Crystal structure of an Xrcc4-DNA ligase IV complex. Nat Struct Biol 8:1015–1019

Sidiropoulos M, Hanna W, Raphael SJ, Ghorab Z (2011) Expression of TdT in Merkel cell carcinoma and small cell lung carcinoma. Am J Clin Pathol 135:831–838

Sosunov VV, Santamaria F, Victorova LS, Gosselin G, Rayner B, Krayevsky AA (2000) Stereochemical control of DNA biosynthesis. Nucleic Acids Res 28:1170–1175

Spigelman Z, Duff R, Beardsley GP, Broder S, Cooney D, Landau NR, Mitsuya H, Ullman B, McCaffrey R (1988) 2′,3′-Dideoxyadenosine is selectively toxic for TdT-positive cells. Blood 71:1601–1608

Steen SB, Gomelsky L, Roth DB (1996) The 12/23 rule is enforced at the cleavage step of V(D)J recombination *in vivo*. Genes Cells 1:543–553

Steitz TA (1999) DNA polymerases: structural diversity and common mechanisms. J Biol Chem 274:17395–17398

Steitz TA, Steitz JA (1993) A general two-metal-ion mechanism for catalytic RNA. Proc Natl Acad Sci USA 90:6498–6502

Sur M, AlArdati H, Ross C, Alowami S (2007) TdT expression in Merkel cell carcinoma: potential diagnostic pitfall with blastic hematological malignancies and expanded immunohistochemical analysis. Mod Pathol 20:1113–1120

Takahara K, Hayashi N, Fujita-Sagawa K, Morishita T, Hashimoto Y, Noda A (1994) Alternative splicing of bovine terminal deoxynucleotidyl transferase cDNA. Biosci Biotechnol Biochem 58:786–787

Takata M, Sasaki MS, Sonoda E, Morrison C, Hashimoto M, Utsumi H, Yamaguchi-Iwai Y, Shinohara A, Takeda S (1998) Homologous recombination and non-homologous end-joining pathways of DNA double-strand break repair have overlapping roles in the maintenance of chromosomal integrity in vertebrate cells. EMBO J 17:5497–5508

Thai TH, Kearney JF (2004) Distinct and opposite activities of human terminal deoxynucleotidyltransferase splice variants. J Immunol 173:4009–4019

Thai TH, Kearney JF (2005) Isoforms of terminal deoxynucleotidyltransferase: developmental aspects and function. Adv Immunol 86:113–136

Traut TW (1994) Physiological concentrations of purines and pyrimidines. Mol Cell Biochem 140:1–22

Trubiani O, Bollum FJ, Di Primio R (1995) Terminal deoxynucleotidyl transferase is a nuclear PKC substrate. FEBS Lett 374:367–370

Urban M, Joubert N, Purse BW, Hocek M, Kuchta RD (2010) Mechanisms by which human DNA primase chooses to polymerize a nucleoside triphosphate. Biochemistry 49:727–735

Venditti A, Del Poeta G, Buccisano F, Tamburini A, Aronica G, Bruno A, Cox-Froncillo MC, Maffei L, Simone MD, Papa G, Amadori S (1997) Biological pattern of AML-M0 versus AML-M1: response. Blood 89:345–346

Yamashita N, Shimazaki N, Ibe S, Kaneko R, Tanabe A, Toyomoto T, Fujita K, Hasegawa T, Toji S, Tamai K, Yamamoto H, Koiwai O (2001) Terminal deoxynucleotidyltransferase directly interacts with a novel nuclear protein that is homologous to p65. Genes Cells 6:641–652

Yamtich J, Sweasy JB (2010) DNA polymerase family X: function, structure, and cellular roles. Biochim Biophys Acta 1804:1136–1150

Yang W (2003) Damage repair DNA polymerases Y. Curr Opin Struct Biol 13:23–30

Yang B, Gathy KN, Coleman MS (1994) Mutational analysis of residues in the nucleotide binding domain of human terminal deoxynucleotidyl transferase. J Biol Chem 269:11859–11868
Zhang Y, Wu X, Yuan F, Xie Z, Wang Z (2001) Highly frequent frameshift DNA synthesis by human DNA polymerase mu. Mol Cell Biol 21:7995–8006
Zhang X, Motea E, Lee I, Berdis AJ (2011) Replication of a universal nucleobase provides unique insight into the role of entropy during DNA polymerization and pyrophosphorolysis. Biochemistry 49:3009–3023
Zuo Z, Lin HK, Trakselis MA (2011) Strand annealing and terminal transferase activities of a B-family DNA polymerase. Biochemistry 50:5379–5390

Chapter 6
Archaeal DNA Polymerases: Enzymatic Abilities, Coordination, and Unique Properties

Michael A. Trakselis and Robert J. Bauer

Abstract DNA polymerases from archaea coordinate both replication and repair activities under extreme conditions. Traditional DNA polymerase families B and Y are represented, but an entire phylum of archaea contains members from the unique D-family. These model archaeal polymerases from each family have be exploited for a variety of biotechnology applications but have also provided great insight into kinetic mechanisms, structural properties, conformational changes, and protein interactions with DNA polymerases from other domains. Interestingly, individual polymerase members have revealed some unique features including template uracil recognition, more open active sites, novel motifs, and direct polymerase interactions that modulate replication and repair in these simpler organisms. How multiple polymerases coordinate synthesis on the leading and lagging strands and replication or repair duties is an ongoing fundamental question in archaea. Nevertheless, their intrinsic enzymatic properties are fascinating and continue to shape and influence research avenues for all DNA polymerases. The ability to regulate robust and accurate DNA synthesis through specific and loosely associate oligomeric states and protein interactions maintains a high degree of genomic stability in spite of harsh environmental conditions.

Keywords DNA replication • DNA repair • Fidelity • Archaea • PCNA • Polymerase • Holoenzyme

M.A. Trakselis (✉) • R.J. Bauer
Department of Chemistry, University of Pittsburgh, 219 Parkman Ave, 801 Chevron, Pittsburgh, PA 15260, USA
e-mail: mtraksel@pitt.edu

K.S. Murakami and M.A. Trakselis (eds.), *Nucleic Acid Polymerases*, Nucleic Acids and Molecular Biology 30, DOI 10.1007/978-3-642-39796-7_6,

Abbreviations

BER	Base excision repair
NER	Nucleotide excision repair
NTP	Nucleotide triphosphate
Pab	*Pyrococcus abyssi*
PCNA	Proliferating cellular nuclear antigen
Pfu	*Pyrococcus furiosus*
Pho	*Pyrococcus horikoshii*
Pol	Polymerase
PP_i	Pyrophosphate
RFC	Replication factor C
Sac	*Sulfolobus acidocaldarius*
Sso	*Sulfolobus solfataricus*
Tgo	*Thermococcus gorgonarius*
UV	Ultraviolet

6.1 Introduction

Efficient DNA replication relies on both precise and uninterrupted DNA synthesis at the replication fork. As such, DNA replication polymerases coordinate accurate synthesis on the leading and lagging strands with error-prone polymerases when damage is encountered. Most organisms contain members from multiple DNA polymerase families including Archaea which provide a relevant model system for assessing the kinetics, dynamics, structure, and interactions of multiple DNA polymerases.

Studies over the past decade of DNA replication machinery have revealed that many components from Eukaryotes have evolved from a common ancestor in Archaea (Yutin et al. 2008). In particular, the archaeal DNA replication machinery is essentially a simplified eukaryotic vestige and provides an excellent experimental system for deciphering mechanisms of enzymatic action and evolutionary relationships. Although there are a number of similarities in sequence, structure, and function between Archaea and Eukaryotes, the link is not absolute, as Archaea also contain bacterial- and archaeal-specific features.

A number of DNA replication systems from divergent archaeal species have been examined enzymatically and structurally to gain insight into fundamentally conserved metabolic processes as well as for the development of a variety of ubiquitous biotechnology tools. In this chapter, we will describe and compare the current findings on the relationship and function of archaeal DNA polymerases, their importance in deciphering DNA replication and repair mechanisms, as well as interactions that promote elevated enzymatic properties.

6.2 Archaeal DNA Polymerase Families and Function

The archaeal domain is subdivided minimally into several phyla with the largest two: Crenarchaeota (crenarchaea) and Euryarchaeota (euryarchaea), containing replication systems with high homology to those found in eukaryotic systems. DNA polymerases have been classified into at least six different families. Compared with the 15 human DNA polymerases, archaeal organisms generally have two to four DNA polymerases from two different families. Crenarchaea contain members from both the B- and Y-families, while euryarchaea contain those belonging to B- and D-families (Table 6.1). DNA polymerases employed by archaeal organisms have high sequence, structural, and functional similarities to those found in eukaryotes, yet they have adapted for optimal function under extreme conditions, most notably high temperatures. The ease of purification, high conservation to eukaryotes, and adaptation for biotechnology applications have made archaeal polymerases models for studying processivity, protein interactions, lesion bypass, polymerase–exonuclease shuttling, and polymerase switching mechanisms essential to all domains of life.

6.2.1 *Crenarchaeal B-Family DNA Replication Polymerases*

B-family polymerases are typically robust and accurate enzymes, containing an N-terminal 3′–5′ exonuclease and a C-terminal polymerase domain (Kim et al. 2008; Savino et al. 2004). The exonuclease domain increases selective nucleotide incorporation efficiencies generally by a factor of 10^2 up to 10^8 in total (Kunkel and Bebenek 2000). The polymerase domain is similar in structure to a right hand with fingers, thumb, and palm subdomains that act to bind the DNA template, orientate the incoming nucleotide, and catalyze polymerization through conformational changes between domains (Fig. 6.1). The enzymes are typically not highly processive on their own, but possess the ability to form complexes with their respective processivity clamps (PCNAs), allowing for the incorporation of >10,000 nucleotides in a single binding event (Jeruzalmi et al. 2002). As a result, B-family enzymes are also thought to be the main replication polymerases in crenarchaea.

Crenarchaea possess three B-family polymerases. PolB1 has robust synthesis activity, high nucleotide fidelity, and an included exonuclease domain (Pisani et al. 1998; Zhang et al. 2009). PolB1 is evolutionary related to eukaryotic B-family polymerases (α, δ, ε, ζ) (Prakash et al. 2005) but seems to have arisen after PolB2 and PolB3 in a gene duplication event (Table 6.1) (Edgell et al. 1998). PolB1 has a typical right-hand conformation but includes two extra α-helices in the N-terminal domain that contact the fingers domain (Fig. 6.1). It is proposed that these helices strengthen the contacts between the N- and C-terminal domains for catalysis at high temperatures (Savino et al. 2004). Alternatively, they could also play a role in promoting protein complex formation as seen for the trimeric PolB1 complex discussed below (Mikheikin et al. 2009).

Table 6.1 Archaeal polymerase family members

	Archaeal phyla				
Family	Crenarchaeota	Euryarchaeota	Nanoarchaeota	Thaumarchaeota	Korachaeota
A					
B	PolB1			PolB	PolBI
C	PolB2				
	PolB3	PolB	PolB		PolBII
D		PolD	PolD	PolD	PolD
X	PriSL[a]	PriSL[a]	PriSL[a]	PriSL[a]	PriSL[a]
Y	PolY			PolY	

[a]AEP family of DNA primases with homology to X-family polymerases

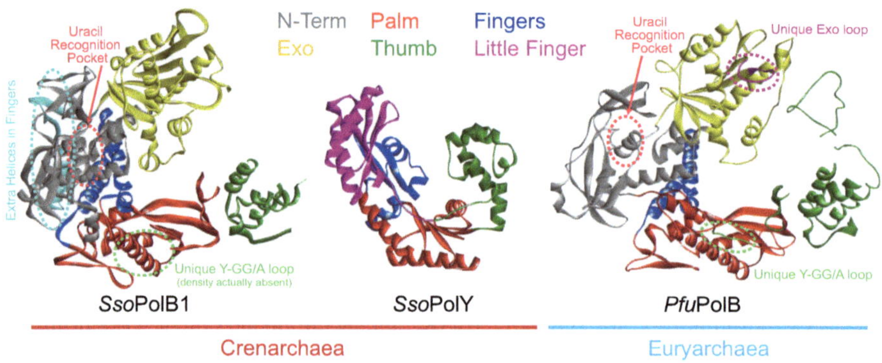

Fig. 6.1 Structures of crenarchaeal, *Sso*PolB1 (PDB: 1S5J), *Sso*PolY (PDB: 1JX4), and euryarchaeal *Pfu*PolB (2JGU) DNA polymerases highlighting conserved N-terminal (*gray*), exonuclease (*yellow*), palm (*red*), thumb (*green*), fingers (*blue*), and little finger (*pink*) domains. The extra α-helices in the fingers of *Sso*PolB1 are highlighted in *cyan*; the unique Y-GG/A loop is highlighted in *green*; the unique exonuclease (Exo) loop is highlighted in *pink*; and the uracil recognition pocket is circled in *red*

The exonuclease domain of *Sso*PolB1 contains an inherent 3′–5′ proofreading ability that enhances the fidelity 14-fold (Zhang et al. 2009) (Fig. 6.2). Although PolB1 utilizes an induced fit mechanism for nucleotide incorporation (Brown and Suo 2009), occasionally, it can make a mistake. The shuttling between the multiple active sites (pol and exo) has been examined in great detail in phage organisms with B-family polymerases and involves multiple steps including long-range movement of the separated primer strand from the polymerase active site to the exonuclease site (Fidalgo da and Reha-Krantz 2007). In crenarchaeal B-family polymerases, control of these two catalytic activities occurs intramolecularly, mediated by a flexible loop (Y-GG/A) in the palm domain (Truniger et al. 1996; Bohlke et al. 2000) (Fig. 6.1). Most likely, the polymerase is able to efficiently achieve this feat by maintaining contact with their respective processivity clamps and allowing alternative holoenzyme conformations (discussed below). Mutation of the conserved aspartates in the exonuclease domain was instrumental in measuring the inherent fidelity of PoB1

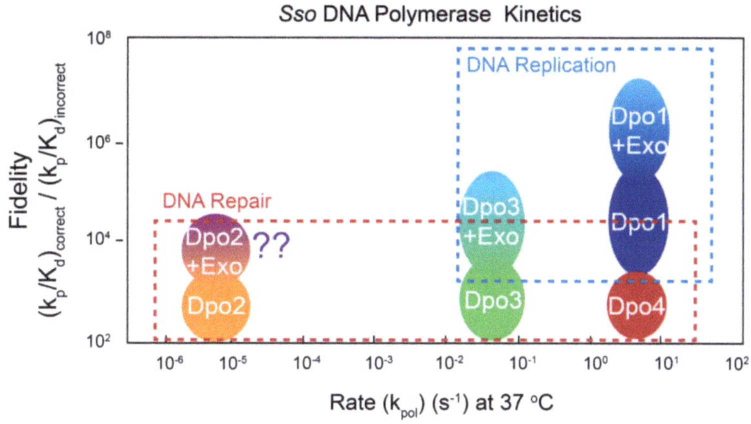

Fig. 6.2 Representation of the kinetics and fidelity of all four DNA polymerases in *Sso*. The fidelities for both the polymerase and exonuclease domains are shown. *Highlighted* are regions required for efficient DNA replication or DNA repair calculated from the total genome size (2.99 Mb) (She et al. 2001), number of origins of replication (3) (Robinson and Bell 2005), and S-phase time (1.5 h) of *Sso* (Duggin et al. 2008)

(Zhang et al. 2009) as well as other archaeal B-family polymerases (Bauer et al. 2012). Surprisingly, without an active exonuclease domain, PolB1exo$^-$ also exhibited a masked ability to extend ssDNA with a template-independent and template-dependent terminal transferase activities (Zuo et al. 2011). Short 20 nucleotide ssDNA templates are extended in a template-independent fashion initially adding three to five bases on the 3′ end. Newly added DNA is then wrapped around and stabilized intramolecularly through incomplete base paring interactions before template-dependent slipping extension creates products greater than 7 kb in a mechanism akin to what has been shown for human X-family DNA polymerase μ and λ (Dominguez et al. 2000; Maga et al. 2005).

Crenarchaeal PolB2 is not as well characterized and has been postulated to be inactive based on bioinformatic analysis of unconserved active site residues (Rogozin et al. 2008). PolB2 has high sequence conservation with PolB1, and it is thought that duplication of PolB2 gave rise to PolB1. Intriguingly, expression of *Sso*PolB2 is upregulated in response to UV exposure prompting many to hypothesize a role in bypassing cyclobutane thymine dimers (Frols et al. 2007, 2009; Gotz et al. 2007). Recently, *Sso*PolB2 has been shown to have some basic polymerase activity, but it is the least active of the four *Sso* polymerases (Fig. 6.2) and actually lacks the in vitro ability to bypass thymine dimers (Choi et al. 2011). Rather, PolB2 was able to bypass uracil, hypoxanthine, and 8-oxoguanine. The binding affinity of PolB2 to DNA is weak compared to PolB1 and limits its polymerase and exonuclease activities. It is speculated that PolB2 participates either in oxidative DNA lesion bypass or in short patch repair of UV-induced DNA damage after excision of the damaged bases, and additional protein interactions may be required to form stable complexes on DNA for efficient activity.

Bioinformatic analysis suggests that crenarchaeal PolB3 and its homolog from euryarchaea, PolB, are actually the original archaeal polymerases (Iwai et al. 2000). Further studies indicate PolB3 may have evolved into the inactive polymerase or exonuclease domain found near the C-terminus of eukaryotic ortholog pol ε (Tahirov et al. 2009). In PolB3, the highly conserved active site YxDTD motif, which is generally responsible for the coordination of the two active site Mg^{2+} ions for catalysis, has diverged significantly. While it had been previously shown that both aspartic acid residues in this motif were required for polymerase activity (Bernad et al. 1990; Copeland and Wang 1993), *Sso*PolB3 has been found to be moderately active despite having a mutated motif (LAN-D) (Choi et al. 2011; Bauer et al. 2012). As a consequence, the polymerase activity and DNA-binding ability of PolB3 are surprisingly low and similar to PolB2. The kinetics and fidelity of PolB3 are moderate, placing it squarely between values for replication or repair polymerases (Fig. 6.2). Although the exact metabolic role of PolB3 has not yet been uncovered, it can bypass cyclobutane dimers in vitro more efficiently than any of the other polymerases in *Sso* including PolY, indicating a potential role in UV-induced damage repair (Choi et al. 2011).

The number, conservation, and biochemistry of B-family polymerases in crenarchaea present the possibility that they are utilized similarly to those of eukaryotes (Edgell et al. 1997; Iwai et al. 2000), with PolB1 acting as the leading strand replicase and either PolB2 or PolB3 as the lagging strand replicase. The fidelities of B-family polymerases are generally greater than Y-family members providing for highly accurate and robust synthesis on the leading and lagging strands. On the other hand, genetic ablation of PolB2 or PolB3 has no effect on viability of *Sulfolobus* (Steve Bell, personal communication). More likely is that PolB1 acts as both the leading and lagging strand polymerases or can compensate for the loss of either PolB2 or PolB3 during replication. It may be that PolB2 and/or PolB3 will have a more specialized unrecognized role in DNA damage repair, potentially complimenting or providing redundancy to the function of the Y-family polymerase. The combination of specificities, fidelities, kinetics, and lesion bypass abilities of the four DNA polymerases in *Sso* encompass a broad range of complimentary activities for efficient replication and repair (Fig. 6.2).

6.2.2 Euryarchaeal B-Family DNA Replication Polymerases

In euryarchaea, there is typically only a single chromosomally encoded PolB. *Pyrococcus furiosus* (*Pfu*) and *Thermococcus gorgonarius* (*Tgo*) PolB have been characterized extensively and are commonly used in PCR applications due to their robust accurate activities and high thermostabilities (Lundberg et al. 1991; Hopfner et al. 1999). Many euryarchaeal PolB polymerases contain inteins which are protein sequences that facilitate their excision from flanking polypeptides while catalytically ligating the remaining segments splicing together the catalytic polymerase active site (Perler et al. 1992). The resulting spliced amino acid sequence is similar between crenarchaeal and euryarchaeal PolBs. The crystal structure of euryarchaeal B-family polymerases shows a right-hand conformation with only slight differences in the loop regions compared to crenarchaeal PolB. One unique loop structure was revealed within the exonuclease domain responsible for regulating the polymerase and nuclease activities of these polymerases (Fig. 6.1) (Bohlke et al. 2000; Hashimoto et al. 2001). This exonuclease loop is not observed in structures of B-family polymerases from other organisms like RB69 and appears to be unique to euryarchaea PolB. Mutation of the exonuclease loop results in a conformational change in the editing cleft caused by altered interactions between the loop and the thumb domain decreasing exonuclease activity and making the polymerase more amenable for PCR reactions (Kuroita et al. 2005). Control and regulation of the polymerase and nuclease activities is important for the accurate and efficient replication necessary for the maintenance of the archaeal genome. While, in crenarchaea, PolB1 is thought to perform both leading and lagging strand synthesis, euryarchaeal PolB was found to function on only the leading strand during chromosomal replication (Henneke et al. 2005), although recent conflicting data indicates that PolB may not be essential to the replication process (Sarmiento et al. 2013; Cubonova et al. 2013).

6.2.3 Archaeal-Specific D-Family Polymerases

The D-family polymerase (PolD), ubiquitously found in all euryarchaeal species, was first discovered in cellular extracts from *P. furiosus* (*Pfu*) (Cann et al. 1998) but is also found in nanoarchaea (Waters et al. 2003), thaumarchaea (Hallam et al. 2006), and korarchaea (Elkins et al. 2008), but not members of the crenarchaeal phyla nor within the genome of any bacterial or eukaryotic organism. The presence of D-family polymerases in so many phyla of Archaea suggests that PolD may have been at least partly responsible for the replication of the genome of the last common ancestor of Archaea (Matsui et al. 2011; Tahirov et al. 2009). Archaeal-specific PolD enzymes are heterotetramers composed of two large (DP2) and two small (DP1) subunits. DP2 is the catalytic subunit of the polymerase, while DP1 is the catalytic subunit of an Mre11 like 3′–5′ exonuclease (Cann et al. 1998). Although DP2 shares some of the common motifs found in other DNA polymerases, it has no global similarities to any other proteins found in the database justifying a new family of DNA polymerases (D-family). The exonuclease region in DP1 is homologous to a C-terminal region in the small noncatalytic subunits (B subunits) of multisubunit eukaryotic B-family polymerases α, δ, and ε (Cann et al. 1998; Ishino et al. 1998; Yamasaki et al. 2010; Shen et al. 2004). Additional homology has been established between zinc fingers found in D-family polymerase and those found near the C-terminus in all eukaryotic B-family polymerases (Tahirov et al. 2009).

Studies performed with *Pyrococcus horikoshii* (*Pho*) PolD illustrated that strong polymerase and exonuclease activities were obtained only when both subunits are present (Uemori et al. 1997). In *Pyrococcus abyssi* (*Pab*), PolD was unable to bypass abasic sites and maintained error rates comparable to *Pab*PolB (Palud et al. 2008). Importantly, D-family polymerases genes have been found to be essential to the survival of the euryarchaeal cell (Berquist et al. 2007; Cubonova et al. 2013). This, in tandem with their high fidelity, implicates them in a primary role in the replication process. *Pab*PolD has been found to be efficient in the extension of RNA-primed DNA, potentially involving it in the initial elongation of RNA-primed DNA prior to replacement by the B-family polymerase (Rouillon et al. 2007). It is possible that B- and D-family polymerases in euryarchaea may perform coordinated DNA synthesis much like the eukaryotic polα primase and pols ε and δ (Hubscher et al. 2002). Intriguingly, *Pab*PolD is also implicated as the lagging strand replicase, due to its ability to displace downstream complementary RNA/DNA duplexes required for Okazaki fragment processing by facilitating excision of the RNA primer by FEN-1 like nucleases (Henneke et al. 2005). In *Thermococcus kodakarensis*, it has been found that only PolD is required for genomic DNA replication, and it is speculated that while the tandem replication by both PolB and PolD may be preferable, PolD itself may be the primary replication in archaea as a whole (Cubonova et al. 2013).

Many of the D-family polymerases also contain inteins. All catalytic elements for intein splicing and subsequent ligation lie within the intein and flanking peptide regions (Paulus 2000; Perler 2005). In *Thermococcus litoralis* splicing was found to control production of the mature form of a polymerase (Perler et al. 1992). In *Pab*, an intein sequence is found within the Dp2 subunit (Mills et al. 2004). It is possible that these sequences serve as a means to regulate polymerase expression levels in vivo. Intein sequences are also present in euryarchaeal B-family polymerases (Perler et al. 1997) but are not common in crenarchaea. Both B- and D-family polymerases were found to contain inteins in *Thermococcus fumicolans* that when expressed separately, possess nuclease activity. The two intein nucleases have different substrate specificities and metal cofactor requirements (Saves et al. 2000), indicating that inteins may not only serve in a regulatory capacity for protein maturation but also potentially play a role in the maintenance of the archaeal genome.

6.2.4 Archaeal Y-Family Lesion Bypass Polymerases

Archaeal Y-family polymerases are found primarily in crenarchaea but are not universally conserved throughout the phyla and are proposed to only be present in those organisms exposed to UV light (Kelman and White 2005). Interestingly, Y-family polymerases do not share sequence identity to any of the other polymerase families (A, B, C, D, X) and also lack the 3′–5′ exonuclease domain present in the archaeal B- and D-families (Ling et al. 2001). Despite these differences, the structure of PolY is similar to that of polymerases in the A- and B-families, possessing the usual palm, fingers, and thumb subdomains (Fig. 6.1). However, in addition to the usual domains, PolY possesses an additional "little finger" subdomain and linker, primarily important for DNA binding. Archaeal Y-family polymerases have served as models for understanding lesion bypass mechanisms and specificities as related to the eukaryotic orthologs of the same family: pol η, pol ι, and pol κ.

As has been observed in bacteria and eukaryotes, archaeal Y-family polymerases have a much more specialized role in the maintenance of the archaeal genome. Y-family polymerases have error rates 100–1,000-fold higher than B-family polymerases (Fig. 6.2) (McCulloch and Kunkel 2008). This is due not only to the lack of a proofreading exonuclease domain but also to a larger, more accommodating active site which allows for binding of incoming nucleotides in additional orientations, prevented by steric clashes with residues in the active sites of polymerases in other high-fidelity polymerase families (Perlow-Poehnelt et al. 2004). Multiple dNTP orientations in the active site allow for potential base pairing with a damaged template base and the concurrent bypass of these sites of damage. The Y-family polymerases from *Sso*(Dpo4) and *Sulfolobus acidocaldarius* (*Sac*Dbh) are two of the best characterized polymerase enzymes and have been found to bypass a large number and variety of DNA lesions including: abasic (Fiala et al. 2007; Ling et al. 2004a), (deoxyguanosin-8-yl)-1-aminopyrene (Sherrer et al. 2009), benzo[*a*]pyrene diol epoxide (Ling et al. 2004b), 8-oxoguanine (Rechkoblit et al. 2006; Zang et al. 2006), methylguanine and benzylguanine (Choi et al. 2011), and thymine dimers (Johnson et al. 2005; Boudsocq et al. 2001). Active site metal ion composition has also been shown to be important in determining which lesions can be bypassed by Y-family polymerases. While typically Mg^{2+} is the metal cofactor most associated with nucleotide binding and incorporation, it has been observed that when replaced by Mn^{2+}, *Sso*Dpo4 exhibits increased catalytic efficiency yet reduced fidelity, with an ability to efficiently bypass otherwise unfavorable substrates such as abasic sites and cyclopyrimidine dimers (Vaisman et al. 2005).

Y-family polymerases are able to bypass lesions through error-free and error-prone means (Fig. 6.3). For example, *Sso*Dpo4 is able to correctly incorporate cytosine across from the aminofluorene adduct of guanine; however, base deletions or substitutions occur directly after the lesion (Rechkoblit et al. 2010). When bypassing an abasic site, Dpo4 is known to employ either the "A-rule" (Strauss 1991), where an adenosine is added opposite a noninstructional template lesion, or a template slippage loop-out mechanism where the template lesion is looped out and replication continues opposite the next base resulting in a −1 frameshift (Fiala and Suo 2007; Wu et al. 2011). Y-family polymerases are also known to be able to induce deletions of a single base through a template slippage mechanism (Wu et al. 2011). While *Sso*Dpo4 is able to bypass most lesions, the ability to bypass cyclopyrimidine dimers is limited and instead speculated to be performed by *Sso*Dpo3 (Choi et al. 2011).

The structural mechanism of lesion bypass involves a large conformational change in the little finger domain of *Sso*Dpo4 upon formation of the polymerase–DNA binary complex, with a 131° rotation relative to the palm, fingers, and thumb domains (Wong et al. 2008). A third little finger orientation,

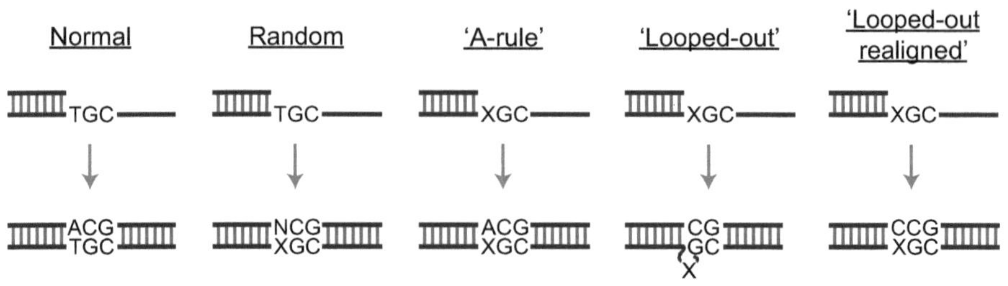

Fig. 6.3 Lesion bypass mechanisms for archaeal Y-family DNA polymerases. Random incorporation includes partial base pairing to the template strand for preferential incorporation of nucleotides dependent on the lesion type. The "A-rule" is utilized primarily when no templating base is available and preferentially incorporates adenine over the other three bases. The "looped-out" mechanism utilizes base pairing interactions at the +1 site of the template strand while excluding the lesion and results in a −1 frameshift. The looped-out mechanism can also realign to avoid the frameshift albeit with much less frequency

distinct from the DNA bound and apo forms, is observed upon polymerase binding to PCNA (Xing et al. 2009). It was found that the linker region between the little finger and thumb domain is important in the interactions between *Sso*Dpo4, DNA, and possibly PCNA (Sherrer et al. 2012). Generally, the activity and lesion bypass ability of *Sso*Dpo4 is much greater than for *Sac*Dbh and has been attributed to the sequence of the little finger linker domain, which contributes to the strength of DNA binding and processivity (Boudsocq et al. 2004; Wilson et al. 2013). Interestingly, knockouts of *Sac*Dbh resulted in no changes to growth rates or sensitivity to a variety of tested DNA damaging agents (Sakofsky et al. 2012). However, an increase in the number of spontaneous mutations was observed, much like that which is seen upon deletion of *E. coli* pol IV (DinB) in cells in a stationary phase (Nowosielska et al. 2004). This indicates that while the Y-family polymerase is not absolutely required for the bypass of lesions, its absence results in the bypass being performed in an even more error-prone fashion possibly by the other archaeal DNA polymerases.

One consequence of the more spacious solvent accessible Y-family polymerase active site is the potential incorporation of ribonucleotides. Selective incorporation of only dNTPs is essential for the DNA replication process, and much like polymerases from other families, Y-family members still maintain a high level of selectivity in spite of their more relaxed active site. This selectivity is preserved through the presence of a highly conserved tyrosine residue termed the "steric gate" (Fig. 6.1) (Delucia et al. 2003). The steric gate blocks binding of rNTPs through steric clash with the 3′OH of the incoming nucleotide (Kirouac et al. 2011). The steric gate also seems to be conserved in a variety of polymerases from the A, B, X, Y, and RT families to provide specificity for deoxyribonucleotides (Brown and Suo 2011).

Another feature identified in *Sso*Dpo4 but not *Sac*Dbh is the ability to perform pyrophosphorolysis or reattachment of the pyrophosphate (PP_i) to the 3′ end of a nucleotide monophosphate, reversing the NTP incorporation mechanism.

Particularly upon misincorporation of dGTP or dCTP across from template dT, pyrophosphorolysis can occur in *Sso*Dpo4 to remove the misincorporated nucleotide (Vaisman et al. 2005). Pyrophosphorolysis is related to the strength of the interaction between the little finger and DNA. Exchange of the little finger from *Sso*Dpo4 with the weaker binding motif from *Sac*Dbh abolished pyrophosphorolytic activity. It may be that *Sso*Dpo4 and other exonuclease-deficient archaeal polymerases utilize pyrophosphorolysis as a means of error checking favored by the slow release of PP_i after nucleotide misincorporation to increase fidelity further.

6.2.5 PriSL Primase, a Possible X-Family Polymerase Homolog

The eukaryotic-like archaeal primase, PriSL, is a heterodimeric protein composed of a large regulatory subunit (PriL) and a small catalytic subunit (PriS) that is conserved across the archaeal domain. It is proposed to be responsible for the *de novo* synthesis of short RNA primers to initiate DNA replication on the leading and lagging strands. PriSL has the remarkable ability to synthesize DNA and RNA in a template-dependent and template-independent manner producing oligonucleotide products from 2 bases to greater than 7,000 (Lao-Sirieix and Bell 2004; Le Breton et al. 2007). The PriS subunit shares significant structural and sequence homology to eukaryotic X-family polymerases, most notably polymerase β (Augustin et al. 2001; Lao-Sirieix et al. 2005). PriSL can also synthesize across discontinuous templates (Hu et al. 2012) in a similar manner to that of eukaryotic pol μ, which is known to be involved in both base excision repair and double-strand break (DSB) repair. Archaea do not have a direct polymerase homolog from the X-family but the biochemical data suggests that PriSL may be a functional homolog (Table 6.1).

6.3 Archaeal DNA Polymerase Holoenzymes

DNA polymerases by themselves are traditionally not very processive and require complexation with their respective clamps to replicate long stretches of DNA without dissociating. In Archaea, the proliferating cell nuclear antigen (PCNA) clamp is loaded onto DNA by the replication factor-C (RFC) clamp loader in an ATP-dependent manner to facilitate recruitment of the DNA polymerase to the holoenzyme complex. This minimal DNA polymerase holoenzyme complex will be active on both the leading and lagging strands during replication as well as participate in lesion bypass mechanisms of exchange at DNA damage sites.

B-family polymerases from both crenarchaea and euryarchaea have been found to interact with their processivity clamps through a motif called the PCNA-interacting peptide (PIP) box (Castrec et al. 2009; Pisani et al. 2000). This motif has a consensus sequence of Qxxhxxaa, where x is any amino acid, h is a hydrophobic residue, and a is an aromatic residue (Warbrick 1998). PCNA loading on DNA allows for the binding of the polymerase and formation of a replicative holoenzyme similar to those seen in both prokaryotes and eukaryotes. The association of the PCNA clamp with PolB results in a highly processive complex that limits dissociation and allows synthesis of greater than 10 kb products in a single binding event. The structure of euryarchaeal (*Pfu*) DNA polymerase holoenzyme has provided significant insight into the holoenzyme assembly mechanism and conformational changes required for both polymerization, editing, and switching (Fig. 6.4a) (Mayanagi et al. 2011; Nishida et al. 2009; Bunting et al. 2003). *Pfu*PolB was found to possess an additional site of contact with PCNA (standby), proposed to be important for pivoting between the editing (locked-down) and polymerization (tethered) modes. This flexibility provided through multiple interactions between PCNA, and the polymerase allows for uninterrupted and dynamic, error-free DNA synthesis and may also be important for polymerase switching during replication.

Euryarchaeal D-family polymerases also interact with the homotrimeric PCNA at the PIP site to increase processivity (Tori et al. 2007; Henneke et al. 2005). Like B-family polymerases, PolD also possesses a PIP box (Castrec et al. 2009); *Pab*PolD has been shown to have two separate sites for interaction with PCNA, one at the C-terminus and a separate palindromic PIP box at the N-terminus of the large subunit. Both sites effect PolD binding to PCNA; however, while mutation of the N-terminal PIP site reduced processivity, mutation of the C-terminal PIP had no effect. It was also shown that both PIP sites interact with the same site on PCNA, raising the question of whether the binding site changes depending on the desired activity between polymerization and editing as seen for euryarchaeal PolB (Castrec et al. 2009).

Unlike the euryarchaeal homotrimeric PCNA, the crenarchaeal PCNA is a heterotrimeric complex comprised of three separate proteins (PCNA1, PCNA2, and PCNA3). PolB1 from *Sulfolobus solfataricus* has been found to interact specifically with PCNA2 subunit within the heterotrimer to increase replication rate (Dionne et al. 2003). We have recently found that the *Sso* DNA polymerase holoenzyme synthesizes DNA somewhat distributively and is weakly processive unlike other well characterized systems (Bauer, Trakselis et al, submitted). Specific contacts between the C-terminal PIP motif in *Sso*PolB1 and *Sso*PCNA2 is responsible for actively re-recruiting the polymerase during replication to maintain a dynamic processivity. Using specific contact points and measured distances (Trakselis, unpublished data) between the *Sso*PCNA123 and *Sso*PolB1, we have created a DNA polymerase holoenzyme model to highlight interactions in the polymerization mode (Fig. 6.4b). By having three separate proteins, crenarchaeal PCNA123 can also specifically interact with additional proteins, such as flap endonuclease (PCNA1) and DNA ligase (PCNA3) in close proximity to the polymerase to efficiently process Okazaki fragments in a "tool-belt" fashion

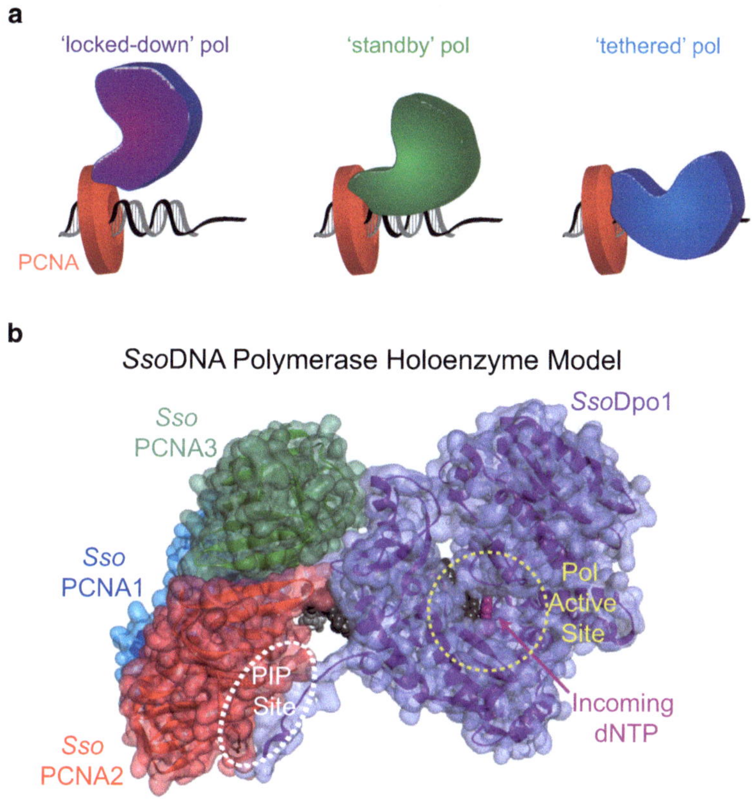

Fig. 6.4 (**a**) Complexation of the DNA polymerase with the PCNA clamp can occur in different conformational states for assembly or activation of exonuclease or polymerase activities. (**b**) DNA polymerase holoenzyme model from Sso showing the specific interactions of the C-terminal PIP domain of SsoDpo1 (*purple*) with the interdomain connecting loop (ICL) of SsoPCNA2 (*red*) as well as adjacent interactions with SsoPCNA3 (*green*) modeled from the PfuPolB1–PCNA cocrystal structure (PDB ID: 3A2F). Model is built from specific detected interactions of PCNA123 (PDB ID: 2IX2) and SsoPolB1 (PDB ID: 1S5J) simulating the tethered complex. The polymerase active site is *highlighted* with the incoming dNTP base (*pink*) pairing to the primer–template junction prior to catalysis. Primer–template DNA was modeled in the active site from analogous interactions with the RB69 polymerase-DNA structure (PDB ID: 1IG9)

(Beattie and Bell 2012; Pascal et al. 2006; Indiani et al. 2005). The retention of additional DNA replication accessory factors as well as multiple copies of DNA polymerases in high local concentration at the replication fork as coordinated by PCNA provides for uninterrupted DNA synthesis ability.

Lesion bypass polymerases from the Y-family are also known to interact with the crenarchaeal PCNA1 subunit (Xing et al. 2009). Much like the B-family polymerases, those of the Y-family exhibit increased processivity and replication rates in the presence of the clamp (Grúz et al. 2001; Dionne et al. 2008). Although much is known about the mechanism of lesion bypass for these Y-family DNA polymerases, the influence of PCNA on this activity is not known. PolY from euryarchaeal *Methanosarcina acetivorans* is unique among Y-family pols as when it is complexed with PCNA; it can synthesize extremely long products, greater than 7.2 kb (Lin et al. 2010). Typically mutagenic lesion bypass polymerases are not able to synthesize such long products presumably due to the effect their low fidelities would have on the genome. Therefore, molecular access to PCNA must be regulated.

The interaction between the polymerases and differing subunits of PCNA has led to the proposal that both B-family and Y-family polymerases can be retained simultaneously for dynamic polymerization in the presence of lesions (Furukohri et al. 2008; Heltzel et al. 2009). If the replicative polymerase (PolB) encounters a DNA damage site, it can shuttle between polymerase and exonuclease sites destabilizing binding in favor of PolY binding (Maxwell and Suo 2013). After switching, the Y-family polymerase bypasses the damage site while maintaining PolB within the holoenzyme. After the lesion has been successfully bypassed, the B-family polymerase can return, reestablishing accurate and normal DNA synthesis. Of course, this would require that multiple polymerases are intimately associated in and around the replisome.

6.4 Polymerase Coordination During DNA Replication and Repair

The DNA-binding specificity for all DNA polymerases is primarily afforded by the 3′OH on the primer strand but also includes significant contacts with both the dsDNA and the ssDNA templates. When multiple DNA polymerases are present in a single organism with similar DNA substrate specificities, a question arises as to how polymerase binding to DNA is regulated to ensure accurate DNA replication and efficient DNA repair. Regulation of polymerase binding and access to DNA in the cell utilizes multiple biophysical strategies including kinetics, thermodynamics, transcriptional and translational regulation, interactions with accessory proteins, and oligomer formation.

6.4.1 Oligomeric DNA Polymerase Complexes

In *E. coli*, three DNA polymerase III cores are coordinated within the replisome by the clamp-loader complex (McInerney et al. 2007). Two of the three polymerase cores are proposed to be involved in the formation of alternating Okazaki fragments, while the other consistently synthesizes DNA on the leading strand. Other polymerases including T4 gp43 (Ishmael et al. 2003), and Klenow (Purohit et al. 2003), have been found to interact with DNA in a dimeric state. In Archaea,

*Sso*PolB1 can assemble into a unique trimeric complex to increase both replication rate and processivity of the enzyme (Mikheikin et al. 2009). Trimeric *Sso*PolB1 is observed at temperatures ranging from 10 to 70 °C where processivity values increase with temperature and routinely exceed 1,000 bases (Lin et al. 2012). As discussed above, DNA polymerases are generally nonprocessive with synthesis of only 20 bases before dissociation from the template and require interactions with their respective clamps to achieve extremely large processivity values. As such, we have suggested that the large processivity value for trimeric *Sso*PolB1 is a consequence of direct interactions between subunits, effectively encircling the DNA template akin to the structural role of the clamp proteins (Fig. 6.5a). Similarly, the Y-family polymerase, *Sso*PolY, also forms oligomeric complexes on DNA in a concentration-dependent manner as highlighted in a variety of crystal structures (Fig. 6.5b). The function of oligomeric PolY complexes is not known and no known enzymatic enhancement has been noted, but this interaction may be used to keep high concentrations of polymerases at the replication fork. In addition, direct interactions between the replication (*Sso*PolB1) and lesion bypass (*Sso*PolY) polymerases have also been detected highlighting another contact point for polymerase exchange (De Felice et al. 2007).

In all, these results suggest that polymerase action may be more complicated than previously thought, utilizing both intimately and loosely bound polymerase molecules in the replisome. In T4 and T7, DNA polymerase molecules can exchange freely at the site of catalysis during replication (Yang et al. 2004; Loparo et al. 2011). This "dynamic polymerase processivity," along with the variety of different possible homo- and heteroligomeric complexes that can form between *Sso*PolB1 and *Sso*PolY, highlights the possibility that the archaeal cell utilizes these complexes in a variety of yet unknown genomic maintenance functions. The regulation of oligomeric polymerase complex formation may be controlled through subtle changes in equilibria at the site of action to afford specific genome maintenance functions.

6.4.2 *Thermodynamic DNA Polymerase Selection*

Although DNA polymerase binding can be quantified using a number of methods, the most complete way to quantify the binding specificity is from a measure of the heat capacity change (ΔC_p^o). For thermophilic DNA polymerases binding to DNA, there is a broad temperature range that can be explored generally giving large and negative ΔC_p^o values. These negative ΔC_p^o values highlight strong structural complementarity between the polymerase and DNA instead of the more traditional explanations of DNA sequence specificity. In comparing the ΔC_p^o values for B- and Y-family polymerases from *Sso*, we have found that *Sso*PolB1 and *Sso*PolY have similar strong enthalpy–entropy compensation in binding over a large temperature range giving rise to smaller fluctuations in free energy (ΔG^o) up to 50 °C

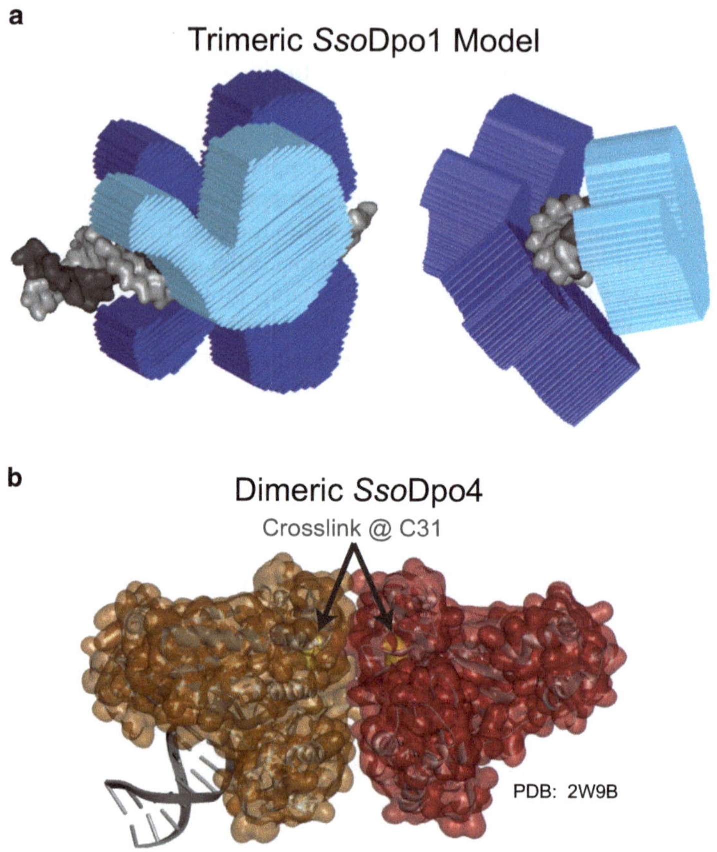

Fig. 6.5 (**a**) Model of the trimeric *Sso*Dpo1 polymerase bound to DNA. The *encircled* conformation is proposed based on binding and footprinting on a small primer–template DNA substrate. (**b**) Crystal structure (PDB: 2W9B) of dimeric *Sso*Dpo4 consistent with chemical cross-linking at cysteine 31

(Lin et al. 2012). Above 50 °C, there is higher affinity of binding for *Sso*PolB1 over *Sso*PolY providing thermodynamic discrimination of monomeric binding at high physiological temperatures. In addition, there are similar sequential binding events dependent on local concentration and temperature that modulate oligomeric *Sso*PolB1 and *Sso*PolY coupled equilibria (Lin, Trakselis et al., unpublished). There are extreme differences in the free energies (ΔG^{o}) of binding oligomeric *Sso*PolB1 over *Sso*PolY that are exaggerated with temperature providing an additional level of thermodynamic selection for oligomeric B-family polymerases on DNA during normal replication in the absence of damage (Lin et al. 2012). The collective results suggest that binding of both polymerases to DNA is not necessarily dictated by specific interaction differences between replication and repair polymerases and DNA, but rather thermodynamic selection of more stable oligomeric DNA polymerase complexes at high temperatures to promote accurate synthesis.

6.4.3 Uracil Read-Ahead Function for Archaeal B-Family Polymerases

Another unique property of archaeal B-family polymerases is their ability to detect uracil in the downstream DNA sequence (Greagg et al. 1999). The presence of uracil in DNA usually arises due to the deamination of cytosine; however, as a result of uracil base pairing to thymine, this deamination results in a transversion to a C-G base pair after replication. A uracil recognition pocket was identified in the N-terminal region of *Tgo*PolB (Hopfner et al. 1999) and *Sso*PolB1 (Savino et al. 2004) that stalls the polymerase four bases from the primer–template junction (Fig. 6.1). Mutations in the recognition pocket of *Pfu*Pol restored the ability to replicate through uracil, confirming that this region is responsible for the uracil "read-ahead" function (Fogg et al. 2002). A Mg^{2+} ion (exchangeable with Mn^{2+}) has recently been identified just below the uracil recognition pocket in *P. abyssi* (*Pab*) PolB that guides the DNA to the binding site (Gouge et al. 2012). In *Pfu*PolB, sensing of uracil (+4 on the template) seems to stimulate the unwinding of the 3′ end of the primer and exonuclease proofreading, trimming the primer to maintain distance from the uracil (Richardson et al. 2013). It is noteworthy that this uracil binding pocket is not present in eukaryotic B-family polymerases, including those with high homology to archaeal B-family polymerases. As a result, it is speculated that the uracil read-ahead function serves as a means to recognize DNA damage prior to replication in organisms with incomplete base excision repair (BER) or nucleotide excision repair (NER) pathways.

6.4.4 Polymerase Participation in DNA Repair

Archaea maintain a level of genomic stability equivalent to or slightly better than other microorganisms. This is somewhat surprising due to the environments that many of these organisms thrive in where oxidations and deaminations would be common. Therefore, Archaea must possess very robust mechanisms by which DNA damage is repaired (Grogan et al. 2001). In bacteria and eukaryotes, DNA is typically repaired through several means including error-prone lesion bypass, BER, NER, and mismatch repair. B-family and D-family polymerases contain exonuclease domains and would shuttle between incorporation and excision events at sites of DNA damage. Most Archaea also possess Y-family polymerases capable of performing the error-prone lesion bypass processes in spite of DNA damage; however, this ability may not be essential as knockouts of PolY are unaffected by DNA damaging agents (Sakofsky et al. 2012).

As the error-prone polymerases are not universally conserved, DNA repair in Archaea must include alternative mechanisms. Traditionally, BER and NER pathways would fulfill this role, but many Archaea are missing many of the proteins homologs to either the eukaryotic or bacterial versions making identification difficult (Kelman and White 2005). Therefore, it is still not understood how Archaea maintain stable genomes even in spite of intense environmental stress, but it is certain that polymerases in Archaea will have both specific and redundant roles in repair processes.

6.5 Conclusion

DNA polymerases from Archaea have both unique and conserved biochemical features compared with polymerases from other domains. Because of their high homology, similar structures, and conserved catalytic features, they are considered to be relevant model systems for understanding mechanisms of action and protein interactions involved in DNA replication and repair. Cells that contain multiple DNA polymerases need a mechanism to regulate individual recruitment to the replication fork or DNA damage site for specialized nucleotide incorporation. As we learn more about the specificities and kinetics of each DNA polymerase, the molecular interactions with accessory proteins, and the dynamics that occur to coordinate their activities, we can better comprehend how their activities are regulated. This is of great importance for understanding how genomes of all organisms are accurately maintained in spite of intense environmental stressors that are mercilessly acting the DNA. Only after revealing how multiple DNA polymerases dynamically orchestrate their specificities can we attempt to modulate this activity to promote evolutionary diversity of new function or restrict cancer cell growth for specialized therapies.

References

Augustin MA, Huber R, Kaiser JT (2001) Crystal structure of a DNA-dependent RNA polymerase (DNA primase). Nat Struct Biol 8(1):57–61

Bauer RJ, Begley MT, Trakselis MA (2012) Kinetics and fidelity of polymerization by DNA polymerase III from *Sulfolobus solfataricus*. Biochemistry 51(9):1996–2007

Beattie TR, Bell SD (2012) Coordination of multiple enzyme activities by a single PCNA in archaeal Okazaki fragment maturation. EMBO J 31(6):1556–1567

Bernad A, Blanco L, Salas M (1990) Site-directed mutagenesis of the YCDTDS amino acid motif of the Φ29 DNA polymerase. Gene 94(1):45–51

Berquist BR, DasSarma P, DasSarma S (2007) Essential and non-essential DNA replication genes in the model halophilic Archaeon, Halobacterium sp. NRC-1. BMC Genet 8(31)

Bohlke K, Pisani FM, Vorgias CE, Frey B, Sobek H, Rossi M, Antranikian G (2000) PCR performance of the B-type DNA polymerase from the thermophilic euryarchaeon *Thermococcus aggregans* improved by mutations in the Y-GG/A motif. Nucleic Acids Res 28(20):3910–3917

Boudsocq F, Iwai S, Hanaoka F, Woodgate R (2001) *Sulfolobus solfataricus* P2 DNA polymerase IV (Dpo4): an archaeal DinB-like DNA polymerase with lesion-bypass properties akin to eukaryotic polη. Nucleic Acids Res 29(22):4607–4616

Boudsocq F, Kokoska RJ, Plosky BS, Vaisman A, Ling H, Kunkel TA, Yang W, Woodgate R (2004) Investigating the role of the little finger domain of Y-family DNA polymerases in low fidelity synthesis and translesion replication. J Biol Chem 279(31):32932–32940

Brown JA, Suo Z (2009) Elucidating the kinetic mechanism of DNA polymerization catalyzed by *Sulfolobus solfataricus* P2 DNA polymerase B1. Biochemistry 48(31):7502–7511

Brown JA, Suo Z (2011) Unlocking the sugar "steric gate" of DNA polymerases. Biochemistry 50(7):1135–1142

Bunting KA, Roe SM, Pearl LH (2003) Structural basis for recruitment of translesion DNA polymerase Pol IV/DinB to the beta-clamp. EMBO J 22(21):5883–5892

Cann IKO, Komori K, Toh H, Kanai S, Ishino Y (1998) A heterodimeric DNA polymerase: evidence that members of Euryarchaeota possess a distinct DNA polymerase. Proc Natl Acad Sci USA 95(24):14250–14255

Castrec B, Rouillon C, Henneke G, Flament D, Querellou J, Raffin J-P (2009) Binding to PCNA in euryarchaeal DNA replication requires two PIP motifs for DNA polymerase D and one PIP motif for DNA polymerase B. J Mol Biol 394(2):209–218

Choi JY, Eoff RL, Pence MG, Wang J, Martin MV, Kim EJ, Folkmann LM, Guengerich FP (2011) Roles of the four DNA polymerases of the crenarchaeon *Sulfolobus solfataricus* and accessory proteins in DNA replication. J Biol Chem 286(36):31180–31193

Copeland WC, Wang TS (1993) Mutational analysis of the human DNA polymerase alpha. The most conserved region in alpha-like DNA polymerases is involved in metal-specific catalysis. J Biol Chem 268(15):11028–11040

Cubonova L, Richardson T, Burkhart BW, Kelman Z, Connolly BA, Reeve JN, Santangelo TJ (2013) Archaeal DNA polymerase D but not DNA polymerase B is required for genome replication in *Thermococcus kodakarensis*. J Bacteriol 195:2322–2328

De Felice M, Medagli B, Esposito L, De Falco M, Pucci B, Rossi M, Gruz P, Nohmi T, Pisani FM (2007) Biochemical evidence of a physical interaction between *Sulfolobus solfataricus* B-family and Y-family DNA polymerases. Extremophiles 11(2):277–282

Delucia AM, Grindley ND, Joyce CM (2003) An error-prone family Y DNA polymerase (DinB homolog from *Sulfolobus solfataricus*) uses a 'steric gate' residue for discrimination against ribonucleotides. Nucleic Acids Res 31(14):4129–4137

Dionne I, Nookala RK, Jackson SP, Doherty AJ, Bell SD (2003) A heterotrimeric PCNA in the hyperthermophilic archaeon *Sulfolobus solfataricus*. Mol Cell 11(1):275–282

Dionne I, Brown NJ, Woodgate R, Bell SD (2008) On the mechanism of loading the PCNA sliding clamp by RFC. Mol Microbiol 68(1):216–222

Dominguez O, Ruiz JF, de Lain LT, Garcia-Diaz M, Gonzalez MA, Kirchhoff T, Martinez A, Bernad A, Blanco L (2000) DNA polymerase mu (Pol mu), homologous to TdT, could act as a DNA mutator in eukaryotic cells. EMBO J 19(7):1731–1742

Duggin IG, McCallum SA, Bell SD (2008) Chromosome replication dynamics in the archaeon *Sulfolobus acidocaldarius*. Proc Natl Acad Sci USA 105(43):16737–16742

Edgell DR, Klenk HP, Doolittle WF (1997) Gene duplications in evolution of archaeal family B DNA polymerases. J Bacteriol 179(8):2632–2640

Edgell DR, Malik SB, Doolittle WF (1998) Evidence of independent gene duplications during the evolution of archaeal and eukaryotic family B DNA polymerases. Mol Biol Evol 15(9):1207–1217

Elkins JG, Podar M, Graham DE, Makarova KS, Wolf Y, Randau L, Hedlund BP, Brochier-Armanet C, Kunin V, Anderson I, Lapidus A, Goltsman E, Barry K, Koonin EV, Hugenholtz P, Kyrpides N, Wanner G, Richardson P, Keller M, Stetter KO (2008) A korarchaeal genome reveals insights into the evolution of the Archaea. Proc Natl Acad Sci USA 105(23):8102–8107

Fiala KA, Suo Z (2007) Sloppy bypass of an abasic lesion catalyzed by a Y-family DNA polymerase. J Biol Chem 282(11):8199–8206

Fiala KA, Hypes CD, Suo Z (2007) Mechanism of abasic lesion bypass catalyzed by a Y-family DNA polymerase. J Biol Chem 282(11):8188–8198

Fidalgo da SE, Reha-Krantz LJ (2007) DNA polymerase proofreading: active site switching catalyzed by the bacteriophage T4 DNA polymerase. Nucleic Acids Res 35(16):5452–5463

Fogg MJ, Pearl LH, Connolly BA (2002) Structural basis for uracil recognition by archaeal family B DNA polymerases. Nat Struct Biol 9(12):922–927

Frols S, Gordon PM, Panlilio MA, Duggin IG, Bell SD, Sensen CW, Schleper C (2007) Response of the hyperthermophilic archaeon *Sulfolobus solfataricus* to UV damage. J Bacteriol 189(23):8708–8718

Frols S, White MF, Schleper C (2009) Reactions to UV damage in the model archaeon *Sulfolobus solfataricus*. Biochem Soc Trans 37(Pt 1):36–41

Furukohri A, Goodman MF, Maki H (2008) A dynamic polymerase exchange with *Escherichia coli* DNA polymerase IV replacing DNA polymerase III on the sliding clamp. J Biol Chem 283(17):11260–11269

Gotz D, Paytubi S, Munro S, Lundgren M, Bernander R, White MF (2007) Responses of hyperthermophilic crenarchaea to UV irradiation. Genome Biol 8(10):R220

Gouge J, Ralec C, Henneke G, Delarue M (2012) Molecular recognition of canonical and deaminated bases by *P. abyssi* family B DNA polymerase. J Mol Biol 423(3):315–336

Greagg MA, Fogg MJ, Panayotou G, Evans SJ, Connolly BA, Pearl LH (1999) A read-ahead function in archaeal DNA polymerases detects promutagenic template-strand uracil. Proc Natl Acad Sci USA 96(16):9045–9050

Grogan DW, Carver GT, Drake JW (2001) Genetic fidelity under harsh conditions: analysis of spontaneous mutation in the thermoacidophilic archaeon *Sulfolobus acidocaldarius*. Proc Natl Acad Sci USA 98(14):7928–7933

Grúz P, Pisani FM, Shimizu M, Yamada M, Hayashi I, Morikawa K, Nohmi T (2001) Synthetic activity of *Sso* DNA polymerase Y1, an archaeal DinB-like DNA polymerase, is stimulated by processivity factors proliferating cell nuclear antigen and replication factor C. J Biol Chem 276(50):47394–47401

Hallam SJ, Konstantinidis KT, Putnam N, Schleper C, Watanabe Y-i, Sugahara J, Preston C, de la Torre J, Richardson PM, DeLong EF (2006) Genomic analysis of the uncultivated marine crenarchaeote *Cenarchaeum symbiosum*. Proc Natl Acad Sci USA 103(48):18296–18301

Hashimoto H, Nishioka M, Fujiwara S, Takagi M, Imanaka T, Inoue T, Kai Y (2001) Crystal structure of DNA polymerase from hyperthermophilic archaeon *Pyrococcus kodakaraensis* KOD1. J Mol Biol 306(3):469–477

Heltzel JM, Maul RW, Scouten Ponticelli SK, Sutton MD (2009) A model for DNA polymerase switching involving a single cleft and the rim of the sliding clamp. Proc Natl Acad Sci USA 106(31):12664–12669

Henneke G, Flament D, Hubscher U, Querellou J, Raffin JP (2005) The hyperthermophilic euryarchaeota *Pyrococcus abyssi* likely requires the two DNA polymerases D and B for DNA replication. J Mol Biol 350(1):53–64

Hopfner KP, Eichinger A, Engh RA, Laue F, Ankenbauer W, Huber R, Angerer B (1999) Crystal structure of a thermostable type B DNA polymerase from *Thermococcus gorgonarius*. Proc Natl Acad Sci USA 96(7):3600–3605

Hu J, Guo L, Wu K, Liu B, Lang S, Huang L (2012) Template-dependent polymerization across discontinuous templates by the heterodimeric primase from the hyperthermophilic archaeon *Sulfolobus solfataricus*. Nucleic Acids Res 40(8):3470–3483

Hubscher U, Maga G, Spadari S (2002) Eukaryotic DNA polymerases. Annu Rev Biochem 71:133–163

Indiani C, McInerney P, Georgescu R, Goodman MF, O'Donnell M (2005) A sliding-clamp toolbelt binds high- and low-fidelity DNA polymerases simultaneously. Mol Cell 19(6):805–815

Ishino Y, Komori K, Cann IKO, Koga Y (1998) A novel DNA polymerase family found in Archaea. J Bacteriol 180(8):2232–2236

Ishmael FT, Trakselis MA, Benkovic SJ (2003) Protein-protein interactions in the bacteriophage T4 replisome. The leading strand holoenzyme is physically linked to the lagging strand holoenzyme and the primosome. J Biol Chem 278(5):3145–3152

Iwai T, Kurosawa N, Itoh YH, Kimura N, Horiuchi T (2000) Sequence analysis of three family B DNA polymerases from the thermoacidophilic crenarchaeon *Sulfurisphaera ohwakuensis*. DNA Res 7(4):243–251

Jeruzalmi D, O'Donnell M, Kuriyan J (2002) Clamp loaders and sliding clamps. Curr Opin Struct Biol 12(2):217–224

Johnson RE, Prakash L, Prakash S (2005) Distinct mechanisms of cis-syn thymine dimer bypass by Dpo4 and DNA polymerase η. Proc Natl Acad Sci USA 102(35):12359–12364

Kelman Z, White MF (2005) Archaeal DNA replication and repair. Curr Opin Microbiol 8(6):669–676

Kim SW, Kim D-U, Kim JK, Kang L-W, Cho H-S (2008) Crystal structure of *Pfu*, the high fidelity DNA polymerase from *Pyrococcus furiosus*. Int J Biol Macromol 42(4):356–361

Kirouac KN, Suo Z, Ling H (2011) Structural mechanism of ribonucleotide discrimination by a Y-family DNA polymerase. J Mol Biol 407(3):382–390

Kunkel TA, Bebenek K (2000) DNA replication fidelity. Annu Rev Biochem 69:497–529

Kuroita T, Matsumura H, Yokota N, Kitabayashi M, Hashimoto H, Inoue T, Imanaka T, Kai Y (2005) Structural mechanism for coordination of proofreading and polymerase activities in archaeal DNA polymerases. J Mol Biol 351(2):291–298

Lao-Sirieix SH, Bell SD (2004) The heterodimeric primase of the hyperthermophilic archaeon *Sulfolobus solfataricus* possesses DNA and RNA primase, polymerase and 3′-terminal nucleotidyl transferase activities. J Mol Biol 344(5):1251–1263

Lao-Sirieix SH, Nookala RK, Roversi P, Bell SD, Pellegrini L (2005) Structure of the heterodimeric core primase. Nat Struct Mol Biol 12(12):1137–1144

Le Breton M, Henneke G, Norais C, Flament D, Myllykallio H, Querellou J, Raffin JP (2007) The heterodimeric primase from the euryarchaeon *Pyrococcus abyssi*: a multifunctional enzyme for initiation and repair? J Mol Biol 374(5):1172–1185

Lin LJ, Yoshinaga A, Lin Y, Guzman C, Chen YH, Mei S, Lagunas AM, Koike S, Iwai S, Spies MA, Nair SK, Mackie RI, Ishino Y, Cann IK (2010) Molecular analyses of an unusual translesion DNA polymerase from *Methanosarcina acetivorans* C2A. J Mol Biol 397(1):13–30

Lin HK, Chase SF, Laue TM, Jen-Jacobson L, Trakselis MA (2012) Differential temperature-dependent multimeric assemblies of replication and repair polymerases on DNA increase processivity. Biochemistry 51(37):7367–7382

Ling H, Boudsocq F, Woodgate R, Yang W (2001) Crystal structure of a Y-family DNA polymerase in action: a mechanism for error-prone and lesion-bypass replication. Cell 107(1):91–102

Ling H, Boudsocq F, Woodgate R, Yang W (2004a) Snapshots of replication through an abasic lesion; structural basis for base substitutions and frameshifts. Mol Cell 13(5):751–762

Ling H, Sayer JM, Plosky BS, Yagi H, Boudsocq F, Woodgate R, Jerina DM, Yang W (2004b) Crystal structure of a benzo[a]pyrene diol epoxide adduct in a ternary complex with a DNA polymerase. Proc Natl Acad Sci USA 101(8):2265–2269

Loparo JJ, Kulczyk AW, Richardson CC, van Oijen AM (2011) Simultaneous single-molecule measurements of phage T7 replisome composition and function reveal the mechanism of polymerase exchange. Proc Natl Acad Sci USA 108(9):3584–3589

Lundberg KS, Shoemaker DD, Adams MW, Short JM, Sorge JA, Mathur EJ (1991) High-fidelity amplification using a thermostable DNA polymerase isolated from *Pyrococcus furiosus*. Gene 108(1):1–6

Maga G, Ramadan K, Locatelli GA, Shevelev I, Spadari S, Hubscher U (2005) DNA elongation by the human DNA polymerase lambda polymerase and terminal transferase activities are differentially coordinated by proliferating cell nuclear antigen and replication protein A. J Biol Chem 280(3):1971–1981

Matsui I, Urushibata Y, Shen Y, Matsui E, Yokoyama H (2011) Novel structure of an N-terminal domain that is crucial for the dimeric assembly and DNA-binding of an archaeal DNA polymerase D large subunit from *Pyrococcus horikoshii*. FEBS Lett 585(3):452–458

Maxwell BA, Suo Z (2013) Single-molecule investigation of substrate binding kinetics and protein conformational dynamics of a B-family replicative DNA polymerase. J Biol Chem 288:11590–11600

Mayanagi K, Kiyonari S, Nishida H, Saito M, Kohda D, Ishino Y, Shirai T, Morikawa K (2011) Architecture of the DNA polymerase B-proliferating cell nuclear antigen (PCNA)-DNA ternary complex. Proc Natl Acad Sci USA 108:1845–1849

McCulloch SD, Kunkel TA (2008) The fidelity of DNA synthesis by eukaryotic replicative and translesion synthesis polymerases. Cell Res 18(1):148–161

McInerney P, Johnson A, Katz F, O'Donnell M (2007) Characterization of a triple DNA polymerase replisome. Mol Cell 27(4):527–538

Mikheikin AL, Lin HK, Mehta P, Jen-Jacobson L, Trakselis MA (2009) A trimeric DNA polymerase complex increases the native replication processivity. Nucleic Acids Res 37 (21):7194–7205

Mills KV, Manning JS, Garcia AM, Wuerdeman LA (2004) Protein splicing of a *Pyrococcus abyssi* intein with a C-terminal glutamine. J Biol Chem 279(20):20685–20691

Nishida H, Mayanagi K, Kiyonari S, Sato Y, Oyama T, Ishino Y, Morikawa K (2009) Structural determinant for switching between the polymerase and exonuclease modes in the PCNA-replicative DNA polymerase complex. Proc Natl Acad Sci USA 106(49):20693–20698

Nowosielska A, Janion C, Grzesiuk E (2004) Effect of deletion of SOS-induced polymerases, pol II, IV, and V, on spontaneous mutagenesis in *Escherichia coli* mutD5. Environ Mol Mutagen 43(4):226–234

Palud A, Villani G, L'Haridon S, Querellou J, Raffin JP, Henneke G (2008) Intrinsic properties of the two replicative DNA polymerases of *Pyrococcus abyssi* in replicating abasic sites: possible role in DNA damage tolerance? Mol Microbiol 70(3):746–761

Pascal JM, Tsodikov OV, Hura GL, Song W, Cotner EA, Classen S, Tomkinson AE, Tainer JA, Ellenberger T (2006) A flexible interface between DNA ligase and PCNA supports conformational switching and efficient ligation of DNA. Mol Cell 24(2):279–291

Paulus H (2000) Protein splicing and related forms of protein autoprocessing. Annu Rev Biochem 69(1):447–496

Perler FB (2005) Protein splicing mechanisms and applications. IUBMB Life 57(7):469–476

Perler FB, Comb DG, Jack WE, Moran LS, Qiang B, Kucera RB, Benner J, Slatko BE, Nwankwo DO, Hempstead SK (1992) Intervening sequences in an archaea DNA polymerase gene. Proc Natl Acad Sci USA 89(12):5577–5581

Perler FB, Olsen GJ, Adam E (1997) Compilation and analysis of intein sequences. Nucleic Acids Res 25(6):1087–1093

Perlow-Poehnelt RA, Likhterov I, Scicchitano DA, Geacintov NE, Broyde S (2004) The spacious active site of a Y-family DNA polymerase facilitates promiscuous nucleotide incorporation opposite a bulky carcinogen-DNA adduct. J Biol Chem 279(35):36951–36961

Pisani FM, De FM, Manco G, Rossi M (1998) Domain organization and biochemical features of *Sulfolobus solfataricus* DNA polymerase. Extremophiles 2(3):171–177

Pisani FM, De FM, Carpentieri F, Rossi M (2000) Biochemical characterization of a clamp-loader complex homologous to eukaryotic replication factor C from the hyperthermophilic archaeon *Sulfolobus solfataricus*. J Mol Biol 301(1):61–73

Prakash S, Johnson RE, Prakash L (2005) Eukaryotic translesion synthesis DNA polymerases: specificity of structure and function. Annu Rev Biochem 74:317–353

Purohit V, Grindley NDF, Joyce CM (2003) Use of 2-aminopurine fluorescence to examine conformational changes during nucleotide incorporation by DNA polymerase I (klenow fragment). Biochemistry 42(34):10200–10211

Rechkoblit O, Malinina L, Cheng Y, Kuryavyi V, Broyde S, Geacintov NE, Patel DJ (2006) Stepwise translocation of Dpo4 polymerase during error-free bypass of an oxoG lesion. PLoS Biol 4(1):e11

Rechkoblit O, Kolbanovskiy A, Malinina L, Geacintov NE, Broyde S, Patel DJ (2010) Mechanism of error-free and semitargeted mutagenic bypass of an aromatic amine lesion by Y-family polymerase Dpo4. Nat Struct Mol Biol 17(3):379–388

Richardson TT, Wu X, Keith BJ, Heslop P, Jones AC, Connolly BA (2013) Unwinding of primer-templates by archaeal family-B DNA polymerases in response to template-strand uracil. Nucleic Acids Res 41(4):2466–2478

Robinson NP, Bell SD (2005) Origins of DNA replication in the three domains of life. FEBS J 272 (15):3757–3766

Rogozin IB, Makarova KS, Pavlov YI, Koonin EV (2008) A highly conserved family of inactivated archaeal B family DNA polymerases. Biol Direct 3:32

Rouillon C, Henneke G, Flament D, Querellou J, Raffin J-P (2007) DNA polymerase switching on homotrimeric PCNA at the replication fork of the euryarchaea *Pyrococcus abyssi*. J Mol Biol 369(2):343–355

Sakofsky CJ, Foster PL, Grogan DW (2012) Roles of the Y-family DNA polymerase Dbh in accurate replication of the *Sulfolobus* genome at high temperature. DNA Repair (Amst) 11(4):391–400

Sarmiento F, Mrazek J, Whitman WB (2013) Genome-scale analysis of gene function in the hydrogenotrophic methanogenic archaeon Methanococcus maripaludis. Proc Natl Acad Sci USA 110(12):4726–4731

Saves I, Ozanne V, Dietrich J, Masson JM (2000) Inteins of *Thermococcus fumicolans* DNA polymerase are endonucleases with distinct enzymatic behaviors. J Biol Chem 275(4): 2335–2341

Savino C, Federici L, Johnson KA, Vallone B, Nastopoulos V, Rossi M, Pisani FM, Tsernoglou D (2004) Insights into DNA replication: the crystal structure of DNA polymerase B1 from the archaeon *Sulfolobus solfataricus*. Structure 12(11):2001–2008

She Q, Singh RK, Confalonieri F, Zivanovic Y, Allard G, Awayez MJ, Chan-Weiher CC, Clausen IG, Curtis BA, De MA, Erauso G, Fletcher C, Gordon PM, Heikamp-de JI, Jeffries AC, Kozera CJ, Medina N, Peng X, Thi-Ngoc HP, Redder P, Schenk ME, Theriault C, Tolstrup N, Charlebois RL, Doolittle WF, Duguet M, Gaasterland T, Garrett RA, Ragan MA, Sensen CW, Van der OJ (2001) The complete genome of the crenarchaeon *Sulfolobus solfataricus* P2. Proc Natl Acad Sci USA 98(14):7835–7840

Shen Y, Tang XF, Yokoyama H, Matsui E, Matsui I (2004) A 21-amino acid peptide from the cysteine cluster II of the family D DNA polymerase from *Pyrococcus horikoshii* stimulates its nuclease activity which is Mre11-like and prefers manganese ion as the cofactor. Nucleic Acids Res 32(1):158–168

Sherrer SM, Brown JA, Pack LR, Jasti VP, Fowler JD, Basu AK, Suo Z (2009) Mechanistic studies of the bypass of a bulky single-base lesion catalyzed by a Y-family DNA polymerase. J Biol Chem 284(10):6379–6388

Sherrer SM, Maxwell BA, Pack LR, Fiala KA, Fowler JD, Zhang J, Suo Z (2012) Identification of an unfolding intermediate for a DNA lesion bypass polymerase. Chem Res Toxicol 25(7): 1531–1540

Strauss BS (1991) The 'A rule' of mutagen specificity: a consequence of DNA polymerase bypass of non-instructional lesions? Bioessays 13(2):79–84

Tahirov TH, Makarova KS, Rogozin IB, Pavlov YI, Koonin EV (2009) Evolution of DNA polymerases: an inactivated polymerase-exonuclease module in Pol ε and a chimeric origin of eukaryotic polymerases from two classes of archaeal ancestors. Biol Direct 4(1):11

Tori K, Kimizu M, Ishino S, Ishino Y (2007) DNA polymerases BI and D from the hyperthermophilic archaeon *Pyrococcus furiosus* both bind to proliferating cell nuclear antigen with their C-terminal PIP-box motifs. J Bacteriol 189(15):5652–5657

Truniger V, Lazaro JM, Salas M, Blanco L (1996) A DNA binding motif coordinating synthesis and degradation in proofreading DNA polymerases. EMBO J 15(13):3430–3441

Uemori T, Sato Y, Kato I, Doi H, Ishino Y (1997) A novel DNA polymerase in the hyperthermophilic archaeon, *Pyrococcus furiosus*: gene cloning, expression, and characterization. Genes Cells 2(8):499–512

Vaisman A, Ling H, Woodgate R, Yang W (2005) Fidelity of Dpo4: effect of metal ions, nucleotide selection and pyrophosphorolysis. EMBO J 24(17):2957–2967

Warbrick E (1998) PCNA binding through a conserved motif. Bioessays 20(3):195–199

Waters E, Hohn MJ, Ahel I, Graham DE, Adams MD, Barnstead M, Beeson KY, Bibbs L, Bolanos R, Keller M, Kretz K, Lin X, Mathur E, Ni J, Podar M, Richardson T, Sutton GG, Simon M, Söll D, Stetter KO, Short JM, Noordewier M (2003) The genome of *Nanoarchaeum equitans*: insights into early archaeal evolution and derived parasitism. Proc Natl Acad Sci USA 100(22):12984–12988

Wilson RC, Jackson MA, Pata JD (2013) Y-family polymerase conformation is a major determinant of fidelity and translesion specificity. Structure 21(1):20–31

Wong JH, Fiala KA, Suo Z, Ling H (2008) Snapshots of a Y-family DNA polymerase in replication: substrate-induced conformational transitions and implications for fidelity of Dpo4. J Mol Biol 379(2):317–330

Wu Y, Wilson RC, Pata JD (2011) The Y-family DNA polymerase Dpo4 uses a template slippage mechanism to create single-base deletions. J Bacteriol 193(10):2630–2636

Xing G, Kirouac K, Shin YJ, Bell SD, Ling H (2009) Structural insight into recruitment of translesion DNA polymerase Dpo4 to sliding clamp PCNA. Mol Microbiol 71(3):678–691

Yamasaki K, Urushibata Y, Yamasaki T, Arisaka F, Matsui I (2010) Solution structure of the N-terminal domain of the archaeal D-family DNA polymerase small subunit reveals evolutionary relationship to eukaryotic B-family polymerases. FEBS Lett 584(15):3370–3375

Yang J, Zhuang Z, Roccasecca RM, Trakselis MA, Benkovic SJ (2004) The dynamic processivity of the T4 DNA polymerase during replication. Proc Natl Acad Sci USA 101(22):8289–8294

Yutin N, Makarova KS, Mekhedov SL, Wolf YI, Koonin EV (2008) The deep archaeal roots of eukaryotes. Mol Biol Evol 25(8):1619–1630

Zang H, Irimia A, Choi J-Y, Angel KC, Loukachevitch LV, Egli M, Guengerich FP (2006) Efficient and high fidelity Incorporation of dCTP opposite 7,8-dihydro-8-oxodeoxyguanosine by *Sulfolobus solfataricus* DNA polymerase Dpo4. J Biol Chem 281(4):2358–2372

Zhang L, Brown JA, Newmister SA, Suo Z (2009) Polymerization fidelity of a replicative DNA polymerase from the hyperthermophilic archaeon *Sulfolobus solfataricus* P2. Biochemistry 48(31):7492–7501

Zuo Z, Lin HK, Trakselis MA (2011) Strand annealing and terminal transferase activities of a B-family DNA polymerase. Biochemistry 50(23):5379–5390

Chapter 7
Engineered DNA Polymerases

Roberto Laos, Ryan W. Shaw, and Steven A. Benner

Abstract This chapter reviews the methods used to generate variants of DNA polymerases that have improved ability, in particular to accept unnatural nucleotides, focusing especially on in vitro and directed evolution methods. Several natural families of DNA polymerases have independently evolved for millions of years to accept their natural nucleotide substrates with high fidelity and the ability to exclude closely related structures, such as ribonucleoside derivatives. However, polymerases that can accept unnatural nucleotide substrates would have many applications in biotechnology. Directed evolution may be an efficient method to produce new DNA polymerases capable to do so. Directed evolution relies on methods to create a library of sequence diverse polymerases starting with a gene for a parent polymerase. These methods are reviewed here, as well as examples of their application to produce variant polymerases. An evolutionary rationalization is offered to explain some mutations produced by directed evolution experiments.

Keywords Nucleic acids • DNA polymerases • AEGIS • PCR • CSR • Phage display • In vitro evolution

Abbreviations

AEGIS	Artificially expanded genetic information system
CSR	Compartmentalized self-replication
ddGTP	Dideoxy guanosine triphosphates
ddNTPs	Dideoxy nucleotides triphosphates
dNTPs	Deoxy nucleotides triphosphates
ePCR	Error-prone PCR

R. Laos (✉) • R.W. Shaw • S.A. Benner
Foundation for Applied Molecular Evolution, Gainesville, FL, USA
e-mail: rlaos@ffame.org

K.S. Murakami and M.A. Trakselis (eds.), *Nucleic Acid Polymerases*, Nucleic Acids and Molecular Biology 30, DOI 10.1007/978-3-642-39796-7_7,

PCR	Polymerase chain reaction
REAP	Reconstructed evolutionary adaptive paths
Sf	Stoffel fragment
Taq	*Thermus aquaticus*
Taq pol I	*Taq* DNA polymerase
TBD	Thioredoxin-binding domain

7.1 Introduction

DNA polymerases are enzymes responsible for catalyzing the template-directed synthesis of DNA. Over billions of years, they have evolved to have the speed, specificity, and accuracy required for them to transmit valuable genetic information from and to living organisms, with just enough infidelity to support Darwinian evolution. These properties allow polymerases to be used routinely in biotechnology. This use increased with the invention of the polymerase chain reaction (PCR) by Kary Mullis, who was awarded the Nobel Prize in chemistry in 1993 for his accomplishment.

PCR iterates a relatively simple cycle of strand-strand dissociation, primer annealing, and polymerase extension to produce, in an exponential amplification, millions of copies of product DNA from just a few molecules of starting nucleic acid (DNA or RNA). This technique was improved dramatically since it was first reported (Saiki et al. 1985). Initially, PCR was performed using *Escherichia coli* DNA polymerase I. However, this enzyme is unstable at the temperature used for duplex strand dissociation, typically 95 °C. This made it necessary to add extra polymerase for each cycle.

Three years later, the introduction of a homolog of DNA polymerase I from the thermophilic bacterium *Thermus aquaticus*, *Taq* polymerase (*Taq* pol I), was reported to overcome this limitation (Saiki et al. 1988). *T. aquaticus* grows naturally at 70 °C. Therefore, *Taq* pol I has evolved to remain active at high temperatures, and this thermostability allows it to survive the repeated heating steps required for PCR. The availability of *Taq* polymerase and of PCR thermal cyclers made PCR automated and routine.

Currently, many DNA polymerases are used to support PCR and other procedures that involve the copying of nucleic acids. These include multiplexed PCR, nested PCR, reverse transcription PCR, and DNA sequencing among others. Polymerases are also used to incorporate modified nucleotides, including those that tag, report, or signal the presence of product DNA. These allow nucleic acids to be amplified from complex samples, including blood, saliva, forensic traces, and fossil remains. The choice of a particular polymerase depends on the specific need, especially for processivity and fidelity, temperature of initiation, or ability to accept unnatural nucleotide analogs.

However, despite the large number of commercially available polymerases with their particular attributes, the need for additional polymerase variants, especially those with specialized attributes, shows no sign of diminishing. This is especially true as new architectures for amplifying and detecting nucleic acids emerge and with the growing interest in the use of unnatural nucleic acids.

7.2 Direct Design of New Polymerases with Novel Properties

The availability of high-quality crystal structures for many polymerases, as well as advanced computational tools, has encouraged many to alter the structure of natural polymerases by design. Two categories of direct design have achieved special importance.

7.2.1 Fully Guided Modifications to the Sequences of DNA Polymerases Themselves

The first efforts to modify a DNA polymerase to obtain variants better able to incorporate unnatural nucleotide substrates were motivated by the need to use dideoxynucleotides triphosphates (ddNTPs) in Sanger-type DNA sequencing. Native *Taq* pol I can incorporate ddNTPs, but at considerably lower rates than it incorporates natural deoxynucleotides triphosphates (dNTPs). In addition, this enzyme incorporates each ddNTP with a different rate, producing uneven DNA sequence signals.

Using a guided approach, Tabor and Richardson started with three DNA polymerases that belong to the same Family A (Braithwaite and Ito 1993). They noticed that bacteriophage T7 DNA polymerase incorporated ddNTPs better than the homologous polymerases from *E. coli* and *T. aquaticus*. Of course, the T7 polymerase has evolved to grow at 37 °C and does not have the thermostability required for PCR. However, Tabor and Richardson noticed that T7 polymerase had a tyrosine at a position (numbered 526) homologous to positions that held a phenylalanine in the *E. coli* and *Taq* polymerases (positions numbered 762 and 667, respectively). They hypothesized that this single amino acid difference was responsible for the different levels of discrimination against ddNTPs among the three polymerases.

Based on this hypothesis, Tabor and Richardson replaced the phenylalanine in the *Taq* polymerase by a tyrosine. The result was a variant *Taq* (F667Y) that retained the thermostability of the *Taq* parent but had improved ability to accept ddNTPs. Similar improvement was seen when the analogous replacement was made in the polymerase from *E. coli*. The mutant *Taq* (F667Y) became one of the first specialized polymerases used for DNA sequencing (Tabor and Richardson 1995).

Subsequently, Li et al. (1999) studied the crystal structures of Klentaq1, a derivative of *Taq* DNA polymerase that lacks an exonuclease domain. In separate structures, protein crystals binding ddNTPs were observed to have closed ternary complexes, where a conformational change upon substrate binding was associated with a large shift in the position of the side chain of residue 660 in the O helix. Comparing the open and closed structures with ddGTP, Li et al. concluded that the selective interaction of arginine 660 with the O6 and N7 atoms of the G nucleobase provided the structural grounds for better incorporation of ddGTP by *Taq* polymerase. Guided by these observations, Li et al. then replaced amino acids at residue 660 in Klentaq1 already holding the Tabor–Richardson (F667Y) replacement and then studied the resulting variants. Among the variants, the double mutant *Taq* (F667Y; R660D) showed superior performance in DNA sequencing architectures that used ddNTPs (Li et al. 1999).

7.2.2 Fused Polymerases for Increased Processivity

These engineering efforts have modified the sequence of the polymerase itself. However, a separate class of modifications appends polymerases to proteins that are not polymerases.

Most DNA polymerases responsible for large-scale DNA copying in their natural environments have evolved to use accessory proteins to enhance their processivity. These are, however, generally not used in biotechnology, because of the complexity of the assembled combination. Indeed, *Taq* polymerase and other enzymes are used by themselves for PCR because of their simplicity, a simplicity arising from their physiological roles in lagging strand replication and DNA repair. For this natural function, *Taq* need not be particularly processive compared to, for example, polymerase III, which carries out most leading strand DNA synthesis.

Thus, it is sensible to consider adding to simple polymerases one or two of the factors used in complex polymerases that makes them processive. This would create a simple system that is processive, with various advantages for in vitro DNA replication.

Following this rationale, Wang and collaborators covalently fused the double-stranded DNA-binding protein *Sso*7d from *Sulfolobus solfataricus* at the N-terminus of *Taq* polymerase (S-*Taq*) and to the fragment of *Taq* polymerase that results from the deletion of the first 289 amino acids which lacks the exonuclease domain (S-*Taq*(Δ289)). The average primer extension of *Taq*(Δ289) was increased from 2.9 to 51 nucleotides in S-*Taq*(Δ289). The full-length *Taq* polymerase which is intrinsically more processive than *Taq*(Δ289) improves its average primer extension from 22 (*Taq*) to 104 (S-*Taq*) nucleotides. Similarly they fused the *Sso*7d domain to the C-terminus of the polymerase from *Pyrococcus furiosus*, *Pfu* polymerase (*Pfu*-S). Just as in the case of *Taq* polymerase, the fusion of the *Sso*d7 domain leads to an increase of the average primer extension, from 22 nucleotides for *Pfu* to 104 for *Pfu*-S (Wang et al. 2004).

Another example of increased processivity of *Taq* polymerase uses the insertion of the T3 bacteriophage DNA polymerase thioredoxin-binding domain (TBD) in the thumb domain of *Taq* DNA polymerase, deleting amino acids 480–485 (Davidson et al. 2003). The rationale behind this modification is that T7 DNA polymerase forms a complex with *E. coli* thioredoxin. Upon binding, the processivity increases from 15 to 2,000 nucleotides as well as increasing the affinity to the primer/template by 80-fold. The TBD of T3 bacteriophage differs only in one amino acid from the TBD of T7 bacteriophage. The resulting polymerase remains thermostable, and its processivity is 20–50 times higher than the original *Taq* pol.

7.3 Protein Engineering Methods for Evolved DNA Polymerases

These successes of "design" in polymerase engineering are not frequent and for good reason. First, the design required crystal structures of enzymes that are closely homologous to the enzyme that starts the engineering. These crystal structures are useful to guide engineering only if they have bound substrates or substrate analogs. With the Tabor–Richardson analysis, the guidance also relied on the fortunate discovery of a homolog having the desired properties. This kind of information and understanding is not always available.

Further, the demands placed on a polymerase to incorporate unnatural triphosphates in Sanger sequencing procedures are much lower than the demands placed on a polymerase to repeatedly copy, as in PCR, DNA molecules containing unnatural nucleotides. In particular, the fidelity required by a DNA polymerase to support PCR with unnatural nucleotides must be very high. Further, the structural differences between a DNA polymerase that makes one error per thousand nucleotides and one error per million can be quite subtle, involving geometric differences that would not be necessarily distinguished even in a high-resolution crystal structure. Thus, the design of a high-fidelity DNA polymerase from a medium-fidelity polymerase is largely beyond current structure theory, making it impossible to get polymerases with the desired high-level behaviors from fully guided protein engineering.

As a consequence, many investigators use protein engineering to select for polymerases with certain properties improved with respect to a desired function, starting from libraries of polymerase variants. This approach, often called "directed evolution," is today considered by many to be the method of choice for protein engineering (Bornscheuer and Pohl 2001; Yuan et al. 2005; Leemhuis et al. 2009; Turner 2009) although other complementary methods also exist.

7.3.1 *Directed Evolution*

In a directed evolution experiment, a "starting" enzyme is first identified as having (at least) some of the properties desired in the enzyme that is desired as the end product. The gene of this parent enzyme is then altered to create a library encoding variant forms of the enzyme, some of which might be able to catalyze the desired transformation better than the parent enzyme. The members of the library are then not screened, a process whereby the members of the library are tested individually to find those having the desired properties. Rather, in directed evolution, the collection of variants is processed in bulk, where this "selection" generates extracts enriched in variants that better allowed them to survive. If the selection is properly constructed, the survivors will have mutations that confer the desired properties. This process is shown schematically in Fig. 7.1. With iteration, the process mimics natural evolution, except that the selective pressures applied come from the bioengineer, rather than nature.

7.3.1.1 Library Creation and Protein Sequence Space

The success of a directed evolution exercise depends on the success of two key steps (a) the generation of a collection of variants of the protein (called a library) that include members having the desired properties and (b) a selection strategy that allows survivors to actually have the properties that are desired in the end product. As we consider the first of these, it is useful to understand the magnitude of the challenge.

The size of the "protein sequence space" that might be explored by a directed evolution experiment is immense. As discussed by Smith (1970), the behavior of all possible proteins of length n with respect to a measurable behavior can be represented by a space in n dimensions, where each dimension can have 1 of 20 values, representing the 20 natural amino acids. Each protein sequence is represented by a point in that space. Two points are neighbors in that space if one can be converted into another by a single amino acid substitution. With 20 amino acids, each point in the sequence space has $19n$ neighbors. The measurable behavior is a real number displayed in the nth + 1 dimension.

Different sequences have different functions, and moving from a sequence having a function to another functional sequence can proceed via intermediates that either have or lack function. This is illustrated in Fig. 7.2 with a word game used by Smith, where functional protein sequences are analogous to strings of letters with a meaning in English. In Smith's (somewhat imperfect) analogy, the sequence of letters in the word "WORD" is converted to the sequence of letters in the word "GENE" by exchanging one letter at the time, with one path having a sequence of letters with a meaning (WORE, GORE, and GONE), illustrated by solid lines. Pathways that proceed via words lacking meaning are illustrated by broken lines (e.g., WOND, GOND, and GEND).

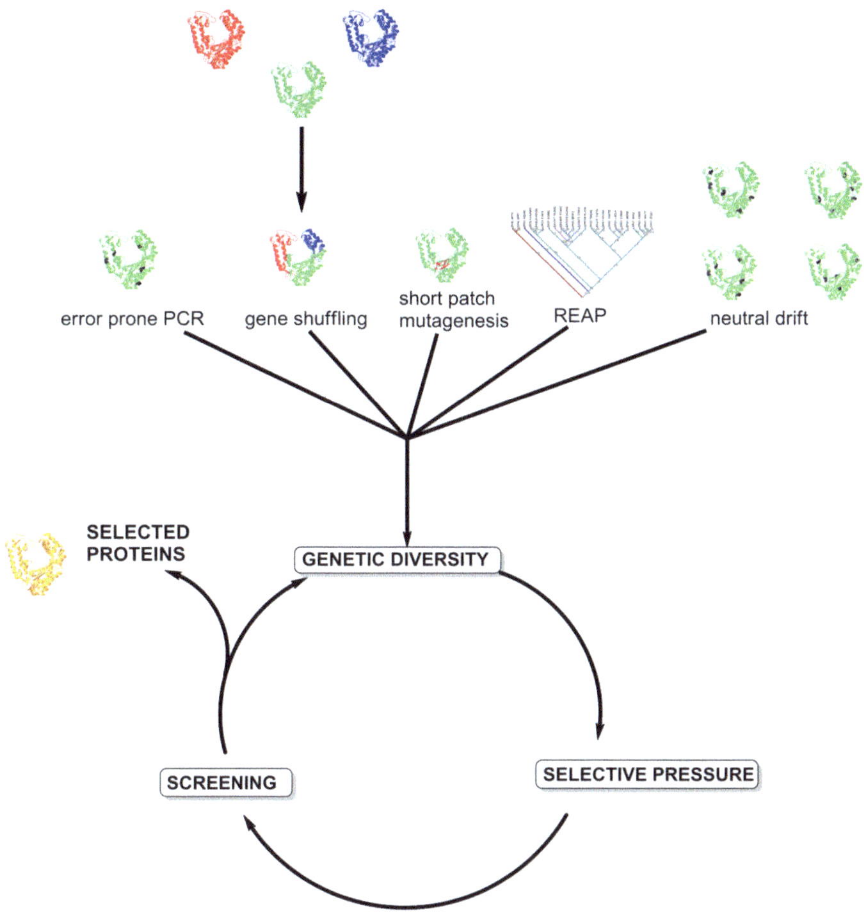

Fig. 7.1 Directed evolution experiments start with genetic diversity created by, for example, error-prone PCR (ePCR); gene shuffling; short-patch mutagenesis or design, including design methods that rely on evolutionary information such as reconstructed evolutionary adaptive paths (REAP); or neutral drift libraries. A selective pressure is imposed on the variants so that only polymerases with desired properties are enriched in a collection of polymerases, which may then be screened. This is most appropriate if the desired behavior is sparsely distributed within the library. Alternatively, genes for the variants that survived the selection can then be the starting points for processes that create new genetic diversity. Iteration of this process eventually leads to enzymes displaying properties produced by the selective pressures imposed on the experiments

In this example, meaningfulness is an analog of fitness, which provides the *n*th +1 dimension to the surface, a "fitness landscape" (Wright 1932). The landscape is portrayed like in a topographic map with peaks, marked with a (+) for optimal sequences. The absence of function is depicted as dips, marked with a (–). Smith proposed that natural evolution proceeds along paths only if all intermediates are functional. Nonfunctional sequences are removed by "purifying" selection. Thus,

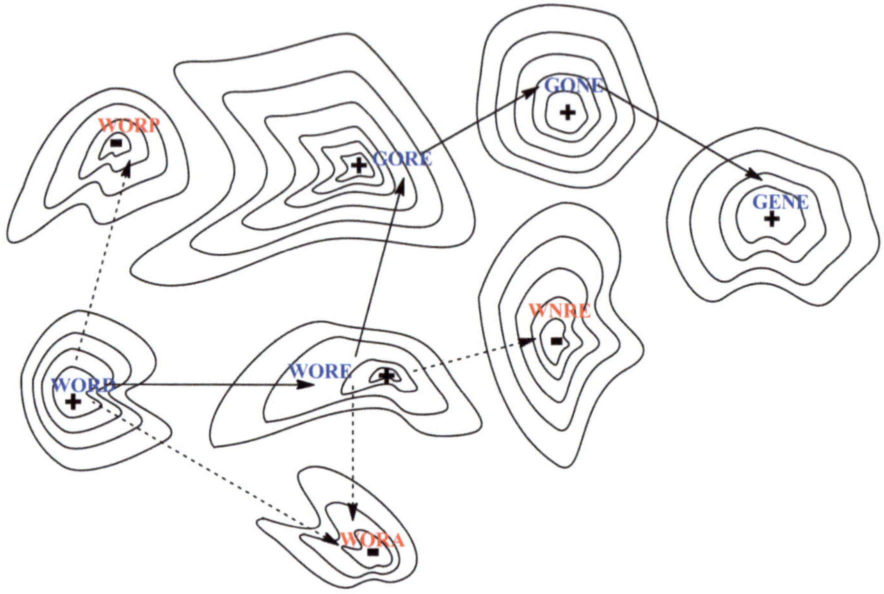

Fig. 7.2 Evolution can be modeled as a walk across a fitness landscape, here presented as a two-dimensional representation of a multiple dimensional hypersurface; analogous to a topographic map, peaks (*plus symbols*) indicate the locations where function exists, while dips (*minus symbols*) represent regions with lack of function. Illustrated through an analogy to a word game, a meaningful (functional) string of letters (here "word") must be reached starting from another string ("gene") via stepwise replacement of single letters, where every intermediate along the path must itself also be a functional word. *Solid arrows* indicate a path of accepted mutations, while *dashed arrows* illustrate deleterious mutations that produce nonfunctional proteins

the only valid pathways to explore a sequence space proceed via functional sequences, just as the evolution of words can proceed only via meaningful words.

Sequence space is vast to the point of being incomprehensible. For example, a 100-amino acid protein can be arranged in 20^{100} different ways. Typical polymerases, eight times longer, are found in a space with 20^{800} points. Both numbers are astronomical. No experiment can sample this space effectively, even if all the atoms of the known universe were available.

Accordingly, a search of sequence space can be successful only if one of three things is true: (a) the fitness landscape is "smooth," meaning that a single useful protein can be obtained starting with any sequence via a path that encounters only other functional proteins or, if not, then (b) useful functional proteins can be obtained no matter where one starts the search, as the surface is littered with many of them or, if not, then (c) the library is guided so as to start the search in a region of the astronomical enormity of sequence space where usefully functional proteins reside. Presumably, selecting a parent invariably does (c), at least in part.

7.3.1.2 Completely Unguided Library Creation by Random Variation

Cast in this way and given our ignorance of the "smoothness" of protein sequence spaces in general, it is clear that the directed evolutionist should give some thought to how a library might be generated. Given that selection of a parent polymerase is likely to start in a region of the fitness landscape that is already quite elevated, the simplest approach is to simply randomly mutate the gene of the parent polymerase. Various tools are available to do so. One of these exploits PCR, under conditions where the polymerase performing the copying makes more mutations than typical. Known as "mutagenic" or "error-prone" PCR (ePCR), this approach takes advantage of the inherent propensity of *Taq* polymerase to introduce mistakes into the copies of DNA under certain conditions. The frequency of mismatching is often increased by introducing manganese Mn^{2+} along with the natural cofactor Mg^{2+} (Vartanian et al. 1996). Other additives such as alcohols or unbalanced concentrations of nucleotides can also be used to introduce mutations through PCR.

Error-prone PCR produces mutations that are "unguided," in the sense used in this review. Nevertheless, they do not produce a truly random set of amino acid replacements, for a variety of reasons:

(i) *Taq* pol tends to replace purines (adenine and guanine) by purines and pyrimidines (thymidine and cytidine) by pyrimidines; these changes are called transitions. The exchanges of a purine by a pyrimidine and vice versa are called transversions. The tendency of the polymerase for transitions leads to produce libraries with amino acid replacements that are nonrandom with respect to the parent protein.
(ii) Even if ePCR introduced transitions and transversions entirely randomly, the resulting amino acid replacements would not be random. Most amino acids are encoded by more than one codon; most of them cannot be interconverted by a single-nucleotide change. Further, similar codons code for amino acids having similar chemical properties (Wong et al. 2007). For instance, the valine codon (GTN[1]) is converted by a single-nucleotide replacement to a phenylalanine (TTY[2]), leucine (CTN), isoleucine (ATN), aspartate (GAY), or glycine (GGN) codon. To gain access to other amino acids and, consequently, more dramatically altered chemical properties, two or three nucleotide replacements are required. For example, with valine, two nucleotides must be replaced to get a codon for proline (CCN). Typical ePCR introduces no more than four to six mutations per 1,000 nucleotides; higher mutation rates tend to disrupt protein function. Under these conditions it is very unlikely to introduce two or three consecutive mutations at the DNA level to fully explore amino acid sequence space.

[1] N is any nucleotide.
[2] Y is a pyrimidine.

Reetz et al. (2008) applied an approach where a subset of the 64 standard codons is introduced in specific sites by one or two degenerated codons: NNK and NDT. Here, D is adenine, guanine, or thymine, K is guanine or thymine, and N is any nucleobase. With the NNK degenerate codon, all 20 amino acids are covered by just 32 ($= 4 \times 4 \times 2$) of the 64 codons possible with standard nucleotides. With the 12 ($= 4 \times 3 \times 1$) codons covered by the NDT degenerate codon, only a sample of the standard amino acids can be encoded. This sample includes, however, a representative set of nonpolar, aromatic hydrophobic, hydrophilic, and charged amino acids.

Further thought generates further considerations about what is meant by "random" when discussing amino acid replacements. With 20 amino acids, a fully random sequence might have, on average, 5 % of each of the natural amino acids in bulk. However, this distribution is not found in natural proteins nor is it expected in proteins encoded by random genes. Because some amino acids (like serine, with six codons) are encoded by more codons than other amino acids (like tryptophan, with just one codon), a gene with a truly random nucleotide sequence would give proteins with a codon-weighed distribution of amino acids. Even this might not be the desired goal of an unguided approach to library generation, since some amino acids (aspartate and glutamate, e.g., each with two codons) appear in natural proteins more abundantly than is expected from their few codons. Thus, an "ideal" library might arguably be one where amino acids are replaced by a process that leaves the naturally observed overall composition of the protein unchanged. Finally, given our ignorance on the shape of function landscapes generally, as well as our ignorance of the local topography around any individual parent sequence, a perfectly random mutagenesis tool need not be the most useful approach for a directed evolution experiment.

7.3.2 Libraries Made by Gene Shuffling or Molecular Breeding

One obvious deficiency of random mutagenesis of a parent gene is that it fails to use all of the available information, information that is especially abundant in a post-genomic world. In particular, nature has already provided many homologs of a parent protein having many amino replacements relative to the parent sequences. Most of these homologs are functional, identifying other points in the vast sequence space that are elevated on the fitness landscape. It would be desirable to use the information that these homologs provide.

Gene shuffling was introduced by Stemmer (1994) more than a decade ago to directly use these homologs. Here, the starting point is a family of genes that share enough sequence similarity that they can be recombined. Using a modified PCR protocol, gene chimeras are produced. These are then often screened for recombinants that have the desired activity, although a collection of recombinants can also be the starting point for a selection step.

Those using shuffling in protein evolution assume, of course, that sequence space is more efficiently searched by combining the outcomes of two historically successful searches of a particular region of sequence space, than a search that simply replaces single amino acids starting from a single parent. These historical searches delivered the two functioning proteins whose genes are being shuffled. Here, the landscape is assumed to be such that specific paths between two elevated points are also similarly elevated.

This would be a more compelling hypothesis if natural evolution were observed to use shuffling. Natural evolution does, of course, have access to mechanisms that shuffle parts of genes. Natural evolution uses these mechanisms to rearrange, for example, the order of independently folded units in multiunit polypeptides. This is famously done in the evolution of multiunit proteins involved in metazoan signal transduction, where a regulatory protein might contain one "src homology domain 1" unit (SH1, a protein kinase), a few SH2 units, and a few SH3 units (Benner et al. 1993). Evolutionary analysis shows that these are all obtained by shuffling, implying that shuffling is an efficient way to search sequence space when no protein folding unit is disrupted.

However, natural evolution does not provide many examples where polypeptide chains within a *single* folded unit are shuffled. This is presumably because the buried contacts binding collections of secondary structural units are finely tuned to permit packing. Changing a single hydrophobic side chain in a packed protein fold often converts a core that is (typically) as densely packed as an organic crystal into a "molten globule." Thus, these biophysical realities would make it surprising to expect that shuffling explores sequence space more effectively than point mutation. Such expectations rely, of course, on the view that natural evolution exploits the most effective ways to search sequence space.

This notwithstanding, various individuals have used shuffling to generate polymerase libraries for directed evolution experiments. For example, d'Abbadie et al. (2007) shuffled the genes of the polymerases from three *Thermus* species (*aquaticus*, *thermophilus* and *flavus*) to generate libraries to start a directed evolution experiment to identify DNA polymerases that can extend single, double, and quadruple mismatches, process noncanonical primer-template duplexes, and bypass hydantoins and abasic sites (d'Abbadie et al. 2007). They applied these to PCR-amplify cave bear DNA from remains ca. 50,000 years old. These experiments showed that the polymerases obtained by directed evolution applied to these libraries outperformed Taq DNA polymerase, and were therefore better able to solve a biotechnological problem, here the sequencing of ancient damaged genomes.

In another example from the Holliger laboratory, Baar et al. (2011) shuffled the genes from eight different polymerase orthologs from the genus *Thermus* to generate libraries for their directed evolution system (Baar et al. 2011). Here, polymerases were sought that were not inhibited by various complex environmental inhibitors. For example, a polymerase resistant to inhibition by a broad spectrum of complex inhibitors (humic acid, bone dust, peat extract, clay-rich soil, cave sediment, and tar) was found that contained parts of the sequence of DNA polymerases from *Thermus aquaticus*, *Thermus oshimai*, *Thermus thermophilus*, and *Thermus brockianus*. Again, these polymerases might be useful in analyzing archaeological samples. However, as before, these experiments do not help us understand whether shuffling is a more effective way to search sequence space within compact single protein folds.

7.3.3 Evolution-Based Approaches to Search Sequence Space Around Parent Sequences

Alternative approaches now exist to create libraries that search sequence space around parent sequences (Lutz and Patrick 2004; Jackel et al. 2008; Lutz 2010). One class of these exploits the divergence of sequences in homologs of the parent enzyme. For example, Gaucher, Benner, and their coworkers introduced an approach, called the reconstructed evolutionary adaptive paths (REAP) approach, to create libraries that were hypothesized to more efficiently explore local sequence space (Chen et al. 2010). REAP begins with a phylogenetic analysis of homologous sequences, seeking signatures of functional divergence (Cole and Gaucher 2011). For example, an amino acid at a site may be entirely conserved in one branch of a phylogenetic tree, while not conserved at all in a second branch. This pattern of divergence, called *heterotachy*[3] (Lopez et al. 2002), indicates that the purifying selective pressures operating in the first branch at this site are different (and stronger) than those in the second branch. This, in turn, means that the function of the proteins within the first phylogenetic branch is different from the function in the second branch.

The variation observed in natural history was, of course, only rarely responsive to the specific adaptive changes desired by today's biotechnologist. Ancient polymerases, for example, were most likely *not* evolving to become, for example, resistant to heparin. The rationale is more subtle. The phylogenetic information used in a REAP analysis identifies sites that have been historically involved in *some* adaptive event. Because *some* changes are involved, the amino acid at the site cannot be *absolutely* required for core function. Conversely, the REAP-identified sites are not likely to be those whose amino acid *never* has a phenotypic impact. The rationale is that sites that have in the past been involved in an adaptive event without losing core function are sites that might be productively examined to identify sites that might adapt the protein to the *new*, biotechnologist-demanded function.

REAP is based on the hypothesis that the most productive sites to replace in a protein engineering experiment are *neither* sites whose amino acids contribute to a

[3] Different speed in Greek.

core function (as indicated by their absolute conservation) *nor* sites where the choice of amino acid is incidental to function (as indicated by their easy variability). By identifying sites where replacement *might* have phenotypic impact *without* destroying core function, REAP is proposed to have an advantage compared to other methods in the generation of libraries with productively altered behaviors. The advantage of the REAP approach relies on the fact that nature has already tested several amino acid sites and these modifications on these sites produce enzymes that retain the original activity. Searching for new variants in a REAP library gives the advantage of having several parent enzymes.

Chen et al. have applied REAP to generate DNA polymerases able to accept unnatural triphosphates modified on their sugar units. Using REAP, they identified 35 sites having heterotachous pattern of divergence, after filtering for sites where additional information from evolutionary history, structural biology, and experiments was exploited. They then asked which replacements improve the ability of *Taq* polymerase to accept reversible terminating triphosphates, where the 3′-OH unit of the nucleoside triphosphate had been replaced by an $-ONH_2$ unit, which prevents continued primer extensions. A single modification (L616A) appears to open space behind Phe-667, allowing the enzyme to accommodate a larger 3′-substituent (Chen et al. 2010).

7.3.4 Neutral Drift Libraries

REAP focuses the bioengineer on sites that have historically been involved in adaptive events. It does, perforce, exclude sites whose amino acids must be specific for core function. Deleterious mutations, those that diminish the fitness of the host by damaging core function, are expected to be removed by "purifying selection."

A conceptually simpler way to focus the bioengineering experiment considers only sites that are able to vary without impact on core function, variation that is often called "neutral." Neutral drift is traditionally defined as a natural process whereby genetic changes accumulate over geological time without having any impact on the fitness of the host organism. More recently, the concept has been expanded to include nucleotide substitutions that have perhaps only small impacts on fitness. It has been argued that most mutations that accumulate during natural history are neutral.

Here, the rationale holds that mutations that do not damage core function might nevertheless allow the enzyme to perform in a biotechnologically more useful way. Thus, a "neutral drift library" might be expected, as a starting point for a directed evolution experiment, to be already depleted in proteins lacking core function. Under this rationale, a bioengineer should start by creating a collection of variants that are all active under (approximately) natural conditions. This smaller but functional library might be more useful than a large randomized library where the large majority of the population lacks of function.

Neutral drift libraries have not been applied to DNA polymerases, but have been successfully used for other enzymes. For instance, Frances Arnold and her collaborators began an engineering experiment by selecting for active cytochrome P450 enzymes among a random library of enzyme variants. Screening the recovered mutants, they found enzymes able to catalyze oxidations of five other substrates different from the original (Bloom et al. 2007b). The use of neutral drift libraries has been further discussed by Tawfik and his coworkers (Gupta and Tawfik 2008; Bershtein et al. 2008). A particularly interesting example applied neutral drift libraries to create a beta-galactosidase from a beta-glucuronidase via two different evolutionary pathways (Smith et al. 2011).

7.4 Directed Evolution Methods Involving Genotype–Phenotype Linkage

The variants generated by shuffling, REAP, and other approaches can be (and have been) the starting points for simple screens, where the behaviors of individual proteins are examined one at a time. This was done in the early papers of Stemmer and in the work with REAP on DNA polymerases.

Screening variants requires a certain level of confidence that the approach used to vary the parent has a high probability of producing a variant that has the desired properties. If that confidence is low, or if the desired behavior is only sparsely distributed within a library of variants, screening is too slow to identify variants with the desired activities. Rather, processes are needed to be applied to all members of a large library that will generate extracts enriched in variants that have those activities. Further, if the first library is unable to deliver variants with improved activities, but not the desired levels of activities, it would be desirable to have the first extract serve as the library for subsequent selections.

In such systems, the protein must remain linked to the gene that encodes it. Further, the protein cannot have an opportunity to become linked with a gene that encodes an inferior enzyme. This linking can be done if, for example, the gene is within a bacterium whose fitness is enhanced by an enzyme that catalyzes the desired transformations. These strategies for linking genotype and phenotype are, however, problematic, as the selection often ends up being focused not on the enzyme of interest, but rather on some of the many other enzymes that are important to the survival of the host bacterium.

Accordingly, a variety of techniques have been developed to provide a physical genotype–phenotype linkage in the laboratory without requiring a complex host taken from natural biology. Two techniques used for polymerase directed evolution are compartmentalized self-replication (CSR) and phage display.

7.4.1 Compartmentalized Self-Replication

Compartmentalized self-replication (CSR) is a method that has been frequently used to link genotype and phenotype in polymerase development. Developed first by Tawfik and Griffiths, CSR holds proteins and genes together in water droplets suspended in oil emulsions. These generally receive the gene–protein pair from a single *E. coli* cell that is encapsulated within individual droplets (Tawfik and Griffiths 1998).

CSR was first adapted a decade ago for the directed evolution of DNA polymerases in the Holliger laboratory (Ghadessy et al. 2001). A library of different polymerase genes was delivered in plasmids to create clones in *E. coli* cells. These cells were dispersed into emulsified water droplets containing the primers and buffers needed to perform a PCR amplification of the polymerase gene. Approximately $\sim 10^8$–10^9 compartments are formed per milliliter of emulsion; ideally each compartment contains a single variant. PCR cycling was then performed, with the first heat step lysing the *E. coli* cell to present its expressed thermostable polymerase and its encoding plasmids to the primers. Polymerases that functioned under the conditions imposed by the experiment were able to make copies of their own genes (Fig. 7.3). After 20 rounds of PCR, the emulsions were broken to give a pool of PCR products enriched in the genes that encoded the polymerase variants able to replicate their own genes. These genes could be directly used, or be introduced in cells for another round of selection.

The Holliger laboratory applied CSR to develop polymerases that remained active in the presence of heparin. Further, as noted above, they applied CSR to polymerase libraries generated by shuffling to avoid inhibition by materials contained in samples of archaeological DNA. In other examples, CSR was applied to obtain DNA polymerases that accept unnatural sugars (Pinheiro et al. 2012). A summary of these and other results is in Table 7.1.

The Benner laboratory has also attempted to use CSR to copy DNA that is constructed from artificially expanded genetic information systems (AEGIS) (Piccirilli et al. 1990; Geyer et al. 2003). AEGIS is a species of DNA having additional nucleobases that present alternative hydrogen bonding patterns, allowing its members to form mutually exclusive pairing schemes. Some members of AEGIS are used in the clinic to monitor the viral load of patients infected with human immunodeficiency virus and hepatitis C virus (Collins et al. 1997).

While many natural polymerases are able to synthesize duplex DNA containing AEGIS nucleobase pairs, the efficiency of the synthesis is typically less than with completely natural DNA. This efficiency includes pausing at sites containing AEGIS nucleotides. This might be expected, given two unnatural features of the AEGIS pyrimidine components:

(a) Some are C-glycosides, having a nucleobase joined to the carbohydrate ring via a C–C bond, rather than the C–N bond joining natural pyrimidine nucleoside. This is expected to distort the conformation of the carbohydrate ring, a distortion that the polymerase might reject as "unnatural."

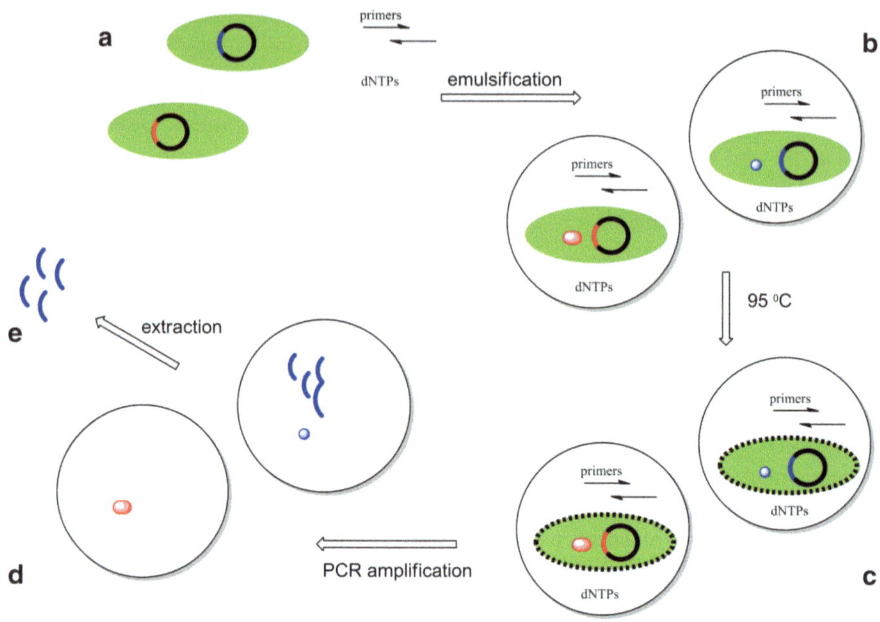

Fig. 7.3 Compartmentalized self-replication system (CSR) experiments start with the creation of a library of genes encoding variants of a polymerase. Members of this library are introduced into *E. coli* cells by electroporation. Here, just two variant genes (*red* and *blue*) are represented. These genes drive the expression of mutant polymerases in each *E. coli* cell, each of which is isolated in its own water-in-oil-emulsion droplet. (**b**) The first cycle of PCR breaks the cell wall of the *E. coli*, exposing the expressed polymerase molecules and their gene to the contents of a water droplet containing all of the necessary components necessary for a PCR amplification (1) primers, (2) dNTPs, (3) a mutated gene of the polymerase, and (4) the enzyme expressed by this gene (**c**). During PCR, any polymerases active under the selective pressure (*blue*) amplify their respective genes, enriching the pool of mutants having the desired properties; inactive polymerases (*red*) fail to do so (**d**). The emulsion is then broken and the amplified genes enriched in those encoding polymerases having the desired behaviors are extracted and inserted in a plasmid vector (circular DNA) (**e**). These then enter the cycle of selection again (**a**). After repeating these cycles, an enriched pool of variants of the original gene are produced

(b) Some lack electron density in the minor groove. This is believed to be a key element for recognition by polymerases (Joyce and Steitz 1994; Morales and Kool 1999).

We applied the water–oil emulsions to select for polymerases that copied a pair (the Z:P pair) (Fig. 7.4) where both components present electron density to the minor groove, but where one component (Z) is a C-glycoside. The library was generated by ePCR and short-patch saturated mutagenesis of the O helix. Two polymerase variants emerged that had particularly interesting properties, including diminished stalling at the point where the Z:P pair is synthesized in in vitro copying. Even more interestingly, several of the amino acid replacements were at sites

Table 7.1 Some DNA polymerases developed for specific purposes

Target/mutation	Pol	Nucleotide or analog(s)	Characteristics	Comments	Reference
F667T	Taq	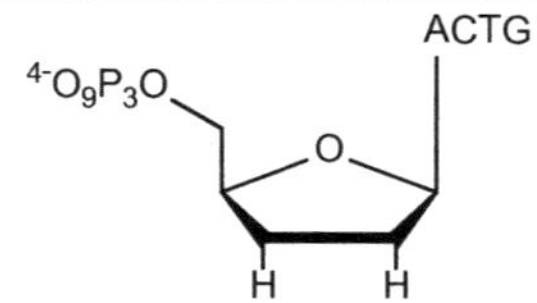	Increased rates of dideoxynucleoside triphosphate (ddNTP) incorporation	Obtained by rational design	Tabor and Richardson (1995)
F667T R660D	Taq	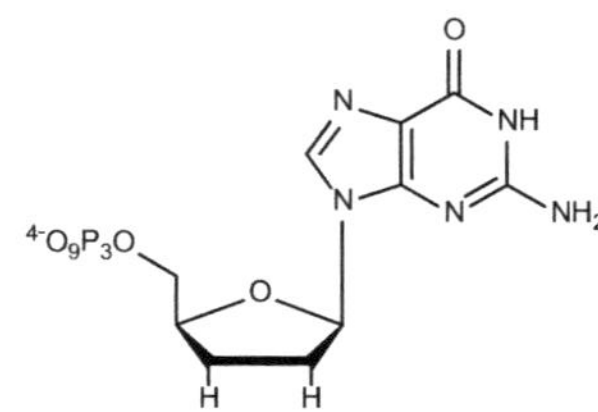	Incorporates ddGTP at rates similar to the other dideoxynucleotides	Obtained by rational design	Li et al. (1999)
R660S	Taq	Standard	Reduces T → C transitions		Yoshida et al. (2001)
T664P	Taq	Standard	Works like WT for correct nucleotides, enlarges catalytic pocket for misincorporation		Tosaka et al. (2001)
E531K A597T A600T W604G A608S L609V I614T E615G	Stoffel fragment (Δ289 *Taq*)	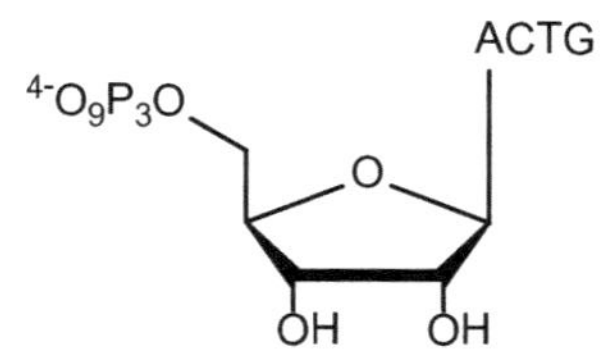	Incorporate ribonucleoside triphosphates	Obtained by phage display	Xia et al. (2002)

(continued)

Table 7.1 (continued)

Target/mutation	Pol	Nucleotide or analog(s)	Characteristics	Comments	Reference
P2: Q489H F598I I614F	Stoffel fragment (Δ289 *Taq*)	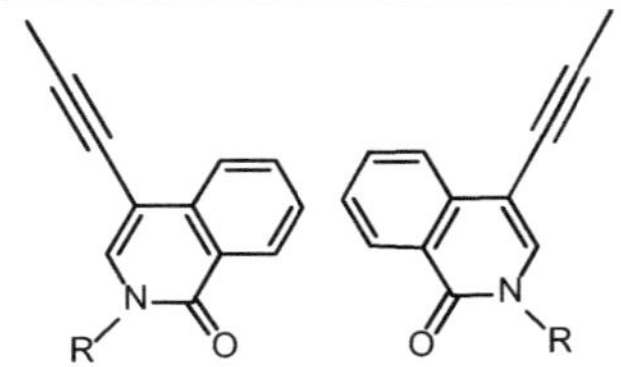	Incorporates hydrophobic nucleobases analog propynylisocarbostyril (PICS) self-pair		Leconte et al. (2005)
V618I L619M V631A I638V T640K M646V M658L A661I T664A I665V L670M F700Y T756S A757G	Taq	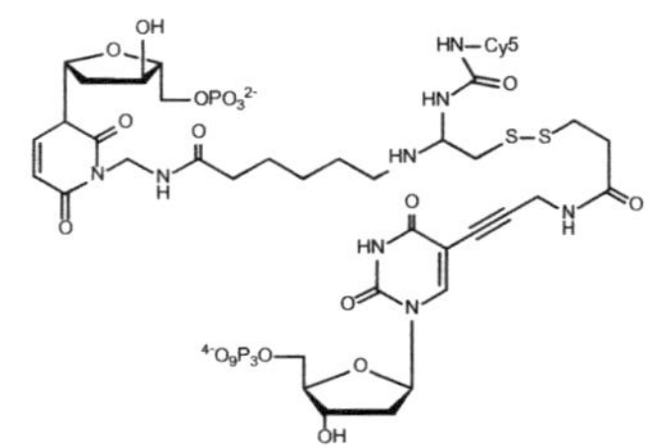	Incorporates deoxynucleotides triphosphates-bearing fluorophore attached through cleavable linkers		Leconte et al. (2010)
V93Q V337I E399D N400D R407I Y546H Chimeric polymerases	Pfu-exo-	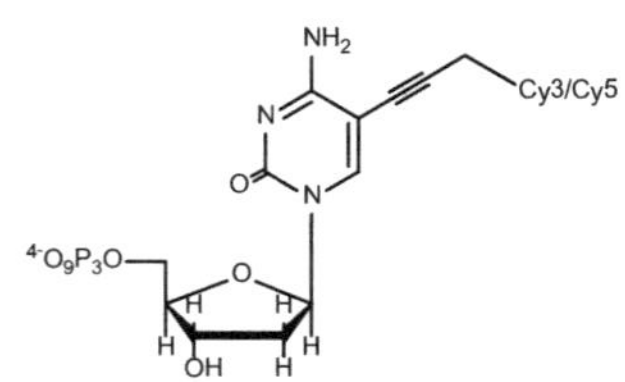	Incorporates Cy3-dCTP and Cy5-dCTP	Obtained by CSR	Ramsay et al. (2010)
N577S L625V E829V	Stoffel fragment (Δ289 *Taq*)	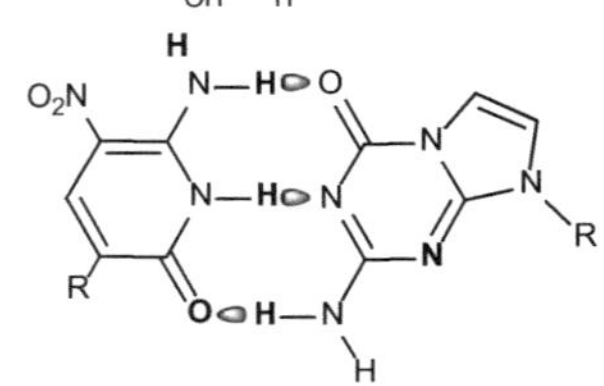	Incorporation of a third nucleobases pair with alternative hydrogen bonding pattern	CSR	Laos R. (unpublished results)

A597T E615G	Stoffel fragment (Δ289 *Taq*)	Ribonucleotides	DNA polymerase mutated into an efficient RNA polymerase	Obtained by phage display	Xia et al. (2002)
A597T W604R L605Q I614T E615G	Stoffel fragment (Δ289 *Taq*)	Ribonucleotides	DNA polymerase mutated into an efficient RNA polymerase	Obtained by phage display	Xia et al. (2002)
N/A	Taq-T3 chimera	Standard	Converts *Taq* DNA pol to a highly processive enzyme	Fused polymerase	Davidson et al. (2003)
L33P E76K D145G P552S E775G M777T	*Thermus thermophilus* (*Tth*)–*Thermus aquaticus* chimera (*Tth*–*Taq*)	Standard	Capable of bypassing DNA lesions	Obtained by CSR	d'Abbadie et al. (2007)
E76K E91Q D145G R336Q A448T I616M V739M E744G	*Tth*–*Taq* chimera	Standard	Capable of bypassing DNA lesions	Obtained by CSR	d'Abbadie et al. (2007)
L616A	*Taq* pol	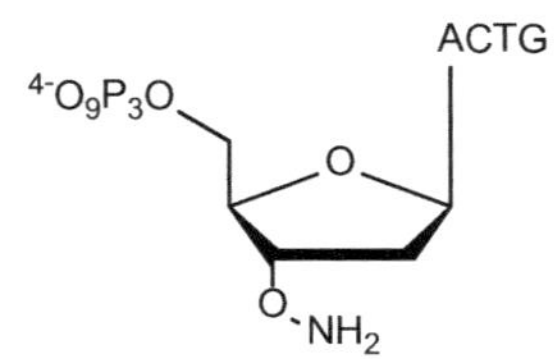	Incorporates reversible terminators having 3′ hydroxyl blocked with a $-ONH_2$ moiety	REAP	Chen et al. (2010)
(81 mutations from the Taq consensus)	Chimera of: *Taq* (1–109), *Tos* (110–388), *Tth* (389–456), *Tbr* (457–471), *Taq* (472–834)	Standard	Resistance to environmental inhibitors: clay-rich soil; tar-impregnated bone material	Obtained by CSR	Baar et al. (2011)
L322M L459M S515R I638F S739G L789F E773G	*Taq*	Standard	Increased reverse transcriptase activity	Randomized library and screening	Kranaster et al. (2010) Sauter and Marx (2006)

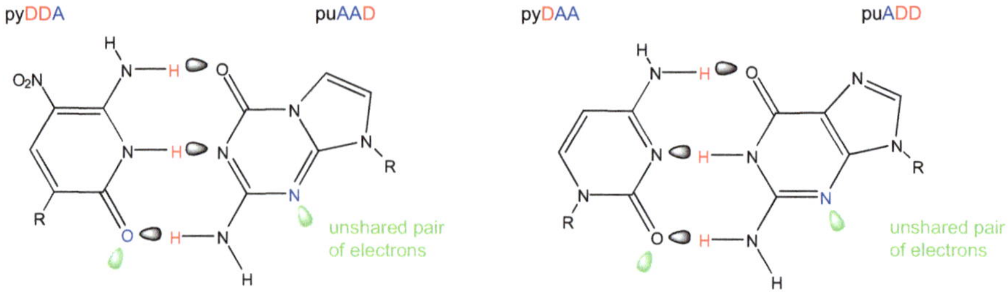

Fig. 7.4 The Z:P pair from an artificially expanded genetic information system (AEGIS, *left*) showing in *green* orbitals containing the unshared electron pairs presented to the minor groove of DNA. The natural C:G pair (*right*) showing these unshared pairs of electrons in the standard C:G pair. The pattern of hydrogen bonding is indicated A, hydrogen bond acceptor, and D, hydrogen bond donor

previously identified as heterotachous by the REAP method on *Taq* polymerase, even though the starting point was a randomized library (Laos et al. 2013).

7.4.2 Phage Display

An alternative way to link a polymerase and its encoding gene through a selection step is to join them in a single viral particle. In this technique, the protein of interest is co-expressed in the coat of a virus, keeping the genotype linked with phenotype. The Romesberg laboratory has been especially active in generating useful polymerases using this technique. They have evolved an RNA polymerase from a DNA polymerase (Xia et al. 2002), found polymerases having an improved ability to incorporate the self-pairing hydrophobic nucleobases analog propynylisocarbostyril (PICS) (Leconte et al. 2005), and more recently found polymerases having an improved ability to incorporate modified dUTP with a fluorophore (dUTP-Fl) that can be used for sequencing by synthesis. Leconte and his coworkers generated a library of Stoffel fragment (Sf) which is *Taq* DNA polymerase minus the first 289 amino acids. This fragment conserves the polymerase activity but lacks the exonuclease domain. The library was done by shuffling the genes of six homologous polymerases: *Thermus aquaticus*, *Thermus thermophilis*, *Thermus caldophilus*, *Thermus filiformis*, *Spirochaeta thermophila, and Thermomicrobium roseum*. The three most active polymerase mutants were as follows: **Sf168** (F598I, I614F, V618I, L619M, L622F, I638V, T640A, A643G, M646V, A661T, T664V, I665V, L670M, A691V, F700Y, F749A, V753I, A757G, Q782H), **Sf197** (V618I, L619M, V631A, I638V, T640K, M646V, M658L, A661I, T664A, I665V, L670M, F700Y, T756S, A757G), and **Sf267** (I614F, L619M, L622F, T640E, M658K, T664A, M751T, V753I, T756S, A757G, L760I). These mutants had between 10- and 50-fold increase in efficiency for dUTP-Fl incorporation compared with

Table 7.2 Some heterotachous sites in *Taq* polymerase reported in the literature from directed evolution experiments

Mutation	Method	Comments on site	Reference
D144G	CSR	Viral clade presents only G	Ghadessy et al. (2004)
A597T	Phage display	Viral clade presents only T	Xia et al. (2002)
F598L	CSR	Viral clade presents only L	Ghadessy et al. (2004)
A600T	Phage display	T is sparsely present in both clades. Viral clade presents also contain V, S	Xia et al. (2002)
I614K	Genetic complementation and screening	Viral clade presents L, Q, I, G, E; nonviral has I	Patel et al. (2001)
I614N	Genetic complementation and screening	Viral clade presents L, Q, I, G, E; nonviral has I	Patel et al. (2001)
I614T	Phage display	Viral clade presents L, Q, I, G, E; nonviral has I	Xia et al. (2002)
I614E	Phage display	Viral clade presents L, Q, I, G, E; nonviral has I	Fa et al. (2004)
E615G	Phage display	Viral clade presents E, D, Q, I, V, T; non-viral has E	Xia et al. (2002)
L616I	Genetic complementation and screening	Viral clade presents I, P, A, V, D, Q; non-viral has L and few M	Patel et al. (2001)

These amino acid positions have been identified by the REAP method as heterotachous sites

wild-type Sf (Leconte et al. 2010). Therefore, phage display seems to be an alternative for directed evolution experiments.

7.5 Outlook

The literature on DNA polymerase engineering includes examples representative of a range of strategies and tactics used more widely to generate enzymes with unnatural behaviors, including the ability to act on unnatural substrates. The demand for engineered polymerases is certain to grow, as many scientists seek to copy DNA constructed from alternative genetic alphabets (Geyer et al. 2003), highly tagged substrates (Hollenstein et al. 2009), modified backbones (Pinheiro et al. 2012), and other unusual structures (Hirao et al. 2007; Fa et al. 2004; Leconte et al. 2005). Further, we (and many others) are seeking to develop living systems that implement a "synthetic biology" based on unnatural DNA analogs. These have the potential for being "biosafe" platforms for artificial metabolisms, fermentations, diagnostics, and therapeutic tools, inter alia (Schmidt 2010).

The literature teaches that in some cases, simple downstream screening can obtain polymerases with the needed properties. This is illustrated in particular with efforts by Tabor and Richardson (1995) and Chen et al. (2010) to create polymerases that accept various 3′-terminating groups. Their combination of structural biology and evolutionary biology analyses was sufficiently powerful to ensure that regions of sequence space small enough to be screened contained polymerases having the desired properties. In each case, screening began with rich sets of variants extracted from the sequence space around a deftly chosen parent, allowing a relatively simple one-at-a-time inspection to get useful enzymes.

For less effectively constructed libraries of variants, including those generated by shuffling and by undirected mutagenesis, various selection tools stand to pick up where screening cannot possibly go. Here, CSR and phage display have been especially useful. These have yielded polymerases that are able to read through highly damaged DNA that might be extracted from forensic and archaeological samples, as well as polymerases that support the copying of entirely different genetic systems. Other approaches are possible. For example, neutral drift libraries (Amitai et al. 2007; Bloom et al. 2007a, b) have yet to be applied as starting points for directed evolution experiments targeted against polymerases. These have, however, been found to be useful for other enzyme systems.

Acknowledgements This work was supported by the US National Human Genome Research Institute (R01HG004647) and the Defense Threat Reduction Agency (through its program in basic research, HDTRA1-08-1-0052).

References

Amitai G, Gupta RD, Tawfik DS (2007) Latent evolutionary potentials under the neutral mutational drift of an enzyme. HFSP J 1(1):67–78. doi:10.2976/1.2739115

Baar C, d'Abbadie M, Vaisman A, Arana ME, Hofreiter M, Woodgate R, Kunkel TA, Holliger P (2011) Molecular breeding of polymerases for resistance to environmental inhibitors. Nucleic Acids Res 39(8):e51. doi:10.1093/nar/gkq1360

Benner SA, Cohen MA, Gerloff D (1993) Predicted secondary structure for the SRC homology-3 domain. J Mol Biol 229(2):295–305. doi:10.1006/jmbi.1993.1035

Bershtein S, Goldin K, Tawfik DS (2008) Intense neutral drifts yield robust and evolvable consensus proteins. J Mol Biol 379(5):1029–1044. doi:10.1016/j.jmb.2008.04.024

Bloom JD, Lu Z, Chen D, Raval A, Venturelli OS, Arnold FH (2007a) Evolution favors protein mutational robustness in sufficiently large populations. BMC Biol 5:29. doi:10.1186/1741-7007-5-29

Bloom JD, Romero PA, Lu Z, Arnold FH (2007b) Neutral genetic drift can alter promiscuous protein functions, potentially aiding functional evolution. Biol Direct 2:17. doi:10.1186/1745-6150-2-17

Bornscheuer UT, Pohl M (2001) Improved biocatalysts by directed evolution and rational protein design. Curr Opin Chem Biol 5(2):137–143. doi:10.1016/s1367-5931(00)00182-4

Braithwaite DK, Ito J (1993) Compilation, alignment, and phylogenetic-relationships of DNA-polymerases. Nucleic Acids Res 21(4):787–802. doi:10.1093/nar/21.4.787

Chen F, Gaucher EA, Leal NA, Hutter D, Havemann SA, Govindarajan S, Ortlund EA, Benner SA (2010) Reconstructed evolutionary adaptive paths give polymerases accepting reversible terminators for sequencing and SNP detection. Proc Natl Acad Sci USA 107(5):1948–1953. doi:10.1073/pnas.0908463107

Cole MF, Gaucher EA (2011) Exploiting models of molecular evolution to efficiently direct protein engineering. J Mol Evol 72(2):193–203. doi:10.1007/s00239-010-9415-2

Collins ML, Irvine B, Tyner D, Fine E, Zayati C, Chang CA, Horn T, Ahle D, Detmer J, Shen LP, Kolberg J, Bushnell S, Urdea MS, Ho DD (1997) A branched DNA signal amplification assay for quantification of nucleic acid targets below 100 molecules/ml. Nucleic Acids Res 25 (15):2979–2984

d'Abbadie M, Hofreiter M, Vaisman A, Loakes D, Gasparutto D, Cadet J, Woodgate R, Paeaebo S, Holliger P (2007) Molecular breeding of polymerases for amplification of ancient DNA. Nat Biotechnol 25(8):939–943. doi:10.1038/nbt1321

Davidson JF, Fox R, Harris DD, Lyons-Abbott S, Loeb LA (2003) Insertion of the T3 DNA polymerase thioredoxin binding domain enhances the processivity and fidelity of Taq DNA polymerase. Nucleic Acids Res 31(16):4702–4709. doi:10.1093/nar/gkg667

Fa M, Radeghieri A, Henry AA, Romesberg FE (2004) Expanding the substrate repertoire of a DNA polymerase by directed evolution. J Am Chem Soc 126(6):1748–1754. doi:10.1021/ja038525p

Geyer CR, Battersby TR, Benner SA (2003) Nucleobase pairing in Watson-Crick-like genetic expanded information systems. Structure 11(12):1485–1498. doi:10.1016/j.str.2003.11.008

Ghadessy FJ, Ong JL, Holliger P (2001) Directed evolution of polymerase function by compartmentalized self-replication. Proc Natl Acad Sci USA 98(8):4552–4557. doi:10.1073/pnas.071052198

Ghadessy FJ, Ramsay N, Boudsocq F, Loakes D, Brown A, Iwai S, Vaisman A, Woodgate R, Holliger P (2004) Generic expansion of the substrate spectrum of a DNA polymerase by directed evolution. Nat Biotechnol 22(6):755–759, http://www.nature.com/nbt/journal/v22/n6/suppinfo/nbt974_S1.html

Gupta RD, Tawfik DS (2008) Directed enzyme evolution via small and effective neutral drift libraries. Nat Methods 5(11):939–942. doi:10.1038/nmeth.1262

Hirao I, Mitsui T, Kimoto M, Yokoyama S (2007) An efficient unnatural base pair for PCR amplification. J Am Chem Soc 129(50):15549–15555. doi:10.1021/ja073830m

Hollenstein M, Hipolito CJ, Lam CH, Perrin DM (2009) A DNAzyme with three protein-like functional groups: enhancing catalytic efficiency of M2+-independent RNA cleavage. Chembiochem 10(12):1988–1992. doi:10.1002/cbic.200900314

Jackel C, Kast P, Hilvert D (2008) Protein design by directed evolution. Annu Rev Biophys 37:153–173. doi:10.1146/annurev.biophys.37.032807.125832

Joyce CM, Steitz TA (1994) Function and structure relationships in DNA-polymerases. Annu Rev Biochem 63:777–822. doi:10.1146/annurev.biochem.63.1.777

Kranaster R, Drum M, Engel N, Weidmann M, Hufert FT, Marx A (2010) One-step RNA pathogen detection with reverse transcriptase activity of a mutated thermostable Thermus aquaticus DNA polymerase. Biotechnol J 5(2):224–231. doi:10.1002/biot.200900200

Laos R, Shaw RW, Leal NA, Gaucher EA, Benner SA (2013) Directed evolution of polymerases to accept nucleotides with nonstandard hydrogen bond patterns. Biochemistry 52(31):5288. doi:10.1021\bi40558c

Leconte AM, Chen L, Romesberg FE (2005) Polymerase evolution: efforts toward expansion of the genetic code. J Am Chem Soc 127(36):12470–12471. doi:10.1021/ja053322h

Leconte AM, Patel MP, Sass LE, McInerney P, Jarosz M, Kung L, Bowers JL, Buzby PR, Efcavitch JW, Romesberg FE (2010) Directed evolution of DNA polymerases for next-generation sequencing. Angew Chem Int Ed Engl 49(34):5921–5924. doi:10.1002/anie.201001607

Leemhuis H, Kelly RM, Dijkhuizen L (2009) Directed evolution of enzymes: library screening strategies. IUBMB Life 61(3):222–228. doi:10.1002/iub.165

Li Y, Mitaxov V, Waksman G (1999) Structure-based design of Taq DNA polymerases with improved properties of dideoxynucleotide incorporation. Proc Natl Acad Sci USA 96 (17):9491–9496. doi:10.1073/pnas.96.17.9491

Lopez P, Casane D, Philippe H (2002) Heterotachy, an important process of protein evolution. Mol Biol Evol 19(1):1–7

Lutz S (2010) Beyond directed evolution-semi-rational protein engineering and design. Curr Opin Biotechnol 21(6):734–743. doi:10.1016/j.copbio.2010.08.011

Lutz S, Patrick WM (2004) Novel methods for directed evolution of enzymes: quality, not quantity. Curr Opin Biotechnol 15(4):291–297. doi:10.1016/j.copbio.2004.05.004

Morales JC, Kool ET (1999) Minor groove interactions between polymerase and DNA: more essential to replication than Watson-Crick hydrogen bonds? J Am Chem Soc 121 (10):2323–2324. doi:10.1021/ja983502+

Patel PH, Kawate H, Adman E, Ashbach M, Loeb LA (2001) A single highly mutable catalytic site amino acid is critical for DNA polymerase fidelity. J Biol Chem 276(7):5044–5051. doi:10.1074/jbc.M008701200

Piccirilli JA, Krauch T, Moroney SE, Benner SA (1990) Enzymatic incorporation of a new base pair into DNA and RNA extends the genetic alphabet. Nature 343(6253):33–37

Pinheiro VB, Taylor AI, Cozens C, Abramov M, Renders M, Zhang S, Chaput JC, Wengel J, Peak-Chew S-Y, McLaughlin SH, Herdewijn P, Holliger P (2012) Synthetic genetic polymers capable of heredity and evolution. Science 336(6079):341–344. doi:10.1126/science.1217622

Ramsay N, Jemth A-S, Brown A, Crampton N, Dear P, Holliger P (2010) CyDNA: synthesis and replication of highly Cy-Dye substituted DNA by an evolved polymerase. J Am Chem Soc 132 (14):5096–5104. doi:10.1021/ja909180c

Reetz MT, Kahakeaw D, Lohmer R (2008) Addressing the numbers problem in directed evolution. Chembiochem 9(11):1797–1804. doi:10.1002/cbic.200800298

Saiki RK, Scharf S, Faloona F, Mullis KB, Horn GT, Erlich HA, Arnheim N (1985) Enzymatic amplification of b-globin genomic sequences and restriction site analysis for diagnosis of sickle cell anemia. Science 230(4732):1350–1354. doi:10.1126/science.2999980

Saiki RK, Gelfand DH, Stoffel S, Scharf SJ, Higuchi R, Horn GT, Mullis KB, Erlich HA (1988) Primer-directed enzymatic amplification of DNA with a thermostable DNA-polymerase. Science 239(4839):487–491. doi:10.1126/science.2448875

Sauter KBM, Marx A (2006) Evolving thermostable reverse transcriptase activity in a DNA polymerase scaffold. Angew Chem Int Ed Engl 45(45):7633–7635. doi:10.1002/anie.200602772

Schmidt M (2010) Xenobiology: a new form of life as the ultimate biosafety tool. Bioessays 32 (4):322–331. doi:10.1002/bies.200900147

Smith JM (1970) Natural selection and concept of a protein space. Nature 225(5232):563–564

Smith WS, Hale JR, Neylon C (2011) Applying neutral drift to the directed molecular evolution of a beta-glucuronidase into a beta-galactosidase: two different evolutionary pathways lead to the same variant. BMC Res Notes 4:138

Stemmer WPC (1994) Rapid evolution of a protein in-vitro by DNA shuffling. Nature 370 (6488):389–391. doi:10.1038/370389a0

Tabor S, Richardson CC (1995) A single residue in DNA polymerases of the *Escherichia coli* DNA polymerase I family is critical for distinguishing between deoxyribonucleotides and dideoxyribonucleotides. Proc Natl Acad Sci USA 92(14):6339–6343. doi:10.1073/pnas.92.14.6339

Tawfik DS, Griffiths AD (1998) Man-made cell-like compartments for molecular evolution. Nat Biotechnol 16(7):652–656

Tosaka A, Ogawa M, Yoshida S, Suzuki M (2001) O-helix mutant T664P of Thermus aquaticus DNA polymerase I. J Biol Chem 276(29):27562–27567. doi:10.1074/jbc.M010635200

Turner NJ (2009) Directed evolution drives the next generation of biocatalysts. Nat Chem Biol 5 (8):568–574. doi:10.1038/nchembio.203

Vartanian JP, Henry M, WainHobson S (1996) Hypermutagenic PCR involving all four transitions and a sizeable proportion of transversions. Nucleic Acids Res 24(14):2627–2631. doi:10.1093/nar/24.14.2627

Wang Y, Prosen DE, Mei L, Sullivan JC, Finney M, Vander Horn PB (2004) A novel strategy to engineer DNA polymerases for enhanced processivity and improved performance in vitro. Nucleic Acids Res 32(3):1197–1207. doi:10.1093/nar/gkh271

Wong TS, Roccatano D, Schwaneberg U (2007) Challenges of the genetic code for exploring sequence space in directed protein evolution. Biocatal Biotransformation 25(2–4):229–241. doi:10.1080/10242420701444280

Wright S (1932) The roles of mutation, inbreeding, crossbreeding and selection in evolution. In: Proceedings of the sixth international congress of genetics, Ithaca, vol 1, pp 356–366

Xia G, Chen L, Sera T, Fa M, Schultz PG, Romesberg FE (2002) Directed evolution of novel polymerase activities: mutation of a DNA polymerase into an efficient RNA polymerase. Proc Natl Acad Sci USA 99(10):6597–6602. doi:10.1073/pnas.102577799

Yoshida K, Tosaka A, Kamiya H, Murate T, Kasai H, Nimura Y, Ogawa M, Yoshida S, Suzuki M (2001) Arg660Ser mutation in Thermus aquaticus DNA polymerase I suppresses T -> C transitions: implication of wobble base pair formation at the nucleotide incorporation step. Nucleic Acids Res 29(20):4206–4214. doi:10.1093/nar/29.20.4206

Yuan L, Kurek I, English J, Keenan R (2005) Laboratory-directed protein evolution. Microbiol Mol Biol Rev 69(3):373–392. doi:10.1128/jmbr.69.3.393-392.2005

Chapter 8
Reverse Transcriptases

Stuart F.J. Le Grice and Marcin Nowotny

Abstract Reverse transcriptases (RTs) catalyze the multi-step process that converts the single-stranded viral RNA genome to double-stranded integration-competent DNA, a process that is essential for the proliferation of retroviruses and retrotransposons. The former are released from cells as infectious particles and include human immunodeficiency virus (HIV) and hepadnaviruses such as hepatitis B virus (HBV). In contrast, retrotransposons are restricted to intracellular proliferation cycle and are very common in eukaryotic genomes. The N-terminal DNA polymerase domain of RTs structurally resembles other nucleic acid polymerases and assumes the topology of a right hand, with subdomains designated fingers, palm, and thumb. RTs exhibit low fidelity and processivity and are capable of both intra- and intermolecular strand displacement synthesis. A distinguishing feature of RTs is the presence of a C-terminal ribonuclease H (RNase H) domain in enzymes from viruses and long terminal repeat (LTR)-containing retrotransposons. The general role of this domain is sequence-independent removal of RNA from the RNA/DNA hybrid intermediates of reverse transcription, although precise cleavage is also required to mediate key steps such as (+) strand primer selection and DNA strand transfer. In LTR retroelements the RNase H domain is located immediately following the thumb subdomain, while in retroviruses an RNase H-like "connection domain" devoid of catalytic activity is followed by the catalytically competent C-terminal RNase H domain. DNA polymerase and RNase H activities are coordinated, but RNA/DNA hybrid hydrolysis is significantly slower than nucleotide

S.F.J. Le Grice (✉)
RT Biochemistry Section, HIV Drug Resistance Program, Frederick National Laboratory for Cancer Research, Frederick, MD 21702, USA
e-mail: legrices@mail.nih.gov

M. Nowotny (✉)
Laboratory of Protein Structure, International Institute of Molecular and Cell Biology, 4 Trojdena Street, 02-109 Warsaw, Poland
e-mail: mnowotny@iimcb.gov.pl

K.S. Murakami and M.A. Trakselis (eds.), *Nucleic Acid Polymerases*, Nucleic Acids and Molecular Biology 30, DOI 10.1007/978-3-642-39796-7_8,

incorporation. Whether this reflects simultaneous or individual active site occupancy of the nucleic acid substrate remains controversial.

Keywords Human immunodeficiency virus • Reverse transcription • DNA polymerase • Ribonuclease H • Antiviral therapy

8.1 Overview

Reverse transcription is a unique process through which single-stranded RNA is converted into double-stranded DNA. During its course, genetic information flows from RNA to DNA which is opposite to the "central dogma" of molecular biology and hence the term "reverse." This process is essential for mobile genetic elements such as viruses and retroelements (also called retrotransposons), for which at some stage of their proliferation, the genetic information is stored in RNA. Reverse transcription is catalyzed by multifunctional enzymes designated as reverse transcriptases (RTs), discovered in 1970 by Temin and Baltimore (Baltimore 1970; Temin and Mizutani 1970). These enzymes possess RNA- and DNA-dependent DNA polymerase activities and a ribonuclease H (RNase H) activity responsible for the hydrolysis of RNA in RNA/DNA hybrid intermediates of reverse transcription (Telesnitsky and Goff 1997).

RT-containing viruses belong to two groups: retroviruses such as human immunodeficiency virus 1 (HIV-1), most of which store their genetic information in single-stranded RNA, and caulimoviruses and hepadnaviruses (an example of the latter is hepatitis B virus, HBV) with a double-stranded DNA (dsDNA) genome and an RNA replication intermediate. The main difference between retroviruses and retrotransposons is that the former contain *env* genes coding for viral envelope proteins that allow the virus to leave the cell and spread, while retrotransposons lack this gene and are restricted to an intracellular life cycle. Retrotransposons thus have to integrate into the germ line to be passed on to the next generation. They are quite diverse and can be divided into two broad classes depending on whether or not their genetic information is flanked by long terminal repeats (LTRs). The diversity of retrotransposons suggests that they are the ancestral form from which the viruses evolved by acquisition of an *env* gene (Eickbush and Jamburuthugoda 2008). The DNA polymerase domain of RT is the only common element of all retrotransposons and related viruses (Eickbush and Jamburuthugoda 2008).

Retrotransposons are ubiquitous and ~42 % of the human genome corresponds to retroelements, mainly non-LTR long interspersed nuclear elements (LINE-1 or L1) (Lander et al. 2001; Cordaux and Batzer 2009). In maize 75 % of the genome reflects multiplication of retroelements, mostly from the LTR group (SanMiguel et al. 1998). Retroelements are also present in bacteria (Simon and Zimmerly 2008) and include retrons which utilize an RT to synthesize chimeric RNA/DNA structures (designated multicopy single-stranded DNA—msDNA) of unknown function (Lampson et al. 2005). The other RT-containing bacterial elements are type II introns (Simon and Zimmerly 2008). Discussion of the bacterial elements is outside of the scope of this chapter and the reader is referred to review articles on this topic.

The most extensively studied RTs are the dimeric enzyme from HIV-1 due to its clinical importance and monomeric RT from the gammaretrovirus Moloney murine leukemia virus (Mo-MLV) (Moelling 1974; Roth et al. 1985; Cote and Roth 2008). Enzymes from retroelements have also been purified and studied, e.g., from LINE-like elements (Gabriel and Boeke 1991; Ivanov et al. 1991), human L1 element (Mathias et al. 1991), yeast LTR transposons Ty1 (Wilhelm et al. 2000) and Ty3 (Bibillo et al. 2005a, b), and ORF from Penelope-like retroelements (Pyatkov et al. 2004). Here we will focus on HIV-1 RT, but will include data for other enzymes where available.

8.2 Reverse Transcription and Viral DNA Synthesis in LTR Elements

Converting the single-stranded (+) RNA genome of retroviruses and LTR-containing retrotransposons to integration-competent dsDNA is a multi-step process catalyzed by the virus-coded RT. However, while the integrated provirus is flanked by the hallmark LTRs, these encode sites for initiation and termination of transcription. The (+) RNA genome from which the provirus is synthesized therefore harbors incomplete LTR copies at its 5′ and 3′ termini, requiring their regeneration during RNA- and DNA-templated DNA synthesis (Fig. 8.1).

(a) tRNA-primed (−) DNA synthesis. Reverse transcription initiates from a host-coded tRNA molecule, whose 3′-terminal 18 nucleotides are complementary to a sequence immediately adjacent to U5 and designated the primer binding site (PBS). Examples of tRNA primer usage include $tRNA^{Lys,3}$ (HIV-1, HIV-2), $tRNA^{Trp}$ (Rouse sarcoma virus), $tRNA^{Pro}$ (Moloney murine leukemia virus), and $tRNA^{iMet}$ (Ty1 and Ty3) (Le Grice 2003). An exception to priming with tRNA is the Gypsy group of retrotransposons. In Tf1 retrotransposon from the fission yeast *Schizosaccharomyces pombe*, the first 11 bases of the (+) RNA genome are used for priming (Lin and Levin 1997). For HIV-1, current evidence suggests $tRNA^{Lys,3}$ is packaged into virions complexed with a component of the host translational machinery, namely, lysyl-tRNA synthetase (LysRS) (Cen et al. 2001). Whether primer annealing occurs pre- or post-virus assembly/budding remains to be determined.

Establishing a productive initiation complex occurs subsequent to RT binding to the tRNA/viral RNA duplex. A combination of chemical and enzymatic probing studies suggest this RNA/protein complex assumes a complex tertiary structure that controls early steps of initiation (Isel et al. 1995, 1998, 1999). Besides the tRNA/PBS duplex, additional interactions with the HIV-1 RNA genome control early events of (−) strand DNA synthesis, with one model suggesting that a primer activation sequence (PAS) on the viral genome downstream of the PBS interacts with an "anti-PAS" sequence of the tRNA TΨC stem (Beerens and Berkhout 2002), and an alternative suggests an A-rich sequence in the U5-IR stem interacts with the U-rich $tRNA^{Lys,3}$ anticodon loop (Isel et al. 1993). Initial steps of (−) strand DNA synthesis in HIV-1 are characterized by pausing and premature termination of the

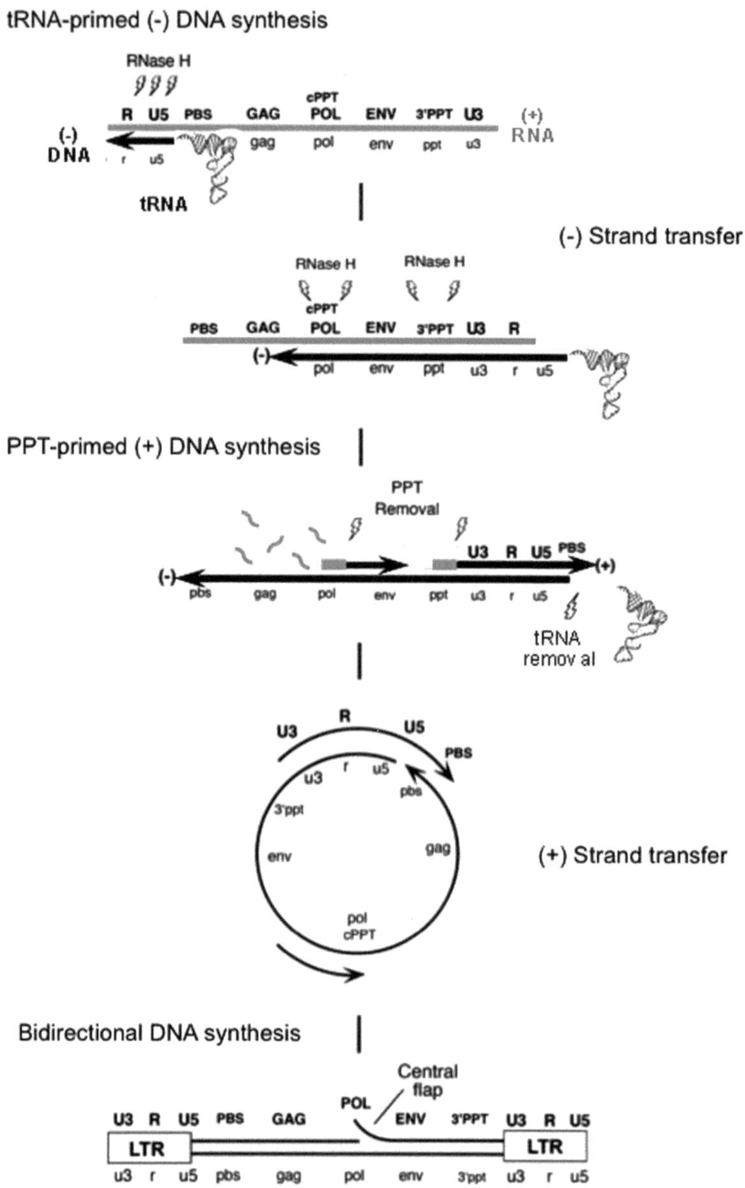

Fig. 8.1 RT-catalyzed synthesis of double-stranded, integration-competent HIV-1 DNA from the (+) strand RNA genome. RNA and DNA strands are depicted in *gray* and *black*, respectively. Note that for HIV-1 and related lentiviruses, the (+) strand of the DNA duplex is discontinuous, reflecting a second site for initiation of DNA synthesis at the center of the genome. Individual steps of DNA synthesis are described in the text

replication machinery, followed by a transition into a processive polymerization mode. This initiation "program," variations of which can be demonstrated for feline immunodeficiency and equine infectious anemia viruses, may represent a control mechanism that prevents premature reverse transcription prior to virus budding.

(b) (−) Strand strong-stop DNA synthesis and strand transfer. tRNA-primed DNA synthesis continues through U5 and R regions of the genome to the 5′ terminus, producing (−) strand strong-stop (−ss) DNA. The template RNA in RNA/DNA hybrid created by this process is hydrolyzed by RNase H activity of both the polymerizing enzyme and additional enzymes that reassociate with the substrate, producing short (+) strand oligoribonucleotides that spontaneously dissociate from nascent (−) strand DNA. Complementarity between the newly synthesized (−) DNA and the R region at the 3′-end of the genome facilitates the first DNA strand transfer event. (−) Strand DNA transfer can be both intra- or intermolecular, while inter-strand DNA transfer facilitates recombination, an essential process that maintains virus diversity and fitness. (−) Strand DNA transfer proceeds via two distinct mechanisms. In one scenario, transfer takes place between the termini of the full-length (−) strand DNA and the 3′-end of the viral RNA genome. An alternative mechanism invokes strand invasion, where an internal stretch of single-stranded DNA is accessed by the corresponding complement in the viral RNA, followed by branch migration and subsequent completion of (−) ss DNA synthesis (Kim et al. 1997).

(c) Polypurine tract-primed (+) strand DNA synthesis. Following (−) strand DNA transfer, RNA-dependent DNA synthesis resumes, accompanied by hydrolysis of (+) RNA of the ensuing RNA/DNA hybrid. However, a polypurine tract (PPT), located near the 3′ terminus, is refractory to hydrolysis, providing the primer for (+) strand, DNA-dependent DNA synthesis. A second, central PPT (cPPT) is a controversial feature of several lentiviruses and will be discussed later.

Since 3′ PPT-primed (+) strand DNA synthesis defines 5′ LTR sequences critical to the successful integration of viral dsDNA, its selection from the RNA/DNA replication intermediate and removal from nascent (+) DNA thus requires a considerable degree of precision. Although the structural basis for PPT selection remains elusive, X-ray crystallography (Sarafianos et al. 2001) and chemical footprinting (Kvaratskhelia et al. 2002) indicate the presence of locally altered base pairing, and, while speculative, the unusual conformation of this RNA/DNA hybrid may promote an "induced" fit to position the scissile PPT/U3 phosphodiester bond in the RNase H active site. Initiation of (+) strand DNA synthesis produces a PPT RNA/(+) DNA chimera hybridized to (−) DNA, and the unique architecture of the RNA/DNA junction, demonstrated by NMR spectroscopy (Fabris et al. 2009), may likewise mediate its recognition for accurate primer removal.

(d) tRNA primer removal and (+) strand DNA transfer. Prior to (+) strand DNA transfer, 3′ PPT-primed (+) strand DNA synthesis is templated by both DNA and RNA, namely, U3, R and U5 DNA, together with 18 terminal of the covalently linked tRNA primer. For HIV, the replication complex pauses at a modified $tRNA^{Lys,3}$ base (1-methyladenosine 58), placing the (−) DNA–tRNA junction in

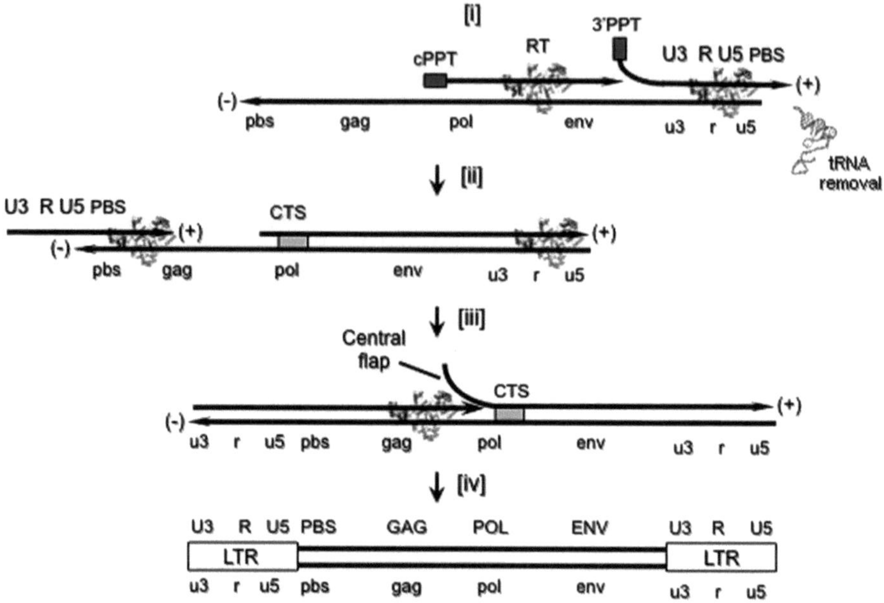

Fig. 8.2 Proposed mechanism for the termination of HIV-1 (+) strand DNA synthesis near the cPPT and generation of the "central flap." cPPT and 3′ PPT RNA primers are depicted in *gray* and (+) DNA in *black*, respectively. HIV-1 RT is represented by the 2-color dimer. *CTS* central termination sequence. Individual steps A–D are described in the text

the RNase H active site to facilitate tRNA removal. Although the tRNA–DNA junction is the logical RNase H target site, HIV-1 RT cleaves the penultimate ribonucleotide bond (Furfine and Reardon 1991a; Pullen et al. 1992). RNase H activity subsequently removes additional ribonucleotides from the tRNA 3′ terminus, promoting its dissociation and freeing homologous (+) and (−) strand PBS sequences for intramolecular (+) strand DNA transfer.

(e) Central termination of DNA synthesis. Following (+) strand DNA transfer, bidirectional DNA-dependent DNA synthesis would suffice to produce double-stranded, integration-competent viral DNA. However, a (+) strand discontinuity mapping near the center of the HIV-1 genome suggested a second initiation site and a unique termination mechanism (Hungnes et al. 1991; Charneau et al. 1992). Since the HIV-1 IN coding region contains a duplication of the 3′ PPT (designated the central, or cPPT), this provided the additional initiation site, and termination of (+) strand DNA synthesis was explained by the nearby *central termination sequence*, or CTS (Fig. 8.2).

3′ PPT-primed (+) strand DNA synthesis creates (+) strong-stop DNA containing the PBS complement necessary for (+) strand DNA transfer. However, cPPT-primed (+) strand DNA lacks a PBS sequence, precluding its involvement in strand transfer and thereby establishing one component of the discontinuous (+) strand. Following (+) strand DNA transfer and resumption of DNA-dependent

DNA synthesis, the replication complex ultimately reaches the duplex product of cPPT-primed DNA synthesis. After displacing ~100 nucleotides, the CTS is encountered in the form of duplex DNA, a prominent feature of which is phased dA:dT tracts that induce minor groove compression and "dislocation" of the replication complex (Berdis et al. 2001). Central termination also produces a "flap" of displaced DNA that is later removed and repaired by host-coded enzymes to create the replication-competent DNA duplex. Although the role of the central flap remains controversial, equivalent elements are present in the equine infectious anemia (Stetor et al. 1999) and feline immunodeficiency virus genomes (Whitwam et al. 2001).

8.3 Retrotransposition Mechanism of Non-LTR Elements

The mechanism of non-LTR retrotransposition is very different from that of LTR retroelements and retroviruses described above and is best characterized for the R2 retroelement from *Bombyx mori*. A key component in this process is the R2 protein, which contains an RT DNA polymerase domain flanked by N-terminal DNA-binding domain and C-terminal DNA-binding/endonuclease domain. In the proposed mechanism of "target-primed reverse transcription," R2 dimer binds ends of R2 transcript and the target DNA sequence in the 28S rRNA gene (Christensen et al. 2006). The R2 subunit bound upstream from the target sequence cleaves the DNA and the liberated 3′-OH terminus is used to prime reverse transcription of R2 element RNA (Luan et al. 1993). After first strand reverse transcription is completed, the second subunit of the R2 dimer cleaves the second DNA strand to initiate DNA synthesis from the other end to produce dsDNA. The RNA is either degraded by cellular RNases or displaced during second strand DNA synthesis (Christensen et al. 2006). R2 RT has higher strand displacement activity (Kurzynska-Kokorniak et al. 2007) and processivity than retroviral RTs which is likely important for the mode of action of this protein.

The large ORF of another non-LTR element, the retrotransposon L1, was demonstrated to possess RT activity (Mathias et al. 1991) and to catalyze target-primed reverse transcription, if nicks are present in the DNA (Cost et al. 2002). The N-terminal nuclease domain fused to L1 RT belongs to apurinic/apyrimidinic endonuclease family, whose role is to create a nick to prime reverse transcription (Christensen et al. 2000; Anzai et al. 2001).

8.4 RT Subunit Organization

The compact genomes of retroviruses exploit a small number of proteins and polyproteins that form repeating structures to enclose space, minimizing the requirement for a large protein-coding capacity. In this context, their RTs share a common property of maturation from larger gag-pol polyprotein precursors by the virus-coded protease. In many retroviruses, synthesis of the gag-pol polyprotein involves a frameshift (−1 frameshifting in the case of lentiviruses) or termination suppression (e.g., for gammaretroviruses), often between gag and pol proteins (Herschhorn and Hizi 2010). Since frameshifting or termination suppression occurs with a defined frequency, this ensures the proper excess of gag protein, which is further processed to structural proteins of the virus. An interesting exception is spumaretroviruses in which the RT is produced from single mRNA (Yu et al. 1996).

RTs exhibit significant diversity in quaternary structure. For HIV-1 and HIV-2 RT, the biologically relevant enzymes are 66 kDa/51 kDa (HIV-1) or 68 kDa/55 kDa heterodimers (HIV-2), the smaller subunit resulting from protease-mediated cleavage of p66 or p68 (Telesnitsky and Goff 1997). Both subunits of the heterodimer thus share similar subdomains, designated (by analogy to a right hand) fingers, palm, thumb, and connection, while p66 and p68 retain the C-terminal RNase H domain. Despite primary sequence identity, subdomains of the heterodimer adopt significantly different conformations, i.e., the p66 and p68 polymerase domains exhibit an open, extended structure with a large active-site cleft, while the equivalent subdomains of the small subunit are closed, compact structures devoid of catalytic activity. The p51 subunit of HIV-1 RT is essentially rigid, having nearly the same conformation in numerous crystal structures of HIV-1 RT, while the p66 subunit undergoes large-scale motions (Bahar et al. 1999).

Proposed roles for the p51 HIV-1 subunit include providing a structural support and facilitating p66 loading onto nucleic acid. The recent structure of HIV-1 RT containing a non-PPT RNA/DNA hybrid (Lapkouski et al. 2013) has highlighted two novel features of the p51 C-terminus that facilitate accommodation of the duplex and correct presentation of the RNA strand in the RNase H active site. The first of these is a peptide spanning connection residues Phe416–Pro421, comprising the β-20–α-M connecting loop, which interacts with the DNA strand, crosses the shallow minor groove (where the hybrid bends), and forms van der Waals contacts with the RNA strand 3 nt from the scissile phosphate. In addition, Tyr427 of α-M is involved in hydrogen bonding with Asn348, a residue whose mutation to Ile has been associated with increased resistance to nucleoside and nonnucleoside RT inhibitors (Radzio and Sluis-Cremer 2011).

RTs of alpharetroviruses (avian leukosis virus (ALV) and Rous sarcoma virus (RSV)) are also heterodimers of p63 and p95 proteins. In this case, the 95 kDa β subunit retains the integrase domain of the gag-pol precursor, and the 63 kDa α subunit specifies the DNA polymerase and RNase H domains. Site-directed mutagenesis studies suggest that both enzymatic activities of the α/β heterodimer are conferred by the α subunit (Werner and Wohrl 2000).

In contrast to alpharetroviral and lentiviral RTs, the counterpart enzymes of gammaretroviruses (Moloney murine leukemia virus and xenotropic murine leukemia viruslike retrovirus—XMRV) and deltaretroviruses (bovine leukemia virus (Perach and Hizi 1999)) are monomeric, with molecular masses of ~70 kDa. Another feature distinguishing gamma- and deltaretrovirus RTs from lentiviral enzymes is the inclusion of an additional α-helix (designated the basic protrusion) in their C-terminal RNase H domain, i.e., they closely resemble RNase H of *Escherichia coli* (see below). One of the main structural differences between monomeric and dimeric RTs is the positioning of their RNase H domains. In dimeric enzymes it is relatively rigidly positioned on the platform of the non-catalytic subunit. In contrast, the RNase H domain of monomeric enzymes is tethered to the polymerase domain through a flexible linker. As shown by recent small-angle X-ray scattering data, this C-terminal domain is very mobile in the absence of the nucleic acid but becomes organized on the substrate when it is bound by polymerase domain (Nowak et al. 2013).

8.5 DNA Polymerase Domain: Structure and Mechanism

RTs constitute a separate class of polymerases and are most closely related to telomerases (see Chap. 9) and viral RNA-dependent RNA polymerases. Although the overall shape and spatial arrangement of subdomains is similar across various classes of polymerases, their order in the primary sequence varies. In RTs the N-terminal region contains the palm subdomain with two insertions forming the fingers, followed by the thumb subdomain (Fig 8.3a).

The only RTs for which structural information is available are the dimeric HIV-1 enzyme (Kohlstaedt et al. 1992; Huang et al. 1998) and two nearly identical monomeric RTs from Mo-MLV and XMRV (Das and Georgiadis 2004; Zhou et al. 2012; Nowak et al. 2013). Although the structural description below is based on the information available for these enzymes, the high degree of amino acid sequence conservation of the palm and fingers subdomains (Xiong and Eickbush 1988, 1990) suggests that for other RTs at least those two subdomains are very similar. The thumb subdomain is more divergent in amino acid sequence and its full structural characterization across different groups of RTs awaits the solution of further structures.

At the center of the DNA polymerase domain of HIV-1 RT is the palm subdomain (Fig 8.3b, c), characteristic features of which are two three-stranded β-sheets and two long α-helices. The fingers subdomain comprises an α-helix and a two-stranded β-sheet, while the thumb contains three or four α-helices. The DNA polymerase active site is located in the palm subdomain and binds two metal ions, termed here A and B, which are essential for catalysis (Fig 8.3c, d). The preferred catalytic ion is Mg^{2+}, while Mn^{2+} can also support DNA synthesis. Metal ion B is coordinated by all three phosphate groups of the incoming nucleotide and a backbone carbonyl (Val111 in HIV-1 RT). Metal ion A is coordinated by two essential aspartates (Asp110 and 185 in HIV-1 RT) by the 3′-OH group of the primer, which is the attacking nucleophile in the nucleotidyl transfer reaction. Two carboxylates (Asp185 and Asp186 in HIV-1 RT) form the active-site motif -Tyr-X-Asp-Asp-, of which X denotes a hydrophobic residue. The two aspartic acid residues are absolutely essential for DNA polymerase activity (Kaushik et al. 1996), also for retrotransposon RTs. For example, for Ty1 RT substituting

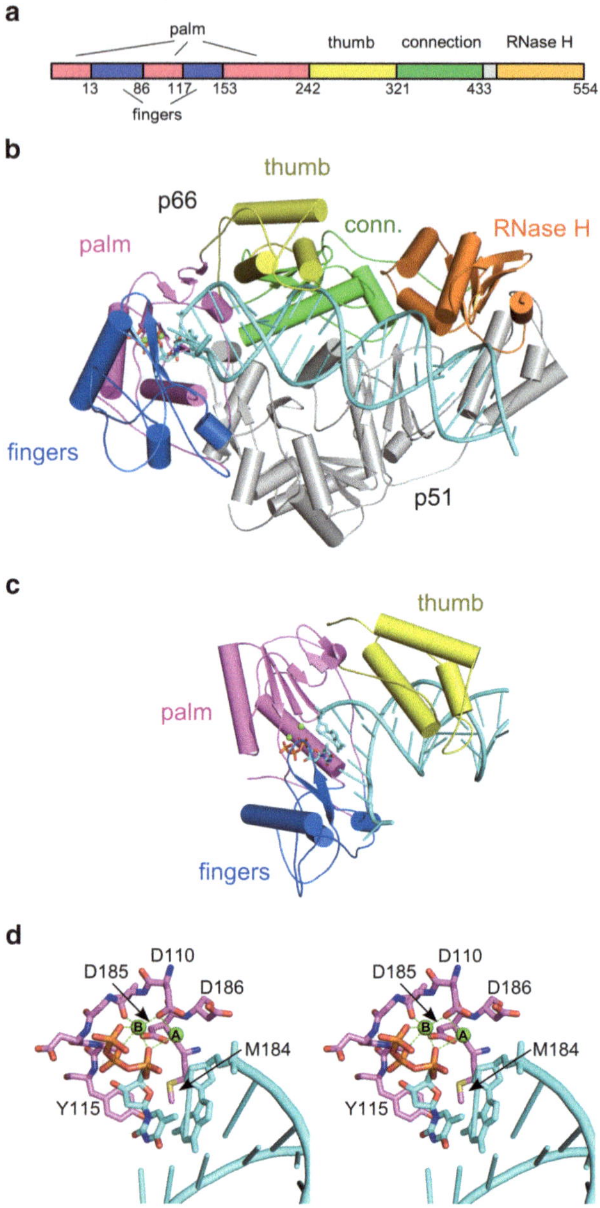

Fig. 8.3 Structure of HIV-1 RT. (**a**) Domain composition of the p66 subunit. The residue numbers at the boundaries of the domains are given. (**b**) Overall structure of HIV-1 RT in complex with dsDNA substrate (PDB ID: 1RTD). p66 subunit is shown in *color* and p51 in *gray*. The nucleic acid substrate is shown in *cyan ladder*. (**c**) Polymerase domain (fingers, palm, and thumb subdomains). The incoming nucleotide at the active site is shown in *sticks* and the metal ions (Mg^{2+}) as *green spheres*. (**d**) Close-up view of the active site (stereoview). Residues forming the active site (from palm domain) are shown in *stick* representation and metal ion coordination in *green dashed lines*. Note that the substrate present in the structure is missing the attacking 3′-OH group to inhibit the polymerization reaction

the first Asp with Asn abolished DNA polymerase function, and for both Ty3 and Ty1 RTs, mutating the second Asp only reduced DNA activity in vitro but abolished retrotransposition in vivo (Uzun and Gabriel 2001; Pandey et al. 2004; Bibillo et al. 2005a).

Another key element of the DNA polymerase active site is the pocket binding the incoming nucleotide, designated the N site. For HIV-1 RT the γ-phosphate of the dNTP interacts with Lys65 and Asp113, while the base stacks with Arg72. With respect to the ribose ring, the 3′-OH group interacts with Gln151 and the backbone amide of Tyr115. The side chain of the latter residue also forms a very important stacking interaction with the ribose ring forming a so-called steric gate in common with other DNA polymerases (Cases-Gonzalez et al. 2000; Brown and Suo 2011). The stacking is only efficient when the 2′-OH group is absent from the ring, selecting for the incoming dNTP and conferring DNA polymerase specificity. To investigate how structural changes of Tyr115 affect nucleotide incorporation by HIV-1 RT, Klarmann et al. substituted this residue with a variety of nonnatural tyrosine analogues (Klarmann et al. 2007). This approach demonstrated that aminomethyl-Phe115 RT incorporated dCTP more efficiently than the WT enzyme and was also resistant to the chain terminator 3TCTP. Molecular modeling suggested that the aminomethyl-Phe115 substitution provided new hydrogen bonds through the minor groove to the incoming dNTP and the template residue of the terminal base pair, which might contribute to the increased efficiency of dCTP incorporation. Substituting the equivalent steric gate residue in Mo-MLV RT, Phe155, with valine resulted in enhanced ribonucleotide incorporation (Gao et al. 1997).

The polymerization rate of RTs compared to other polymerases is slow—between 1 and 100 nucleotides per second in pre-steady state when released from a stalled state (Kati et al. 1992) and approximately 170 nucleotides per second during processive incorporation (Ortiz et al. 2005). The reaction starts with the binding of primer-template duplex such that the primer 3′ terminus is located at the active site (Jacobo-Molina et al. 1993; Huang et al. 1998; Sarafianos et al. 2001). Substrate binding is accompanied by a change in position of the thumb subdomain, which moves away from the substrate binding cleft (Rodgers et al. 1995; Hsiou et al. 1996). The incoming dNTP is next bound, leading to re-positioning of the fingers subdomain to align the attacking 3′-OH of the primer and the α-phosphate of the dNTP (Kati et al. 1992; Reardon 1992). The movement of the fingers subdomain is thought to be the rate-limiting step of nucleotide incorporation. Following dNTP binding, the reaction follows an S_N2-like mechanism shared by all polymerases, which involves the two divalent metal ions (Steitz and Steitz 1993; Steitz 1998; Yang et al. 2006). The primer 3′-OH, which is positioned and activated by metal ion A, catalyzes nucleophilic attack on the α phosphorus atom of the incoming dNTP, releasing pyrophosphate, and nascent DNA subsequently translocates to position the next 3′-OH group at the active site. Studies with pyrophosphate analogue foscarnet indicate that translocation occurs through a thermal ratchet mechanism in which the enzyme spontaneously switches between pre- and post-translocated states (Marchand et al. 2007).

RTs have a higher error rate than other replicative DNA polymerases, due in part to the fact that they lack proofreading exonuclease activity. HIV-1 RT is in fact among the least accurate RTs, making one error in 1,700 to 4,000 polymerized nucleotides (Preston et al. 1988; Roberts et al. 1988). Mo-MLV MuLV RT, for example, has significantly higher fidelity, making 1 error per 30,000 nucleotides (Roberts et al. 1988, 1989). A critical residue that contributes to fidelity is the hydrophobic amino acid of the ^{183}-Tyr-X-Asp-Asp-^{186} active-site motif. In HIV-1 RT this is Met184, whose substitution to Val or Leu resulted in increased fidelity (Bakhanashvili et al. 1996; Pandey et al. 1996; Wainberg et al. 1996; Oude Essink et al. 1997), while replacing it with Ala caused a four- to eightfold reduction in fidelity (Pandey et al. 1996). In Mo-MLV and Ty1 RTs the equivalent hydrophobic residue is Val which could explain their higher fidelity (Kaushik et al. 2000; Boutabout et al. 2001). Interestingly, substitutions of HIV-1 Met184 which increase fidelity lead to resistance to nucleoside analogue inhibitors of polymerase activity (NRTIs) which is likely caused by higher discrimination against incorporation of incorrect incoming nucleotides (Wainberg et al. 1996; Oude Essink et al. 1997).

RTs are low-processivity polymerases and can synthesize up to several hundred nucleotides in a single binding event depending greatly on the nucleic acid sequence (Huber et al. 1989; Avidan et al. 2002). An element important for DNA synthesis processivity is the "pin" structure guiding the template strand opposite of the active site (Kim et al. 1999; Gu et al. 2001; Nowak et al. 2013) (see below). A unique feature of RTs is DNA synthesis with concurrent displacement of nucleic acid hybridized to the template upstream from the active site (Huber et al. 1989; Fuentes et al. 1996). Such strand displacement synthesis is important at several steps during viral DNA synthesis, for example, during removal of RNA fragments left over by RNase H activity or during the termination of reverse transcription and creation of the central flap (see above). For non-LTR enzymes, which lack RNase H activity, displacement synthesis must be essential for the completion of retrotransposition. Displacement synthesis is approximately threefold slower than non-displacement synthesis (Kelleher and Champoux 1998; Whiting and Champoux 1998). It has been shown that for retroviral RTs, an aromatic residue in the fingers subdomain (Trp24 in HIV-1 and Tyr64 in Mo-MLV) is important for displacement synthesis (Agopian et al. 2007; Paulson et al. 2007). This residue forms a stacking interaction with bases in the single-stranded template region ahead of the active site (Winshell and Champoux 2001; Nowak et al. 2013).

HIV-1 RT DNA polymerase activity, along with viral integrase and protease, is the prime target for development of antiretroviral drugs. In fact, the first drug used to treat AIDS was DNA polymerase inhibitor—the nucleoside analogue azidothymidine (AZT). Since then multiple nucleoside analogues have been developed. Nonnucleoside RT inhibitors (NNRTIs) are a second class of HIV-1 DNA polymerase activity inhibitors which bind to an allosteric pocket in the vicinity of the active site. The description of polymerase inhibitors and evolution of drug resistance is outside of the scope of this chapter and the reader is referred to reviews on this topic (Ilina and Parniak 2008).

8.6 The Connection and RNase H Domains

A unique feature present only in retroviral RTs is the connection subdomain between the DNA polymerase and RNase H domains. The connection adopts a structure similar to the RNase H domain, but the catalytic residues are absent. Based on amino acid sequence, the RNase H domain from retrotransposons and retroviruses are different, with retroviral RNases H resembling more closely the cellular enzymes. It was therefore proposed that retroviruses acquired a new RNase H domain from the host, while the ancestral one became the connection domain and lost catalytic activity (Malik and Eickbush 2001).

RNase H activity of RTs is essential for reverse transcription and inactivating mutations of this domain inhibit virus infectivity (Tisdale et al. 1991). Cellular RNases H are present in all eukaryotes and are divided into two classes: type 1 and type 2. The RNase H domain from RTs is a type 1 enzyme. Cellular RNases H1 are implicated in the (1) maintenance of genomic stability by processing R-loops which arise when the mRNA hybridizes with the DNA from which it was transcribed (Broccoli et al. 2004), (2) removal of RNA/DNA hybrids which can promote genomic instability (Wahba et al. 2011), and (3) removal of ribonucleotides from DNA, which is essential for genome integrity and development (Reijns et al. 2012). The general role of RNase H activity of RTs is to nondiscriminately degrade (+) genomic RNA during (−) strand, DNA-dependent DNA synthesis. RNase H also performs specific cuts at several stages of reverse transcription, namely, excising the tRNA primer and generating and removing the PPT primer of (+) strand DNA synthesis (Rausch and Le Grice 2004; Schultz and Champoux 2008).

The RT-associated RNase H domain adopts a typical RNase H fold with the central β-sheet of five strands (Fig. 8.4a). The first three strands in the primary sequence run antiparallel to each other and the last two are shorter and parallel to the first. The central β-sheet is flanked by α-helices. The single C-terminal helix is located on one side of the sheet and three or four helices are on the other. If four helices are present, two of them, together with the following loop, form an element designated the basic protrusion, which participates in substrate binding (Kanaya et al. 1991; Keck and Marqusee 1996; Haruki et al. 1997; Nowotny et al. 2007). RNase H both containing and lacking a basic protrusion are evident in RTs, most likely depending on the architecture of the enzyme. For example, dimeric HIV-1 RT lacks this element but it is present in the monomeric Mo-MLV and XMRV enzymes.

Currently no structures are available for retroviral RNases H interacting with nucleic acid substrate. The mechanism of RNase H-mediated RNA hydrolysis has been elucidated from crystal structures of substrate complexes of cellular RNases H1 from the bacterium *Bacillus halodurans* (Nowotny et al. 2005) as well as the human enzyme (Nowotny et al. 2007). The minor groove of the hybrid is bound by conserved carboxylate and amide side chains (equivalents in HIV-1 RNase H are Glu449, Asn474, and Gln475). The RNA backbone forms a network of interactions with the protein, many mediated by the main chain. In particular 2′-OH groups of

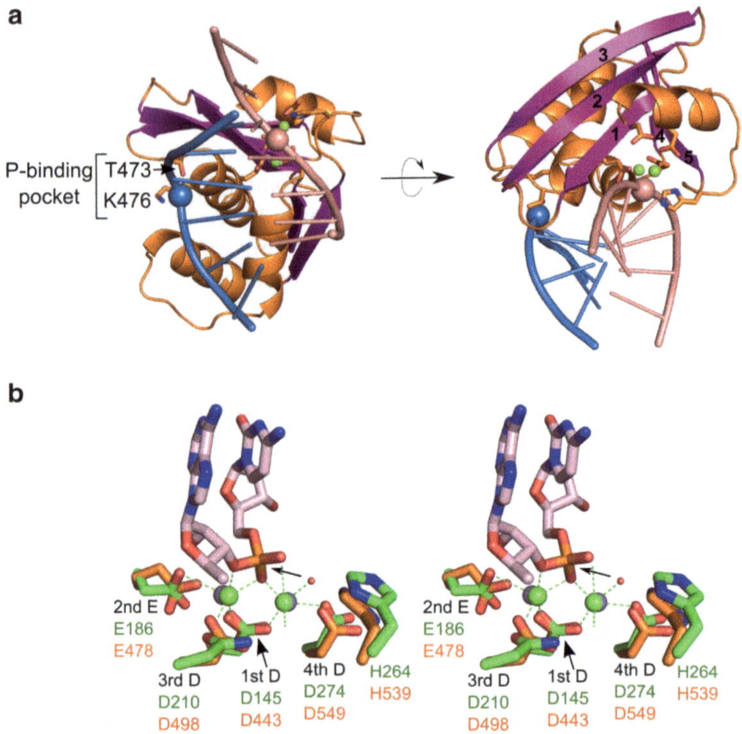

Fig. 8.4 HIV-1 RNase H domain. (**a**) Two views of the model of RNase H domain of HIV-1 RT interacting with RNA/DNA hybrid substrate and metal ions. The model was created by superimposing the structure of HIV-1 RNase H onto the structure of the substrate complex structures of very similar human and bacterial (*B. halodurans*) RNases H1. HIV-1 RNase H structure is shown in *cartoon* representation with the central β-sheet of the RNase H fold shown in *purple* and the strands labeled as they appear in the sequence. RNA and DNA from the human RNase H1 structure are shown in *light blue* and *pink ladder* representations, respectively. Two Mg^{2+} ions modeled from the bacterial structure are shown as *green spheres*. Residues forming the active site and the phosphate-binding pocket are shown as *sticks* and the latter are labeled. Phosphate group interacting with the active site and the P-binding pocket are indicated with *pink* and *light blue spheres*, respectively. (**b**) Close-up view of the active site (stereoview). The figure shows a superposition of the active sites of human RNase H1 (PDB ID:3K2P) and HIV-1 RNase H in complex with β-thujaplicinol and manganese ions. Protein residues forming the active site are shown as *sticks* (*orange* for HIV-1 and *green* for human) and a fragment of substrate RNA from human RNase H1 structure in *pink*. Calcium ions from human RNase H1 structure are shown as *green spheres* and manganese ions from HIV-1 structure as *gray spheres*. Metal ion coordination is shown as *green dashed lines*. *Small red sphere* represents the attacking nucleophilic water and the direction of the attack on the scissile phosphate is indicated with an *arrow*

two nucleotides on each side of the scissile phosphate form hydrogen bonds with RNase H which facilitates recognition of the RNA strand. The backbone of the DNA interacts with two elements. The first is the "phosphate-binding pocket," which forms a tight interaction with the phosphate group of DNA located two base pairs from the active site (Fig 8.4a). In most RNases H1 this motif comprises 3–4 residues, while in HIV-1 RNase H domain, there are only two residues: Thr473 and Lys476. The importance of the former is further underscored by the fact that its substitution in HIV-1 RT inhibits virus replication (Julias et al. 2002). The tight interaction between the phosphate-binding pocket and DNA results in deformation of the nucleic acid manifested by unusual torsion angles. This deformation also requires that the strand interacting with the pocket adopts B-form conformation with a narrow minor groove and B-form-like sugar puckers allowed only for DNA. This serves for the recognition of DNA as the non-cleaved strand. When the basic protrusion is present, it promotes additional DNA binding. A tight channel is formed between the core of the enzyme and the protrusion through which the DNA strand passes. In human RNase H, the channel contains an aromatic residue (Trp) which forms a stacking interaction with the ribose ring of the substrate. This interaction is efficient only when 2′-OH group is absent, thus further selecting for DNA. Equivalents of this Trp are not found in retroviral RNase H domains, i.e., perhaps this feature is unique to cellular enzymes.

Analogous to the mechanism of DNA synthesis, RNase H-mediated hydrolysis proceeds via two metal ion-assisted catalysis (Nowotny et al. 2005; Nowotny and Yang 2006), but the attacking nucleophile is water rather than the primer 3′-OH, resulting in hydrolysis and not phosphoryl transfer. The preferred catalytic metal ion is Mg^{2+}. Mn^{2+} can also support the reaction but for HIV-1 RNase H, Mn^{2+} induces nonspecific dsRNA cleavage termed RNase H* activity, which is considered nonphysiological (Ben-Artzi et al. 1992; Hostomsky et al. 1994). Ca^{2+} ions inhibit the RNase H activity (Nowotny and Yang 2006). Similar to the DNA polymerase active site, the metal ions are termed A and B, where A positions and activates the nucleophilic water and B stabilizes the transition state and the leaving group (Fig 8.4b). The reaction product contains 5′-phosphate and 3′-OH groups (Miller et al. 1973; Krug and Berger 1989).

The active site of RNases H is composed of carboxylate residues forming an -Asp-Glu-Asp-Asp- motif and coordinating the two metal ions (Fig 8.4b). The first three residues of this motif are absolutely conserved and their substitutions in HIV-1, Mo-MLV, and Ty3 RTs led to a complete loss of RNase H activity (Schatz et al. 1989; Mizrahi et al. 1990, 1994; Lener et al. 2002). The first Asp is located in the middle of the first β-strand of the central β-sheet, the second Asp, at the C-terminus of the fourth strand and Glu in the first α-helix.

The nucleic acid substrate contributes important coordination contacts for metal ions and only in the presence of the substrate is proper positioning and coordination of metal ions observed (Nowotny et al. 2005). Since, as mentioned above, catalytic complex structures are not available for RNases H from RTs, the configuration of the active site has been inferred from the structures of the cellular counterparts. From this comparison, an architecture can be proposed in which both metal ions are coordinated by the first carboxylate of the -Asp-Glu-Asp-Asp- motif which is the central element of the active site (Asp443 in HIV-1) and the non-bridging oxygen of the scissile phosphate (Fig 8.4b). In addition metal ion A is coordinated by the last carboxylate of the motif and coordinates two water molecules, one of which is the attacking nucleophile. Metal ion B is also coordinated by the O3′ the scissile phosphate and Glu of the -Asp-Glu-Asp-Asp- motif. This residue also forms a hydrogen bond with 2′-OH group of the nucleotide on the 3′ side of the scissile phosphate, coupling RNA recognition to catalysis. This model is supported by crystal structures of complexes of HIV-1 RT with RNase H inhibitors interacting with the active site through coordination of metal ions. These inhibitors mimic the scissile phosphate and lead to proper positioning of the metal ions nearly identical to bona fide substrate complexes of cellular RNases H1 (Himmel et al. 2009).

In retroviral RNases H, but not in the domains from retrotransposons, an additional His (His539 in HIV-1 and His638 in Mo-MLV) is located in the proximity of the active site and has been shown to be important for catalysis. This residue located in a flexible loop before the C-terminal α-helix. The mobility of this residue is probably exploited by the enzyme to promote dissociation of the reaction product (Nowotny et al. 2007).

8.7 Substrate Binding and Coordination of DNA Polymerase and RNase H Activities

The details of substrate binding by retroviral RTs have been elucidated from crystal structures of HIV-1 RT in complex with nucleic acid (dsDNA and RNA/DNA) (Jacobo-Molina et al. 1993; Huang et al. 1998; Sarafianos et al. 2001) and recent structure of polymerase-connection fragment of monomeric XMRV RT in complex with RNA/DNA (Nowak et al. 2013). For HIV-1 RT the nucleic acid substrates span the entire length of the enzyme (Fig 8.3b) and the distance between the DNA polymerase and RNase H active sites is ~60 Å or 17–18 bp. The interactions between the substrate and DNA polymerase domain of HIV-1 and XMRV RTs can be divided into several segments. If an overhang is present in the template, it is stabilized by interactions with the fingers subdomain. The template nucleotide pairing with the incoming nucleotides is stabilized by a "pin"—an Arg from the fingers subdomain (Arg78 in HIV-1 and Arg116 in Mo-MLV)—to guide the nucleic acid. For an RNA template, after the "pin" there is a region of interactions between the protein and 2′-OH groups. This region is followed by a positively charged patch comprising Lys and/or Arg residues interacting with the nucleic acid backbone. An important element for substrate binding is an α-helix of the thumb subdomain inserted into the minor groove of the duplex and forming interactions predominantly with the primer strand. These interactions of the thumb subdomain are also conserved for RT from the *Saccharomyces cerevisiae* LTR-retrotransposon Ty3, as demonstrated by site-directed mutagenesis (Bibillo et al. 2005b).

Although interactions of substrate with polymerase domain are conserved between HIV-1 RT, XMRV RT, and likely Ty3 RT, especially around the active site, further towards the connection and RNase H domains, these are quite different. For example, HIV-1 RT contains in this region a unique and important element designated the "RNase H primer grip," which comprises (i) Gly359, Ala360, and

His361 from the p66 connection subdomain (ii) Thr473, Asn474, Gln475, Lys476, Tyr501, and Ile505 from the RNase H domain of p66 and (iii) Lys395 and Glu396 from p51 subunit. This region is important for both the DNA polymerase and RNase H activities of HIV-1 (Arion et al. 2002; Julias et al. 2002, 2003; Rausch et al. 2002; McWilliams et al. 2006).

In many available crystal structures of HIV-1 RT, the nucleic acid substrate forms productive interactions with the active site of DNA polymerase domain but is bound by the RNase H domain without contacting its active site. This is also true for the complex between HIV-1 RT and the PPT-containing RNA/DNA hybrid (Sarafianos et al. 2001). The hybrid in this structure, however, has a conformation very similar to dsDNA and contains a region of anomalous base pairing, prominent among which is a G:T mispair. However, NMR analysis of HIV-1 RT bound to the PPT (Turner et al. 2008) and related studies with PPT variants (Le Grice et al., unpublished observations) together failed to confirm the presence of this mispair. It remains to be seen whether such a configuration will be confirmed for other RNA/DNA substrates. Nevertheless, the lack of interaction between the substrate and the active site of RNase H clearly does not allow for RNase H cleavage. This is in agreement with the biochemical data that show that the interaction of the substrate with the two active sites is mutually exclusive (e.g., see Delviks-Frankenberry et al. 2008). Recent crystal structures of HIV-1 RT in complex with RNA/DNA hybrids showed that a unique substrate conformation stabilized by interactions with the connection domain is required for the interaction with the RNase H active site. The duplex RNA/DNA in those structures is underwound in the middle with a widened major groove (Lapkouski et al. 2013). This changes the trajectory of the hybrid, so that the RNA strand moves closer to the RNase H active site and the 3′ end of the primer is removed from the polymerase active site precluding DNA polymerization.

The situation is very different for monomeric RTs which may not require a deformation of the substrate to allow RNase H cleavage. Instead, the RNase H is tethered to the connection domain via a flexible linker and very mobile, requiring organization on the substrate to catalyze hydrolysis (Nowak et al. 2013). The weak and transient interaction between the RNase H domain and the substrate that requires either substrate deformation (for dimeric HIV-1 RT) or RNase H domain organization (for monomeric XMRV RT) appears to emerge as a common feature of RTs. Its likely function is to allow for the regulation of the RNase H domain in cleavages that require a greater degree of precision—removal of tRNA primer and generation and removal of PPT primers.

Even though the DNA polymerase and RNase H activities are not simultaneous, the extensive interactions between the RNA/DNA substrate and the polymerase domain result in its strong influence on RNase H activity. It has been shown, for example, that substitutions in the DNA polymerase domain affect RNase H activity (Boyer et al. 1992a, b, 1994; Gao et al. 1998; Powell et al. 1999; Mandal et al. 2006). In fact, the isolated HIV-1 RNase H domain has very low activity (Hostomsky et al. 1991; Smith and Roth 1993) which can be restored by fusing it with artificial extensions, e.g., poly-His tags (Evans et al. 1991; Smith and Roth

1993), including the basic protrusion from cellular enzymes (Stahl et al. 1994; Keck and Marqusee 1995) or addition of p66 fragments N-terminal to the RNase H domain (Smith et al. 1994). Addition of the p51 subunit or the isolated connection domain has also been demonstrated to restore activity (Evans et al. 1991; Hostomsky et al. 1991; Smith and Roth 1993; Smith et al. 1994). In contrast to HIV-1 RNase H, the isolated domain of Mo-MLV RT retains some activity (Tanese and Goff 1988; Telesnitsky et al. 1992; Schultz and Champoux 1996; Zhan and Crouch 1997). This may be due to the fact that Mo-MLV RNase H contains a basic protrusion and its phosphate-binding pocket is comprised of four residues, which may promote stronger substrate binding, required for the organization of this domain on the RNA/DNA duplex.

The DNA polymerase domain also strongly affects the sites of RNase H cleavage. RNase H cleavage is defined as DNA 3′-end-directed, RNA 5′-end-directed, and internal (Fig. 8.5). The first occurs when the recessed 3′-end of the primer is positioned for extension at the DNA polymerase active site (Furfine and Reardon 1991b; Gopalakrishnan et al. 1992; Kati et al. 1992). For HIV-1 RT this results in RNase H cuts 15–20 bp from the primer 3′ terminus. For RTs from retrotransposons, Ty3 (Lener et al. 2002) and Ty1 (Wilhelm et al. 2000), the 3′-end-directed cleavages occur 13–14 bp from the primer terminus, most likely a consequence of the lack of a connection subdomain in retrotransposon RTs and a shorter distance between the two active sites.

3′-end-directed cleavages are further subdivided into polymerization independent, when the substrate is statically bound at the DNA polymerase active site, and polymerization dependent, which are simultaneous with DNA synthesis. For HIV-1 RT the RNase H activity is approximately sevenfold slower than polymerase activity (Kati et al. 1992), which results in incomplete degradation of the RNA strand. During processive DNA synthesis, 20 % of the RNA remained in fragments longer than 7 nt which can remain stably associated the DNA template.

5′-end-directed cleavages occur when recessed 5′-end of the RNA template is bound at the DNA polymerase active site (Fig 8.5) (Palaniappan et al. 1996; Wisniewski et al. 2000a, b; DeStefano et al. 2001). Primary and secondary cuts are observed, the former occurring 13–19 nt from the 5′-end of RNA and the latter 7–10 nt from the 5′-end and 5 nt from the 3′-end of the RNA (Wisniewski et al. 2000a, 2002). On long substrates RNase H can also perform internal cuts that do not depend of duplex ends (Schultz et al. 2004).

8.8 Conclusions and Perspectives

RTs are among the most extensively studied and best characterized enzymes, due to the clinical significance of HIV-1 RT. However, RTs are also very important tools that have allowed tremendous advancements of molecular biology, finding use in converting mRNA to complementary DNAs (cDNA) applied, among other things, in cloning of protein-coding regions of genes. RTs are intriguing enzymes able to

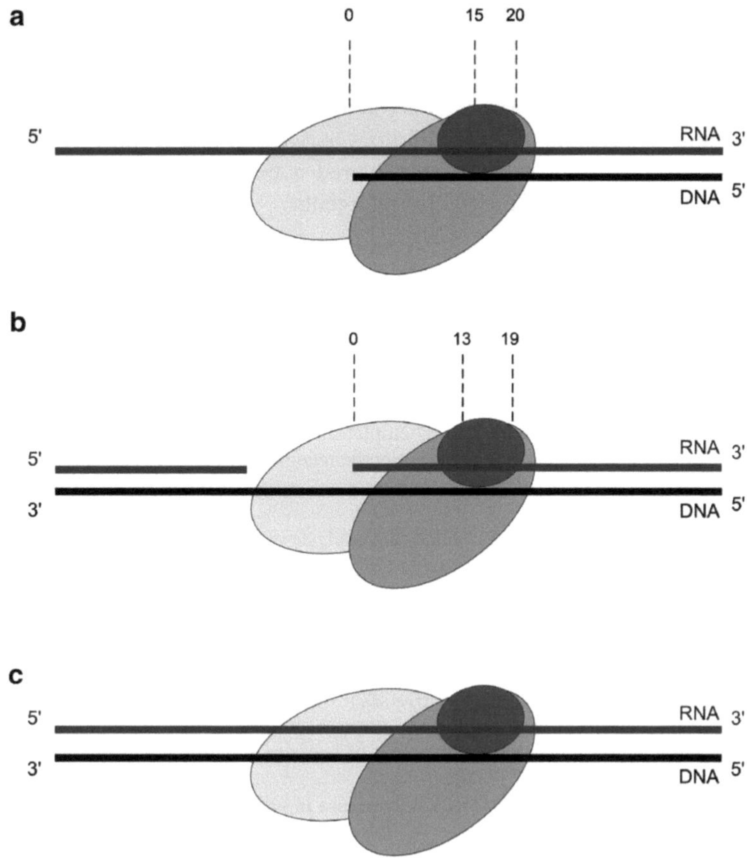

Fig. 8.5 Polymerase-guided RNase H cleavage modes. (**a**) 3′-DNA-directed cleavage. *Darker oval* represents the RNase H domain and *dashed lines* show the region in which the cleavages occur. (**b**) 5′-RNA-directed cleavage. (**c**) Internal cleavage

catalyze complicated multi-step reverse transcription reaction involving an elaborate protein-nucleic acid choreography.

As a final "chapter" of the HIV-1 RT journey, agents targeting this enzyme are showing promise as vaginal and rectal microbicides (Lewi et al. 2012). Prominent among these is tenofovir (TFV), shown in clinical trials to be safe and well tolerated in a study on HIV-negative women with a vaginal gel applied during 24 weeks. Repeated application of TFV intravaginal gel was well tolerated, produced low plasma levels, and, importantly, failed to select for resistance-conferring mutations, a continuing challenge of antiretroviral therapy. NNRTIs such as Dapivirine (DPV) have also displayed promising virucidal properties. When applied intravaginally, DPV is absorbed by the outer mucosal layers, while plasma concentrations reportedly remained low. Long-term constant DPV release has been obtained from a variety of intravaginal rings. However, HIV microbicide development still faces formidable challenges, including conclusive demonstration of efficacy, selection of drug-resistant virus in clinical settings, cultural acceptability, and affordability. These issues notwithstanding, advances in HIV RT research over that last 25 years, ranging from expression of active recombinant enzyme for high throughput screening to the potential of inhibitors as vaginal and rectal microbicides in resource-limited settings, must be considered a bench-to-bedside success and a model for development of future antiviral agents.

References

Agopian A, Depollier J et al (2007) p66 Trp24 and Phe61 are essential for accurate association of HIV-1 reverse transcriptase with primer/template. J Mol Biol 373(1):127–140

Anzai T, Takahashi H et al (2001) Sequence-specific recognition and cleavage of telomeric repeat (TTAGG)(n) by endonuclease of non-long terminal repeat retrotransposon TRAS1. Mol Cell Biol 21(1):100–108

Arion D, Sluis-Cremer N et al (2002) Mutational analysis of Tyr-501 of HIV-1 reverse transcriptase. Effects on ribonuclease H activity and inhibition of this activity by N-acylhydrazones. J Biol Chem 277(2):1370–1374

Avidan O, Meer ME et al (2002) The processivity and fidelity of DNA synthesis exhibited by the reverse transcriptase of bovine leukemia virus. Eur J Biochem 269(3):859–867

Bahar I, Erman B et al (1999) Collective motions in HIV-1 reverse transcriptase: examination of flexibility and enzyme function. J Mol Biol 285(3):1023–1037

Bakhanashvili M, Avidan O et al (1996) Mutational studies of human immunodeficiency virus type 1 reverse transcriptase: the involvement of residues 183 and 184 in the fidelity of DNA synthesis. FEBS Lett 391(3):257–262

Baltimore D (1970) RNA-dependent DNA polymerase in virions of RNA tumour viruses. Nature 226(5252):1209–1211

Beerens N, Berkhout B (2002) The tRNA primer activation signal in the human immunodeficiency virus type 1 genome is important for initiation and processive elongation of reverse transcription. J Virol 76(5):2329–2339

Ben-Artzi H, Zeelon E et al (1992) Double-stranded RNA-dependent RNase activity associated with human immunodeficiency virus type 1 reverse transcriptase. Proc Natl Acad Sci USA 89 (3):927–931

Berdis AJ, Stetor SR et al (2001) Molecular mechanism of sequence-specific termination of lentiviral replication. Biochemistry 40(40):12140–12149

Bibillo A, Lener D et al (2005a) Functional roles of carboxylate residues comprising the DNA polymerase active site triad of Ty3 reverse transcriptase. Nucleic Acids Res 33(1):171–181

Bibillo A, Lener D et al (2005b) Interaction of the Ty3 reverse transcriptase thumb subdomain with template-primer. J Biol Chem 280(34):30282–30290

Boutabout M, Wilhelm M et al (2001) DNA synthesis fidelity by the reverse transcriptase of the yeast retrotransposon Ty1. Nucleic Acids Res 29(11):2217–2222

Boyer PL, Ferris AL et al (1992a) Cassette mutagenesis of the reverse transcriptase of human immunodeficiency virus type 1. J Virol 66(2):1031–1039

Boyer PL, Ferris AL et al (1992b) Mutational analysis of the fingers domain of human immunodeficiency virus type 1 reverse transcriptase. J Virol 66(12):7533–7537

Boyer PL, Ferris AL et al (1994) Mutational analysis of the fingers and palm subdomains of human immunodeficiency virus type-1 (HIV-1) reverse transcriptase. J Mol Biol 243(3):472–483

Broccoli S, Rallu F et al (2004) Effects of RNA polymerase modifications on transcription-induced negative supercoiling and associated R-loop formation. Mol Microbiol 52 (6):1769–1779

Brown JA, Suo Z (2011) Unlocking the sugar "steric gate" of DNA polymerases. Biochemistry 50 (7):1135–1142

Cases-Gonzalez CE, Gutierrez-Rivas M et al (2000) Coupling ribose selection to fidelity of DNA synthesis. The role of Tyr-115 of human immunodeficiency virus type 1 reverse transcriptase. J Biol Chem 275(26):19759–19767

Cen S, Khorchid A et al (2001) Incorporation of lysyl-tRNA synthetase into human immunodeficiency virus type 1. J Virol 75(11):5043–5048

Charneau P, Alizon M et al (1992) A second origin of DNA plus-strand synthesis is required for optimal human immunodeficiency virus replication. J Virol 66(5):2814–2820

Christensen S, Pont-Kingdon G et al (2000) Target specificity of the endonuclease from the Xenopus laevis non-long terminal repeat retrotransposon, Tx1L. Mol Cell Biol 20 (4):1219–1226

Christensen SM, Ye J et al (2006) RNA from the 5′ end of the R2 retrotransposon controls R2 protein binding to and cleavage of its DNA target site. Proc Natl Acad Sci USA 103 (47):17602–17607

Cordaux R, Batzer MA (2009) The impact of retrotransposons on human genome evolution. Nat Rev Genet 10(10):691–703

Cost GJ, Feng Q et al (2002) Human L1 element target-primed reverse transcription in vitro. EMBO J 21(21):5899–5910

Cote ML, Roth MJ (2008) Murine leukemia virus reverse transcriptase: structural comparison with HIV-1 reverse transcriptase. Virus Res 134(1–2):186–202

Das D, Georgiadis MM (2004) The crystal structure of the monomeric reverse transcriptase from Moloney murine leukemia virus. Structure 12(5):819–829

Delviks-Frankenberry KA, Nikolenko GN et al (2008) HIV-1 reverse transcriptase connection subdomain mutations reduce template RNA degradation and enhance AZT excision. Proc Natl Acad Sci USA 105(31):10943–10948

DeStefano JJ, Cristofaro JV et al (2001) Physical mapping of HIV reverse transcriptase to the 5′ end of RNA primers. J Biol Chem 276(35):32515–32521

Eickbush TH, Jamburuthugoda VK (2008) The diversity of retrotransposons and the properties of their reverse transcriptases. Virus Res 134(1–2):221–234

Evans DB, Brawn K et al (1991) A recombinant ribonuclease H domain of HIV-1 reverse transcriptase that is enzymatically active. J Biol Chem 266(31):20583–20585

Fabris D, Marino JP et al (2009) Revisiting plus-strand DNA synthesis in retroviruses and long terminal repeat retrotransposons: dynamics of enzyme: substrate interactions. Viruses 1 (3):657–677

Fuentes GM, Rodriguez-Rodriguez L et al (1996) Strand displacement synthesis of the long terminal repeats by HIV reverse transcriptase. J Biol Chem 271(4):1966–1971

Furfine ES, Reardon JE (1991a) Human immunodeficiency virus reverse transcriptase ribonuclease H: specificity of tRNA(Lys3)-primer excision. Biochemistry 30(29):7041–7046

Furfine ES, Reardon JE (1991b) Reverse transcriptase. RNase H from the human immunodeficiency virus. Relationship of the DNA polymerase and RNA hydrolysis activities. J Biol Chem 266(1):406–412

Gabriel A, Boeke JD (1991) Reverse transcriptase encoded by a retrotransposon from the trypanosomatid Crithidia fasciculata. Proc Natl Acad Sci USA 88(21):9794–9798

Gao G, Orlova M et al (1997) Conferring RNA polymerase activity to a DNA polymerase: a single residue in reverse transcriptase controls substrate selection. Proc Natl Acad Sci USA 94 (2):407–411

Gao HQ, Boyer PL et al (1998) Effects of mutations in the polymerase domain on the polymerase, RNase H and strand transfer activities of human immunodeficiency virus type 1 reverse transcriptase. J Mol Biol 277(3):559–572

Gopalakrishnan V, Peliska JA et al (1992) Human immunodeficiency virus type 1 reverse transcriptase: spatial and temporal relationship between the polymerase and RNase H activities. Proc Natl Acad Sci USA 89(22):10763–10767

Gu J, Villanueva RA et al (2001) Substitution of Asp114 or Arg116 in the fingers domain of moloney murine leukemia virus reverse transcriptase affects interactions with the template-primer resulting in decreased processivity. J Mol Biol 305(2):341–359

Haruki M, Noguchi E et al (1997) Kinetic and stoichiometric analysis for the binding of Escherichia coli ribonuclease HI to RNA-DNA hybrids using surface plasmon resonance. J Biol Chem 272(35):22015–22022

Herschhorn A, Hizi A (2010) Retroviral reverse transcriptases. Cell Mol Life Sci 67 (16):2717–2747

Himmel DM, Maegley KA et al (2009) Structure of HIV-1 reverse transcriptase with the inhibitor beta-Thujaplicinol bound at the RNase H active site. Structure 17(12):1625–1635

Hostomsky Z, Hostomska Z et al (1991) Reconstitution in vitro of RNase H activity by using purified N-terminal and C-terminal domains of human immunodeficiency virus type 1 reverse transcriptase. Proc Natl Acad Sci USA 88(4):1148–1152

Hostomsky Z, Hughes SH et al (1994) Redesignation of the RNase D activity associated with retroviral reverse transcriptase as RNase H. J Virol 68(3):1970–1971

Hsiou Y, Ding J et al (1996) Structure of unliganded HIV-1 reverse transcriptase at 2.7 A resolution: implications of conformational changes for polymerization and inhibition mechanisms. Structure 4(7):853–860

Huang H, Chopra R et al (1998) Structure of a covalently trapped catalytic complex of HIV-1 reverse transcriptase: implications for drug resistance. Science 282(5394):1669–1675

Huber HE, McCoy JM et al (1989) Human immunodeficiency virus 1 reverse transcriptase. Template binding, processivity, strand displacement synthesis, and template switching. J Biol Chem 264(8):4669–4678

Hungnes O, Tjotta E et al (1991) The plus strand is discontinuous in a subpopulation of unintegrated HIV-1 DNA. Arch Virol 116(1–4):133–141

Ilina T, Parniak MA (2008) Inhibitors of HIV-1 reverse transcriptase. Adv Pharmacol 56:121–167

Isel C, Marquet R et al (1993) Modified nucleotides of tRNA(3Lys) modulate primer/template loop-loop interaction in the initiation complex of HIV-1 reverse transcription. J Biol Chem 268 (34):25269–25272

Isel C, Ehresmann C et al (1995) Initiation of reverse transcription of HIV-1: secondary structure of the HIV-1 RNA/tRNA(3Lys) (template/primer). J Mol Biol 247(2):236–250

Isel C, Keith G et al (1998) Mutational analysis of the tRNA3Lys/HIV-1 RNA (primer/template) complex. Nucleic Acids Res 26(5):1198–1204

Isel C, Westhof E et al (1999) Structural basis for the specificity of the initiation of HIV-1 reverse transcription. EMBO J 18(4):1038–1048

Ivanov VA, Melnikov AA et al (1991) Authentic reverse transcriptase is coded by jockey, a mobile Drosophila element related to mammalian LINEs. EMBO J 10(9):2489–2495

Jacobo-Molina A, Ding J et al (1993) Crystal structure of human immunodeficiency virus type 1 reverse transcriptase complexed with double-stranded DNA at 3.0 A resolution shows bent DNA. Proc Natl Acad Sci USA 90(13):6320–6324

Julias JG, McWilliams MJ et al (2002) Mutations in the RNase H domain of HIV-1 reverse transcriptase affect the initiation of DNA synthesis and the specificity of RNase H cleavage in vivo. Proc Natl Acad Sci USA 99(14):9515–9520

Julias JG, McWilliams MJ et al (2003) Mutation of amino acids in the connection domain of human immunodeficiency virus type 1 reverse transcriptase that contact the template-primer affects RNase H activity. J Virol 77(15):8548–8554

Kanaya S, Katsuda-Nakai C et al (1991) Importance of the positive charge cluster in Escherichia coli ribonuclease HI for the effective binding of the substrate. J Biol Chem 266 (18):11621–11627

Kati WM, Johnson KA et al (1992) Mechanism and fidelity of HIV reverse transcriptase. J Biol Chem 267(36):25988–25997
Kaushik N, Rege N et al (1996) Biochemical analysis of catalytically crucial aspartate mutants of human immunodeficiency virus type 1 reverse transcriptase. Biochemistry 35 (36):11536–11546
Kaushik N, Chowdhury K et al (2000) Valine of the YVDD motif of moloney murine leukemia virus reverse transcriptase: role in the fidelity of DNA synthesis. Biochemistry 39 (17):5155–5165
Keck JL, Marqusee S (1995) Substitution of a highly basic helix/loop sequence into the RNase H domain of human immunodeficiency virus reverse transcriptase restores its Mn(2+)-dependent RNase H activity. Proc Natl Acad Sci USA 92(7):2740–2744
Keck JL, Marqusee S (1996) The putative substrate recognition loop of Escherichia coli ribonuclease H is not essential for activity. J Biol Chem 271(33):19883–19887
Kelleher CD, Champoux JJ (1998) Characterization of RNA strand displacement synthesis by Moloney murine leukemia virus reverse transcriptase. J Biol Chem 273(16):9976–9986
Kim JK, Palaniappan C et al (1997) Evidence for a unique mechanism of strand transfer from the transactivation response region of HIV-1. J Biol Chem 272(27):16769–16777
Kim B, Ayran JC et al (1999) New human immunodeficiency virus, type 1 reverse transcriptase (HIV-1 RT) mutants with increased fidelity of DNA synthesis. Accuracy, template binding, and processivity. J Biol Chem 274(39):27666–27673
Klarmann GJ, Eisenhauer BM et al (2007) Investigating the "steric gate" of human immunodeficiency virus type 1 (HIV-1) reverse transcriptase by targeted insertion of unnatural amino acids. Biochemistry 46(8):2118–2126
Kohlstaedt LA, Wang J et al (1992) Crystal structure at 3.5 A resolution of HIV-1 reverse transcriptase complexed with an inhibitor. Science 256(5065):1783–1790
Krug MS, Berger SL (1989) Ribonuclease H activities associated with viral reverse transcriptases are endonucleases. Proc Natl Acad Sci USA 86(10):3539–3543
Kurzynska-Kokorniak A, Jamburuthugoda VK et al (2007) DNA-directed DNA polymerase and strand displacement activity of the reverse transcriptase encoded by the R2 retrotransposon. J Mol Biol 374(2):322–333
Kvaratskhelia M, Budihas SR et al (2002) Pre-existing distortions in nucleic acid structure aid polypurine tract selection by HIV-1 reverse transcriptase. J Biol Chem 277(19):16689–16696
Lampson BC, Inouye M et al (2005) Retrons, msDNA, and the bacterial genome. Cytogenet Genome Res 110(1–4):491–499
Lander ES, Linton LM et al (2001) Initial sequencing and analysis of the human genome. Nature 409(6822):860–921
Lapkouski M, Tian L et al (2013) Complexes of HIV-1 RT, NNRTI and RNA/DNA hybrid reveal a structure compatible with RNA degradation. Nat Struct Mol Biol 20(2):230–236
Le Grice SF (2003) "In the beginning": initiation of minus strand DNA synthesis in retroviruses and LTR-containing retrotransposons. Biochemistry 42(49):14349–14355
Lener D, Budihas SR et al (2002) Mutating conserved residues in the ribonuclease H domain of Ty3 reverse transcriptase affects specialized cleavage events. J Biol Chem 277 (29):26486–26495
Lewi P, Heeres J et al (2012) Reverse transcriptase inhibitors as microbicides. Curr HIV Res 10 (1):27–35
Lin JH, Levin HL (1997) A complex structure in the mRNA of Tf1 is recognized and cleaved to generate the primer of reverse transcription. Genes Dev 11(2):270–285
Luan DD, Korman MH et al (1993) Reverse transcription of R2Bm RNA is primed by a nick at the chromosomal target site: a mechanism for non-LTR retrotransposition. Cell 72(4):595–605
Malik HS, Eickbush TH (2001) Phylogenetic analysis of ribonuclease H domains suggests a late, chimeric origin of LTR retrotransposable elements and retroviruses. Genome Res 11 (7):1187–1197

Mandal D, Dash C et al (2006) Analysis of HIV-1 replication block due to substitutions at F61 residue of reverse transcriptase reveals additional defects involving the RNase H function. Nucleic Acids Res 34(10):2853–2863
Marchand B, Tchesnokov EP et al (2007) The pyrophosphate analogue foscarnet traps the pre-translocational state of HIV-1 reverse transcriptase in a Brownian ratchet model of polymerase translocation. J Biol Chem 282(5):3337–3346
Mathias SL, Scott AF et al (1991) Reverse transcriptase encoded by a human transposable element. Science 254(5039):1808–1810
McWilliams MJ, Julias JG et al (2006) Combining mutations in HIV-1 reverse transcriptase with mutations in the HIV-1 polypurine tract affects RNase H cleavages involved in PPT utilization. Virology 348(2):378–388
Miller HI, Riggs AD et al (1973) Ribonuclease H (hybrid) in Escherichia coli. Identification and characterization. J Biol Chem 248(7):2621–2624
Mizrahi V, Usdin MT et al (1990) Site-directed mutagenesis of the conserved Asp-443 and Asp-498 carboxy-terminal residues of HIV-1 reverse transcriptase. Nucleic Acids Res 18 (18):5359–5363
Mizrahi V, Brooksbank RL et al (1994) Mutagenesis of the conserved aspartic acid 443, glutamic acid 478, asparagine 494, and aspartic acid 498 residues in the ribonuclease H domain of p66/p51 human immunodeficiency virus type I reverse transcriptase. Expression and biochemical analysis. J Biol Chem 269(30):19245–19249
Moelling K (1974) Characterization of reverse transcriptase and RNase H from friend-murine leukemia virus. Virology 62(1):46–59
Nowak E, Potrzebowski W et al (2013) Structural analysis of monomeric retroviral reverse transcriptase in complex with an RNA/DNA hybrid. Nucleic Acids Res 41(6):3874–3887
Nowotny M, Yang W (2006) Stepwise analyses of metal ions in RNase H catalysis from substrate destabilization to product release. EMBO J 25(9):1924–1933
Nowotny M, Gaidamakov SA et al (2005) Crystal structures of RNase H bound to an RNA/DNA hybrid: substrate specificity and metal-dependent catalysis. Cell 121(7):1005–1016
Nowotny M, Gaidamakov SA et al (2007) Structure of human RNase H1 complexed with an RNA/DNA hybrid: insight into HIV reverse transcription. Mol Cell 28(2):264–276
Ortiz TP, Marshall JA et al (2005) Stepping statistics of single HIV-1 reverse transcriptase molecules during DNA polymerization. J Phys Chem B 109(33):16127–16131
Oude Essink BB, Back NK et al (1997) Increased polymerase fidelity of the 3TC-resistant variants of HIV-1 reverse transcriptase. Nucleic Acids Res 25(16):3212–3217
Palaniappan C, Fuentes GM et al (1996) Helix structure and ends of RNA/DNA hybrids direct the cleavage specificity of HIV-1 reverse transcriptase RNase H. J Biol Chem 271(4):2063–2070
Pandey VN, Kaushik N et al (1996) Role of methionine 184 of human immunodeficiency virus type-1 reverse transcriptase in the polymerase function and fidelity of DNA synthesis. Biochemistry 35(7):2168–2179
Pandey M, Patel S et al (2004) Insights into the role of an active site aspartate in Ty1 reverse transcriptase polymerization. J Biol Chem 279(46):47840–47848
Paulson BA, Zhang M et al (2007) Substitution of alanine for tyrosine-64 in the fingers subdomain of M-MuLV reverse transcriptase impairs strand displacement synthesis and blocks viral replication in vivo. Virology 366(2):361–376
Perach M, Hizi A (1999) Catalytic features of the recombinant reverse transcriptase of bovine leukemia virus expressed in bacteria. Virology 259(1):176–189
Powell MD, Beard WA et al (1999) Residues in the alphaH and alphaI helices of the HIV-1 reverse transcriptase thumb subdomain required for the specificity of RNase H-catalyzed removal of the polypurine tract primer. J Biol Chem 274(28):19885–19893
Preston BD, Poiesz BJ et al (1988) Fidelity of HIV-1 reverse transcriptase. Science 242 (4882):1168–1171

Pullen KA, Ishimoto LK et al (1992) Incomplete removal of the RNA primer for minus-strand DNA synthesis by human immunodeficiency virus type 1 reverse transcriptase. J Virol 66 (1):367–373
Pyatkov KI, Arkhipova IR et al (2004) Reverse transcriptase and endonuclease activities encoded by Penelope-like retroelements. Proc Natl Acad Sci USA 101(41):14719–14724
Radzio J, Sluis-Cremer N (2011) Subunit-specific mutational analysis of residue N348 in HIV-1 reverse transcriptase. Retrovirology 8:69
Rausch JW, Le Grice SF (2004) 'Binding, bending and bonding': polypurine tract-primed initiation of plus-strand DNA synthesis in human immunodeficiency virus. Int J Biochem Cell Biol 36(9):1752–1766
Rausch JW, Lener D et al (2002) Altering the RNase H primer grip of human immunodeficiency virus reverse transcriptase modifies cleavage specificity. Biochemistry 41(15):4856–4865
Reardon JE (1992) Human immunodeficiency virus reverse transcriptase: steady-state and pre-steady-state kinetics of nucleotide incorporation. Biochemistry 31(18):4473–4479
Reijns MA, Rabe B et al (2012) Enzymatic removal of ribonucleotides from DNA is essential for mammalian genome integrity and development. Cell 149(5):1008–1022
Roberts JD, Bebenek K et al (1988) The accuracy of reverse transcriptase from HIV-1. Science 242(4882):1171–1173
Roberts JD, Preston BD et al (1989) Fidelity of two retroviral reverse transcriptases during DNA-dependent DNA synthesis in vitro. Mol Cell Biol 9(2):469–476
Rodgers DW, Gamblin SJ et al (1995) The structure of unliganded reverse transcriptase from the human immunodeficiency virus type 1. Proc Natl Acad Sci USA 92(4):1222–1226
Roth MJ, Tanese N et al (1985) Purification and characterization of murine retroviral reverse transcriptase expressed in Escherichia coli. J Biol Chem 260(16):9326–9335
SanMiguel P, Gaut BS et al (1998) The paleontology of intergene retrotransposons of maize. Nat Genet 20(1):43–45
Sarafianos SG, Das K et al (2001) Crystal structure of HIV-1 reverse transcriptase in complex with a polypurine tract RNA:DNA. EMBO J 20(6):1449–1461
Schatz O, Cromme FV et al (1989) Point mutations in conserved amino acid residues within the C-terminal domain of HIV-1 reverse transcriptase specifically repress RNase H function. FEBS Lett 257(2):311–314
Schultz SJ, Champoux JJ (1996) RNase H domain of Moloney murine leukemia virus reverse transcriptase retains activity but requires the polymerase domain for specificity. J Virol 70 (12):8630–8638
Schultz SJ, Champoux JJ (2008) RNase H activity: structure, specificity, and function in reverse transcription. Virus Res 134(1–2):86–103
Schultz SJ, Zhang M et al (2004) Recognition of internal cleavage sites by retroviral RNases H. J Mol Biol 344(3):635–652
Simon DM, Zimmerly S (2008) A diversity of uncharacterized reverse transcriptases in bacteria. Nucleic Acids Res 36(22):7219–7229
Smith JS, Roth MJ (1993) Purification and characterization of an active human immunodeficiency virus type 1 RNase H domain. J Virol 67(7):4037–4049
Smith JS, Gritsman K et al (1994) Contributions of DNA polymerase subdomains to the RNase H activity of human immunodeficiency virus type 1 reverse transcriptase. J Virol 68 (9):5721–5729
Stahl SJ, Kaufman JD et al (1994) Construction of an enzymatically active ribonuclease H domain of human immunodeficiency virus type 1 reverse transcriptase. Protein Eng 7(9):1103–1108
Steitz TA (1998) A mechanism for all polymerases. Nature 391(6664):231–232
Steitz TA, Steitz JA (1993) A general two-metal-ion mechanism for catalytic RNA. Proc Natl Acad Sci USA 90(14):6498–6502
Stetor SR, Rausch JW et al (1999) Characterization of (+) strand initiation and termination sequences located at the center of the equine infectious anemia virus genome. Biochemistry 38(12):3656–3667

Tanese N, Goff SP (1988) Domain structure of the Moloney murine leukemia virus reverse transcriptase: mutational analysis and separate expression of the DNA polymerase and RNase H activities. Proc Natl Acad Sci USA 85(6):1777–1781

Telesnitsky A, Goff SP (1997) Reverse transcriptase and the generation of retroviral DNA. In: Coffin JM, Hughes SH, Varmus HE (eds) Retroviruses. Cold Spring Harbor Laboratory, Ney York

Telesnitsky A, Blain SW et al (1992) Defects in Moloney murine leukemia virus replication caused by a reverse transcriptase mutation modeled on the structure of Escherichia coli RNase H. J Virol 66(2):615–622

Temin HM, Mizutani S (1970) RNA-dependent DNA polymerase in virions of Rous sarcoma virus. Nature 226(5252):1211–1213

Tisdale M, Schulze T et al (1991) Mutations within the RNase H domain of human immunodeficiency virus type 1 reverse transcriptase abolish virus infectivity. J Gen Virol 72(Pt 1):59–66

Turner KB, Brinson RG et al (2008) Structural probing of the HIV-1 polypurine tract RNA:DNA hybrid using classic nucleic acid ligands. Nucleic Acids Res 36(8):2799–2810

Uzun O, Gabriel A (2001) A Ty1 reverse transcriptase active-site aspartate mutation blocks transposition but not polymerization. J Virol 75(14):6337–6347

Wahba L, Amon JD et al (2011) RNase H and multiple RNA biogenesis factors cooperate to prevent RNA:DNA hybrids from generating genome instability. Mol Cell 44(6):978–988

Wainberg MA, Drosopoulos WC et al (1996) Enhanced fidelity of 3TC-selected mutant HIV-1 reverse transcriptase. Science 271(5253):1282–1285

Werner S, Wohrl BM (2000) Asymmetric subunit organization of heterodimeric Rous sarcoma virus reverse transcriptase alphabeta: localization of the polymerase and RNase H active sites in the alpha subunit. J Virol 74(7):3245–3252

Whiting SH, Champoux JJ (1998) Properties of strand displacement synthesis by Moloney murine leukemia virus reverse transcriptase: mechanistic implications. J Mol Biol 278(3):559–577

Whitwam T, Peretz M et al (2001) Identification of a central DNA flap in feline immunodeficiency virus. J Virol 75(19):9407–9414

Wilhelm M, Boutabout M et al (2000) Expression of an active form of recombinant Ty1 reverse transcriptase in Escherichia coli: a fusion protein containing the C-terminal region of the Ty1 integrase linked to the reverse transcriptase-RNase H domain exhibits polymerase and RNase H activities. Biochem J 348(Pt 2):337–342

Winshell J, Champoux JJ (2001) Structural alterations in the DNA ahead of the primer terminus during displacement synthesis by reverse transcriptases. J Mol Biol 306(5):931–943

Wisniewski M, Balakrishnan M et al (2000a) The sequential mechanism of HIV reverse transcriptase RNase H. J Biol Chem 275(48):37664–37671

Wisniewski M, Balakrishnan M et al (2000b) Unique progressive cleavage mechanism of HIV reverse transcriptase RNase H. Proc Natl Acad Sci USA 97(22):11978–11983

Wisniewski M, Chen Y et al (2002) Substrate requirements for secondary cleavage by HIV-1 reverse transcriptase RNase H. J Biol Chem 277(32):28400–28410

Xiong Y, Eickbush TH (1988) Similarity of reverse transcriptase-like sequences of viruses, transposable elements, and mitochondrial introns. Mol Biol Evol 5(6):675–690

Xiong Y, Eickbush TH (1990) Origin and evolution of retroelements based upon their reverse transcriptase sequences. EMBO J 9(10):3353–3362

Yang W, Lee JY et al (2006) Making and breaking nucleic acids: two-Mg2+-ion catalysis and substrate specificity. Mol Cell 22(1):5–13

Yu SF, Baldwin DN et al (1996) Human foamy virus replication: a pathway distinct from that of retroviruses and hepadnaviruses. Science 271(5255):1579–1582

Zhan X, Crouch RJ (1997) The isolated RNase H domain of murine leukemia virus reverse transcriptase. Retention of activity with concomitant loss of specificity. J Biol Chem 272 (35):22023–22029

Zhou D, Chung S et al (2012) Crystal structures of the reverse transcriptase-associated ribonuclease H domain of xenotropic murine leukemia-virus related virus. J Struct Biol 177(3):638–645

Chapter 9
Telomerase: A Eukaryotic DNA Polymerase Specialized in Telomeric Repeat Synthesis

Andrew F. Brown, Joshua D. Podlevsky, and Julian J.-L. Chen

Abstract Telomerase is an RNA-dependent DNA polymerase (reverse transcriptase) specialized in the synthesis of short DNA repeats onto chromosome ends, a function essential for chromosome stability and cellular immortality in eukaryotes. Unlike conventional polymerases, telomerase is a ribonucleoprotein (RNP) enzyme, minimally composed of the catalytic telomerase reverse transcriptase (TERT) and the telomerase RNA (TR) subunit. While the TERT catalytic core is well conserved and shares several motifs with conventional reverse transcriptases, the TR subunit is highly divergent and has evolved species-specific structural elements essential for telomerase RNP assembly and biogenesis. Telomerase is unique among polymerases, capable of producing a DNA product vastly longer than the RNA template. This unique polymerization reaction relies on repeatedly regenerating and reusing the short TR template during DNA synthesis, producing the characteristic repetitive telomeric DNA sequence. Processive telomeric repeat synthesis is dependent on "template translocation" for template regeneration, a complex mechanism that is only partially understood. Correspondingly, telomerase-specific domains within telomerase TERT-TR core function cooperatively with telomerase accessory proteins to coordinate template translocation during processive telomeric DNA repeat synthesis. Telomerase is thus a fascinating polymerase, singular in function and unrivaled in complexity.

Keywords Telomerase • Telomere • Reverse transcriptase • Template translocation • Repeat addition processivity

---

A.F. Brown • J.D. Podlevsky • J.J.-L. Chen (✉)
Department of Chemistry & Biochemistry, Arizona State University, Tempe, AZ 85287, USA
e-mail: jlchen@asu.edu

K.S. Murakami and M.A. Trakselis (eds.), *Nucleic Acid Polymerases*, Nucleic Acids and Molecular Biology 30, DOI 10.1007/978-3-642-39796-7_9,

Abbreviations

CAB	Cajal body localization
CP2	Ciliate counterpart
CTE	C-terminal extension
IFD	Insertion in fingers domain
LINE-1	Long interspersed element-1
PLE	Penelope-like element
RID1	RNA-interacting domain 1
RNP	Ribonucleoprotein
RT	Reverse transcriptase
scaRNA	Small Cajal body RNA
snoRNA	Small nucleolar RNA
TASC	Telomere adaptor sub-complex
TBE	Template boundary element
TEN	Telomerase essential N-terminal
TERT	Telomerase reverse transcriptase
TR	Telomerase RNA
TRBD	Telomerase RNA-binding domain
TWJ	Three-way junction
VSR	Vertebrate-specific region

9.1 Introduction

Eukaryotic cells evolved linear chromosomes with special end caps, termed telomeres, for distinction from double-stranded DNA breaks. These telomere-capped ends prevent DNA fusions that often occur after double-stranded DNA breaks and lead to chromosomal rearrangements (Muller 1938; McClintock 1941). Linear chromosomes also inherit an "end replication problem." Conventional DNA polymerases cannot initiate DNA synthesis de novo, requiring an RNA primer for extension, and are thus unable to fully replicate the ends of linear DNA (Watson 1972; Olovnikov 1973). Incomplete replication of chromosome ends results in the progressive loss of terminal DNA with each cell division, endangering genome integrity. The elusive nature of the telomere cap began to be understood by the discovery of repetitive $(TTGGGG)_n$ sequences in *Tetrahymena* telomeric DNA by Blackburn in 1978 (Blackburn and Gall 1978). Shortly after, a linear DNA with terminally flanked TTGGGG repeats was not only found to be stably maintained in budding yeast cells, but was even extended with yeast telomeric DNA repeats (Szostak and Blackburn 1982). This finding predicted the presence of a telomeric DNA-extending enzyme. This enzyme, responsible for synthesis of these unusual telomeric DNA repeats, was later discovered and termed "telomerase" (Greider and Blackburn 1985).

Telomerase is an RNA-dependent DNA polymerase that contains an intrinsic and essential RNA component, thus functioning as a ribonucleoprotein (RNP) enzyme (Greider and Blackburn 1987). The telomerase RNP core is composed of the telomerase reverse transcriptase (TERT) and the integral telomerase RNA (TR). The TR provides the template for DNA synthesis by TERT as well as essential structural domains for enzymatic activity. In addition to the catalytic core, the telomerase holoenzyme consists of a variety of telomerase-associated proteins. These accessory proteins are dispensable for enzymatic activity, while crucial for telomerase biogenesis, localization, and regulation (Fu and Collins 2003; Venteicher and Artandi 2009; Egan and Collins 2010; Kiss et al. 2010). The biogenesis pathway of telomerase RNA holoenzyme has been extensively reviewed elsewhere (Egan and Collins 2012a) and will only be discussed briefly in this chapter.

Distinct from all known DNA polymerases, telomerase synthesizes a DNA product that is vastly larger than the RNA template. The telomeric DNA product is composed of repetitive copies of a simple DNA sequence, TTGGGG in *Tetrahymena* and TTAGGG in human. Identification of the *Tetrahymena thermophila* TR revealed a template sequence complementary for 1.5 copies of telomeric sequence (Greider and Blackburn 1989). This implied a "template translocation" mechanism where the half repeat of the template sequence serves as a realignment site for repositioning the DNA primer on the RNA template after each repeat synthesis (Fig. 9.1). Such a mechanism would explain the processive synthesis of long stretches of telomeric DNA repeats from a short RNA template (Greider 1991). Similar to conventional reverse transcriptase (RT) reactions, the telomerase reaction is initiated by the DNA primer annealing to the RNA template, and DNA synthesis terminates upon reaching the end of template (Fig. 9.1). Remarkably, telomerase regenerates and reuses the same RNA template for successive telomeric DNA repeat synthesis. However, the detailed mechanism underlying telomerase repeat addition processivity and template translocation is only beginning to emerge in the last few years, impeded by the complex interactions of multiple domains within TERT, TR, and accessory proteins.

9.2 Telomerase Reverse Transcriptase

The catalytic TERT protein is well conserved among most known species. It comprises four structural domains: the telomerase essential N-terminal (TEN) domain, the telomerase RNA-binding domain (TRBD), the RT domain, and the C-terminal extension (CTE) (Fig. 9.2a). The RT domain contains motifs that are conserved in all RTs and constitutes the active site for RNA-dependent DNA polymerization (Lingner et al. 1997b). The other three domains are telomerase specific and function cooperatively with TR and telomerase accessory proteins for processive telomeric repeat addition.

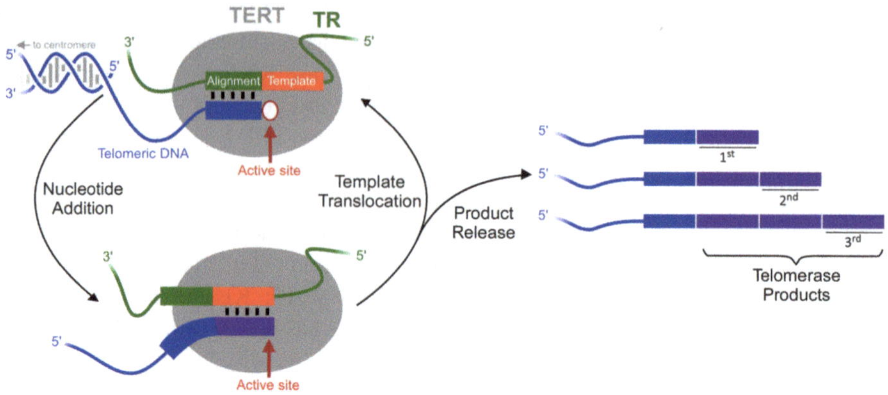

Fig. 9.1 The unique telomerase catalytic cycle. Telomerase functions as a conventional reverse transcriptase by synthesizing single-stranded DNA onto the ends of a telomeric DNA primer (*blue*) from an intrinsic RNA template (*orange*), *nucleotide addition*. Apart from conventional polymerases, upon reaching the end of the template, telomerase has the capacity to regenerate the template by a complex mechanism of *template translocation*. Following this, an additional round of *nucleotide addition* then proceeds generating longer telomerase products (*violet*). Unsuccessful *template translocation* terminates the synthesis of additional telomeric repeats by *product release*

The TEN domain of TERT has two important functions: binding the TR subunit and the single-stranded telomeric DNA (Fig. 9.2b). Binding the single-stranded telomeric DNA by the TEN domain is vital for enhancing processive repeat synthesis. TEN contains a high-affinity DNA "anchor" site for specifically binding single-stranded telomeric DNA (Jacobs et al. 2006; Lue and Li 2007; Romi et al. 2007; Sealey et al. 2010; Finger and Bryan 2008). This binding restrains the telomeric DNA within close proximity of the active site, delaying product release which in turn increases processive telomeric repeat synthesis (Wyatt et al. 2010). A conserved leucine residue outside the DNA anchor site was found to also enhance processive telomeric repeat synthesis (Zaug et al. 2008; Eckert and Collins 2012). This residue was first identified in *T. thermophila*, with putatively homologous residues found within yeast and human TEN domains. While mutating this residue decreased processive telomeric repeat synthesis, this mutation failed to decrease telomeric DNA retention by the DNA anchor site. Instead, this leucine is believed to function as a molecular switch. However, the mechanism is poorly understood at this time. Outside the DNA anchor site and the important leucine residue, TEN contains a low-affinity RNA-binding domain for binding the TR pseudoknot (Lai et al. 2001; Moriarty et al. 2004). However, the mechanistic significance of this TERT-pseudoknot interaction is also poorly understood. Interestingly, while TEN contains elements that enhance processive telomeric repeat synthesis and a low-affinity RNA-binding site, this domain is not essential for catalytic activity (Eckert and Collins 2012). Certain insect species, including *Tribolium castaneum*, have a truncated TERT N-terminus, seeming to lack the TEN domain entirely (Fig. 9.2) (Gillis et al. 2008).

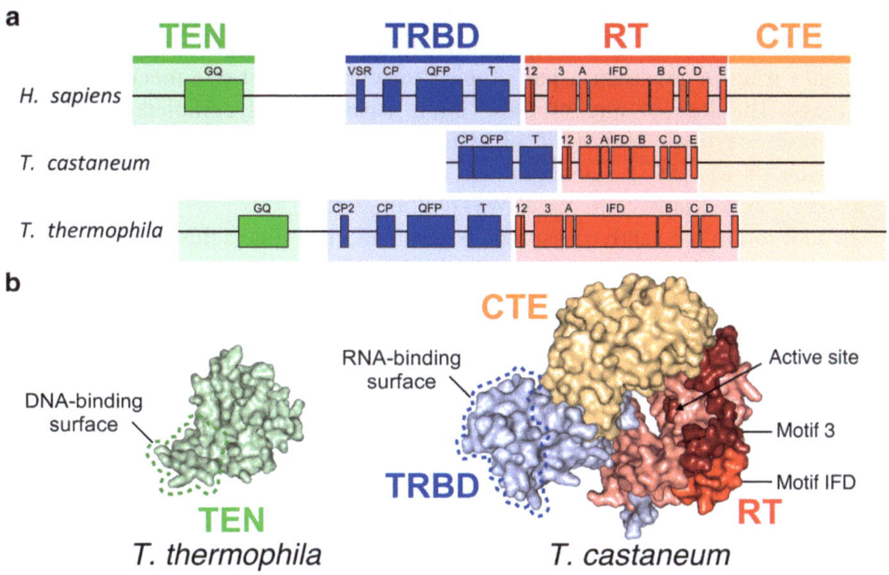

Fig. 9.2 Structural organization of the catalytic TERT protein. (**a**) TERT is composed of four structural domains: telomerase essential N-terminal (TEN, *green*) domain, telomerase RNA-binding domain (TRBD, *blue*), reverse transcriptase (RT, *red*), and the C-terminal extension (CTE, *orange*). The TEN and TRBD are telomerase-specific, essential for *template translocation*, and not found among conventional polymerases. Important motifs are colored similarly in the encompassing domains. (**b**) The crystallized *Tetrahymena thermophila* TEN and *Tribolium castaneum* TRBD crystal fragments present important nucleotide-binding surfaces. The *Tribolium castaneum* TERT protein forms a ring structure unseen in other polymerases. The TERT active site, motif 3 (*dark red*) and motif IFD (*red-orange*) in the RT domain are denoted

Unlike the TEN domain, TRBD is more conserved and found in all known TERT proteins to date (Podlevsky et al. 2008). This domain contains a high-affinity RNA interacting domain essential for RNP assembly (Moriarty et al. 2004). The highly helical TRBD structure comprises three universal motifs: CP, QFP, and T (Fig. 9.2a) (Rouda and Skordalakes 2007; Gillis et al. 2008). Separate from RNA binding, motif T has been implicated in processive telomeric repeat synthesis through proposed interactions between motif T and the RNA backbone of the TR template (Drosopoulos and Prasad 2010; Mitchell et al. 2010). In addition to the universally conserved motifs, TRBD has been shown to contain a vertebrate-specific region (VSR) and a ciliate counterpart (CP2). VSR and CP2 both bind the TR, and CP2 is involved with delineating the template boundary within the ciliate TR (Harley 2002; Moriarty et al. 2002).

The catalytic RT domain is centrally located within the primary sequence of the TERT protein (Fig. 9.2). As a reverse transcriptase, TERT contains all the hallmark motifs of DNA polymerases: 1, 2, and A through E. TERT is a right-handed polymerase containing the “fingers” and “palm” domains similar to conventional RTs. The TERT fingers bind incoming nucleotides and positions the RNA template

(Wyatt et al. 2010; Gillis et al. 2008; Bosoy and Lue 2001), while the TERT palm forms the catalytic site for DNA polymerization. Within the palm, motifs A and C contain a triad of invariant aspartic acids for conventional two-metal nucleotide synthesis. The loss of any of these three aspartic acids completely abolishes telomerase enzymatic activity in vitro and results in telomere shortening in vivo (Weinrich et al. 1997; Harrington et al. 1997; Bryan et al. 2000b; Counter et al. 1997; Nakayama et al. 1998; Wyatt et al. 2010).

In addition to the catalytic aspartic acids, flanking amino acids in motif C have been shown to play a role in repeat addition rate and processivity as well as nucleotide addition fidelity, though the effects appear to be species specific (Peng et al. 2001; Drosopoulos and Prasad 2007; Bryan et al. 2000a). TERT also contains an invariant lysine residue in motif D which is believed to function as an acid, activating the pyrophosphate generated from nucleotide addition. The loss of this lysine in TERT severely reduces telomerase activity (Miller et al. 2000; Bryan et al. 2000b; Sekaran et al. 2010). Motif E, in the TERT palm domain, functions as a primer grip for positioning the DNA primer (Peng et al. 2001; Wyatt et al. 2007). This function is further supported by the *Tribolium castaneum* TERT crystal structure, in which a coiled loop is proximal to the end of the DNA primer (Mitchell et al. 2010).

Although the TERT RT domain is well conserved among RTs, several telomerase-specific motifs have evolved within this domain for telomerase-specific functions. A large insertion found in the fingers domain, aptly termed the "insertion in fingers domain" (IFD), influences processive telomeric repeat synthesis in vitro and telomere maintenance in vivo (Lue et al. 2003). In addition to IFD, another telomerase-specific motif, motif 3—so named for its location immediately following motif 2 (Fig. 9.2a)—is directly involved in repeat addition processivity (Xie et al. 2010). Interestingly, alanine screening revealed that specific residues within motif 3 could alter the rate of telomeric repeat synthesis independent of altering the processivity of telomeric repeat synthesis. The helix-coil-helix structure of motif 3 lies atop the active site and appears to be in close contact with the RNA/DNA duplex (Gillis et al. 2008; Mitchell et al. 2010).

The C-terminal extension (CTE), comprising the C-terminus of TERT, contains little-to-no sequence homology to conventional RTs. However, the overall structure and function of CTE is similar to the "thumb" domain of retroviral RTs, specifically the HIV1 RT (Nakamura et al. 1997; Gillis et al. 2008). The TERT CTE affects telomeric DNA binding, telomerase activity, and processive telomeric repeat synthesis (Hossain et al. 2002; Huard 2003). The crystal structure of *Tribolium castaneum* TERT revealed interactions between CTE and TRBD (Fig. 9.2b) (Gillis et al. 2008). These interactions shape the global architecture of the TERT protein, forming a ring structure instead of the commonly seen horseshoe shape of conventional RTs. It has been proposed that CTE may interact with TR when bound to TRBD (Bley et al. 2011). Thus the TR could function as a brace to strengthen the CTE-TRBD interactions and maintain the TERT structure.

9.3 Telomerase RNA

Unique among RTs, telomerase contains an integral RNA component. The RNA associated with conventional RTs provides merely a template for reverse transcription. In contrast, TR is a multifaceted RNA with unique structural elements crucial for telomerase enzymatic activity in addition to providing the template for nucleotide addition. TR is remarkably divergent in size, sequence, and structure with distinct structural domains and motifs emerging along separate phylogenetic lineages (Fig. 9.3a–d). The inherent differentiation in TR is associated with a plethora of species-specific TR-binding proteins (Podlevsky and Chen 2012). However, despite this immense variation, two structural elements outside of the template are universal to all known TRs: the pseudoknot and a stem-loop moiety (Fig. 9.3a–d) (Chen et al. 2000, 2002; Chen and Greider 2004; Lin et al. 2004; Brown et al. 2007; Qi et al. 2012a). These two ubiquitous TR elements are sufficient for reconstituting telomerase activity in vitro when added to TERT either as a truncated TR or excised as two physically separated RNA fragments and combined in trans (Tesmer et al. 1999; Mitchell and Collins 2000; Qi et al. 2012a).

The pseudoknot element from human TR was found to contain a triple helix formed from Hoogsteen base pairings (Theimer et al. 2005). Additional TR pseudoknots from other species have either been shown, or predicted, to contain a similar triple helix (Qiao and Cech 2008; Shefer et al. 2007). While the pseudoknot—and apparently the triple helix—is well conserved and essential for telomerase activity, the precise function of either structure is not well known (Ly et al. 2003; Chen and Greider 2005; Qiao and Cech 2008). While the template is distant from the pseudoknot in the primary sequence, it is located proximal to the pseudoknot within the secondary structure (Fig. 9.3a–d). From this, the pseudoknot has been postulated to function for template positioning or retention of the template in proximity to the TERT active site. Furthermore, NMR structures revealed human TR has a sharp kink located between the pseudoknot and the template (Zhang et al. 2010, 2011). This kink is believed to facilitate proper positioning of the template and may be ubiquitous to TR.

In addition to the pseudoknot structure, the other universal TR element is a stem-loop moiety located downstream of the template-pseudoknot region (Fig. 9.3a–d). Discovered independently within three major phylogenetic groups, this element has been separately termed: CR4/5 in vertebrates (composed of P6 and P6.1), three-way junction (TWJ) in yeasts, and helix IV in ciliates (Brown et al. 2007; Blackburn and Collins 2011; Chen et al. 2000, 2002; Qi et al. 2012a). The vertebrate CR4/5 and yeast TWJ are both a junction of three stems, two of which are capped by an apical loop (Fig. 9.3b, c). It was found that the vertebrate CR4/5 was necessary for telomerase activity in vitro and telomere maintenance in vivo. Among the three stems in the vertebrate CR4/5, P6.1—composed of a 4 bp stem together with a 5 nt loop—is highly conserved and essential for telomerase activity (Chen et al. 2002). Mutations either disrupting the stem or altering the conserved residues in the loop abolished telomerase activity. Recent cross-linking studies mapped the essential

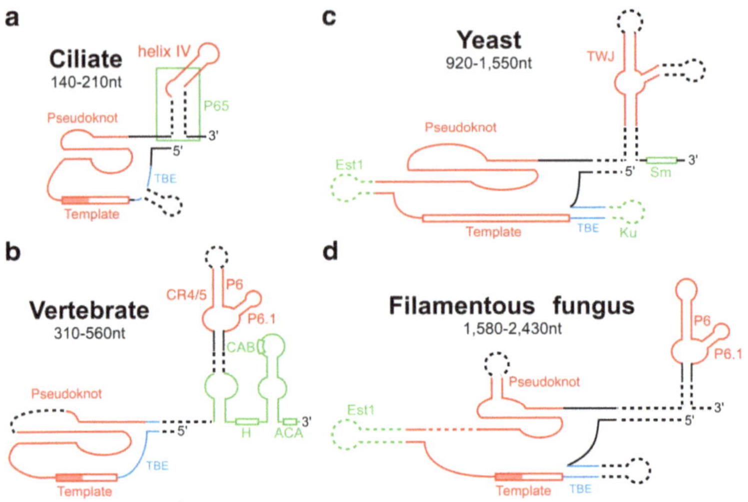

Fig. 9.3 The conserved core of the highly divergent TR. (**a**) Ciliate TRs include the smallest TR identified to date. The binding site for p65 (*green box*) is important for ciliate telomerase activity and RNP assembly. (**b**) Vertebrate TRs contain a CR4/5 domain (*red*) composed of P6 and P6.1. The 3′-proximal H/ACA domain with a CAB box in the apical loop (*green*) is crucial for TR biogenesis and telomerase localization. (**c**) Yeast TRs contain a three-way-junction element (TWJ, *red*) which is functionally dissimilar from vertebrate TRs. Additionally, there are Est1, Ku, and Sm protein-binding sites (*green*) important for telomerase activity in vivo. (**d**) Filamentous fungal TRs contain the largest known TR found and include a CR4/5-like domain (*red*) composed of P6 and P6.1 and an Est1-binding site (*green*). Two major structural elements are common to all known TRs: a template-proximal pseudoknot (*red*) and template-distal three-helical junction [P6/P6.1 and TWJ (*red*)]. Ciliates appear to have a homologous structure, helix IV (*red*). Additionally, there is a template boundary element (TBE, *blue*) which is functionally similar yet structurally divergent

vertebrate P6/6.1 stem-loop element onto the TRBD surface with single-residue resolution (Bley et al. 2011). The close proximity of the P6.1 cross-linking site with the CTE suggests the P6.1 stem-loop could interact with CTE and regulate the CTE-TRBD interactions and telomerase RNP assembly.

Yeast TWJ lacks the vertebrate conserved P6.1 loop and is dispensable for telomerase function in vitro (Brown et al. 2007; Zappulla et al. 2005). However, recent studies of fungal TRs from the fission yeast *Schizosaccharomyces pombe* and Pezizomycotina, such as the filamentous fungi *Neurospora crassa* and *Aspergillus nidulans*, identified a vertebrate-like P6.1 in this region (Fig. 9.3d). The filamentous fungal P6 and P6.1 were also found to be essential for telomerase activity in vitro (Qi et al. 2012a). While ciliates do not have a three-way junction, there is a helix moiety—helix IV—which is weakly bound by TERT and is necessary for telomerase activity (Mason et al. 2003; Lai et al. 2003). The ciliate helix IV has been postulated to be functionally analogous to P6.1 which is conserved in both vertebrates and filamentous fungi (Blackburn and Collins 2011).

Since the TR is vastly larger than the template sequence, a physical boundary is needed to define where reverse transcription must terminate. This template boundary element (TBE) prevents reverse transcription into the 5′ flanking region and the synthesis of non-telomeric DNA sequences. The mechanisms and structural elements of TBE appear to be as divergent as the rest of TR (Fig. 9.3a–d). In ciliates, the 5′ TBE includes conserved sequences at the base of the template adjacent helix II (Autexier and Greider 1995; Lai et al. 2002). The ciliate TBE also serves as a high-affinity TERT binding site, suggesting that the TBE is defined by TERT binding which blocks DNA synthesis from this region of the RNA (Fig. 9.3a). In budding yeast, the TBE is a stable helix located 0–2 nt 5′ of the template (Tzfati et al. 2000; Seto et al. 2003). This helix provides a physical barrier by limiting single-stranded RNA in this region that could function as template (Fig. 9.3c). In fission yeasts, the TBE partially overlaps with the template. This duality in function for specific nucleotides in the TR is responsible for the telomeric repeat heterogeneity found in *S. pombe* telomeres (Box et al. 2008). Filamentous fungal TRs appear to have a similar TBE structure to yeast, composed of a template adjacent helix (Qi et al. 2012a). This suggests that the filamentous fungal TBE is functionally similar to the yeast TBE as well (Fig. 9.3d). In human and most vertebrate TRs, the TBE is located 6–8 nt 5′ of the template and is also composed of a helical structure (Fig. 9.3b). The vertebrate TBE restrains the template in the active site during reverse transcription, rather than steric hindrance by protein–RNA or RNA–RNA interactions as seen in ciliates and yeasts (Moriarty et al. 2005; Chen and Greider 2003b). Surprisingly in vertebrates, certain rodent TRs lack this template-proximal helix and have only two remaining residues 5′ of the template (Hinkley et al. 1998). In these rodents, the lack of any significant sequence 5′ of template serves functionally as TBE (Chen and Greider 2003b).

Outside the universal template-pseudoknot and distal stem-loop moiety domains are species-specific TR structural elements that are essential for biogenesis, localization, and accumulation (Fig. 9.3a–d). Within vertebrate TR, the 3′ end contains an H/ACA domain (Fig. 9.3b). The H/ACA domain is composed of two stem-loops separated by a box H motif and followed by a box ACA motif, a structure also conserved in small nucleolar (sno) and small Cajal body (sca) RNAs (Kiss et al. 2010; Li 2008; Mitchell et al. 1999; Vulliamy et al. 2006). In accordance with the conservation of this moiety with sno- and scaRNAs, two copies of a protein complex composed of dyskerin, NOP10, NHP2, and GAR1 bind to each stem of the H/ACA domain for 3′-end processing and RNA accumulation (Egan and Collins 2012b; Mitchell et al. 1999). Additionally, the terminal loop of the H/ACA domain contains a Cajal body localization (CAB box) motif important for proper RNA localization and a biogenesis promoting (BIO box) motif (Reichow et al. 2007; Venteicher and Artandi 2009; Theimer et al. 2007). Furthermore, the 5′ end of the TR contains a guanosine-rich track which is postulated to form a G-quadruplex structure (Lattmann et al. 2011; Sexton and Collins 2011). The HEXH box RNA helicase RHAU is a part of the active telomerase holoenzyme complex which increases TR accumulation and would resolve the putative G-quadruplex structure.

Fungal TRs are significantly larger than vertebrate TRs and appear to function as a flexible scaffold for binding a separate set of telomerase accessory proteins (Fig. 9.3) (Zappulla et al. 2005). Between the template and the pseudoknot is an extended helix which binds a telomerase accessory protein called ever shorter telomere protein 1 (Est1p), which is essential for telomere maintenance in vivo (Seto et al. 2002; Evans and Lundblad 2002). Also found within all known yeast TRs is an Sm protein-binding site (Fig. 9.3c), which is essential for TR 3′ end processing (Seto et al. 1999). Yeast TRs also contain a template adjacent extended helix, and in most yeast TRs harbor a binding site for the Ku heterodimer, yKu70 and yKu80 (Stellwagen et al. 2003; Fisher and Zakian 2005). *Kluyveromyces lactis* TR includes this extended template-proximal helix; however it lacks Ku heterodimer association (Kabaha et al. 2008). TR is a functional RNA component of telomerase. Rather than merely providing the template for reverse transcription—as is true for all other known RTs—this RNA is an integral component, essential for telomerase activity, with functions yet to be fully understood.

9.4 Evolutionary Aspects of Telomerase RNP

In the transition from circular to linear chromosomes, ancestral eukaryotes required a means to counterbalance progressive terminal DNA loss after chromosome replication. The solution that emerged presumably was to extend the 3′ ends of genomic DNA by a primitive reverse transcriptase, the precursor to TERT, which used a simple RNA as the template. Over time, this TERT precursor evolved specific RNA-binding domains for internalizing this simple RNA template that later became the integral TR component. While little remains of this ancestral TR, the discovery of TRs from several phylogenetic groups is slowly increasing our understanding of TR evolution (Qi et al. 2012a; Sandhu et al. 2013).

The overarching structural and functional conservation within the TRs from fungi and vertebrates suggests that the pseudoknot and TWJ were important features of a common ancestor TR before the branching of fungi and vertebrate lineages. Similar to the TWJ in vertebrates and fungi, stem-loop IV in ciliates is a binding site for TERT and is required for telomerase RNP assembly (Robart et al. 2010). It is unclear if the TWJ structure in yeasts and vertebrates emerged independently or evolved from an ancestral structure homologous to the ciliate stem-loop IV.

Unlike TR, TERT is highly conserved across all explored taxa, although many adaptations have specialized TERT for DNA repeat synthesis. Phylogenetic comparative analysis indicates that TERT shares sequence attributes with eukaryotic retrotransposon RTs, a large and diverse group of self-replicating genetic elements found throughout eukaryotic genomes. The similarities between TERT and retrotransposon RTs provide insight into the evolutionary origins of the telomerase RNP. While the majority of retrotransposons employ endonuclease activity prior to nucleotide synthesis, several retrotransposons—such as Penelope-like element

(PLE) and long interspersed element-1 (LINE-1)—are endonuclease independent and require a free DNA 3′ end for retrotransposition (Kopera et al. 2011). These observations suggest a common ancestral RT in early eukaryotes diverged and specialized into various retrotransposons, including the unique telomerase enzyme.

9.5 Telomerase Mechanism

The mechanism by which telomerase functions is astonishingly complex. In addition to catalyzing nucleotide addition—similar to other polymerases—telomerase evolved a novel mechanism for the processive synthesis of telomeric repeats. Conventional RTs bind pre-annealed RNA/DNA duplex as substrate. While the substrate for telomerase is the single-stranded telomeric DNA, once bound to telomerase, the 3′ end of the DNA base-pairs with the RNA template to form a DNA/RNA duplex which is then bound by the TERT catalytic core. Thus the substrate for the TERT catalytic core is similar to the substrate of conventional RTs, while the cellular substrate for telomerase is only single-stranded DNA at chromosome ends. The complex and highly regulated recruitment of telomerase to telomeres in vivo has been reviewed elsewhere (Nandakumar and Cech 2013) and is not discussed in detail here.

Telomerase employs an incredibly short region within its integral TR component as template, while conventional RTs employ vastly longer RNA templates for the synthesis of complementary DNA. Remarkably, from this short template, TERT is capable of synthesizing a vastly larger DNA product within a single uninterrupted reaction. This is accomplished by regeneration of the template through a template translocation mechanism that readies the template for an additional cycle of DNA polymerization. This repeated DNA synthesis produces the characteristic telomeric repeat sequence. While the telomerase catalytic cycle is complex and remains poorly understood, still much has been discerned in recent years—sufficient for a working model. In this model, the template translocation cycle is minimally composed of five distinct steps: (1) nucleotide addition, (2) duplex disassociation, (3) strand separation, (4) template realignment, and (5) duplex binding, which are discussed below in much detail.

9.5.1 Nucleotide Addition

Similar to all known RTs, telomerase catalyzes nucleotide addition onto the 3′-OH of a given DNA strand employing the RNA strand as a template (Fig. 9.4a, *step 1*). The mechanism for DNA polymerization of incoming nucleotides is similar to other RTs—a two-metal ion system involving three invariant aspartic acids in the TERT palm domain which coordinate two magnesium ions (Lingner et al. 1997a; Gillis et al. 2008). Free nucleotides are bound and positioned by the nucleotide-binding pocket between the fingers and palm domain in TERT (Gillis et al. 2008).

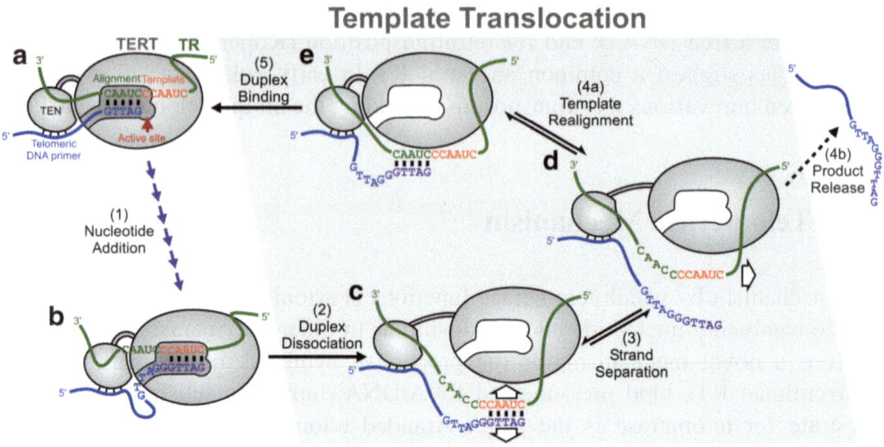

Fig. 9.4 A working model for the telomerase template translocation mechanism. (**a**) Schematic of human telomerase bound to a telomeric DNA primer (*blue*). The TR alignment region (*green*) is base-paired with the 3′-end of the telomeric DNA primer (*blue*) to form 5 base pairs adjacent to the active site (*red arrow*). The TR template (*orange*) is constrained by flanking sequences bound to the TERT protein, while the 5′ region of the telomeric DNA is bound to the TEN domain (*gray*). TERT (*gray*) catalyzes the addition of six deoxyribonucleotides (*violet arrows*) to the 3′ end of the DNA primer by reverse transcribing the TR template sequence (*step 1*). (**b**) After nucleotide addition, a new repeat (*violet*) is generated which retains only 5 base pairs. After reaching the end of the template, the duplex dissociates from the active site (*step 2*). (**c**) Outside the active site, the RNA/DNA duplex undergoes template translocation (*light gray box*) involving strand separation (*step 3*) and (**d**) template realignment (*step 4a*) to reform 5 base pairs. (**e**) Further nucleotide addition can proceed with the regenerated template. Unsuccessful realignment of the DNA primer to the RNA template eventually results in complete dissociation of the DNA product from the enzyme (*step 4b*). Strand separation and template realignment (*steps 3* and *4a*) are postulated as reversible, with multiple binding/separation steps possible

Upon base-pairing the telomeric DNA primer with the TR template, the RNA/DNA duplex is positioned within the active site, presumably by the cooperation of motif 3, IFD, and CTE which surround the central cavity in the *Tribolium castaneum* TERT crystal structure (Gillis et al. 2008; Mitchell et al. 2010; Xie et al. 2010).

After each nucleotide addition, the RNA/DNA duplex advances away from the catalytic site so as to position the 3′ end of the newly incorporated nucleotide adjacent to the catalytic site for further nucleotide addition. This "duplex translocation" during nucleotide addition is believed to be facilitated by both the flanking single-stranded DNA and RNA through a possible DNA secondary structure formation and/or the stretch and compression of the bound RNA (Jarstfer and Cech 2002; Berman et al. 2011). Weak interactions between the flanking single-stranded DNA and TERT surface could also promote duplex translocation during nucleotide addition.

During nucleotide addition, the active site-bound RNA/DNA duplex remains at a constant length. Evidence from both human and yeast telomerase reactions indicates that the extended RNA/DNA duplex is maintained at 5–7 base pairs by the disruption of a base pair at the 5′ end of the DNA for each new base pair formed at the 3′ end (Qi et al. 2012b; Förstemann and Lingner 2005). Interestingly, in the *T. castaneum* TERT crystal structure, the central cavity of the protein is estimated to accommodate a duplex of 7–8 base pairs (Gillis et al. 2008). Moreover, the diameter of the central cavity decreases by 2–3 Å upon binding an RNA/DNA duplex, possibly an induced-fit conformational change (Mitchell et al. 2010). Disrupting base pairs at the 5′ end of the DNA also appears to be functional significant for duplex translocation through the active site during processive nucleotide addition. When supplied with pre-annealed five base-paired RNA/DNA duplex substrates, a template-free telomerase that has the template region deleted from TR can processively add multiple nucleotides to only duplexes with a 5′ single-stranded DNA overhang (Qi et al. 2012b). This suggests that TERT binds the unpaired single-stranded DNA for duplex translocation through the active site during nucleotide addition.

9.5.2 Duplex Dissociation

Upon reaching the end of the template, the RNA/DNA duplex must separate, realign, and re-anneal so that the 3′ end of the DNA is again at the start of the template for further nucleotide addition (Fig. 9.4b, *step 2*). Recent evidence suggests that RNA/DNA duplex dissociation from the active site is the first step of template translocation, occurring prior to strand separation and template realignment and duplex formation (Fig. 9.4c–e). In a pulse-chase assay, a five-base-pair RNA/DNA duplex significantly reduced telomerase repeat addition processivity (Qi et al. 2012b). Thus the active site is temporarily unoccupied during template translocation, supporting template translocation occurring outside the active site.

Separation of the duplex without dissociation from the active site seems less feasible. It would require a significant conformation change in the telomerase RNP to break the hydrogen bonds of the RNA/DNA duplex as well as the protein-nucleic acid contacts restraining the duplex within the active site. This is not consistent with the relatively minor changes observed in the crystal structures of *T. castaneum* TERT with and without an RNA/DNA duplex bound in the active site (Mitchell et al. 2010). Furthermore, the correlation between the repeat addition rate in telomerase mutants and the enzyme turnover rate with duplex substrates indicates that duplex dissociation is the rate-limiting step of template translocation (Qi et al. 2012b). This could explain the observation that changes in template sequence can drastically affect telomerase repeat addition rate, which cannot be explained by the thermodynamic stability of the RNA/DNA duplex (Drosopoulos et al. 2005). Under this model, changes in the dissociation rate of the duplexes with altered template sequences could be the cause of the changes in repeat addition rate.

9.5.3 Strand Separation

Once the RNA/DNA duplex dissociates from the active site, the duplex must dissociate into separate DNA and RNA strands for template realignment and re-annealing the RNA/DNA duplex so that the 3′-OH of the DNA is adjacent to the unoccupied template for further nucleotide addition (Fig. 9.4, *step 3*). The underlying mechanism for strand separation has yet to be determined. However, a 5–6-base-pair duplex—maintained during nucleotide addition—once released from the active site, could spontaneously separate in milliseconds at 37 °C based on thermodynamic predictions (Qi et al. 2012b).

Proceeding separation from the RNA template, the 3′ end of the DNA is no longer bound to the active site (Fig. 9.4d). To prevent complete DNA dissociation and the termination of repeat synthesis, the telomerase holoenzyme evolved several binding sites for DNA retention proximal to the active site, which greatly enhances repeat addition processivity (Fig. 9.4, *step 4a, 4b*). An upstream region of the DNA remains bound to TEN, even when the 3′ end of the DNA has dissociated from the active site (Fig. 9.2b) (Jacobs et al. 2006; Wyatt et al. 2007; Finger and Bryan 2008; Zaug et al. 2008). This DNA–protein interaction tethers the telomeric DNA to the TERT protein during template translocation so that only the DNA 3′ end dissociates from the active site for template realignment.

Additionally in human telomerase, the accessory proteins POT1 and TPP1 have been shown to greatly increase telomerase repeat addition processivity by decreasing DNA dissociation through synergetic interactions with single-stranded DNA and TERT (Wang et al. 2007; Latrick and Cech 2010). As part of the telomeric protein-nucleic acid complex, the POT1–TPP1 complex recruits telomerase to telomeric DNA in vivo in a cell cycle-dependent manner (Nandakumar et al. 2012; Sexton et al. 2012; Zhong et al. 2012; Zhang et al. 2013). While the DNA is constrained within proximity to the active site, the unbound 3′ end can re-anneal either with the template region, producing an un-extendable duplex, or with the alignment region ready for another repeat synthesis (Fig. 9.4, *step 4a*).

In ciliates, a functional equivalent to the POT1–TPP1 complex has been found, Teb1 and the three-subunit telomere adaptor sub-complex (TASC). The Teb1–TASC complex dramatically enhances *T. thermophila* telomerase repeat addition processivity (Min and Collins 2009, 2010). It has been postulated that there are multiple low-affinity DNA-binding sites on TERT and Teb1–TASC complex allowing single-stranded DNA to thread through the enzyme complex, while high-affinity binding sites on Teb1 are hypothesized to prevent secondary structure formation of the upstream DNA that could induce complete DNA dissociation from the enzyme (Min and Collins 2010; Collins 2011).

9.5.4 Template Realignment

Following strand separation, the DNA primer must re-anneal to the alignment region of the RNA template for further nucleotide addition (Fig. 9.4, *step 5*). However, the DNA could completely dissociate from the protein and terminate repeat addition (Fig. 9.4, *step 4b*), or the DNA could re-anneal and occupy the template, resulting in an unproductive template translocation event (Fig. 9.4d). The probability of successful template translocation over complete DNA dissociation defines the "translocation efficiency." Since the DNA must re-anneal to the alignment region for successful template translocation, the length of the alignment region of adjoining the template has been found to greatly influence repeat addition processivity (Chen and Greider 2003a). Steric forces from stretching and compressing the TR template-flanking linkers during nucleotide addition may assist in repositioning the DNA toward the alignment region (Berman et al. 2011).

9.5.5 Duplex Binding

After the DNA has re-annealed with the alignment region of the RNA template, the duplex is then ready to rebind to the TERT active site for another round of nucleotide addition (Fig. 9.4e). The newly formed duplex is incredibly short, at only 5 base pairs for the human telomerase reaction. This diminutive duplex is rather unstable at 37 °C. Rapid binding by the TERT active site is needed to stabilize the duplex with protein-nucleic acid contacts and result in a successful template translocation event (Fig. 9.4, *step 5*). Experiments with template-free telomerase and short duplex substrates show a strong correlation between TERT duplex binding affinity and the repeat addition processivity (Xie et al. 2010; Qi et al. 2012b). TERT mutations in motif 3, IFD, and CTE that reduced repeat addition processivity increased the Km for RNA/DNA duplex substrates and decreased template translocation efficiency (Qi et al. 2012b). This suggests that the TERT binding affinity for RNA/DNA duplex is a major determinant of telomerase repeat addition processivity.

9.6 Concluding Remarks

Over the past three decades, the importance of telomerase for telomere maintenance, cellular immortality, cancer, and human health has become clear. In humans, telomerase expression is tightly regulated: downregulated in healthy human somatic cells, reactivated in most cancer cells, and persistently expressed in germline and stem cells—with the loss or reduction of activity eventually bringing about age-related human diseases (Hiyama and Hiyama 2007). Germline and stem cell defects which reduce telomerase activity result in a variety of human diseases, including dyskeratosis congenita, aplastic anemia, and idiopathic pulmonary fibrosis (Armanios 2009).

Telomerase has become a target for anticancer therapeutics as well as the promise of extended cellular replicative potential. Currently, a variety of telomerase-related cancer therapies have shown promise in preclinical studies, and a few have even moved into clinical trials (Agrawal et al. 2012). While telomerase inhibitors are of great interest, to suppress cancer cell progression, telomerase activators are also of growing interest, in the hope of delaying cellular and potentially even organismal aging. Despite detectable levels of telomerase activity, adult stem cells age with corresponding telomere shortening, telomere dysfunction, and limited cellular renewal capacities (Sahin and Depinho 2010). While telomerase-based therapeutics are still at inception, it has become increasingly clear that still greater knowledge of the inner workings of telomerase is necessary to design new, highly specific therapies that target individual aspects of the telomerase RNP. Further elucidation of telomerase mechanism will thus have direct and broad impacts on biomedical research and human health.

Acknowledgments This work was supported by National Institutes of Health (NIH) Grant R01GM094450 to J.J-L.C.

References

Agrawal A, Dang S, Gabrani R (2012) Recent patents on anti-telomerase cancer therapy. Recent Pat Anticancer Drug Discov 7(1):102–117

Armanios M (2009) Syndromes of telomere shortening. Annu Rev Genomics Hum Genet 10:45–61

Autexier C, Greider CW (1995) Boundary elements of the Tetrahymena telomerase RNA template and alignment domains. Genes Dev 9(18):2227–2239

Berman AJ, Akiyama BM, Stone MD, Cech TR (2011) The RNA accordion model for template positioning by telomerase RNA during telomeric DNA synthesis. Nat Struct Mol Biol 18:1371–1375

Blackburn EH, Collins K (2011) Telomerase: an RNP enzyme synthesizes DNA. Cold Spring Harb Perspect Biol 3(5):a003558

Blackburn EH, Gall JG (1978) A tandemly repeated sequence at the termini of the extrachromosomal ribosomal RNA genes in Tetrahymena. J Mol Biol 120(1):33–53

Bley CJ, Qi X, Rand DP, Borges CR, Nelson RW, Chen JJ-L (2011) RNA-protein binding interface in the telomerase ribonucleoprotein. Proc Natl Acad Sci USA 108:20333–20338

Bosoy D, Lue NF (2001) Functional analysis of conserved residues in the putative "finger" domain of telomerase reverse transcriptase. J Biol Chem 276(49):46305–46312

Box JA, Bunch JT, Zappulla DC, Glynn EF, Baumann P (2008) A flexible template boundary element in the RNA subunit of fission yeast telomerase. J Biol Chem 283(35):24224–24233

Brown Y, Abraham M, Pearl S, Kabaha MM, Elboher E, Tzfati Y (2007) A critical three-way junction is conserved in budding yeast and vertebrate telomerase RNAs. Nucleic Acids Res 35:6280–6289

Bryan TM, Goodrich KJ, Cech TR (2000a) A mutant of Tetrahymena telomerase reverse transcriptase with increased processivity. J Biol Chem 275(31):24199–24207

Bryan TM, Goodrich KJ, Cech TR (2000b) Telomerase RNA bound by protein motifs specific to telomerase reverse transcriptase. Mol Cell 6(2):493–499
Chen J-L, Greider CW (2003a) Determinants in mammalian telomerase RNA that mediate enzyme processivity and cross-species incompatibility. EMBO J 22:304–314
Chen J-L, Greider CW (2003b) Template boundary definition in mammalian telomerase. Genes Dev 17(22):2747–2752
Chen J-L, Greider CW (2004) An emerging consensus for telomerase RNA structure. Proc Natl Acad Sci USA 101:14683–14684
Chen J-L, Greider CW (2005) Functional analysis of the pseudoknot structure in human telomerase RNA. Proc Natl Acad Sci USA 102:8080–8085
Chen J-L, Blasco MA, Greider CW (2000) Secondary structure of vertebrate telomerase RNA. Cell 100:503–514
Chen J-L, Opperman K, Greider C (2002) A critical stem-loop structure in the CR4-CR5 domain of mammalian telomerase RNA. Nucleic Acids Res 30:592–597
Collins K (2011) Single-stranded DNA repeat synthesis by telomerase. Curr Opin Chem Biol 15 (5):643–648
Counter CM, Meyerson M, Eaton EN, Weinberg RA (1997) The catalytic subunit of yeast telomerase. Proc Natl Acad Sci USA 94(17):9202–9207
Drosopoulos WC, Prasad VR (2007) The active site residue Valine 867 in human telomerase reverse transcriptase influences nucleotide incorporation and fidelity. Nucleic Acids Res 35 (4):1155–1168
Drosopoulos WC, Prasad VR (2010) The telomerase-specific T motif is a restrictive determinant of repetitive reverse transcription by human telomerase. Mol Cell Biol 30:447–459
Drosopoulos WC, Direnzo R, Prasad VR (2005) Human telomerase RNA template sequence is a determinant of telomere repeat extension rate. J Biol Chem 280:32801–32810
Eckert B, Collins K (2012) Roles of the telomerase reverse transcriptase N-terminal domain in the assembly and activity of Tetrahymena telomerase holoenzyme. J Biol Chem 287:12805–12814
Egan ED, Collins K (2010) Specificity and stoichiometry of subunit interactions in the human telomerase holoenzyme assembled in vivo. Mol Cell Biol 30:2775–2786
Egan ED, Collins K (2012a) Biogenesis of telomerase ribonucleoproteins. RNA 18 (10):1747–1759
Egan ED, Collins K (2012b) An enhanced H/ACA RNP assembly mechanism for human telomerase RNA. Mol Cell Biol 32(13):2428–2439
Evans SK, Lundblad V (2002) The Est1 subunit of Saccharomyces cerevisiae telomerase makes multiple contributions to telomere length maintenance. Genetics 162(3):1101–1115
Finger SN, Bryan TM (2008) Multiple DNA-binding sites in Tetrahymena telomerase. Nucleic Acids Res 36:1260–1272
Fisher TS, Zakian VA (2005) Ku: a multifunctional protein involved in telomere maintenance. DNA Repair 4:1215–1226
Förstemann K, Lingner J (2005) Telomerase limits the extent of base pairing between template RNA and telomeric DNA. EMBO Rep 6:361–366
Fu D, Collins K (2003) Distinct biogenesis pathways for human telomerase RNA and H/ACA small nucleolar RNAs. Mol Cell 11:1361–1372
Gillis AJ, Schuller AP, Skordalakes E (2008) Structure of the Tribolium castaneum telomerase catalytic subunit TERT. Nature 455:633–637
Greider CW (1991) Telomerase is processive. Mol Cell Biol 11(9):4572–4580
Greider CW, Blackburn EH (1985) Identification of a specific telomere terminal transferase activity in Tetrahymena extracts. Cell 43:405–413
Greider CW, Blackburn EH (1987) The telomere terminal transferase of Tetrahymena is a ribonucleoprotein enzyme with two kinds of primer specificity. Cell 51:887–898
Greider CW, Blackburn EH (1989) A telomeric sequence in the RNA of Tetrahymena telomerase required for telomere repeat synthesis. Nature 337(6205):331–337
Harley CB (2002) Telomerase is not an oncogene. Oncogene 21(4):494–502

Harrington L, Zhou W, McPhail T, Oulton R, Yeung DS, Mar V, Bass MB, Robinson MO (1997) Human telomerase contains evolutionarily conserved catalytic and structural subunits. Genes Dev 11(23):3109–3115

Hinkley CS, Blasco MA, Funk WD, Feng J, Villeponteau B, Greider CW, Herr W (1998) The mouse telomerase RNA 5'-end lies just upstream of the telomerase template sequence. Nucleic Acids Res. 26: 532–536.

Hiyama E, Hiyama K (2007) Telomere and telomerase in stem cells. Br J Cancer 96:1020–1024

Hossain S, Singh S, Lue NF (2002) Functional analysis of the C-terminal extension of telomerase reverse transcriptase. A putative "thumb" domain. J Biol Chem 277:36174–36180

Huard S (2003) The C terminus of the human telomerase reverse transcriptase is a determinant of enzyme processivity. Nucleic Acids Res 31:4059–4070

Jacobs SA, Podell ER, Cech TR (2006) Crystal structure of the essential N-terminal domain of telomerase reverse transcriptase. Nat Struct Mol Biol 13:218–225

Jarstfer MB, Cech TR (2002) Effects of nucleotide analogues on Euplotes aediculatus telomerase processivity: evidence for product-assisted translocation. Biochemistry 41:151–161

Kabaha MM, Zhitomirsky B, Schwartz I, Tzfati Y (2008) The 5′ arm of Kluyveromyces lactis telomerase RNA is critical for telomerase function. Mol Cell Biol 28(6):1875–1882

Kiss T, Fayet-Lebaron E, Jády BE (2010) Box H/ACA small ribonucleoproteins. Mol Cell 37:597–606

Kopera HC, Moldovan JB, Morrish TA, Garcia-Perez JL, Moran JV (2011) Similarities between long interspersed element-1 (LINE-1) reverse transcriptase and telomerase. Proc Natl Acad Sci USA 108(51):20345–20350

Lai CK, Mitchell JR, Collins K (2001) RNA binding domain of telomerase reverse transcriptase. Mol Cell Biol 21:990–1000

Lai CK, Miller MC, Collins K (2002) Template boundary definition in Tetrahymena telomerase. Genes Dev 16(4):415–420

Lai CK, Miller MC, Collins K (2003) Roles for RNA in telomerase nucleotide and repeat addition processivity. Mol Cell 11(6):1673–1683

Latrick CM, Cech TR (2010) POT1-TPP1 enhances telomerase processivity by slowing primer dissociation and aiding translocation. EMBO J 29:924–933

Lattmann S, Stadler MB, Vaughn JP, Akman SA, Nagamine Y (2011) The DEAH-box RNA helicase RHAU binds an intramolecular RNA G-quadruplex in TERC and associates with telomerase holoenzyme. Nucleic Acids Res 39(21):9390–9404

Li H (2008) Unveiling substrate RNA binding to H/ACA RNPs: one side fits all. Curr Opin Struct Biol 18:78–85

Lin J, Ly H, Hussain A, Abraham M, Pearl S, Tzfati Y, Parslow TG, Blackburn EH (2004) A universal telomerase RNA core structure includes structured motifs required for binding the telomerase reverse transcriptase protein. Proc Natl Acad Sci USA 101:14713–14718

Lingner J, Hughes TR, Shevchenko A, Mann M, Lundblad V, Cech TR (1997) Reverse transcriptase motifs in the catalytic subunit of telomerase. Science 276:561–567

Lue NF, Li Z (2007) Modeling and structure function analysis of the putative anchor site of yeast telomerase. Nucleic Acids Res 35:5213–5222

Lue NF, Lin Y-C, Mian IS (2003) A conserved telomerase motif within the catalytic domain of telomerase reverse transcriptase is specifically required for repeat addition processivity. Mol Cell Biol 23:8440–8449

Ly H, Blackburn EH, Parslow TG (2003) Comprehensive structure-function analysis of the core domain of human telomerase RNA. Mol Cell Biol 23:6849–6856

Mason DX, Goneska E, Greider CW (2003) Stem-loop IV of tetrahymena telomerase RNA stimulates processivity in trans. Mol Cell Biol 23(16):5606–5613

McClintock B (1941) The stability of broken ends of chromosomes in zea mays. Genetics 26:234–282

Miller MC, Liu JK, Collins K (2000) Template definition by Tetrahymena telomerase reverse transcriptase. EMBO J 19:4412–4422

Min B, Collins K (2009) An RPA-related sequence-specific DNA-binding subunit of telomerase holoenzyme is required for elongation processivity and telomere maintenance. Mol Cell 36 (4):609–619

Min B, Collins K (2010) Multiple mechanisms for elongation processivity within the reconstituted tetrahymena telomerase holoenzyme. J Biol Chem 285(22):16434–16443

Mitchell JR, Collins K (2000) Human telomerase activation requires two independent interactions between telomerase RNA and telomerase reverse transcriptase. Mol Cell 6:361–371

Mitchell JR, Cheng J, Collins K (1999) A box H/ACA small nucleolar RNA-like domain at the human telomerase RNA 3′ end. Mol Cell Biol 19(1):567–576

Mitchell M, Gillis A, Futahashi M, Fujiwara H, Skordalakes E (2010) Structural basis for telomerase catalytic subunit TERT binding to RNA template and telomeric DNA. Nat Struct Mol Biol 17:513–518

Moriarty TJ, Huard S, Dupuis S, Autexier C (2002) Functional multimerization of human telomerase requires an RNA interaction domain in the N terminus of the catalytic subunit. Mol Cell Biol 22:1253–1265

Moriarty TJ, Marie-Egyptienne DT, Autexier C (2004) Functional organization of repeat addition processivity and DNA synthesis determinants in the human telomerase multimer. Mol Cell Biol 24:3720–3733

Moriarty TJ, Marie-Egyptienne DT, Autexier C (2005) Regulation of 5′ template usage and incorporation of noncognate nucleotides by human telomerase. RNA 11(9):1448–1460

Muller HJ (1938) The remaking of chromosomes. Collecting Net 13(8):181–198

Nakamura TM, Morin GB, Chapman KB, Weinrich SL, Andrews WH, Lingner J, Harley CB, Cech TR (1997) Telomerase catalytic subunit homologs from fission yeast and human. Science 277: 955–959.

Nakayama J, Tahara H, Tahara E, Saito M, Ito K, Nakamura H, Nakanishi T, Ide T, Ishikawa F (1998) Telomerase activation by hTRT in human normal fibroblasts and hepatocellular carcinomas. Nat Genet 18(1):65–68

Nandakumar J, Cech TR (2013) Finding the end: recruitment of telomerase to telomeres. Nat Rev Mol Cell Biol 14(2):69–82. doi:10.1038/nrm3505

Nandakumar J, Bell CF, Weidenfeld I, Zaug AJ, Leinwand LA, Cech TR (2012) The TEL patch of telomere protein TPP1 mediates telomerase recruitment and processivity. Nature 492:285–289

Olovnikov AM (1973) A theory of marginotomy. The incomplete copying of template margin in enzymic synthesis of polynucleotides and biological significance of the phenomenon. J Theor Biol 41:181–190

Peng Y, Mian IS, Lue NF (2001) Analysis of telomerase processivity: mechanistic similarity to HIV-1 reverse transcriptase and role in telomere maintenance. Mol Cell 7:1201–1211

Podlevsky JD, Chen JJ-L (2012) It all comes together at the ends: telomerase structure, function, and biogenesis. Mutat Res 730:3–11

Podlevsky JD, Bley CJ, Omana RV, Qi X, Chen JJ-L (2008) The telomerase database. Nucleic Acids Res 36:D339–D343

Qi X, Li Y, Honda S, Hoffmann S, Marz M, Mosig A, Podlevsky JD, Stadler PF, Selker EU, Chen JJ-L (2012a) The common ancestral core of vertebrate and fungal telomerase RNAs. Nucleic Acids Res 41:450–462

Qi X, Xie M, Brown AF, Bley CJ, Podlevsky JD, Chen JJ-L (2012b) RNA/DNA hybrid binding affinity determines telomerase template-translocation efficiency. EMBO J 31:150–161

Qiao F, Cech TR (2008) Triple-helix structure in telomerase RNA contributes to catalysis. Nat Struct Mol Biol 15:634–640

Reichow SL, Hamma T, Ferre-D'Amare AR, Varani G (2007) The structure and function of small nucleolar ribonucleoproteins. Nucleic Acids Res 35(5):1452–1464

Robart AR, O'Connor CM, Collins K (2010) Ciliate telomerase RNA loop IV nucleotides promote hierarchical RNP assembly and holoenzyme stability. RNA 16:563–571

Romi E, Baran N, Gantman M, Shmoish M, Min B, Collins K, Manor H (2007) High-resolution physical and functional mapping of the template adjacent DNA binding site in catalytically active telomerase. Proc Natl Acad Sci USA 104:8791–8796

Rouda S, Skordalakes E (2007) Structure of the RNA-binding domain of telomerase: implications for RNA recognition and binding. Structure 15:1403–1412

Sahin E, Depinho RA (2010) Linking functional decline of telomeres, mitochondria and stem cells during ageing. Nature 464(7288):520–528

Sandhu R, Sanford S, Basu S, Park M, Pandya UM, Li B, Chakrabarti K (2013) A trans-spliced telomerase RNA dictates telomere synthesis in Trypanosoma brucei. Cell Res 23(4):537–551. doi:10.1038/cr.2013.1035

Sealey DC, Zheng L, Taboski MA, Cruickshank J, Ikura M, Harrington LA (2010) The N-terminus of hTERT contains a DNA-binding domain and is required for telomerase activity and cellular immortalization. Nucleic Acids Res 38(6):2019–2035

Sekaran VG, Soares J, Jarstfer MB (2010) Structures of telomerase subunits provide functional insights. Biochim Biophys Acta 1804:1190–1201

Seto AG, Zaug AJ, Sobel SG, Wolin SL, Cech TR (1999) Saccharomyces cerevisiae telomerase is an Sm small nuclear ribonucleoprotein particle. Nature 401:177–180

Seto AG, Livengood AJ, Tzfati Y, Blackburn EH, Cech TR (2002) A bulged stem tethers Est1p to telomerase RNA in budding yeast. Genes Dev 16:2800–2812

Seto AG, Umansky K, Tzfati Y, Zaug AJ, Blackburn EH, Cech TR (2003) A template-proximal RNA paired element contributes to Saccharomyces cerevisiae telomerase activity. RNA 9 (11):1323–1332

Sexton AN, Collins K (2011) The 5′ guanosine tracts of human telomerase RNA are recognized by the G-quadruplex binding domain of the RNA helicase DHX36 and function to increase RNA accumulation. Mol Cell Biol 31:736–743

Sexton AN, Youmans DT, Collins K (2012) Specificity requirements for human telomere protein interaction with telomerase holoenzyme. J Biol Chem 287(41):34455–34464

Shefer K, Brown Y, Gorkovoy V, Nussbaum T, Ulyanov NB, Tzfati Y (2007) A triple helix within a pseudoknot is a conserved and essential element of telomerase RNA. Mol Cell Biol 27:2130–2143

Stellwagen AE, Haimberger ZW, Veatch JR, Gottschling DE (2003) Ku interacts with telomerase RNA to promote telomere addition at native and broken chromosome ends. Genes Dev 17:2384–2395

Szostak JW, Blackburn EH (1982) Cloning yeast telomeres on linear plasmid vectors. Cell 29 (1):245–255

Tesmer VM, Ford LP, Holt SE, Frank BC, Yi X, Aisner DL, Ouellette M, Shay JW, Wright WE (1999) Two inactive fragments of the integral RNA cooperate to assemble active telomerase with the human protein catalytic subunit (hTERT) in vitro. Mol Cell Biol 19:6207–6216

Theimer CA, Blois CA, Feigon J (2005) Structure of the human telomerase RNA pseudoknot reveals conserved tertiary interactions essential for function. Mol Cell 17:671–682

Theimer CA, Jády BE, Chim N, Richard P, Breece KE, Kiss T, Feigon J (2007) Structural and functional characterization of human telomerase RNA processing and cajal body localization signals. Mol Cell 27:869–881

Tzfati Y, Fulton TB, Roy J, Blackburn EH (2000) Template boundary in a yeast telomerase specified by RNA structure. Science 288(5467):863–867

Venteicher AS, Artandi SE (2009) A human telomerase holoenzyme protein required for Cajal body localization and telomere synthesis. Science 323:644–648

Vulliamy TJ, Marrone A, Knight SW, Walne A, Mason PJ, Dokal I (2006) Mutations in dyskeratosis congenita: their impact on telomere length and the diversity of clinical presentation. Blood 107:2680–2685

Wang F, Podell ER, Zaug AJ, Yang Y, Baciu P, Cech TR, Lei M (2007) The POT1-TPP1 telomere complex is a telomerase processivity factor. Nature 445:506–510

Watson JD (1972) Origin of concatemeric T7 DNA. Nat New Biol 239:197–201

Weinrich SL, Pruzan R, Ma L, Ouellette M, Tesmer VM, Holt SE, Bodnar AG, Lichtsteiner S, Kim NW, Trager JB et al (1997) Reconstitution of human telomerase with the template RNA component hTR and the catalytic protein subunit hTRT. Nat Genet 17(4):498–502

Wyatt HDM, Lobb DA, Beattie TL (2007) Characterization of physical and functional anchor site interactions in human telomerase. Mol Cell Biol 27:3226–3240

Wyatt HDM, West SC, Beattie TL (2010) InTERTpreting telomerase structure and function. Nucleic Acids Res 38:5609–5622

Xie M, Podlevsky JD, Qi X, Bley CJ, Chen JJ-L (2010) A novel motif in telomerase reverse transcriptase regulates telomere repeat addition rate and processivity. Nucleic Acids Res 38:1982–1996

Zappulla DC, Goodrich KJ, Cech TR (2005) A miniature yeast telomerase RNA functions in vivo and reconstitutes activity in vitro. Nat Struct Mol Biol 12:1072–1077

Zaug AJ, Podell ER, Cech TR (2008) Mutation in TERT separates processivity from anchor-site function. Nat Struct Mol Biol 15:870–872

Zhang Q, Kim N-K, Peterson RD, Wang Z, Feigon J (2010) Structurally conserved five nucleotide bulge determines the overall topology of the core domain of human telomerase RNA. Proc Natl Acad Sci USA 107:18761–18768

Zhang Q, Kim N-K, Feigon J (2011) Telomerase and retrotransposons: reverse transcriptases that shaped genomes special feature Sackler colloquium: architecture of human telomerase RNA. Proc Natl Acad Sci USA 108:20325–20332

Zhang Y, Chen LY, Han X, Xie W, Kim H, Yang D, Liu D, Songyang Z (2013) Phosphorylation of TPP1 regulates cell cycle-dependent telomerase recruitment. Proc Natl Acad Sci USA 110:5457–5462. doi:10.1073/pnas.1217733110

Zhong FL, Batista LF, Freund A, Pech MF, Venteicher AS, Artandi SE (2012) TPP1 OB-fold domain controls telomere maintenance by recruiting telomerase to chromosome ends. Cell 150 (3):481–494

Chapter 10
Bacteriophage RNA Polymerases

Ritwika S. Basu and Katsuhiko S. Murakami

Abstract Bacteriophage-encoded RNA polymerase (RNAP) was first discovered in T7 phage infected *Escherichia coli* cells. It was known that phage infection on host bacterial cells led to redirection of the host gene expression towards generation of progeny phage particles, but a previously uncharacterized "switching event" leading to the expression of late bacteriophage genes was first attributed to a phage-encoded RNAP. This phage RNAP could recognize promoters on the phage genome and express phage genes using a single-polypeptide polymerase of ~100 kDa molecular weight, which is ~4 times smaller than bacterial RNAPs. This was a substantial simplification from the previously known RNAPs from bacteria (5 subunits) and eukaryotes (more than 12 subunits); nonetheless, the single-unit T7 RNAP is able to recognize promoter DNA and unwind double-stranded (ds) DNA to form open complex, and after abortive initiation, it proceeds to processive RNA elongation. The simplicity of T7 phage RNAP made it an ideal model system to study the transcription mechanism and an ideal tool for protein expression system in bacterial cells. In this chapter, we will review the current state of knowledge of transcription mechanism in single-unit bacteriophage RNAPs from the two deeply studied T7 and the N4 phage RNAPs.

Keywords RNA polymerase • Transcription • Bacteriophage • T7 • N4

R.S. Basu • K.S. Murakami (✉)
Department of Biochemistry and Molecular Biology, The Pennsylvania State University, University Park, PA 16802, USA

The Center for RNA Molecular Biology, The Pennsylvania State University, University Park, PA 16802, USA
e-mail: kum14@psu.edu

K.S. Murakami and M.A. Trakselis (eds.), *Nucleic Acid Polymerases*, Nucleic Acids and Molecular Biology 30, DOI 10.1007/978-3-642-39796-7_10,

10.1 T7 RNAP Structure: The Prototype of Single-Unit RNAP

The first crystal structure of RNAP was determined in 1993 from the single-unit bacteriophage T7 RNAP (Sousa et al. 1993). The domain organization of the T7 RNAP was found similar to the bacterial DNA polymerase (DNAP) I (Arnold et al. 1995), e.g., Klenow fragment (KF) of DNA Pol I (Fig. 10.1). The structure resembles the anatomy of a right hand comprising of palm, fingers, and thumb subdomains that are arranged around a DNA-binding cleft. In addition, an N-terminal domain (NTD) constitutes the front wall of the DNA-binding cleft, making the DNA-binding cleft deeper and narrower in RNAP, and also plays roles in promoter recognition and DNA unwinding for making transcription-competent open complex.

The palm subdomain (Fig. 10.1a) forms the base of the DNA-binding cleft with the fingers and the tall thumb subdomains forming either sidewalls of the channel. Invariant residues from motifs A, B, and C cluster around this catalytic cleft. Aspartate residues, conserved in all nucleic acid polymerases, bear the most important catalytic function of chelating two divalent metals (Mg^{2+}) at the active site. The catalytic metal A (Me_A) generates the nucleophile at the 3′-end RNA for the S_N2 nucleotidyl transfer reaction, and a nucleotide-binding metal (Me_B) stabilizes the charge distribution of the incoming nucleotide at the reaction transition state (Steitz et al. 1994; Sect. 10.3.3; Fig. 10.4).

10.1.1 Promoter Binding

The T7 promoter sequence is conserved from the −17 to +6 position with a highly AT-rich region centered around −17. The upstream duplex-form DNA from −17 to −5 binds to the NTD, and the DNA bases downstream are melted and a single-stranded template DNA is directed into the active site. T7 RNAP recognizes the promoter through three main interactions (Fig. 10.2a): (1) DNA bases are recognized by an antiparallel β-hairpin of the fingers, the specificity loop, from the major groove; (2) an AT-rich recognition motif in the NTD recognizes AT-rich (−17) region by inserting a flexible surface loop into the DNA minor groove; (3) the intercalating β-hairpin in the NTD melts the promoter DNA and marks the upstream edge of the transcription bubble. This precise location of the transcription bubble ensures correct positioning of the transcription start site DNA base at the active site. At this point, RNAP is ready to accept the two nucleotides that form base pairs with the +1 and +2 template DNA bases to initiate RNA synthesis.

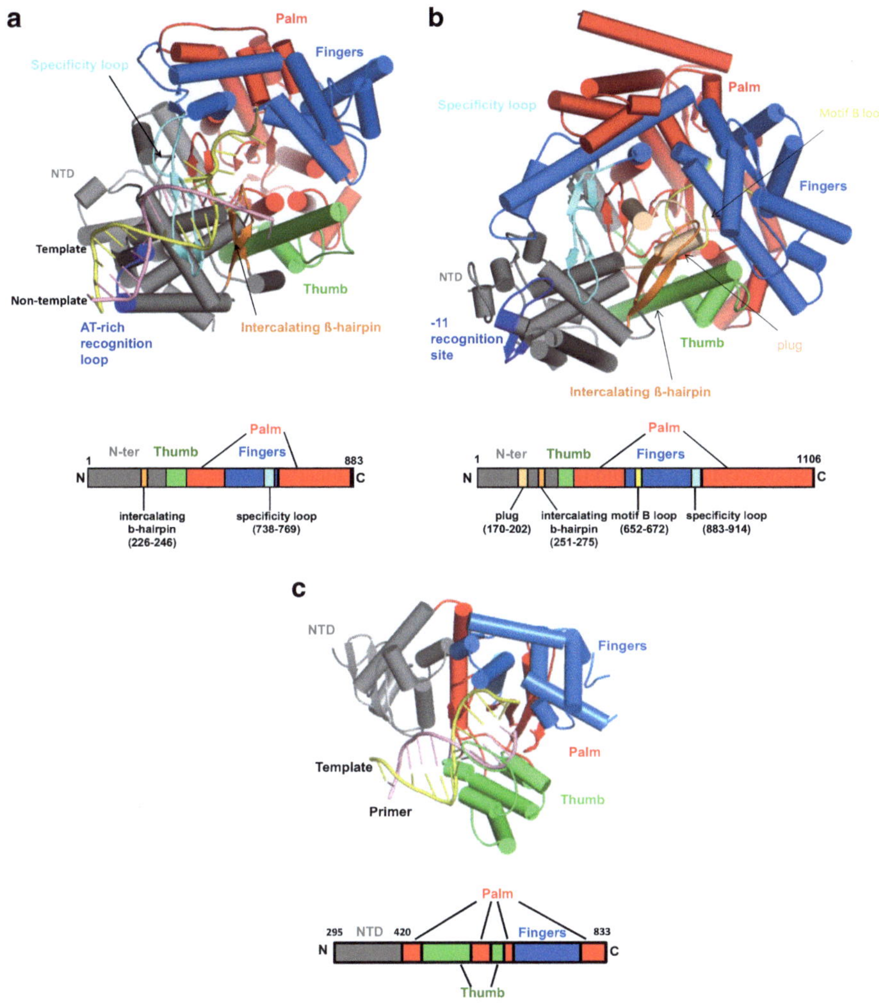

Fig. 10.1 Structures of single-unit DNA-dependent polymerases. Right hand-like organization domains in T7 RNAP complexed with promoter DNA (**a**), N4 mini-vRNAP (**b**), and KlenTaq DNAP I (**c**) are shown. Same orientation of the structures were obtained by superposing the palm domains. The palm (*red*), fingers (*blue*), thumb (*green*) domains, and NTD (*gray*) are shown as *cylinders* (α-helix) and *arrows* (β-strands). Double-stranded promoter DNA containing template (*yellow*) and non-template (*pink*) in T7 RNAP and the primer (*pink*) template (*yellow*) duplex in KlenTaq reach the active site cleft formed by three β-strands. Active site of N4 mini-vRNAP (**b**) is blocked by the plug (*wheat*) and the motif B loop (*yellow*). The primary structures of the polymerases are shown below each 3-D structure with the same color code

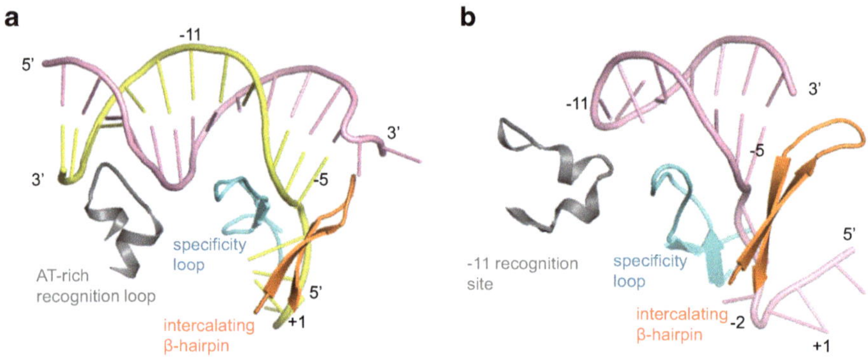

Fig. 10.2 Structural motifs for promoter recognition. (**a**) AT-rich recognition loop (*gray*), specificity loop (*cyan*), and intercalating hairpin (*orange*) in T7 RNAP binary complex. Template (*yellow*) and non-template (*pink*) are shown. (**b**) −11 recognition loop (*gray*), specificity loop (*cyan*), and intercalating hairpin (*orange*) in the N4 RNAP binary complex recognize the hairpin promoter (*pink*). The figure has been adapted from Gleghorn et al. (2008)

10.1.2 Transcript Initiation

The structure of T7 RNAP transcription initiation complex (Cheetham and Steitz 1999) showed how RNAP positions the template DNA bases at the active site at every subsequent step of NTP addition. In this structure, RNAP was bound to a 17-bp duplex promoter and a 3-mer RNA transcript base-paired with the single-stranded tailed template.

The presence of RNA in this polymerase structure offered an insight into the rNTP-specific RNA synthesis. In contrast to DNAP, a bulky glutamate "steric gate" near the active site is replaced to a glycine in RNAP that makes space for the 2′-OH of incoming NTP. Secondly, a carbonyl group of the active site amino acid residues hydrogen bonds with the 2′-OH of the 3′ primer end allowing only 3′-endo ribose conformation of the base. Further, the DNA-binding pocket is also complementary to the A-form DNA/RNA heteroduplex thus favoring the formation of DNA/RNA heteroduplex at the stage of RNA extension.

The promoter contacts are maintained, while the DNA/RNA heteroduplex accumulates in the active site, positioning the growing primer end at the active site. Accordingly, the −1 template base that stacked with the +1 template for proper positioning during *de novo* initiation assumes a flipped out conformation allowing transcript extension to RNA 3-mer stage. This observation served the first structural evidence for the "DNA scrunching" mechanism during initiation. The primer DNA scrunches into the active site through the initial synthesis phase until it chooses one of the two fates, abortive or productive transcription. During initiation, abortive transcripts are displaced from the template by collapsing the newly formed bubble due to small, weak DNA/RNA hybrids (Brieba and Sousa 2001). Alternatively, it could extend the DNA/RNA hybrid and scrunch it until it reaches the threshold length of 8 nt, after which the entire complex undergoes a phenomenal conformational change into a processive elongation complex (Sect. 10.2).

10.2 Transcription Elongation

Two structures of T7 RNAP with 8-mer (Tahirov et al. 2002) and 11-mer (Yin and Steitz 2002) RNAs provide a holistic view of how the features of a transcription elongation complex emerge from the initiation complex after extensive reorganizations of the RNAP and DNA structures. A transcribing complex is committed to elongation when it has successfully accommodated numerous concurrent events including promoter release, partial collapse of the bubble, construction of an RNA exit channel, and peeling of the 5′ end of the nascent RNA from the template DNA. The threshold length of DNA/RNA hybrid determining this transition to elongation had been only predicted from biochemical observations, but in both the transcription elongation structures, the nascent RNA forms a heteroduplex with the template DNA of only 8 bp, upstream of which is peeled off the template and directed into a new formed exit channel.

10.2.1 Promoter Release and Processivity

The NTD being most closely involved in interaction with the promoter undergoes major rearrangement. On comparing the RNAP in the initiation and elongation stages, three subdomains in the NTD show independent rearrangement movements (Fig. 10.3). (1) A six-helix bundle called the promoter-binding domain (PBD) undergoes a massive rigid body rotation of 140° to a position previously occupied by the promoter, thereby destroying its promoter interaction sites. Along with PBD, the adjacent intercalating β-hairpin, an important motif for promoter melting in initiation, also becomes disordered in this promoter release event. (2) An α-helix (named C-helix) at the N-terminus of NTD nearly doubles in length by stacking of two smaller helices seen during initiation, forming part of the binding site for DNA/RNA hybrid. This helix protrudes into the region occupied by the PBD in initiation suggesting the concerted nature of the two motions. (3) Subdomain H undergoes extensive refolding into two antiparallel helices, paired with a large translation of 70 Å to the opposite side of the polymerase, forming the rim of the RNA exit channel on one side, and interacts with the non-template DNA from its opposite surface.

The formation of the RNA exit channel is the most important determinant of transcription processivity. Along with the subdomain H, two important motifs are involved in its formation including the thumb and the specificity loop. The interaction of the channel wall with RNA is only through salt bridges between the phosphate backbone and basic residues of specificity loop and subdomain H.

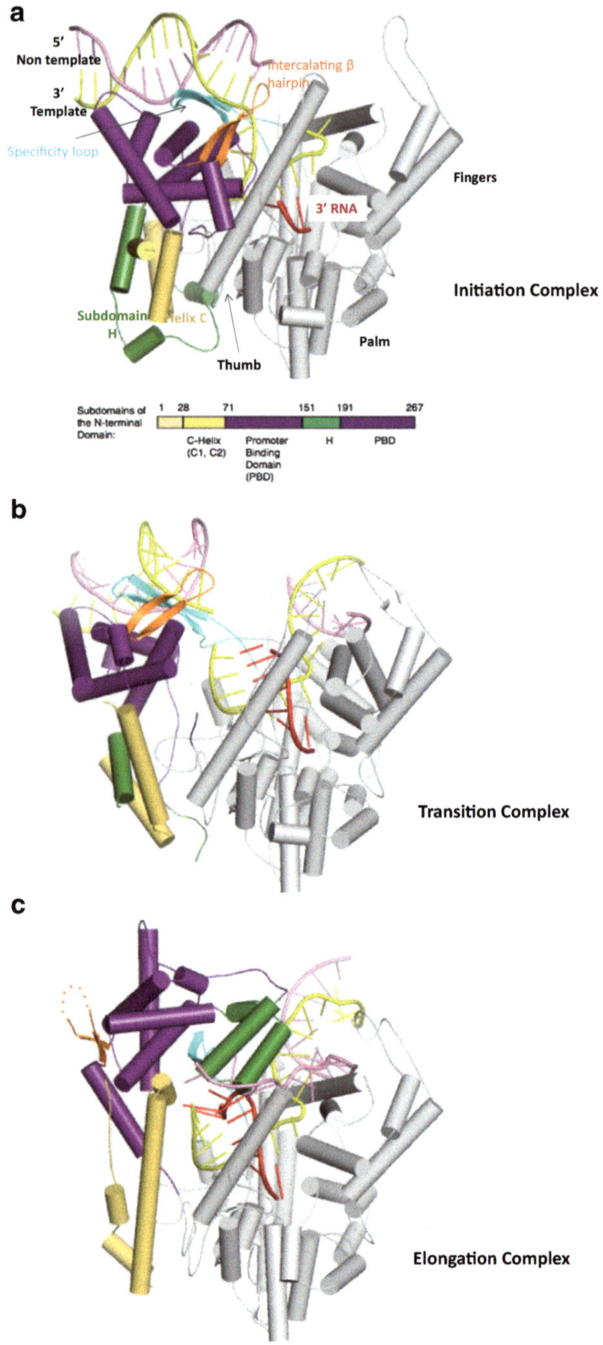

Fig. 10.3 Comparison of the structures of initiation complex (**a**), intermediate complex (**b**), and elongation complex (**c**). α-helices are represented by *cylinders* and β-sheets by *arrows*, while disordered regions are shown in *dotted lines*. The structures are similarly oriented by superposing the palm domains including the active site. Transition involves major conformational changes in the NTD colored as independently moving regions: promoter-binding domain (*purple*), intercalating hairpin (*orange*), helix C motif (*pale yellow*), and subdomain H (*green*). The CTD (*gray*) remains mostly unchanged except for movements in the long thumb helix and the specificity loop (*cyan*). The template strand is *yellow*, non-template is *pink*, and RNA is *red*. The downstream DNA is highly bent with respect to the upstream regions to help bubble formation (**c**). The primary structure of the NTD subdomains is also shown with the same color code (**a**). These figures have been adapted from Yin and Steitz (2002)

Processivity is also favored over abortive transcription due to the extensive interactions of the 7 bp DNA/RNA heteroduplex with its binding site.

Transcription processivity is coupled with the continuous downstream progress of the transcription bubble. At the onset of transcription, the bubble is generated from the unwinding of the downstream duplex DNA, approximately through 146° with respect to the upstream promoter. The template strand also plunges deep into the active site and comes out by bending about 80° at the upstream and downstream end of the bubble (Fig. 10.3c).

10.2.2 Transition to Elongation Complex

Transition from the initiation to elongation complexes should involve metastable complexes that not only form the basis of abortive cycling but also assume the conformation of an expanded active site to accumulate a growing DNA/RNA hybrid of about 8 nt length (Huang and Sousa 2000; Temiakov et al. 2000) while still bound to the promoter. Biochemical studies proposed that the transition was two steps (Bandwar et al. 2007; Guo et al. 2005; Ma et al. 2005), where RNA extended to about 8 nt followed by major refolding events accompanying synthesis of 9–14 nt (Tang et al. 2008). One of the intermediate structures with a 17-mer promoter DNA and 7-mer RNA transcript in a bubble showed the nature of transition between the vastly different initiation and elongation complexes (Durniak et al. 2008; Fig. 10.3b). This structure was captured using a mutant (P266L) in a loop connecting the polymerase NTD and C-terminal domains.

In the first stage of transition, NTD movements including PBD and helix C appear to have proceeded halfway, leading to the enlargement of the active site to accommodate the 7-bp DNA/RNA hybrid. Subdomain H of NTD, however, remains in its initiation orientation. The second stage of transition involves a final movement of the NTD, specificity loop, and subdomain H that loses promoter contact, completes the exit channel formation, and also forces the downstream duplex to its bent position.

10.2.3 Nucleotide Addition Cycle

In every nucleotide addition cycle, RNAP sieves through the pool of NTPs for the correct substrate through its intricate mechanism of nucleotide selection. The catalytic-competent nucleotide-binding N-site elicits the nucleotidyl transfer reaction between the RNA primer 3′-end base at P-site and the incoming NTP to extend the RNA through one base. The extended RNA then translocates upstream relative to the active site, opening the N-site for the next round of cycle. T7 RNAP conducts this harmonized process through fine regulation by elements mainly from the fingers and palm subdomain. Crystal structures of T7 RNAP complexes with the

DNA, RNA, and nucleotide provide great insights into its nucleotide addition cycle (Yin and Steitz 2004; Temiakov et al. 2004).

10.2.3.1 Substrate Selection in Pre-insertion Site

Structure of a ternary elongation complex with non-hydrolyzable ATP analog identified a "pre-insertion site" for a substrate binding prior to catalytically competent nucleotide binding at the "insertion" N-site (Temiakov et al. 2004; Fig. 10.4a). The pre-insertion site is linked to the "open" conformation of a conserved O-helix of the fingers. In this conformation, the templating base faces away from its accepting position, and the conserved tyrosine residue at the end of the O-helix meanwhile blocks the nucleotide insertion site. The tyrosine hydroxyl group interacts with the 2′-OH of the substrate, forming the primary discrimination of incoming rNTPs against dNTPs. The substrate is bound along the O-helix but is not Watson–Crick paired with the template DNA base, implying that the pre-insertion site is an early fidelity checkpoint.

10.2.3.2 Substrate Loading at the Catalytic Site

Substrate loading from the pre-insertion site to the catalytic insertion site is achieved by "closing of the fingers," wherein rotation of the O-helix is the most significant. Structure of a pre-catalytic substrate complex trapped with the use of a nonreactive nucleotide analog showed the nature of O-helix movement when substrate was loaded onto the catalytic insertion site (N-site) and thereby defined the closed conformation of the fingers (Yin and Steitz 2004). The O-helix rotates about a pivot point at its middle, causing the amino end of the helix to close onto the substrate triphosphate moiety, stabilized in this position through positive charge from lysine and arginine residues. Simultaneously, the opposite end of the O-helix, including the important tyrosine residue, moves away to make space for the base moiety of the incoming substrate. The base specific and ribose discriminating interactions of the pre-insertion site are maintained in the closed conformation to face a final round of fidelity check.

The incoming substrate alignment for catalysis, in the active site, is maintained not only by the O-helix residues but also by the accompanying nucleotide-binding metal Me_B, which is, in turn, positioned by chelating the conserved active site aspartates. Further, the octahedral coordination of catalytic metal ion Me_A maintains the critical alignment of the reactive groups, 3′ OH of primer terminus and 5′ αP of incoming substrate. The mechanism of nucleotidyl transfer reaction, which extends one RNA base and produces a pyrophosphate (PPi) by-product, will be described in Sect. 10.3.3.

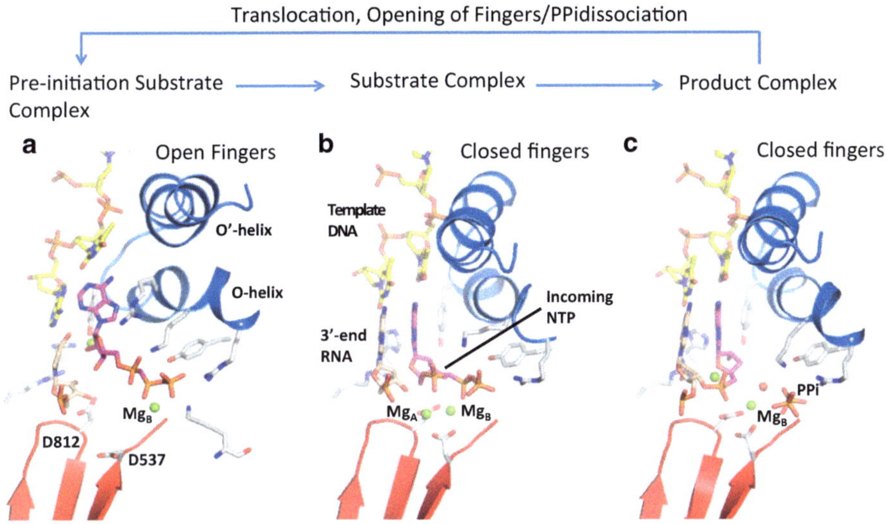

Fig. 10.4 Active site views during nucleotide addition cycle. Structures of the preinitiation complex (PDB: 1S07) (**a**), substrate complex (PDB: 1S76) (**b**), and product complex (PDB: 1S77) (**c**) are aligned similarly by superposing the palm. Important motifs (*ribbons*) and amino acid residues (*sticks*) are shown. Helix O and O′ from fingers (*blue*) and motifs A and C (*red*) from the palm provide amino acid side chains to bind the incoming nucleotide (*magenta* carbons), magnesium atoms (*green spheres*), or transcript primer end (*light pink* carbons). The template DNA is shown as *sticks* with *yellow* carbon atoms

10.2.3.3 Translocation

After the addition of a base to the transcript, the final step of the nucleotide addition cycle is the translocation of the DNA/RNA hybrid through one base distance such that the 3′-end RNA positions in the P-site. The structure of a post-catalytic, pre-translocated product complex, isolated right after phosphodiester bond formation but before dissociation of nascent PPi, showed that the phosphodiester bond does not cause any change in the RNAP or DNA/RNA hybrid (Fig. 10.4c). In the post-translocated state, the only difference lies in the dissociation of PPi along with the coordinated Me_B, which breaks the interactions with the O-helix residues and thus favors the open state. Moreover, the rNTP discriminator tyrosine at the O-helix moves towards the heteroduplex and stacks with the primer end base preventing backtracking of the hybrid, while opening the triphosphate site due to the pivoted helix motion. It is thus proposed that translocation of DNA/RNA hybrid in single-unit RNAP is coupled with PPi dissociation which leads to the opening of O-helix for the next round of nucleotide addition, often called the power-stroke mechanism of translocation (Jiang and Sheetz 1994).

10.3 N4 vRNAP: Factor-Dependent Single-Unit Polymerase

While the characterization of T7-related RNAP was progressing rapidly, the discovery of a unique phage-encoded, virion-encapsidated RNAP (vRNAP) isolated from a lytic coliphage N4 (Falco et al. 1977) added breadth to the studies of single-unit RNAPs. In contrast to other phage RNAPs, N4 vRNAP is encapsulated within the virion to being injected into the bacterial cell at the onset of infection. Upon injection of the N4 double-stranded DNA genome, host proteins DNA gyrase and single-stranded DNA-binding protein (EcoSSB) prepare the N4 vRNAP specific promoter comprising a DNA hairpin with a 5–7-base pair stem and 3-base loop (Glucksmann-Kuis et al. 1996; Haynes and Rothman-Denes 1985). This DNA hairpin promoter is a key to activate the co-injected vRNAP from transcription-inactive state. After transcription initiation, N4 vRNAP depends on EcoSSB for transcript elongation, which displaces nascent RNA transcripts from the template DNA for recycling (Davydova and Rothman-Denes 2003). Thus the part of NTD responsible for RNA separation in T7 transcription elongation (see Sect. 10.3.1) was absent in N4 vRNAP. EcoSSB could thus be termed as a transcription factor, and N4 vRNAP has been emerged as an important model for studying the structural basis of transcription activation as well as factor-dependent transcription of the single-unit RNAP.

Much larger than other phage RNAPs, the 320 kDa N4 vRNAP can be divided into three domains, and a central polymerase domain of 1100 amino acid (called mini-vRNAP) exhibits transcription initiation, elongation, and termination properties identical to full-length vRNAP (Kazmierczak et al. 2002). Sequence alignment classified the N4 mini-vRNAP as the most divergent member of the single-unit RNAP family. There is only small sequence similarity with other members except the catalytically important motifs A, B, C, and T/DxxGR.

10.3.1 Structure of N4 vRNAP

Studies of the structure and function of N4 mini-vRNAP advanced the scope of single-unit RNAP studies. Despite a lower sequence similarity, the N4 mini-vRNAP structure was highly similar to the T7 RNAP structure (Murakami et al. 2008) (Fig. 10.1b). It reiterated the modular organization of the right hand-like structure, with the same subdomains surrounding the DNA-binding cleft and the same structural motifs. However, the promoter recognition motifs in the fingers and NTD, although similar, had adapted interactions specific for the special hairpin N4 promoter (Sect. 10.3.2).

The two major differences between the N4 vRNAP and T7 RNAP structures are the presence of a plug module insertion in the NTD and a loop inserted in the middle of motif B (motif B loop) found in the N4 vRNAP. The plug and motif B loop interact with the motifs A and C of palm that cover all catalytically essential amino acid residues of N4 vRNAP active site until the hairpin-form promoter DNA interacts and activates the polymerase.

10.3.2 Unique Hairpin Promoter DNA Recognition

The N4 vRNAP recognizes hairpin-form promoter DNA by using (1) −11 base recognition motif in the NTD, which recognizes the tip of hairpin loop, (2) specificity loop from the fingers that contact the hairpin from the major groove, and (3) intercalating β-hairpin to maintain the junction of the double- and single-stranded DNA (Fig. 10.2b). After binding the hairpin-form promoter DNA, the N4 vRNAP changes its conformation including a rigid body movement of the plug away from the active site and the motif B loop refolding outwards from the active site, which turns the residues on the O-helix for incoming NTP binding. The hairpin promoter-bound N4 vRNAP conformation is competent for transcription.

10.3.3 Nucleotidyl Transfer Reaction

The nucleotidyl transfer reaction by DNA and RNA polymerases follows a generalized mechanism of two-metal catalysis (Steitz et al. 1994), wherein the catalytic metal (Me_A) is known to work as a Lewis acid to enhance the nucleophilicity of the 3′ oxygen attacking group of the primer 3′ end, initiating the S_N2 reaction onto the 5′ αP of the incoming nucleotide. The nucleotide-binding metal (Me_B) is coordinated by the triphosphates of the incoming NTP, stabilizing the pentacovalent transition state.

In case of the transcription initiation, two NTP substrates bind to the empty P- and N-sites concurrently to form the first phosphodiester bond formation at the 5′-end RNA. The structural snapshots (Gleghorn et al. 2011), Raman crystallography (Chen et al. 2011), and time-resolved trigger-freeze crystallography (Basu and Murakami 2013) studies of ternary complexes of N4 mini-vRNAP during transcript initiation provide the most updated and complete knowledge of the reaction mechanism (Fig. 10.5). Two forms of pre-catalytic substrate complexes could be isolated; while both contained the two initiating nucleotides at the P- and N-sites, the catalytic metal (Me_A) was absent in one of them. This intermediate showed the important, final molecular rearrangements in the active site elicited by the critical Me_A binding to allow catalysis. In the absence of Me_A, the 3′ O from GTP (+1) and the 5′ αP of GTP (+2) were beyond reacting distance (4.1 Å) and were bought closer only with the binding of Me_A (Fig. 10.5b, c). This observation proposed that the catalytic metal binding, which is sensitive to its octahedral coordination requirements, serves as a final fidelity checkpoint where a fine misalignment of

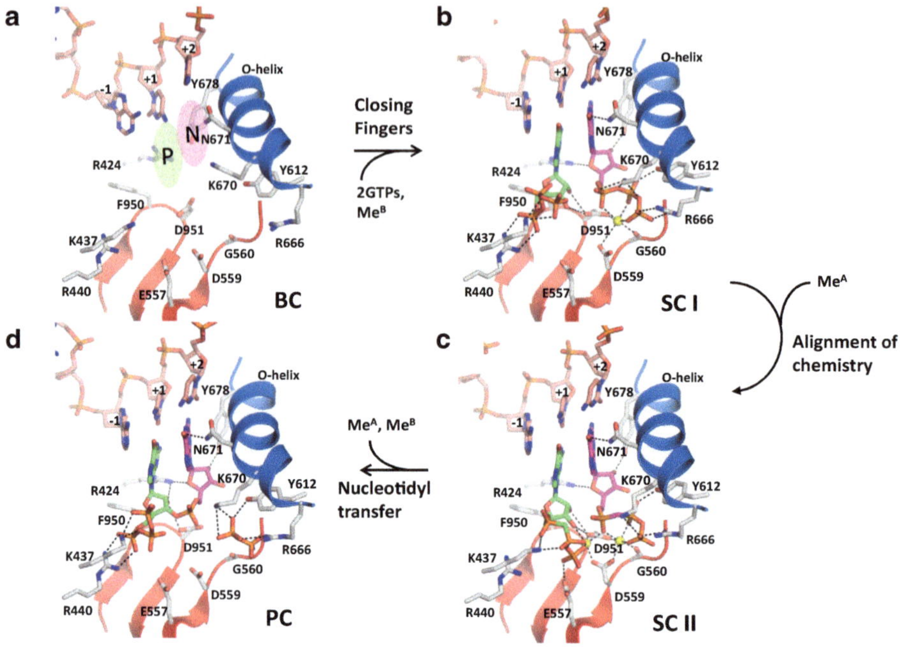

Fig. 10.5 Structures of active site, DNA and nucleotides during nucleotidyl transfer reaction. The main chains (*ribbon models*) of motifs A and C (*red*) and of the O-helix (*blue*) and the main and side chains (*stick models*) involved in nucleotide and metal binding in the promoter binary complex (**a**), substrate complex I (**b**), substrate complex II (**c**), and product complex (**d**). NTP binding P- and N-sites are indicated as *green* and *magenta circles* in **a**. DNA template (from −1 to +2, *pink*) and nucleotides at +1 (*green*) and +2 (*magenta*) positions are shown as *stick models*. Divalent metals are depicted by *yellow spheres*. Hydrogen bonds and salt bridges are depicted by *black dashed lines* (Gleghorn et al. 2011)

the 5′ αP of a mispaired incoming nucleotide disallows the Me_A binding and thereby prevents the catalysis.

Consistent with the steps in transcript elongation, the O-helix in the initiation complexes was in a closed state, providing its basic residues for stable interactions with the nucleotide at N-site. Additionally, the stability of the nucleotide binding at the P-site, an initiation-specific event, could be explained by electrostatic interactions of basic residues of the palm with the triphosphates and the partial base stacking with a purine at −1 template DNA position. Furthermore, the γP group of GTP (+1) participates in the Me_A octahedral coordination. Loss of affinity and catalytic activity with GDP (+1) compared with GTP (+1), led to the proposition of substrate-assisted catalysis for the first phosphodiester bond formation mediated by the γP group and Me_A binding.

A time course soak-trigger-freeze crystallographic study on this transcript initiation process provided the most direct real-time trace of the unperturbed events in the nucleotidyl transfer reaction (Basu and Murakami 2013). The observation of polymerase reaction in real time by high-resolution X-ray crystallography showed that the nucleotide binding, the O-helix closure, and template DNA rearrangement were completed at early stages of reaction. Subsequently, the catalytic metal binding rearranges the reactive groups just prior to the phosphodiester bond formation. Owing to the sensitivity of Me_A for its octahedral coordination, its binding is subject to correct Watson–Crick pairing of the incoming nucleotide, and for the same reason it leaves the active site right after a phosphodiester bond formation, thereby also preventing the backward cleavage reaction.

References

Arnold E, Ding J, Hughes SH, Hostomsky Z (1995) Structures of DNA and RNA polymerases and their interactions with nucleic acid substrates. Curr Opin Struct Biol 5:27–38

Bandwar RP, Ma N, Emanuel SA, Anikin M, Vassylyev DG, Patel SS, McAllister WT (2007) The transition to an elongation complex by T7 RNA polymerase is a multistep process. J Biol Chem 282:22879–22886

Basu RS, Murakami KS (2013) Watching the bacteriophage N4 RNA polymerase transcription by time-dependent soak-trigger-freeze X-ray crystallography. J Biol Chem 288:3305–3311

Brieba LG, Sousa R (2001) T7 promoter release mediated by DNA scrunching. EMBO J 20:6826–6835

Cheetham GM, Steitz TA (1999) Structure of a transcribing T7 RNA polymerase initiation complex. Science 286:2305–2309

Chen Y, Basu R, Gleghorn ML, Murakami KS, Carey PR (2011) Time-resolved events on the reaction pathway of transcript initiation by a single-subunit RNA polymerase: Raman crystallographic evidence. J Am Chem Soc 133:12544–12555

Davydova EK, Rothman-Denes LB (2003) Escherichia coli single-stranded DNA-binding protein mediates template recycling during transcription by bacteriophage N4 virion RNA polymerase. Proc Natl Acad Sci USA 100:9250–9255

Durniak KJ, Bailey S, Steitz TA (2008) The structure of a transcribing T7 RNA polymerase in transition from initiation to elongation. Science 322:553–557

Falco SC, Laan KV, Rothman-Denes LB (1977) Virion-associated RNA polymerase required for bacteriophage N4 development. Proc Natl Acad Sci USA 74:520–523

Gleghorn ML, Davydova EK, Rothman-Denes LB, Murakami KS (2008) Structural basis for DNA-hairpin promoter recognition by the bacteriophage N4 virion RNA polymerase. Mol Cell 32:707–717

Gleghorn ML, Davydova EK, Basu R, Rothman-Denes LB, Murakami KS (2011) X-ray crystal structures elucidate the nucleotidyl transfer reaction of transcript initiation using two nucleotides. Proc Natl Acad Sci USA 108:3566–3571

Glucksmann-Kuis MA, Dai X, Markiewicz P, Rothman-Denes LB (1996) E. coli SSB activates N4 virion RNA polymerase promoters by stabilizing a DNA hairpin required for promoter recognition. Cell 84:147–154

Guo Q, Nayak D, Brieba LG, Sousa R (2005) Major conformational changes during T7RNAP transcription initiation coincide with, and are required for, promoter release. J Mol Biol 353:256–270

Haynes LL, Rothman-Denes LB (1985) N4 virion RNA polymerase sites of transcription initiation. Cell 41:597–605

Huang J, Sousa R (2000) T7 RNA polymerase elongation complex structure and movement. J Mol Biol 303:347–358

Jiang MY, Sheetz MP (1994) Mechanics of myosin motor: force and step size. Bioessays 16:531–532

Kazmierczak KM, Davydova EK, Mustaev AA, Rothman-Denes LB (2002) The phage N4 virion RNA polymerase catalytic domain is related to single-subunit RNA polymerases. EMBO J 21:5815–5823

Ma K, Temiakov D, Anikin M, McAllister WT (2005) Probing conformational changes in T7 RNA polymerase during initiation and termination by using engineered disulfide linkages. Proc Natl Acad Sci USA 102:17612–17617

Murakami KS, Davydova EK, Rothman-Denes LB (2008) X-ray crystal structure of the polymerase domain of the bacteriophage N4 virion RNA polymerase. Proc Natl Acad Sci USA 105:5046–5051

Sousa R, Chung YJ, Rose JP, Wang BC (1993) Crystal structure of bacteriophage T7 RNA polymerase at 3.3 A resolution. Nature 364:593–599

Steitz TA, Smerdon SJ, Jager J, Joyce CM (1994) A unified polymerase mechanism for nonhomologous DNA and RNA polymerases. Science 266:2022–2025

Tahirov TH, Temiakov D, Anikin M, Patlan V, McAllister WT, Vassylyev DG, Yokoyama S (2002) Structure of a T7 RNA polymerase elongation complex at 2.9 A resolution. Nature 420:43–50

Tang GQ, Roy R, Ha T, Patel SS (2008) Transcription initiation in a single-subunit RNA polymerase proceeds through DNA scrunching and rotation of the N-terminal subdomains. Mol Cell 30:567–577

Temiakov D, Mentesana PE, Ma K, Mustaev A, Borukhov S, McAllister WT (2000) The specificity loop of T7 RNA polymerase interacts first with the promoter and then with the elongating transcript, suggesting a mechanism for promoter clearance. Proc Natl Acad Sci USA 97:14109–14114

Temiakov D, Patlan V, Anikin M, McAllister WT, Yokoyama S, Vassylyev DG (2004) Structural basis for substrate selection by t7 RNA polymerase. Cell 116:381–391

Yin YW, Steitz TA (2002) Structural basis for the transition from initiation to elongation transcription in T7 RNA polymerase. Science 298:1387–1395

Yin YW, Steitz TA (2004) The structural mechanism of translocation and helicase activity in T7 RNA polymerase. Cell 116:393–404

Chapter 11
Mitochondrial DNA and RNA Polymerases

Y. Whitney Yin

Abstract Mitochondria are vital to cellular activities, both cell growth and programmed death. The organelle contains its own DNA, which codes for a subset of genes oxidative phosphorylation electron transfer chain that couples oxidation–reduction reactions with ATP synthesis. Maintenance of mitochondrial DNA (mtDNA) requires a designated group of proteins that are nuclear encoded and transported to mitochondria. Both DNA and RNA polymerases have bacteriophage origins but have clearly deviated from their ancestors and have combined features of prokaryotic and eukaryotic enzymes. Mitochondrial DNA polymerase, Pol γ performs all DNA synthesis activities in replication and repair. Mammalian Pol γ holoenzyme consists of a catalytic subunit Pol γA and an accessory subunit Pol γB that regulates all activities of the catalytic subunit. Mitochondrial DNA replication and transcription are intimately related, as the mitochondrial transcription machinery provides transcripts for RNA primers for mtDNA replication as well as gene expression. Human mitochondrial DNA and RNA polymerases are adverse reaction targets for antiviral reagents against HIV and hepatitis viruses, and human Pol γ mutations have been implicated in multisystem clinical disorders. Studies of mitochondrial polymerases are therefore of immediate human health importance.

Keywords DNA polymerase gamma • mtDNA replication and repair • antiretroviral drug toxicity • Mitochondrial RNA transcription

Mitochondria are eukaryotic organelles enclosed by double membranes. Mitochondria are termed the "power plant" of cells, because they are the locations for metabolism of glucose, fatty acids, and some amino acids. The oxidation–reduction reactions generate protons that accumulate between the membranes; an electron transfer chain couples dissipation of this proton potential with synthesis

Y.W. Yin (✉)
University of Texas Medical Branch, Galveston, TX 77555, USA
e-mail: ywyin@utmb.edu

K.S. Murakami and M.A. Trakselis (eds.), *Nucleic Acid Polymerases*, Nucleic Acids and Molecular Biology 30, DOI 10.1007/978-3-642-39796-7_11,

of ATP. This important reaction, named oxidative phosphorylation (OX-PHOS), is critical in sustaining energy for many cellular activities. In addition, mitochondria are involved in signaling, cellular differentiation, cell death, as well as control of the cell cycle and cell growth (McBride et al. 2006).

Mitochondria share many features common with bacteria. They are elongated in shape, with dimension of 500–1,000 nm. The shape and number of mitochondria per cell vary widely by organism and cell type. Mitochondria are thought to have derived from endosymbiotic proteobacteria. Symbiosis resulted in the majority of mitochondrial genes migrating to the nucleus, so the size of mtDNA is only 1–5 % of free-living bacteria. For example, human mitochondrial DNA (mtDNA) is 16.9 kb, and yeast mtDNA is 80 kb. Regardless, all mtDNA code for a subset of OX-PHOS components, as well as tRNAs and rRNAs needed for mitochondrial protein synthesis. To assemble functional OX-PHOS electron transfer chain, gene expression in the nucleus and mitochondria has to be well coordinated. Proteins involved in mtDNA maintenance are exclusively nuclear coded. These proteins have a unique mitochondrial localization sequence (MLS) that is recognized by a specific translocation apparatus for their internalization into mitochondria.

11.1 Mitochondrial DNA Replication

Mitochondrial function is directly correlated with the integrity of its DNA. Unlike nuclear DNA (nDNA), human mitochondrial DNA (mtDNA) is in a relaxed circular form, free of histone proteins, and lacks introns. The DNA contains a 1 kb noncoding region, called the D-loop, containing a replication origin and two transcription promoters (Fig. 11.1). There are 10–1,000 copies of mtDNA per mitochondrion; their replication is not coordinated to the cell cycle. The precise control of mtDNA replication is not completely understood. The two strands of mtDNA are named H (heavy)-strand and L(light)-strand based on their density in poly(UG) CsCl gradients (Attardi and Attardi 1969). Replication on each strand initiates at a unique origin, O_H and O_L, and proceeds unidirectionally. O_H is located in the D-loop, whereas O_L is two-thirds of the way around the genome. The origin recognition proteins are unknown.

There are two competing models for mtDNA leading and lagging strand replication: a conventional mode where leading and lagging strand synthesis occur simultaneously (Aloni and Attardi 1971b; Yang et al. 2002; Yasukawa et al. 2005) and a displacement mode where synthesis initiating from the O_H origin displaces the parental H-strand to form a D-loop. Only when the newly synthesized H-strand DNA crosses a second origin (O_L) does initiation of L-strand synthesis occur. The nascent H and L strands are therefore extended asymmetrically (Clayton 1982; Xu and Clayton 1996). This model was recently modified to allow initiation of L-strand synthesis from a number of origins in addition to O_L (Brown et al. 2005). Supporting evidence for an asymmetrical synthesis mechanism includes that mtSSB can be found coating extensive region the human

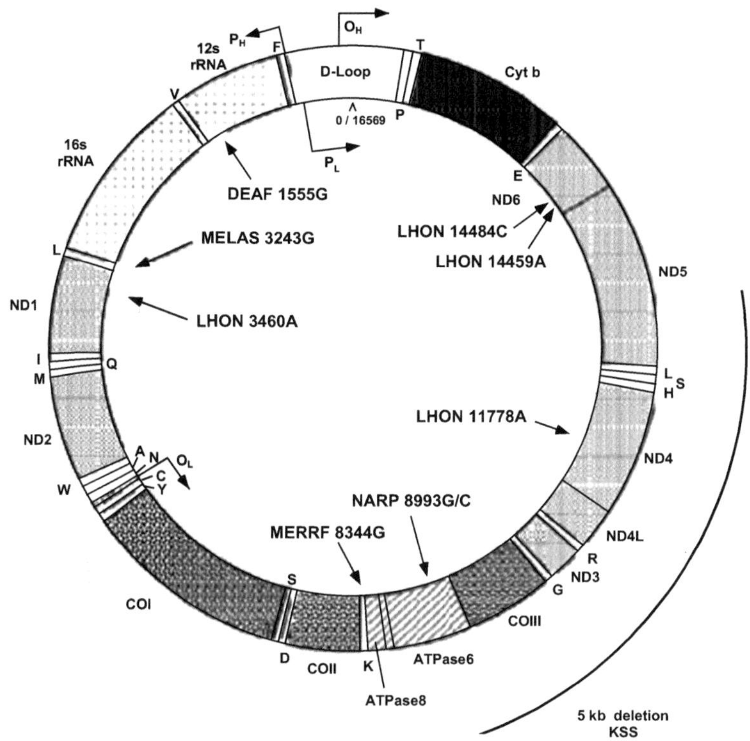

Fig. 11.1 Human mitochondrial DNA. It codes for 13 protein genes, 22 tRNAs, and 2 rRNA

mitochondrial genome and that O_L is important for DNA replication in vivo (Brown et al. 2005; Wanrooij et al. 2012) (Fig. 11.2).

No primase has been identified in mitochondria; maintenance of mtDNA replication therefore requires not only replisomal proteins but also topoisomerase and RNA polymerase with its transcription factors MTFA and MTFB in order to provide primers. The human mtDNA replisome consists of the helicase TWINKLE, DNA polymerase gamma, Pol γ, and a single-stranded DNA-binding protein (mtSSB). TWINKLE is hexameric and unwinds DNA in the 5′–3′ direction.

Human Pol γ is a two-subunit holoenzyme: a catalytic subunit Pol γA and a dimeric accessory subunit Pol γB. Pol γA possesses all enzymatic activities of the holoenzyme, including polymerase, 3′–5′ exonuclease for proofreading, and 5′ deoxyribose phosphate (5′-dRP) lyase activities. Pol γB has no enzymatic activity by itself, but upon association to form holoenzyme, it regulates all Pol γA activities. Holoenzyme exhibits markedly increased processivity of synthesis, reduced exonuclease activity, and elevated 5′-dRP lyase activity (Bogenhagen et al. 2001; Carrodeguas et al. 1999; Lim et al. 1999; Longley et al. 1998). From a crystallographic study, Pol γB can perform all accessory functions on a singly primed

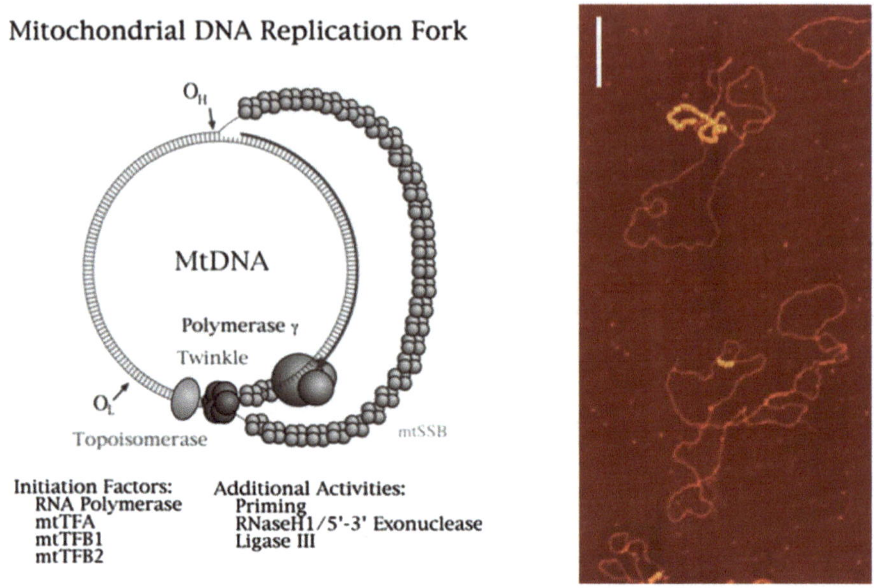

Fig. 11.2 (*Right*) Mitochondrial replisome components. The displaced strand is coated by mitochondrial single-stranded DNA-binding proteins (mtSSB). (*Left*) AFM images of two mtDNA molecules with single-stranded regions coated with SSB (Brown et al. 2005)

template without direct physical contact with DNA, acting entirely through interactions with Pol γA.

Mitochondrial DNA replication in lower eukaryotes is simpler. In fungi, there is no known DNA helicase. To compensate for the lack of helicase, mitochondrial DNAP must perform strand displacement DNA synthesis, i.e., unwind the downstream duplex concurrently with DNA synthesis. It has been shown that yeast (*Saccharomyces cerevisiae*) mtDNAP, Mip1, is indeed able to conduct strand displacement synthesis—an activity that is usually missing in purified replicases, which are normally associated with a helicase (Viikov et al. 2010).

Pol γA belongs to the polymerase A-family. Sequence comparisons with other family members aid identification of the exo and pol active sites and reveal ~400 aa spacer between the exo and pol domains. Although it has no counterpart in other family members, the importance of the spacer is manifested clinically: Pol γ containing mutations in this domain have been found in patients with several distinct mitochondrial diseases. The spacer domain presents a unique fold that can be further divided into two subdomains. The globular IP subdomain is located opposite the palm subdomain, and together they almost completely encircle the template. The IP subdomain thus potentially both enhances affinity for DNA and the intrinsic processivity of Pol γA. The second subdomain is AID, an elongated, flexible segment that extends away from the main body of the enzyme in order to interact with Pol γB (Fig. 11.3). The AID subdomain is likely to be disordered in the absence of Pol γB. The AID subdomain is amphipathic, with one surface being

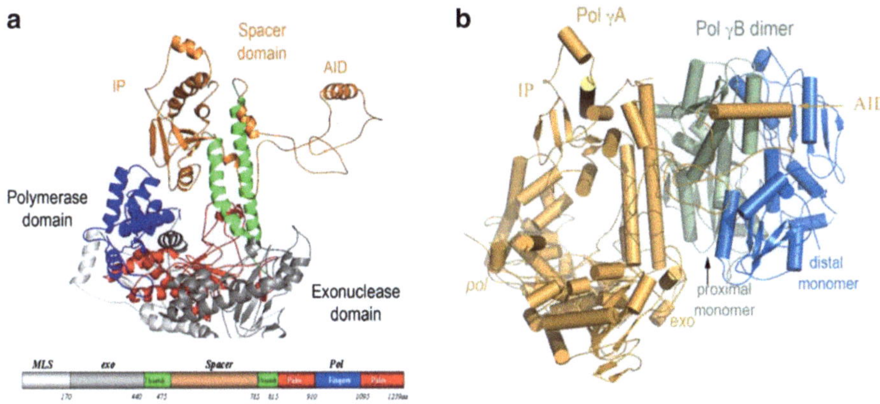

Fig. 11.3 Structure of human Pol γ. (**a**) Structure of Pol γA. The pol domain shows a canonical "right-hand" configuration with thumb (*green*), palm (*red*), and fingers (*blue*) subdomains and the exo domain (*gray*). The spacer domain (*orange*) presents a unique structure and is divided into two subdomains. Domains are shown in a linear form where the N-terminal domain contains residues 1–170; exo, 171–440; spacer, 476–785; and pol, 441–475 and 786–1239. (**b**) Structure of the heterotrimeric Pol γ holoenzyme containing one catalytic subunit Pol γA (*orange*) and the proximal (*green*) and distal (*blue*) monomers of Pol γB. Pol γA primarily interacts with the proximal monomer of the dimeric Pol γB (Lee et al. 2009)

hydrophobic and the opposite face highly positively charged (termed the K-tract, ^{496}KQKKAKKVKK^{505}).

In the trimeric holoenzyme, Pol γA asymmetrically interacts with the homodimeric Pol γB (Lee et al. 2009). This manner of subunit interaction raised the possibility that Pol γ could be a tetrameric enzyme with two Pol γA and a dimer Pol γB (A_2B_2). However, model building, using the twofold axis of Pol γB, indicates that the AID subdomains of two Pol γA monomers in a holoenzyme would result in steric clashes. Human Pol γ holoenzyme is therefore a trimer.

The structural arrangement of Pol γ suggests a mechanism explaining how Pol γB simultaneously affects exo and pol activities. When Pol γB binds to the Pol γA AID subdomain, it positions the positively charged K-tract to interact with upstream DNA, thereby increasing the affinity of holoenzyme for DNA by doubling the area of interaction. Secondly, the location of the primer 3′-OH determines the relative activities of pol and exo. Because of the mode of Pol γA binding to Pol γB, holoenzyme preferentially positions the 3′-OH in pol and away from exo, resulting in increased polymerase but reduced exonuclease activities. It should be noted that even though exonuclease activity is suppressed, after an incorrect nucleotide is incorporated, the synthesis reaction slows, providing the time necessary for the primer terminus to transfer to exo. The fidelity of holoenzyme is not therefore compromised by the interactions of Pol γB with Pol γA in forming holoenzyme.

Despite the strong circumstantial evidence of a bacterial origin of mitochondria, Pol γA shows homology to bacteriophage T7 DNA polymerase, especially in the palm subdomains that contains the catalytic active site; the rmsd of the subdomain

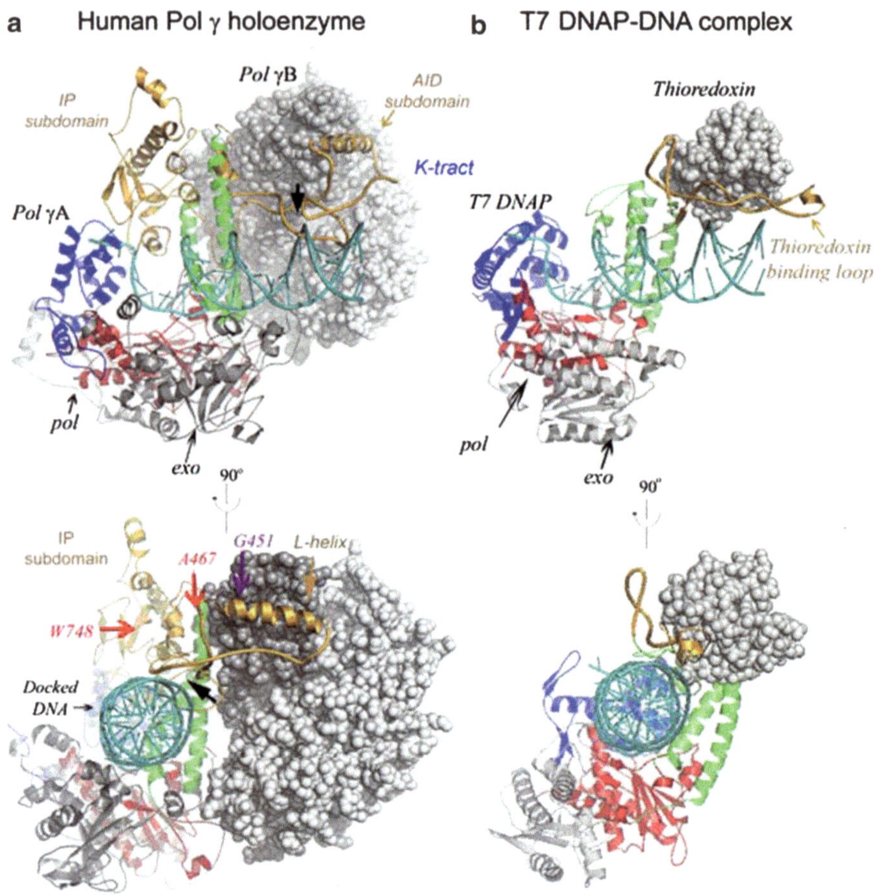

Fig. 11.4 Comparison of a Modeled Pol γ–DNA. (**a**) Modeled Pol γ–DNA complex containing Pol γA (shown in *ribbons*), Pol γB (*gray* CPK), and a docked DNA (*blue ribbons*) shows that IP and AID subdomains enhance DNA binding. Mutations and the region protected by DNA from proteolytic digestion (*black arrow*) are indicated. (**b**) Crystal structure of T7 DNAP–DNA complex containing gp5 (*ribbons*), thioredoxin (*gray* CPK), and a primer-template DNA (*blue ribbons*) (Lee et al. 2009)

between the two polymerases is merely 2.3 Å. A comparable structure to the AID subdomain is also seen, where the flexible thioredoxin-binding domain of T7 gp5 makes the only contact with the processivity factor and the upstream primer-template (Fig. 11.4).

Like the catalytic subunit of many replicases, human Pol γA exhibits limited processivity in synthesis, incorporating ~100 nt per binding event. In the presence of Pol γB, processivity of the holoenzyme increases to ~1,500 nt. Pol γB has a different mode of increasing processivity than other accessory proteins for replicases: it simultaneously enhances DNA binding and the reaction rate. In contrast, thioredoxin only increases the binding affinity of T7 gp5 for DNA.

Nonetheless, because thioredoxin and Pol γB are tightly associated with their catalytic subunit in forming a holoenzyme, they both function differently than loosely bound processivity factors, such as sliding clamps. The ring-shaped sliding clamp, after being loaded on the template, binds to polymerase and reduces its dissociation rate during synthesis. There are mutually beneficial interactions in forming the trimeric Pol γ holoenzyme: Pol γA enhances dimerization of Pol γB, and the interaction between Pol γA and Pol γB is enhanced by the presence of primer-template DNA (Lee et al. 2010b).

Despite an overall similarity, both Pol γA and Pol γB exhibit species-dependent structural variations: for the catalytic subunit, these are focused in the subdomain that interacts with Pol γB and in the C-terminus. In contrast to mammals, Drosophila and mosquitoes have a simpler, monomeric Pol γB, and fungal mtDNAPs are single polypeptides, with no known accessory subunit (Kaguni 2004). They also lack an AID subdomain, yet biochemical characterization of yeast Mip1 indicates the polymerase is still highly processive (Viikov et al. 2010). Relative to the human enzyme, fungal polymerases have C-terminal extensions. *S. cerevisiae* polymerase, Mip1, has the longest known extension: 279 aa, which has been hypothesized to function as an intrinsic processivity factor that provides the same function as Pol γB. Interestingly, deleting 175 aa of the 279 aa extension yields an enzyme with higher processivity than wild type; further deletion (216 aa) results in reduced processivity but increased 3′–5′ exonuclease activity. Deletion of the entire extension abrogates all polymerase activity, but the enzyme retains higher 3′–5′ exonuclease activity than wild type (Viikov et al. 2012).

Pol γB also displays species-dependent variation: mammalian Pol γB is a homodimer of two 55 kDa subunits. It structurally resembles class II aminoacyl-tRNA synthetases, however, the residues for catalytic activity are not conserved, so that Pol γB lacks synthetase activity (Carrodeguas et al. 2001). Nevertheless, the overall tRNA synthetase fold led to examination of Pol γB binding to tRNA-like or stem-loop DNA. This is significant because the single-stranded O_L may also adopt a stem-loop structure. Indeed, Pol γB has high affinity for double-stranded DNAs longer than 45 bp (Carrodeguas et al. 2002). DNA binding by Pol γB is associated with two positively charged regions, termed RK and RKK. Alanine substitutions abolished DNA-binding activity. The substitutions have no effect on holoenzyme Pol γ DNA synthesis on single-stranded templates. This result agrees well the structural studies, where Pol γB appears to enhance DNA binding indirectly. However, when tested on duplex templates using the entire replisome, i.e., in the presence of TWINKLE and SSB, both the RK and RKK mutants result in defective DNA synthesis (Farge et al. 2007). This suggests that the RK and RKK regions only function in synthesis at a replication fork, either by directly contacting the branched DNA or perhaps other components of the replisome. Pol γB therefore appears to perform an additional, as yet unknown, function during replication on duplex templates.

In the trimeric human Pol γ holoenzyme, there are extensive hydrophobic interactions between the AID and thumb subdomains of Pol γA and one Pol γB monomer. In contrast, very limited interactions with the distal Pol γB are observed, with only one salt bridge formed between Pol γA and the distal Pol γB monomer. This asymmetrical provides a structural explanation for the insect monomeric Pol γB can be functional. The limited interaction with the distal Pol γB is of importance in humans, as a mutation that disrupts the only Pol γA and Pol γB distal monomer interaction has severe consequences (Ferrari et al. 2005). Biochemical studies indicate that the mutants abolished the function of Pol γB without physical dissociation of the holoenzyme. The Pol γA–distal Pol γB monomer contact appears to regulate both pol and exo activity; disruption of this interaction elevates exonuclease activity while decreases DNA polymerization (Lee et al. 2010a).

11.2 Pol γ in Mitochondrial DNA Repair

The unique mitochondria environments make mtDNA extremely vulnerable to oxidative damage. Radicals escaping from the oxidative phosphorylation electron transfer chain and reactive oxygen species generated from redox reactions are particularly damaging to DNA. Exogenous radiation sources and the lack of histones in mitochondria further exacerbate mtDNA damage. Consequently, mtDNA suffers much higher levels of oxidative DNA damage than genomic DNA (Bohr et al. 2002; Richter et al. 1988; Wiseman and Halliwell 1996). Oxidative stress plays a large role not only in tumor pathologies but in other diseases of aging, such as Alzheimer's, Parkinson's, and Huntington's diseases (Shoffner et al. 1993; Wang et al. 2005; Yang et al. 2008).

Base excision repair (BER) is the main mechanism for repairing damage to mtDNA. Single-nucleotide BER (SN-BER) begins with damaged base excision by a glycosylase, AP-endonuclease cleavage to generate a 3′-OH, removal of the 5′-deoxyribose by the lyase activity of Pol γ, followed by gap-filling synthesis, and then nick sealing by DNA ligase (Bohr et al. 2002; Hegde et al. 2010; Liu and Demple 2010). However, the 5′-dRP lyase activity of Pol γ is weaker than that of the nuclear repair DNAP Pol β (Longley et al. 1998; Pinz and Bogenhagen 2006), and Pol γ cannot efficiently recognize single-nucleotide gaps (He and Yin, unpublished data). In principle, the latter problem could be overcome by another BER protein actively recruiting Pol γ to the gap, but it is possible that in mitochondria most repair is biased towards the long-patch BER (LP-BER) pathway. Indeed, if the 5′-ribose is oxidatively damaged, often resulting in a deoxyribonolactone, the 5′-dRP lyase activity of Pol γ cannot function to provide a downstream 5′-phosphate for ligation. In this situation, LP-BER must be activated to remove the damaged sugar together with some downstream DNA. The resulting gap is then efficiently filled-in by Pol γ.

The key difference between SN-BER and LP-BER is that the repair extends beyond the lesion site. After excision of the damaged nucleotide, it is replaced along with the strand downstream from the lesion with a nascent DNA. This fashion of DNA synthesis, called strand displacement, requires coordination of Pol γ and a nuclease to synthesize and cleave the displaced strand. The mechanism of removal of downstream DNA has not been resolved, in particular, which nucleases participate in mitochondrial LP-BER. Nucleases known to be active in mitochondria include DNA2, a 5′–3′ helicase and exonuclease; EXOG, a 5′ exonuclease; and FEN1, a flap DNA nuclease (Liu et al. 2008; Tann et al. 2011; Zheng et al. 2008). Both DNA2 and EXOG can be co-immunoprecipitated with Pol γ, suggesting they either directly or indirectly interact. However, both DNA2 and EXOG removed exactly 2-nt from a 5′-flap even though the flap was much longer. FEN1 can remove the entire flap, but as yet there is no evidence of an interaction with Pol γ.

Additionally, Pol γ is able to perform strand displacement synthesis when its 3′–5′ exonuclease activity is abolished, even when the downstream DNA contains a 5′ oxidized ribose (He and Yin, unpublished data). This observation suggests yet another possibility for LP-BER that is suppression of Pol γ exo activity. Whether a major pathway exists or these enzymes can all be involved in redundant pathways requires future studies.

11.3 Pol γ as an Adverse Reaction Target for Antiviral Drugs

The nucleoside/nucleotide reverse transcriptase inhibitors (NRTIs) ushered in a new era of antiviral treatment for human immunodeficiency virus (HIV) and hepatitis virus infections. NRTIs are substrate mimics for DNAPs, but they lack a 3′-OH and therefore function as chain terminators that stop viral replication. The success of highly active antiretroviral therapy (HAARP), which includes inhibitors for HIV reverse transcriptase and protease, also brought attention to the problem of drug toxicity. This is of special importance with long-term usage as the life span of patients extends. Management of toxicity has been ameliorated with varying therapeutic regimes but remains a major challenge. Clinical manifestations of NRTI toxicity are often multisystem, including muscular, cardiovascular, and neurological system dysfunctions, i.e., the systems relying on high-energy output from mitochondria. Histological examinations coupled to electron microscopy show swollen, misshapen mitochondria that have decreased numbers of cristae and non-membrane-bound fat droplets of various sizes, indicative of decreased mitochondrial function for fatty acid β-oxidation (Lewis et al. 2003). The causes of mitochondrial toxicity include the active transport of NRTIs into the organelle and inhibition of major adverse target, human Pol γ.

Studies of isolated human DNA polymerases, Pol α, Pol β, Pol γ and Pol ε showed that Pol γ is most susceptible to inhibition by NRTIs (Hart et al. 1992). This may be attributed to biochemical and structural similarities of Pol γ with the viral enzyme. The structural basis for Pol γ susceptibility is thought to reside in the pol active site and the mechanism for discriminating (or lack thereof) against nucleotide analogs.

The reason for inhibition of Pol γ by drugs designed against both HIV and HBV polymerases may be related to the fact that both viral proteins are reverse transcriptases (RT) that use DNA or RNA templates. Pol γ also possesses RT activity, although synthesis is less processive on RNA templates. This activity is relevant to the presence of RNA in human mtDNA, which is estimated to contain 1–3 % ribonucleotides. The RNA has been hypothesized to be due to incompletely removed primers. Persistence of RNA requires an RT activity of Pol γ to replicate over the RNA segments. The inhibition of Pol γ by NRTIs suggests structural/functional similarities between the human and viral proteins.

Toxicities of NRTIs can be correlated with the kinetics of incorporation and exonucleolytic removal by the mitochondrial DNA polymerase (Fig. 11.5) (Lee and Johnson 2006), supporting the idea that Pol γ is the adverse reaction target. Nonetheless, some NRTIs have high efficacy and low toxicity, suggesting differences exist between the viral and human enzymes (Table 11.1). Kinetic analyses show that Pol γ could not distinguish substrate dNTP from NRTIs, thereby causing cytotoxicity. These NRTIs either have high affinity to Pol γ (Low K_d) that is either comparable to higher substrates, or efficiently incorporated (high k_{pol}). For example, HIV RT inhibitors dideoxy cytosine (ddC or zalcitabine) and 3′-fluroro-3′-deoxythymidine (FLP) (Fig. 11.5) bind to Pol γ with affinity equal or higher than native nucleotides, which could explain their high toxicities. However, the drug toxicity is complicated, because subtle change in NRTI structures can alter their effect on Pol γ drastically. For example, Pol γ discriminates FLP only 8.3-fold from the native nucleotide dTTP; however, it can discriminate a structurally related NRTI, 2′3-didehydro-2′3′-dideoxythymidine (Ed4T), from dTTP greater than 12,000-fold (Sohl et al. 2012). Another NRTI, (−)-2′,3′-dideoxy-3′-thiacytidine [(−)SddC or 3TC] is a potent low-toxic HIV RT inhibitor, but its D(+) isomer, (+) SddC that has the natural nucleoside configuration, is less potent but more toxic (Feng and Anderson 1999a; Johnson et al. 2001). To completely understand drug toxicity will require combinatory studies of enzyme kinetics and computational and structural biology.

Structural comparison of human Pol γ and HIV RT structures reveal differences between the two enzymes that may be exploited in future drug design. For example, the distinct subunit interactions result in substrate DNA being bound in the active site of Pol γ at an angle of 45° to that in HIV RT (Fig. 11.6). More importantly, while the catalytic aspartates of the HIV RT p66 subunit and Pol γA have a similar spatial arrangement, the incoming nucleotide-binding sites formed between the palm and fingers subdomains are structurally distinct, being α-helical in Pol γ but β-sheet in HIV RT.

NRTIs have also been most widely used therapy for HBV. About 350 million people worldwide are living with hepatitis infection (WHO 2003). Hepatitis can be caused by hepatitis virus types A, B, C, D, or E. Of all viral causes of human hepatitis, hepatitis B virus (HBV) is of global importance, because of the sheer number of viral carriers; persistent HBV infection can lead to chronic hepatic insufficiency, cirrhosis, and liver cancer. About one million people die each year

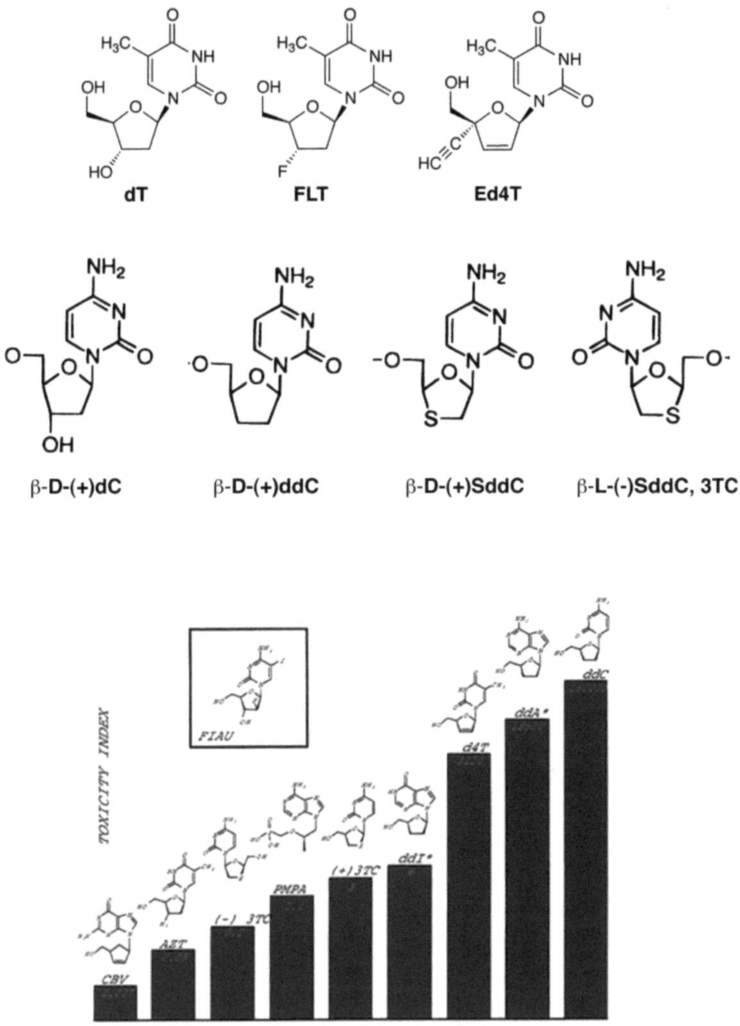

Fig. 11.5 NRTIs drug toxicity correlates with human Pol γ inhibition and structures of selective NRTIs (Lee and Johnson 2007)

of these complications. In the United States, complications from chronic HBV infection count for 5–10 % of cirrhosis and liver transplant (McKenzie et al. 1995).

Clinical treatment of HBV includes administration of Adenine arabinoside (vidarabine), acyclovir, didanosine, zidovudine, and ribavirin (Fried et al. 1992; Garcia et al. 1987; Katsuragi et al. 1997). Unfortunately these drugs are ineffective or too toxic for prolonged use. A clinical trial used the second-generation nucleoside analog fialuridine (1-(2-deoxy-2-fluoro-b-D-arabinofuranosyl)-5-iodouracil, or FIAU) as a treatment for chronic hepatitis B virus infection (McKenzie et al. 1995).

Table 11.1 Comparison of efficiency of human Pol γ and HIV RT incorporation of nucleoside analogs NRTIs that are FDA approved and under development

	Discrimination		
dNTP analog	Human Pol γ	HIV RT	Reference
ddC-TP	2.9	10	Feng and Anderson (1999b), Feng et al. (2001), Ray et al. (2003), Ray and Anderson (2001)
ddA-TP	4.0	5	Johnson et al. (2001)
d4T-TP	7.4	0.56	Vaccaro et al. (2000), Johnson et al. (2001)
KP1212-TP	26	14	Murakami et al. (2005)
FLT-TP	35	4.2	Sohl et al. (2012)[a]
(−)-3TC-TP	2,900	40	Feng and Anderson (1999a), Feng et al. (2001), Ray et al. (2003)
EFdA-TP	4,300	N.D.	Sohl et al. (2011)
Ed4T-TP	6,200	0.51	Sohl et al. (2012)
PMPApp	11,400	6.1	Johnson et al. (2001)
AZT-TP	37,000	2.7	Vaccaro and Anderson (1998), Johnson et al. (2001)
(−)-FTC-TP	290,000	16	Feng et al. (2004)
CBV-TP	900,000	34	Johnson et al. (2001)

N.D. no data, *KP1212* 5-aza-5,6-dihydro-2′-deoxycytidine, *3TC* lamivudine, *PMPApp* tenofovir diphosphate, *FTC* emtricitabine, *CBV* carbovir
[a]Steady-state studies indicated an efficiency of 0.5 (Michailidis et al. 2009)

While nontoxic in animal studies, humans treated with FIAU developed severe hepatotoxicity, with progressive lactic acidosis, and liver failure, even after discontinuation of the drug. Several patients also had pancreatitis, neuropathy, or myopathy. Of the seven patients with severe hepatotoxicity, five died and two survived after a liver transplant.

Extensive investigations revealed that toxicity was caused by FIAU inhibition of human Pol γ. Toxicity predominantly affected organs and tissues that have a slow turnover of cells and a major dependence on mitochondrial function. Triphosphate forms of FIAU and its metabolites, FMAU (1-(2-deoxy-2-fluoro-b-D-arabinofuranosyl)-5-methyluracil) and FAU (1-(2-deoxy-2-fluoro-b-D-arabinofuranosyl)-5-uracil) compete with dTTP for incorporation by human Pol γ. FIAUTP, FMAUTP, and FAUTP can be incorporated into singly primed DNA opposite to adenosine effectively without affecting chain elongation. However, multiple FIAUTP incorporation in adjacent positions dramatically impaired chain elongation by Pol γ (Colacino 1996; Lewis et al. 1996). Consequently, mtDNA content became severely decreased in patients.

11.4 Pol γ Mutations Are Implicated in Clinical Disorders

Mitochondrial diseases are devastating disorders for which there is no cure or proven treatment. About 1 in 4,000 individuals is at risk of developing a mitochondrial disease sometime in their lifetime. Half of those affected are children who

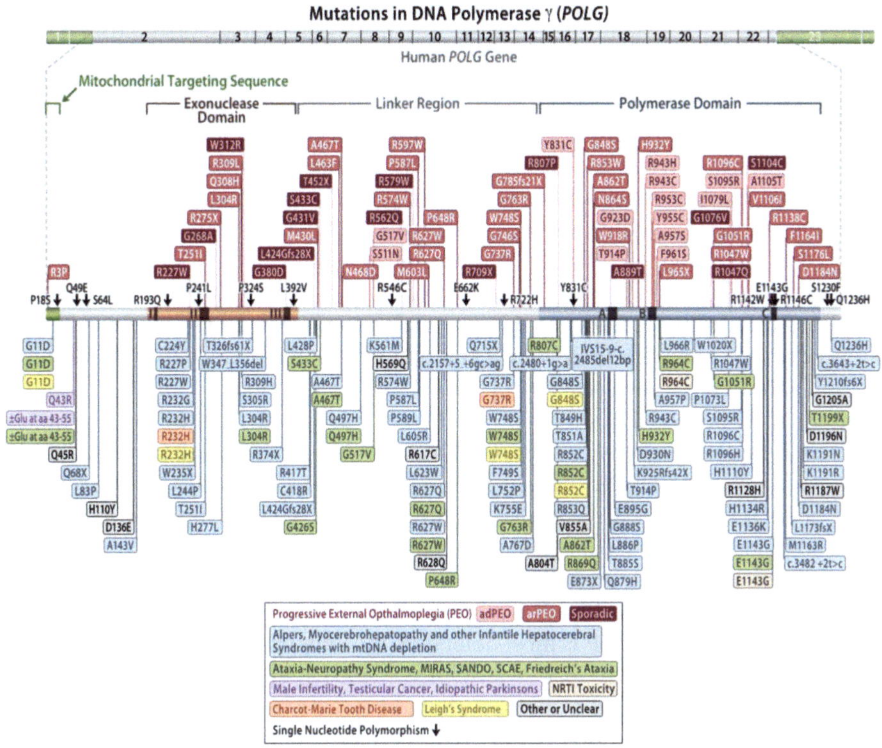

Fig. 11.6 Mutations on human Pol γA associated with diseases (Chan and Copeland 2009)

show symptoms before age 5, and approximately 80 % of them will die before age 20. The mortality rate is roughly that of cancer (Longley et al. 2005).

More than 90 Pol γ mutations have been reported associated with multisystem diseases, spanning cardiovascular, neurological, and muscular systems (Chan and Copeland 2009). Progressive external ophthalmoplegia (PEO) is a late onset mitochondrial disease, with bilateral ptosis and progressive weakening of the external eye muscle, proximal muscle weakness, and wasting as well as exercise intolerance. Skeletal muscles of PEO patients present red ragged fibers and lowered activity of respiratory chain enzymes. Muscle biopsies show that mtDNA contains multiple large-scale deletions (Zeviani et al. 1989). To date, with the exception of one mutation, the autosomal dominant *POLG* mutations responsible for PEO development have been mapped to the pol domain of Pol γ (Fig. 11.6).

Alpers syndrome is a rare but severe, heritable, autosomal recessive disease that afflicts young children. Within the first few years of life, patients develop seizures, cortical blindness, deafness, liver failure, and eventual death. Many Pol γ mutations have been found in Alpers patients. Among them, the Pol γA A467T substitution is most common, accounting for 36 % of all mutations. The mutation is located in the Pol γ thumb subdomain, which interacts directly with the template DNA and guides the primer terminus between exo and pol active sites. The substitution therefore influences the relative activities of both activities; holoenzyme containing Pol γA A467T has only 4 % polymerase activity and a reduced k_{cat} and $K_{m(dNTP)}$ (Chan et al. 2005).

Two other Alpers substitutions, R943H and Y955C, retain less than 1 % of wild-type polymerase activity and display a severe decreased processivity (Graziewicz et al. 2004). The mutations are located on a positively charge helix—O helix—that, based on structural homology modeling with T7DNAP, is critical for dNTP binding. R943 is predicted to interact with the triphosphate moiety of the dNTP, where Y955 facilitates translocation by serving as a pawl to prevent DNA duplex backtracking. The R943H substitution reduces the enzyme's affinity for dNTP, whereas Y955C allows excessive backtracking, which could therefore both reduce the DNA synthesis rate and increase the frequency of insertions or deletions in mtDNA. Using a mouse transgenic model in which Y955C *POLG* was targeted to the heart, the mouse developed cardiomyopathy, loss of mtDNA, an enlarged heart, and increased levels of 8-oxo-deoxyguanine (8oxoG) in its mtDNA (Lewis et al. 2007). Collectively, these phenotypes suggest that patients harboring the Y955C mutation may suffer elevated oxidative damage and should benefit from antioxidant therapy.

Substitutions A957P, A957S, R1096C, and R1096H are also found in patients suffering from aggressive Alpers syndrome to mild PEO. Kinetics analyses of these mutants indicate that A957P is most defective, the deficiency of other point mutants following the order of R1096H > R1096C > A957S, an order that matches well with the severity of clinical symptoms (Sohl et al. 2013).

Substitution reducing Pol γ fidelity will increase mtDNA mutation, which have been associated with numerous other clinical multisystem human disorders, including mitochondrial encephalopathy, lactate acidosis, and stroke-like syndrome (MELAS); cardiovascular, skeletal muscular, endocrinology (Type II diabetes); and neurodegenerative disorders (Alpers syndrome, Parkinson-like, and Alzheimer). Accumulated mutation on mtDNA is correlated with natural process of aging. Because reactive oxygen species (ROS) generated inside the cell will lead, with time, to increasing amounts of oxidative damage to various cell components. Impaired respiratory chain function causes degenerative diseases and accelerates aging. Transgenetic mice harboring exonuclease-deficient Pol γA shows elevated mutation rate; the animals show premature aging phenotype and have significantly reduced life span (Trifunovic et al. 2005).

11.5 Mitochondrial RNA Transcription

11.5.1 Mitochondrial Gene Structure

Human mitochondrial DNA encodes 13 genes that are components of the oxidative phosphorylation electron transport chain, 2 rRNAs (12S and 14S), and 22 tRNAs required for mitochondrial translation. These genes are distributed asymmetrically on both strands of dsDNA: the H-strand encodes the 2 rRNAs, 14 tRNAs, and 12 mRNAs and the L-strand encodes 8 tRNAs and 1 mRNA. As both strands of human mitochondrial DNA code for essential genes, transcription of both strands is necessary. There are two strong promoters, HSP1 and LSP, located in the D-loop, that drive transcription on the H- and L-strand, respectively (Montoya et al. 1982, 1983) (Fig. 11.1). Another promoter, HSP2, located downstream from HSP1 between rRNA and $tRNA^{phe}$, is a weak H-strand promoter (Litonin et al. 2010; Lodeiro et al. 2012). A transcription terminator between 12S rRNA and $tRNA^{leu}$, mediated by a termination factor (mTERF), is an important regulator for the quantity of rRNA and tRNA (Fernandez-Silva et al. 1997; Kruse et al. 1989).

The mitochondrial transcription machinery not only generates mRNAs, tRNAs and rRNAs, but is also an integral component of DNA replication. Mitochondria lack a specialized primase and the RNA polymerase (mtRNAP) must generate RNA primers for DNA synthesis. This suggests that the mitochondrial transcription system has nonspecific as well as promoter-specific initiation abilities. On the H-strand, the HSP promoter is close to the replication origin O_H, so transcripts can also serve as a replication primer. Significantly, transcription from HSP1 pauses or terminates at a conserved sequence (CSBII) near O_H (Pham et al. 2006), which may facilitate primer transfer to DNAP. If replication of mtDNA follows the canonical symmetrical synthesis and the L-strand is replicated as a lagging strand, RNA primers must be produced by mtRNAP in a sequence-independent manner. Indeed, mtRNAP transcribes on a promoter-free, single-stranded template to yield 20–50 nt RNA (Wanrooij et al. 2008).

In contrast to the typically increased complexity in gene structure in higher eukaryotes, yeast mtDNA appears more complicated than its human counterpart. Unlike the circular dsDNA of human mtDNA, that of yeast is a mixture of linear and circular DNA. Although yeast mtDNA codes for the same set of genes, 80 % of the genome codes for type II introns (Tabak et al. 1984), making it five times the size. Interestingly, these group II introns exhibit an extraordinary mobility mechanism in which the excised intron RNA reverse splices directly into a DNA target site and is then reverse transcribed by the intron-encoded protein (Lambowitz and Zimmerly 2004).

11.5.2 Mitochondrial RNA Polymerase and Transcription Factors

All components of mitochondrial transcription machinery are nuclear encoded and are transported into mitochondria via their mitochondrial localization sequences. Early mitochondrial transcription studies were conducted in yeast, which consists of a RNA polymerase, Rpo41, and a single transcription factor, Mtf1 (Greenleaf et al. 1986; Riemen and Michaelis 1993). Rpo41 displays a clear bacteriophage origin, as Rpo41 is homologous to T7 RNAP (Masters et al. 1987), but has an

N-terminal domain that is absent in the phage enzyme. However, unlike T7 RNAP, Rpo41 has lost independent transcription initiation activity and is functional only in the presence of the transcription factor Mtf1. Mitochondrial transcription therefore presents a unique opportunity to understand the evolution of complex transcription machinery. Mtf1 was originally thought to be a σ factor-like protein but, surprisingly, a crystal structure of Mtf1 revealed that Mtf1 is similar to bacterial ribosomal rRNA dimethyltransferase, an enzyme that confers kasugamycin resistance by modifying an adenosine residues in a rRNA stem loop (Seidel-Rogol et al. 2003). Interestingly, even though the yeast Rpo41 is larger than T7 RNAP, the minimal yeast mitochondrial promoters is only 9 bp, about half of a T7 promoter (Marczynski et al. 1989; Tracy and Stern 1995).

Human mitochondrial transcription activity was first detected using partially purified mitochondrial extracts, establishing that mitochondria have a transcription system independent of the nucleus (Aloni and Attardi 1971a, 1972; Pica-Mattoccia and Attardi 1971). However, the purified RNAP showed no activity, which led to the discovery of its transcription factors. The first transcription factor was named TFAM (Fisher and Clayton 1988), a member of the HMG (high-mobility group) family that includes many transcription factors, including LEF-1, TCF-1, and SRY (Rubio-Cosials et al. 2011). Mitochondrial DNA lacks the protection of histones; however, in certain cells, TFAM is so abundant so that it can coat the entire mtDNA (Takamatsu et al. 2002). In additional to its role in transcription, TFAM is also important for mtDNA maintenance, as mutations affecting TFAM reduce mtDNA content.

A common feature of HMG proteins is their high affinity for, and their ability to bend, DNA. An HMG-box domain bends dsDNA by about 90°, forming a superhelical protein–DNA complex. The crystal structure of TFAM complexed to the LSP promoter reveals how this small 25 kDa protein generates a large footprint. The two HMG-box domains of TFAM bind to dsDNA almost symmetrically, forcing the DNA into a 180° U-turn (Rubio-Cosials and Sola 2013). This extreme bending is reminiscent of the pre-initiation transcription complex of TBP to TATA box-containing DNA region (Nikolov et al. 1996) or the bacterial integration host factor (IHF) and HU or "DNABII" family of nucleoid-associated proteins—e.g., Hbb (Mouw and Rice 2007). The two HMG boxes of TFAM make sequence-specific interaction with DNA, suggesting its contribution to promoter specificity.

However, in the presence of TFAM, mtRNAP did not achieve maximal transcriptional activity in vitro. Further studies revealed two homologous proteins, TFB1M and TFB2M, both of which share sequence similarity with the yeast mitochondrial transcription factor, Mtf1, and were predicted by bioinformatics to be methyltransferases (Falkenberg et al. 2002; Seidel-Rogol et al. 2003).

TFB1M complements a *ksg* bacterial mutant, indicating that the human TFB1M can methylate the ksg stem loop in bacterial rRNA, confirming its methyltransferase activity (Seidel-Rogol et al. 2003). However, TFB1M was later shown functioning in translation, and only TFB2M is a transcription factor (Matsushima et al. 2005), interacting with mtRNAP in the absence of any promoter. The minimal human transcription system in vitro consists of an mtRNAP (POLRM) and TFB2M. TFAM differentially affects transcription from HSP and LSP. In the minimal transcription system, HSP1 is a stronger promoter than LSP. However, in the presence of TFAM, equal numbers of transcripts are obtained from the two promoters. At equimolar concentration of TFAM and template, full activation of transcription from LSP is achieved without changing transcription levels from HSP1. When TFAM is in molar excess, transcription from LSP is substantially inhibited, while HSP1 remains active (Lodeiro et al. 2012). These observations support the hypothesis that a TFAM-regulated, two-component transcription system is operative in human mitochondria. The transcription initiation complex is structured that only mtRNAP and TFMA directly contact the promoter DNA, whereas TFBM2 is sandwiched between mtRNAP and TFAM, making only protein–protein interactions (McCulloch and Shadel 2003).

Like yeast Rpo41, human mtRNAP is homologous to T7 RNAP. The similarity allowed an accurate delineation of domains. The various functions of T7 RNAP are neatly divided into two domains: an N-terminal domain (NTD) for promoter specificity and a C-terminal domain for catalysis. Human mtRNAP exhibits 30 and 50 % sequence similarity with the phage NTD and CTD domains; however, as in yeast, the human enzyme also contains an N-terminal extension (Fig. 11.7). Interestingly, an N-terminal-truncated mtRNAP exists naturally in humans as a product of alternative splicing. The truncated enzyme lacks the N-terminal 262 amino acid extension but still functions as an RNAP (Kravchenko et al. 2005) and is confined to the nucleus. Transcription of certain mRNAs in humans and rodents is mediated by this single polypeptide nuclear enzyme.

11.5.3 Mitochondrial Transcription Reaction and Regulation

General mitochondrial transcription follows the same scheme for all promoter-specific transcription:

$$\underbrace{\mathrm{E} + \mathrm{PDc} \leftrightarrow \mathrm{E} \bullet 2\mathrm{PDc} \leftrightarrow \mathrm{E} \bullet \mathrm{PDo}}_{\text{promoter melting}} \xrightarrow[\text{initiation}]{\mathrm{NTP}} \mathrm{E} \bullet \mathrm{D} \bullet R_2 \xrightarrow[\text{elongation}]{\mathrm{NTP_s}} \rightarrow \rightarrow \mathrm{E} \bullet \mathrm{D} \bullet R_n$$

As the N-terminal domain (NTD) of T7 RNAP is responsible for promoter recognition, the Rpo41 NTD was tested for promoter specificity. Rpo41 alone is incapable of transcribing on duplex DNA, and promoter-specific transcription is only seen after Rpo41 associates with Mtf1, suggesting that Mtf1 is responsible for promoter unwinding. The first step of transcription is unwinding the double-stranded promoter; many RNAPs bend DNA ~90° to facilitate the process. However, fluorescence energy transfer (FRET) studies show that Rpo41 by itself bends not only promoter DNA by ~52° but also non-promoter DNA to the same extent (Tang et al. 2010). This level of bending is insufficient for T7 RNAP to open its promoter (Tang and Patel 2006). Remarkably, the addition of Mtf1 to Rpo41

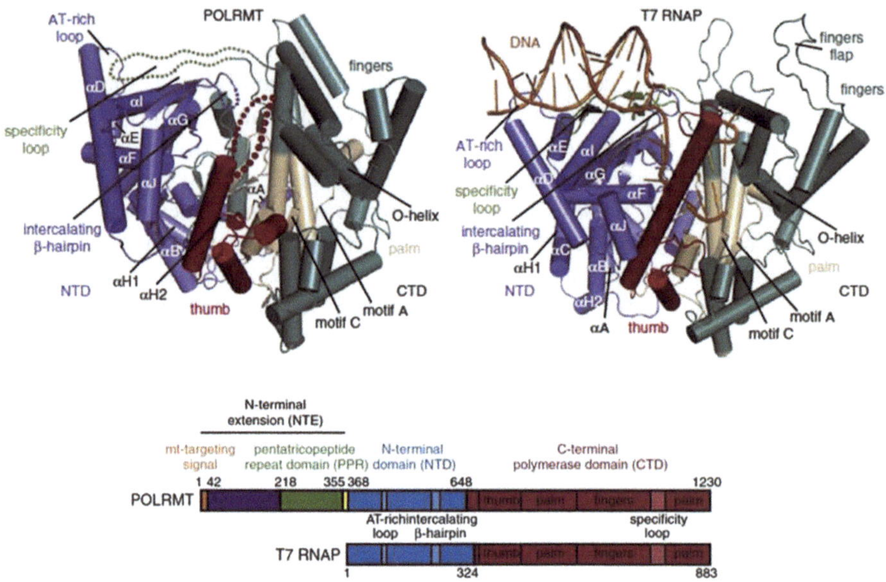

Fig. 11.7 Structural and sequence comparison of human mitochondrial RNA polymerase and T7 RNAP (Ringel et al. 2011)

increases bending ~90°, and the increase is specific for promoter DNA. It appears that Rpo41–Mtf1 catalyzes promoter-specific initiation by an induced-fit mechanism, where the two proteins jointly bend and melt the promoter DNA. This suggests that either both proteins bind the promoter or Mtf1 changes the conformation of Rpo41 to expose additional promoter-binding sites.

Rpo41 displays strong activity on a pre-unwound (bubble) promoter, indicating it contains full catalytic activity, and its inability to transcribe on a closed promoter is therefore due to its inability to separate the DNA strands (Matsunaga and Jaehning 2004). Removal of the N-terminal extension of Rpo41 does not result in Mtf1-independent activity; rather, Rpo41 shows even greater dependency on Mtf1 for activity. This observation, together with that Mtf1 does not bind to DNA, implies that Mtf1 induces significant conformational change on Rpo41 upon formation of holoenzyme on a promoter.

A crystal structure of human mtRNAP provides a clear molecular basis for the unique mechanism of mitochondrial transcription. The NTD of the mtRNAP resembles T7 RNAP, but the N-terminal extension consists of nine α-helices, four of which comprise two pentatricopeptide repeat (PPR) motifs found in both plant and mitochondrial proteins (Delannoy et al. 2007; Lightowlers and Chrzanowska-Lightowlers 2008). The PPR motif appears to be involved in RNA processing and editing. The structure explains the necessity of transcription factors for mtRNAP, as a region in human mtRNAP, equivalent to the promoter recognition motif of T7 RNAP (the AT-rich loop), instead of binding to promoter, is seen to interact with TFAM. Whether promoter interaction is mediated by TFAM has to wait for the crystal mtRNA–TFAM complex structure.

A modeled mtRNAP initiation complex was constructed by docking DNA from a T7 RNAP-promoter complex after superimposing the two active site domains (Cheetham and Steitz 1999; Ringel et al. 2011). Although tentative, the model initiation complex illustrates how promoter recognition may be achieved. Biochemical analyses indicate that the specificity loop confers promoter specificity, interacting with positions -6 and -11 of the promoter (Nayak et al. 2009). The mtRNAP transcription initiation model supports the idea that the specificity loop of mtRNAP, albeit shorter than that of T7 RNAP, could recognize the same region of the promoter. The AT-rich recognition loop is situated in a similar location as T7 RNAP, but it is sequestered by the N-terminal extension and thus does not contribute to promoter binding. In fact, the mtRNAP AT-rich loop appears to have lost promoter-binding activity, because deletion of the extension does not restore factor-independent, promoter-specific transcription (Chang and Yin, unpublished data). It is puzzling why mtRNAP possess the same promoter recognition elements as T7 RNAP but lacks intrinsic promoter specificity. Perhaps substitution of residues in critical regions causes subtle structural changes so that certain elements are less than optimally situated for promoter binding. However, the structure of mtRNAP allows testable hypotheses regarding factor-dependent transcription initiation: TFB2M may reposition the misaligned intercalating hairpin to allow promoter melting or it may directly melt the promoter.

Combined structural and biochemical studies describe the transcription complex on the 38 bp LSP promoter: TFAM binds 23 bp from −38 to −15 at LSP and that the mtRNAP–TFB2M complex covers the region immediately downstream, from bp −14 to +10, with TFB2M putatively positioned close to TFAM (Fig. 11.7) (Dairaghi et al. 1995; Gaspari et al. 2004; Litonin et al. 2010).

Mitochondrial gene expression is regulated by cellular ATP levels. Regulation is best understood in yeast: cells grown in glucose have high levels of ATP, which suppresses transcription. Conversely, cells grown anaerobically produce less ATP and exhibit elevated transcription (Amiott and Jaehning 2006). The increase in transcription is mainly due to increased synthesis rates, as opposed to an increase in mtDNA copy number or elevated Rpo41 and Mtf1 levels. Moreover, different yeast promoters respond to changes in ATP levels to varying degrees. Rpo41–Mtf1 itself acts as the in vivo ATP sensor that couples RNA abundance to respiration (Amiott and Jaehning 2006). In addition, exogenous factors, such as NRTIs, inhibit human mtRNAP activity (Arnold et al. 2012), thus directly contributing to drug toxicity.

Mitochondrial function requires control of mtDNA content, and both are intimately related to human health. Oxidative mtDNA damage has been associated with aging; mutations on mtDNA are implicated in cancers and multisystem clinical disorders. Mitochondrial polymerases, as adverse reaction targets for nucleoside based inhibitors, are important for antiviral drug design. Although much progress has been made in the past decades, a complete understanding of mitochondrial structure and function requires substantial support for further research.

References

Aloni Y, Attardi G (1971a) Expression of the mitochondrial genome in HeLa cells. II. Evidence for complete transcription of mitochondrial DNA. J Mol Biol 55:251–267

Aloni Y, Attardi G (1971b) Symmetrical in vivo transcription of mitochondrial DNA in HeLa cells. Proc Natl Acad Sci USA 68:1757–1761

Aloni Y, Attardi G (1972) Expression of the mitochondrial genome in HeLa cells. XI. Isolation and characterization of transcription complexes of mitochondrial DNA. J Mol Biol 70:363–373

Amiott EA, Jaehning JA (2006) Mitochondrial transcription is regulated via an ATP "sensing" mechanism that couples RNA abundance to respiration. Mol Cell 22:329–338

Arnold JJ, Sharma SD, Feng JY, Ray AS, Smidansky ED, Kireeva ML, Cho A, Perry J, Vela JE, Park Y et al (2012) Sensitivity of mitochondrial transcription and resistance of RNA polymerase II dependent nuclear transcription to antiviral ribonucleosides. PLoS Pathog 8:e1003030

Attardi B, Attardi G (1969) Sedimentation properties of RNA species homologous to mitochondrial DNA in HeLa cells. Nature 224:1079–1083

Bogenhagen DF, Pinz KG, Perez-Jannotti RM (2001) Enzymology of mitochondrial base excision repair. Prog Nucleic Acid Res Mol Biol 68:257–271

Bohr VA, Stevnsner T, de Souza-Pinto NC (2002) Mitochondrial DNA repair of oxidative damage in mammalian cells. Gene 286:127–134

Brown TA, Cecconi C, Tkachuk AN, Bustamante C, Clayton DA (2005) Replication of mitochondrial DNA occurs by strand displacement with alternative light-strand origins, not via a strand-coupled mechanism. Genes Dev 19:2466–2476

Carrodeguas JA, Kobayashi R, Lim SE, Copeland WC, Bogenhagen DF (1999) The accessory subunit of *Xenopus laevis* mitochondrial DNA polymerase gamma increases processivity of the catalytic subunit of human DNA polymerase gamma and is related to class II aminoacyl-tRNA synthetases. Mol Cell Biol 19:4039–4046

Carrodeguas JA, Theis K, Bogenhagen DF, Kisker C (2001) Crystal structure and deletion analysis show that the accessory subunit of mammalian DNA polymerase gamma, Pol gamma B, functions as a homodimer. Mol Cell 7:43–54

Carrodeguas JA, Pinz KG, Bogenhagen DF (2002) DNA binding properties of human pol gammaB. J Biol Chem 277:50008–50014

Chan SS, Copeland WC (2009) DNA polymerase gamma and mitochondrial disease: understanding the consequence of POLG mutations. Biochim Biophys Acta 1787:312–319

Chan SS, Longley MJ, Copeland WC (2005) The common A467T mutation in the human mitochondrial DNA polymerase (POLG) compromises catalytic efficiency and interaction with the accessory subunit. J Biol Chem 280:31341–31346

Cheetham GM, Steitz TA (1999) Structure of a transcribing T7 RNA polymerase initiation complex. Science 286:2305–2309

Clayton DA (1982) Replication of animal mitochondrial DNA. Cell 28:693–705

Colacino JM (1996) Mechanisms for the anti-hepatitis B virus activity and mitochondrial toxicity of fialuridine (FIAU). Antiviral Res 29:125–139

Dairaghi DJ, Shadel GS, Clayton DA (1995) Human mitochondrial transcription factor A and promoter spacing integrity are required for transcription initiation. Biochim Biophys Acta 1271:127–134

Delannoy E, Stanley WA, Bond CS, Small ID (2007) Pentatricopeptide repeat (PPR) proteins as sequence-specificity factors in post-transcriptional processes in organelles. Biochem Soc Trans 35:1643–1647

Falkenberg M, Gaspari M, Rantanen A, Trifunovic A, Larsson NG, Gustafsson CM (2002) Mitochondrial transcription factors B1 and B2 activate transcription of human mtDNA. Nat Genet 31:289–294

Farge G, Pham XH, Holmlund T, Khorostov I, Falkenberg M (2007) The accessory subunit B of DNA polymerase gamma is required for mitochondrial replisome function. Nucleic Acids Res 35:902–911

Feng JY, Anderson KS (1999a) Mechanistic studies comparing the incorporation of (+) and (-) isomers of 3TCTP by HIV-1 reverse transcriptase. Biochemistry 38:55–63
Feng JY, Anderson KS (1999b) Mechanistic studies examining the efficiency and fidelity of DNA synthesis by the 3TC-resistant mutant (184V) of HIV-1 reverse transcriptase. Biochemistry 38:9440–9448
Feng JY, Johnson AA, Johnson KA, Anderson KS (2001) Insights into the molecular mechanism of mitochondrial toxicity by AIDS drugs. J Biol Chem 276:23832–23837
Feng JY, Murakami E, Zorca SM, Johnson AA, Johnson KA, Schinazi RF, Furman PA, Anderson KS (2004) Relationship between antiviral activity and host toxicity: comparison of the incorporation efficiencies of 2′,3′-dideoxy-5-fluoro-3′-thiacytidine-triphosphate analogs by human immunodeficiency virus type 1 reverse transcriptase and human mitochondrial DNA polymerase. Antimicrob Agents Chemother 48:1300–1306
Fernandez-Silva P, Martinez-Azorin F, Micol V, Attardi G (1997) The human mitochondrial transcription termination factor (mTERF) is a multizipper protein but binds to DNA as a monomer, with evidence pointing to intramolecular leucine zipper interactions. EMBO J 16:1066–1079
Ferrari G, Lamantea E, Donati A, Filosto M, Briem E, Carrara F, Parini R, Simonati A, Santer R, Zeviani M (2005) Infantile hepatocerebral syndromes associated with mutations in the mitochondrial DNA polymerase-gammaA. Brain 128:723–731
Fisher RP, Clayton DA (1988) Purification and characterization of human mitochondrial transcription factor 1. Mol Cell Biol 8:3496–3509
Fried MW, Korenman JC, Di Bisceglie AM, Park Y, Waggoner JG, Mitsuya H, Hartman NR, Yarchoan R, Broder S, Hoofnagle JH (1992) A pilot study of 2′,3′-dideoxyinosine for the treatment of chronic hepatitis B. Hepatology 16:861–864
Garcia G, Smith CI, Weissberg JI, Eisenberg M, Bissett J, Nair PV, Mastre B, Rosno S, Roskamp D, Waterman K et al (1987) Adenine arabinoside monophosphate (vidarabine phosphate) in combination with human leukocyte interferon in the treatment of chronic hepatitis B. A randomized, double-blinded, placebo-controlled trial. Ann Intern Med 107:278–285
Gaspari M, Larsson NG, Gustafsson CM (2004) The transcription machinery in mammalian mitochondria. Biochim Biophys Acta 1659:148–152
Graziewicz MA, Longley MJ, Bienstock RJ, Zeviani M, Copeland WC (2004) Structure-function defects of human mitochondrial DNA polymerase in autosomal dominant progressive external ophthalmoplegia. Nat Struct Mol Biol 11:770–776
Greenleaf AL, Kelly JL, Lehman IR (1986) Yeast RPO41 gene product is required for transcription and maintenance of the mitochondrial genome. Proc Natl Acad Sci USA 83:3391–3394
Hart GJ, Orr DC, Penn CR, Figueiredo HT, Gray NM, Boehme RE, Cameron JM (1992) Effects of (-)-2′-deoxy-3′-thiacytidine (3TC) 5′-triphosphate on human immunodeficiency virus reverse transcriptase and mammalian DNA polymerases alpha, beta, and gamma. Antimicrob Agents Chemother 36:1688–1694
Hegde ML, Hazra TK, Mitra S (2010) Functions of disordered regions in mammalian early base excision repair proteins. Cell Mol Life Sci 67:3573–3587
Johnson AA, Ray AS, Hanes J, Suo Z, Colacino JM, Anderson KS, Johnson KA (2001) Toxicity of antiviral nucleoside analogs and the human mitochondrial DNA polymerase. J Biol Chem 276:40847–40857
Kaguni LS (2004) DNA polymerase gamma, the mitochondrial replicase. Annu Rev Biochem 73:293–320
Katsuragi K, Kitagishi K, Mizuguchi T, Nagashima T, Kinoshita M, Kumada H (1997) Method for detection of epsilon-secondary structure in the precore region of human hepatitis B virus DNA using a fluorescence-based polymerase chain reaction-single-strand-conformation polymorphism technique with capillary electrophoresis. J Chromatogr A 781:307–314
Kravchenko JE, Rogozin IB, Koonin EV, Chumakov PM (2005) Transcription of mammalian messenger RNAs by a nuclear RNA polymerase of mitochondrial origin. Nature 436:735–739

Kruse B, Narasimhan N, Attardi G (1989) Termination of transcription in human mitochondria: identification and purification of a DNA binding protein factor that promotes termination. Cell 58:391–397

Lambowitz AM, Zimmerly S (2004) Mobile group II introns. Annu Rev Genet 38:1–35

Lee HR, Johnson KA (2006) Fidelity of the human mitochondrial DNA polymerase. J Biol Chem 281:36236–36240

Lee HR, Johnson KA (2007) Fidelity and processivity of reverse transcription by the human mitochondrial DNA polymerase. J Biol Chem 282:31982–31989

Lee YS, Kennedy WD, Yin YW (2009) Structural insights into human mitochondrial DNA replication and disease-related polymerase mutations. Cell 139:312–324

Lee YS, Johnson KA, Molineux IJ, Yin YW (2010a) A single mutation in human mitochondrial DNA polymerase Pol gammaA affects both polymerization and proofreading activities of only the holoenzyme. J Biol Chem 285:28105–28116

Lee YS, Lee S, Demeler B, Molineux IJ, Johnson KA, Yin YW (2010b) Each monomer of the dimeric accessory protein for human mitochondrial DNA polymerase has a distinct role in conferring processivity. J Biol Chem 285:1490–1499

Lewis W, Levine ES, Griniuviene B, Tankersley KO, Colacino JM, Sommadossi JP, Watanabe KA, Perrino FW (1996) Fialuridine and its metabolites inhibit DNA polymerase gamma at sites of multiple adjacent analog incorporation, decrease mtDNA abundance, and cause mitochondrial structural defects in cultured hepatoblasts. Proc Natl Acad Sci USA 93:3592–3597

Lewis W, Day BJ, Copeland WC (2003) Mitochondrial toxicity of NRTI antiviral drugs: an integrated cellular perspective. Nat Rev Drug Discov 2:812–822

Lewis W, Day BJ, Kohler JJ, Hosseini SH, Chan SS, Green EC, Haase CP, Keebaugh ES, Long R, Ludaway T et al (2007) Decreased mtDNA, oxidative stress, cardiomyopathy, and death from transgenic cardiac targeted human mutant polymerase gamma. Lab Invest 87:326–335

Lightowlers RN, Chrzanowska-Lightowlers ZM (2008) PPR (pentatricopeptide repeat) proteins in mammals: important aids to mitochondrial gene expression. Biochem J 416:e5–e6

Lim SE, Longley MJ, Copeland WC (1999) The mitochondrial p55 accessory subunit of human DNA polymerase gamma enhances DNA binding, promotes processive DNA synthesis, and confers N-ethylmaleimide resistance. J Biol Chem 274:38197–38203

Litonin D, Sologub M, Shi Y, Savkina M, Anikin M, Falkenberg M, Gustafsson CM, Temiakov D (2010) Human mitochondrial transcription revisited: only TFAM and TFB2M are required for transcription of the mitochondrial genes in vitro. J Biol Chem 285:18129–18133

Liu P, Demple B (2010) DNA repair in mammalian mitochondria: much more than we thought? Environ Mol Mutagen 51:417–426

Liu P, Qian L, Sung JS, de Souza-Pinto NC, Zheng L, Bogenhagen DF, Bohr VA, Wilson DM 3rd, Shen B, Demple B (2008) Removal of oxidative DNA damage via FEN1-dependent long-patch base excision repair in human cell mitochondria. Mol Cell Biol 28:4975–4987

Lodeiro MF, Uchida A, Bestwick M, Moustafa IM, Arnold JJ, Shadel GS, Cameron CE (2012) Transcription from the second heavy-strand promoter of human mtDNA is repressed by transcription factor A in vitro. Proc Natl Acad Sci USA 109:6513–6518

Longley MJ, Prasad R, Srivastava DK, Wilson SH, Copeland WC (1998) Identification of 5′-deoxyribose phosphate lyase activity in human DNA polymerase gamma and its role in mitochondrial base excision repair in vitro. Proc Natl Acad Sci USA 95:12244–12248

Longley MJ, Graziewicz MA, Bienstock RJ, Copeland WC (2005) Consequences of mutations in human DNA polymerase gamma. Gene 354:125–131

Marczynski GT, Schultz PW, Jaehning JA (1989) Use of yeast nuclear DNA sequences to define the mitochondrial RNA polymerase promoter in vitro. Mol Cell Biol 9:3193–3202

Masters BS, Stohl LL, Clayton DA (1987) Yeast mitochondrial RNA polymerase is homologous to those encoded by bacteriophages T3 and T7. Cell 51:89–99

Matsunaga M, Jaehning JA (2004) Intrinsic promoter recognition by a "core" RNA polymerase. J Biol Chem 279:44239–44242

Matsushima Y, Adan C, Garesse R, Kaguni LS (2005) Drosophila mitochondrial transcription factor B1 modulates mitochondrial translation but not transcription or DNA copy number in Schneider cells. J Biol Chem 280:16815–16820

McBride HM, Neuspiel M, Wasiak S (2006) Mitochondria: more than just a powerhouse. Curr Biol 16:R551–R560

McCulloch V, Shadel GS (2003) Human mitochondrial transcription factor B1 interacts with the C-terminal activation region of h-mtTFA and stimulates transcription independently of its RNA methyltransferase activity. Mol Cell Biol 23:5816–5824

McKenzie R, Fried MW, Sallie R, Conjeevaram H, Di Bisceglie AM, Park Y, Savarese B, Kleiner D, Tsokos M, Luciano C et al (1995) Hepatic failure and lactic acidosis due to fialuridine (FIAU), an investigational nucleoside analogue for chronic hepatitis B. N Engl J Med 333:1099–1105

Michailidis E, Marchand B, Kodama EN, Singh K, Matsuoka M, Kirby KA, Ryan EM, Sawani AM, Nagy E, Ashida N et al (2009) Mechanism of inhibition of HIV-1 reverse transcriptase by 4′-ethynyl-2-fluoro-2′-deoxyadenosine triphosphate, a translocation-defective reverse transcriptase inhibitor. J Biol Chem 284:35681–35691

Montoya J, Christianson T, Levens D, Rabinowitz M, Attardi G (1982) Identification of initiation sites for heavy-strand and light-strand transcription in human mitochondrial DNA. Proc Natl Acad Sci USA 79:7195–7199

Montoya J, Gaines GL, Attardi G (1983) The pattern of transcription of the human mitochondrial rRNA genes reveals two overlapping transcription units. Cell 34:151–159

Mouw KW, Rice PA (2007) Shaping the Borrelia burgdorferi genome: crystal structure and binding properties of the DNA-bending protein Hbb. Mol Microbiol 63:1319–1330

Murakami E, Basavapathruni A, Bradley WD, Anderson KS (2005) Mechanism of action of a novel viral mutagenic covert nucleotide: molecular interactions with HIV-1 reverse transcriptase and host cell DNA polymerases. Antiviral Res 67:10–17

Nayak D, Guo Q, Sousa R (2009) A promoter recognition mechanism common to yeast mitochondrial and phage t7 RNA polymerases. J Biol Chem 284:13641–13647

Nikolov DB, Chen H, Halay ED, Hoffman A, Roeder RG, Burley SK (1996) Crystal structure of a human TATA box-binding protein/TATA element complex. Proc Natl Acad Sci USA 93:4862–4867

Pham XH, Farge G, Shi Y, Gaspari M, Gustafsson CM, Falkenberg M (2006) Conserved sequence box II directs transcription termination and primer formation in mitochondria. J Biol Chem 281:24647–24652

Pica-Mattoccia L, Attardi G (1971) Expression of the mitochondrial genome in HeLa cells. V. Transcription of mitochondrial DNA in relationship to the cell cycle. J Mol Biol 57:615–621

Pinz KG, Bogenhagen DF (2006) The influence of the DNA polymerase gamma accessory subunit on base excision repair by the catalytic subunit. DNA Repair (Amst) 5:121–128

Ray AS, Anderson KS (2001) Mechanistic studies to understand the inhibition of wild type and mutant HIV-1 reverse transcriptase by Carbovir-triphosphate. Nucleosides Nucleotides Nucleic Acids 20:1247–1250

Ray AS, Schinazi RF, Murakami E, Basavapathruni A, Shi J, Zorca SM, Chu CK, Anderson KS (2003) Probing the mechanistic consequences of 5-fluorine substitution on cytidine nucleotide analogue incorporation by HIV-1 reverse transcriptase. Antivir Chem Chemother 14:115–125

Richter C, Park JW, Ames BN (1988) Normal oxidative damage to mitochondrial and nuclear DNA is extensive. Proc Natl Acad Sci USA 85:6465–6467

Riemen G, Michaelis G (1993) A point mutation in the core subunit gene of yeast mitochondrial RNA polymerase is suppressed by a high level of specificity factor MTF1. Mol Gen Genet 237:49–57

Ringel R, Sologub M, Morozov YI, Litonin D, Cramer P, Temiakov D (2011) Structure of human mitochondrial RNA polymerase. Nature 478:269–273

Rubio-Cosials A, Sola M (2013) U-turn DNA bending by human mitochondrial transcription factor A. Curr Opin Struct Biol 23:116–124

Rubio-Cosials A, Sidow JF, Jimenez-Menendez N, Fernandez-Millan P, Montoya J, Jacobs HT, Coll M, Bernado P, Sola M (2011) Human mitochondrial transcription factor A induces a U-turn structure in the light strand promoter. Nat Struct Mol Biol 18:1281–1289

Seidel-Rogol BL, McCulloch V, Shadel GS (2003) Human mitochondrial transcription factor B1 methylates ribosomal RNA at a conserved stem-loop. Nat Genet 33:23–24

Shoffner JM, Brown MD, Torroni A, Lott MT, Cabell MF, Mirra SS, Beal MF, Yang CC, Gearing M, Salvo R et al (1993) Mitochondrial DNA variants observed in Alzheimer disease and Parkinson disease patients. Genomics 17:171–184

Sohl CD, Singh K, Kasiviswanathan R, Copeland WC, Mitsuya H, Sarafianos SG, Anderson KS (2011) Mechanism of interaction of human mitochondrial DNA polymerase gamma with the novel nucleoside reverse transcriptase inhibitor 4′-ethynyl-2-fluoro-2′-deoxyadenosine indicates a low potential for host toxicity. Antimicrob Agents Chemother 56:1630–1634

Sohl CD, Kasiviswanathan R, Kim J, Pradere U, Schinazi RF, Copeland WC, Mitsuya H, Baba M, Anderson KS (2012) Balancing antiviral potency and host toxicity: identifying a nucleotide inhibitor with an optimal kinetic phenotype for HIV-1 reverse transcriptase. Mol Pharmacol 82:125–133

Sohl CD, Kasiviswanathan R, Copeland WC, Anderson KS (2013) Mutations in human DNA polymerase gamma confer unique mechanisms of catalytic deficiency that mirror the disease severity in mitochondrial disorder patients. Hum Mol Genet 22:1074–1085

Tabak HF, Van der Horst G, Osinga KA, Arnberg AC (1984) Splicing of large ribosomal precursor RNA and processing of intron RNA in yeast mitochondria. Cell 39:623–629

Takamatsu C, Umeda S, Ohsato T, Ohno T, Abe Y, Fukuoh A, Shinagawa H, Hamasaki N, Kang D (2002) Regulation of mitochondrial D-loops by transcription factor A and single-stranded DNA-binding protein. EMBO Rep 3:451–456

Tang GQ, Patel SS (2006) T7 RNA polymerase-induced bending of promoter DNA is coupled to DNA opening. Biochemistry 45:4936–4946

Tang GQ, Anand VS, Patel SS (2010) Fluorescence-based assay to measure the real-time kinetics of nucleotide incorporation during transcription elongation. J Mol Biol 405:666–678

Tann AW, Boldogh I, Meiss G, Qian W, Van Houten B, Mitra S, Szczesny B (2011) Apoptosis induced by persistent single-strand breaks in mitochondrial genome: critical role of EXOG (5′-EXO/endonuclease) in their repair. J Biol Chem 286:31975–31983

Tracy RL, Stern DB (1995) Mitochondrial transcription initiation: promoter structures and RNA polymerases. Curr Genet 28:205–216

Trifunovic A, Hansson A, Wredenberg A, Rovio AT, Dufour E, Khvorostov I, Spelbrink JN, Wibom R, Jacobs HT, Larsson NG (2005) Somatic mtDNA mutations cause aging phenotypes without affecting reactive oxygen species production. Proc Natl Acad Sci USA 102:17993–17998

Vaccaro JA, Anderson KS (1998) Implication of the tRNA initiation step for human immunodeficiency virus type 1 reverse transcriptase in the mechanism of 3′-azido-3′-deoxythymidine (AZT) resistance. Biochemistry 37:14189–14194

Vaccaro JA, Parnell KM, Terezakis SA, Anderson KS (2000) Mechanism of inhibition of the human immunodeficiency virus type 1 reverse transcriptase by d4TTP: an equivalent incorporation efficiency relative to the natural substrate dTTP. Antimicrob Agents Chemother 44:217–221

Viikov K, Valjamae P, Sedman J (2010) Yeast mitochondrial DNA polymerase is a highly processive single-subunit enzyme. Mitochondrion 11:119–126

Viikov K, Jasnovidova O, Tamm T, Sedman J (2012) C-terminal extension of the yeast mitochondrial DNA polymerase determines the balance between synthesis and degradation. PLoS One 7:e33482

Wang J, Xiong S, Xie C, Markesbery WR, Lovell MA (2005) Increased oxidative damage in nuclear and mitochondrial DNA in Alzheimer's disease. J Neurochem 93:953–962

Wanrooij S, Fuste JM, Farge G, Shi Y, Gustafsson CM, Falkenberg M (2008) Human mitochondrial RNA polymerase primes lagging-strand DNA synthesis in vitro. Proc Natl Acad Sci USA 105:11122–11127

Wanrooij S, Miralles Fuste J, Stewart JB, Wanrooij PH, Samuelsson T, Larsson NG, Gustafsson CM, Falkenberg M (2012) In vivo mutagenesis reveals that OriL is essential for mitochondrial DNA replication. EMBO Rep 13:1130–1137

WHO (2003) The World Health Report 2003: shaping the future. World Health Organization, Geneva

Wiseman H, Halliwell B (1996) Damage to DNA by reactive oxygen and nitrogen species: role in inflammatory disease and progression to cancer. Biochem J 313(Pt 1):17–29

Xu B, Clayton DA (1996) RNA-DNA hybrid formation at the human mitochondrial heavy-strand origin ceases at replication start sites: an implication for RNA-DNA hybrids serving as primers. EMBO J 15:3135–3143

Yang MY, Bowmaker M, Reyes A, Vergani L, Angeli P, Gringeri E, Jacobs HT, Holt IJ (2002) Biased incorporation of ribonucleotides on the mitochondrial L-strand accounts for apparent strand-asymmetric DNA replication. Cell 111:495–505

Yang JL, Weissman L, Bohr VA, Mattson MP (2008) Mitochondrial DNA damage and repair in neurodegenerative disorders. DNA Repair (Amst) 7:1110–1120

Yasukawa T, Yang MY, Jacobs HT, Holt IJ (2005) A bidirectional origin of replication maps to the major noncoding region of human mitochondrial DNA. Mol Cell 18:651–662

Zeviani M, Servidei S, Gellera C, Bertini E, DiMauro S, DiDonato S (1989) An autosomal dominant disorder with multiple deletions of mitochondrial DNA starting at the D-loop region. Nature 339:309–311

Zheng L, Zhou M, Guo Z, Lu H, Qian L, Dai H, Qiu J, Yakubovskaya E, Bogenhagen DF, Demple B et al (2008) Human DNA2 is a mitochondrial nuclease/helicase for efficient processing of DNA replication and repair intermediates. Mol Cell 32:325–336

Chapter 12
Eukaryotic RNA Polymerase II

David A. Bushnell and Roger D. Kornberg

Abstract Structures of yeast RNA polymerase II, alone and in the act of transcription with associated DNA and RNA, have been determined to near-atomic resolution. The structures illuminate the basis for the fidelity of transcription, for translocation on the DNA template, and for release of the product RNA. Structures of an RNA polymerase II-general transcription factor complex have led to a model for a closed transcription initiation complex and have suggested a possible basis for promoter escape.

Keywords Yeast • RNA polymerase II structure • X-ray crystallography • Trigger loop • Nucleotide addition • TFIIB

Structural studies of cellular RNA polymerases, both bacterial and eukaryotic, began in the 1980s, with the formation of 2D crystals on lipid layers, electron microscopy, and 3D reconstruction (Darst et al. 1988). This approach is rapid, requires very little material, and can reveal even the slightest degree of ordering. It resulted in low resolution structures of Escherichia coli RNA polymerase, yeast RNA polymerase II (pol II), and a yeast pol II transcribing complex (Darst et al. 1989; Poglitsch et al. 1999; Fu et al. 1999). These studies had three implications for further work: the first was the possibility of crystallization of cellular RNA polymerases, alone and in the act of transcription (crystallization of such large molecules had rarely been attempted, and the size and complexity of these RNA polymerases placed them beyond reach of X-ray structure determination at the time); the studies also revealed that a substoichiometric level of two pol II subunits, Rpb4 and Rpb7, caused heterogeneity that was an impediment to crystallization; and finally, with the removal of Rpb4 and Rpb7 by genetic means, large, well-ordered 2D crystals were obtained

D.A. Bushnell • R.D. Kornberg (✉)
Stanford University School of Medicine, Stanford, CA 94305, USA
e-mail: kornberg@stanford.edu

K.S. Murakami and M.A. Trakselis (eds.), *Nucleic Acid Polymerases*, Nucleic Acids and Molecular Biology 30, DOI 10.1007/978-3-642-39796-7_12,

that showed a tendency towards epitaxial growth (Darst et al. 1991a), leading to the formation of large single crystals for X-ray diffraction analysis.

12.1 Structure of Pol II

A 5 Å electron density map of pol II was obtained by X-ray diffraction with the use of tungsten (W_{18}), tantalum (Ta_6), and iridium (Ir_2) cluster compounds for phase determination (Fu et al. 1999). Close agreement of the map with the results from 2D crystallography (Darst et al. 1991b) provided key validation of the 5 Å phase set. These phases could be used to locate individual heavy atoms in derivative crystals at high resolution. There were two major obstacles. First, the crystals were sensitive to oxidation, resulting in a loss of diffraction to high resolution. The problem was overcome by the growth and maintenance of the crystals in an inert atmosphere or by the use of a high concentration of reductant, such as dithiothreitol. Secondly, the crystals were polymorphic, varying by as much as 10 Å along one unit cell direction. This problem was solved by dehydration of the crystals, resulting in shrinkage of the unit cell volume by 5 % and consistent diffraction to about 3 Å resolution.

Structure determination was finally made possible by developments at synchrotrons, especially increased photon flux and more rapid detectors. For these reasons, the structures of pol II and bacterial RNA polymerase, and also those of ribosomal subunits that had been crystallized long before, were all solved at about the same time. Photon physicists were the unsung heroes of the work. For our studies of pol II, we owe a special debt to the beamline scientists at the Stanford Synchrotron Radiation Lightsource, who gave us extraordinary access during a period of intensive search for heavy atom derivatives.

The structure of ten-subunit pol II (lacking Rpb4 and Rpb7) was solved at 2.8 Å resolution (Fig. 12.1a) (Cramer et al. 2001). The structure reveals the two largest subunits, Rpb1 and Rpb2 on either side of a central cleft, with a magnesium ion marking the location of the active center. The remaining subunits are arrayed around the periphery, with Rpb5 and Rpb9 at the front end or leading edge of the enzyme, on opposite sides of the cleft, and Rpb3, 10, 11, and 12 forming a cluster on the backside of the enzyme.

12.2 Structure of Pol II Transcribing Complex

The 2.8 Å structure was used to solve that of an actively transcribing enzyme by molecular replacement (Gnatt et al. 2001). The transcribing complex was formed by the initiation of transcription on a DNA template with a "tail" of twelve residues at one 3′-end. Initiation on a tailed template avoids the requirement for Rpb4, Rpb7, and a large set of general transcription factors. Transcription on the tailed template was stalled after the synthesis of a 14-residue transcript by the omission of one of the four nucleoside triphosphates (NTPs). The reaction mixture was passed over a

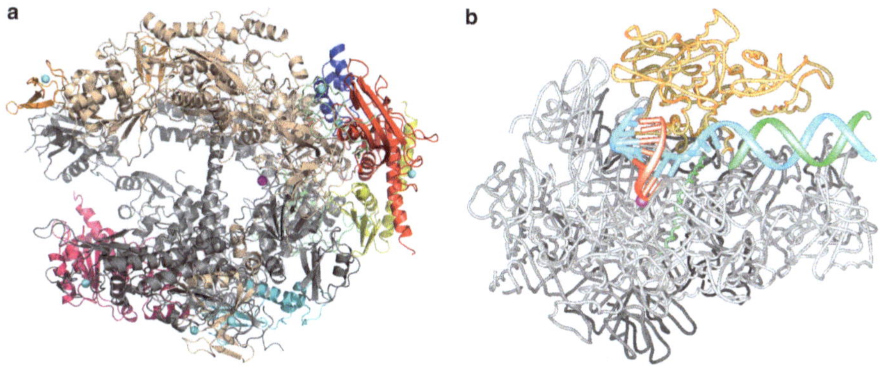

Fig. 12.1 (**a**) Ribbon diagram of 10-subunit yeast RNA pol II. Each subunit is shown in a different color. A *pink sphere* represents a Mg^{2+} ion, marking the location of the active center. *Cyan spheres* represent structural zinc atoms (Cramer et al. 2001). (**b**) Chain traces of polypeptides and nucleic acids in a pol II transcribing complex, formed by initiation of transcription on a tailed template. Protein is *gray* except for the "bridge helix" extending across the central cleft (*green*) and the clamp domain (*gold*). Template DNA strand is *blue*, nontemplate strand is *green*, and RNA is *red*. The view is obtained from (**a**) by a 90° rotation about a horizontal axis (upward rotation) and a 180° rotation about a vertical axis (flip from left to right). Some protein has been removed from the front to expose the nucleic acids in the central cleft. The direction of transcription is to the right; DNA enters from the right (downstream) and translocates during transcription to the left (upstream) (Gnatt et al. 2001)

heparin column, to separate inactive enzyme (which bound to the column, due to interaction of heparin with the empty DNA-binding cleft) from actively transcribing complexes (which flowed through the column, because template DNA filled the cleft). This step was crucial for obtaining homogeneous transcribing complexes for crystallization.

The first transcribing complex structure, at 3.1 Å resolution, differs from that of pol II alone by a major conformational change. A massive protein element, formed by parts of two subunits, swings up to 30 Å over the DNA and RNA in the course of forming a transcribing complex (Fig. 12.1b). This mobile element, referred to as a "clamp," appears to hold the DNA and RNA in place and may be important for the processivity of transcription. The structure reveals the course of nucleic acids through the transcribing complex. Duplex DNA enters the cleft between Rpb5 and Rpb9 and remains in duplex form over a length of 15 bp. The DNA unwinds three residues before the active site, after which there is a bend of about 90° in the template single strand, in consequence of which the next base is flipped and points down towards the floor of the cleft for readout in transcription. This base is paired with that of the ribonucleotide just added to the growing RNA strand. There are eight more DNA–RNA hybrid base pairs in the structure, but no evidence of the corresponding nontemplate strand nor of the duplex DNA beyond the unwound region ("transcription bubble"), due to motion or disorder. The DNA–RNA hybrid interacts with a "wall" of protein density that prevents straight passage of nucleic acids through the cleft.

Because the ribonucleotide just added to the RNA chain is still in the active center, in the nucleotide addition or "A" site, paired with the coding base in the DNA, this structure is referred to as a "pre-translocation complex." The next structure of a transcribing complex was obtained by direct binding of pol II to a nucleic acid "scaffold," consisting of a DNA–RNA hybrid and downstream duplex DNA, rather than by initiation and arrest of RNA synthesis on a tailed template (Westover et al. 2004a). This simplified approach was based on biochemical studies demonstrating the efficient elongation of RNA chains on such scaffolds (Kireeva et al. 2000). The structure represents a "post-translocation complex," in which the nucleic acids have advanced by one residue along the pol II surface, bringing the ribonucleotide just added to the RNA chain to the next position of the DNA–RNA hybrid and leaving the "A" site open for binding NTP. This structure, and others based upon it, reveals the following structural correlates of transcription (Fig. 12.2a, b).

12.2.1 Magnesium Ions and NTP Entry Site

A single metal ion is identified in the active center of pol II. This ion is presumed to be Mg^{2+}, and its location could be confirmed by substitution with Mn^{2+} or Pb^{2+} and anomalous difference diffraction (Cramer et al. 2000, 2001). Upon soaking with NTP and difference diffraction analysis, a second metal ion is observed. In the case of NTP matched with the coding base in the DNA template and with the use of a chain terminator on the RNA strand to prevent nucleotide addition, the NTP is seen to occupy the A site. In the case of NTPs not matched with the coding base, binding with a metal ion is also observed, but with the NTP in an adjacent location, referred to as the entry or "E" site (Westover et al. 2004b). The occurrence of such a second NTP-binding site was anticipated by biochemical studies and may serve to increase NTP concentration in the active center and thereby to facilitate transcription (Sosunov et al. 2003). The series of post-translocation complex structures, with no bound NTP, with NTP in the E site, and with NTP in the A site, along with the pre-translocation complex, recapitulate a round of the transcription cycle (Fig. 12.2b).

12.2.2 Trigger Loop and the Fidelity of Transcription

The first structures of post-translocation complexes with NTP in the A site revealed only interactions of the NTP with DNA, RNA, and Mg^{2+}. Improved structures disclosed additional interactions with a protein element termed the "trigger loop" (Wang et al. 2006). In the absence of NTP, the trigger loop is located some 30 Å from the A site and is partially disordered (Fig. 12.3a). In the presence of NTP, the trigger loop swings into position for contact with the nucleotide base, through a

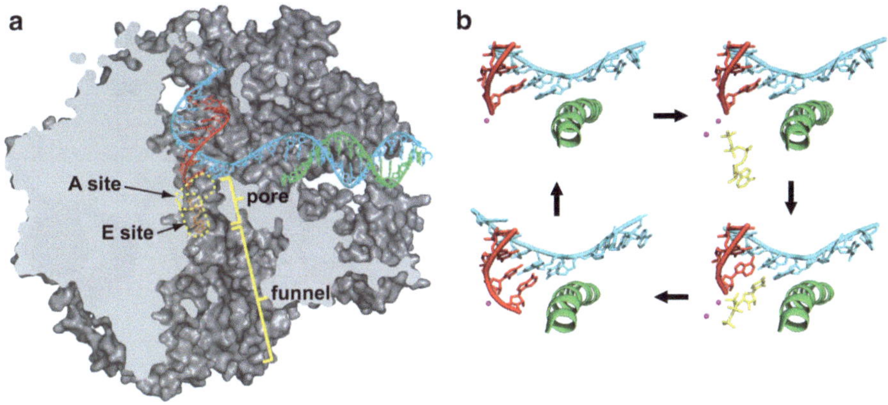

Fig. 12.2 (**a**) Pol II transcribing complex, identical to Fig. 12.1b except shown as a surface representation rather than chain traces. *Solid light gray* areas are cut surfaces where protein was removed to expose the nucleic acids in the central cleft. Also are indicated the locations of the NTP entry (E) and addition (A) sites, as well as the presumed pathway for passage of NTPs to the active center, through a funnel-shaped opening at the bottom of the polymerase, narrowing to a pore beneath the active center (Westover et al. 2004b). (**b**) A transcription cycle, from structures of transcribing complexes formed by initiation on a tailed template (pre-translocation complex, *lower left*) or through binding a nucleic acid scaffold alone (post-translocation complex, *upper left*), soaked with unmatched NTP (E site occupied by NTP, *lower right*) or soaked with matched NTP (A site occupied by NTP, *upper right*). The structures are shown as in Fig. 12.1b, with all protein removed except for the bridge helix (*green*) and with nontemplate DNA strand removed, leaving only the template (*blue*) and RNA strands (*red*) and the magnesium ion (*pink sphere*) in the immediate vicinity of the active center (Gnatt et al. 2001; Westover et al. 2004a, b)

leucine residue, and for contact with the β-phosphate, through a histidine residue (His1085). The imidazole side chain of the histidine can trigger phosphodiester bond formation by promoting the flow of electrons from the 3′-hydroxyl group of the RNA chain terminus to the α-phosphate of the NTP to the pyrophosphate leaving group (Fig. 12.3b), a mechanism common to nucleic acid polymerases (Castro et al. 2009). The imidazole side chain serves as a protein donor to the pyrophosphate leaving group. The alignment of the leucine and histidine residues with the NTP must be precise for this mechanism. A mismatched NTP will be misaligned, resulting in failure of the mechanism. Even a correctly matched dNTP will likely be misaligned, due to the difference in diameter of DNA–DNA and DNA–RNA double helices. Biochemical studies have confirmed a role of the trigger in discrimination between NTP and dNTP (Kaplan et al. 2008; Zhang et al. 2009). Deletion of the trigger loop in the bacterial system leads to a reduction of about 60,000-fold in the rate of RNA chain elongation (Yuzenkova et al. 2010). The trigger loop couples nucleotide specificity to catalysis (Kaplan et al. 2008, 2012; Zhang et al. 2009; Yuzenkova et al. 2010; Vassylyev et al. 2007; Temiakov et al. 2005).

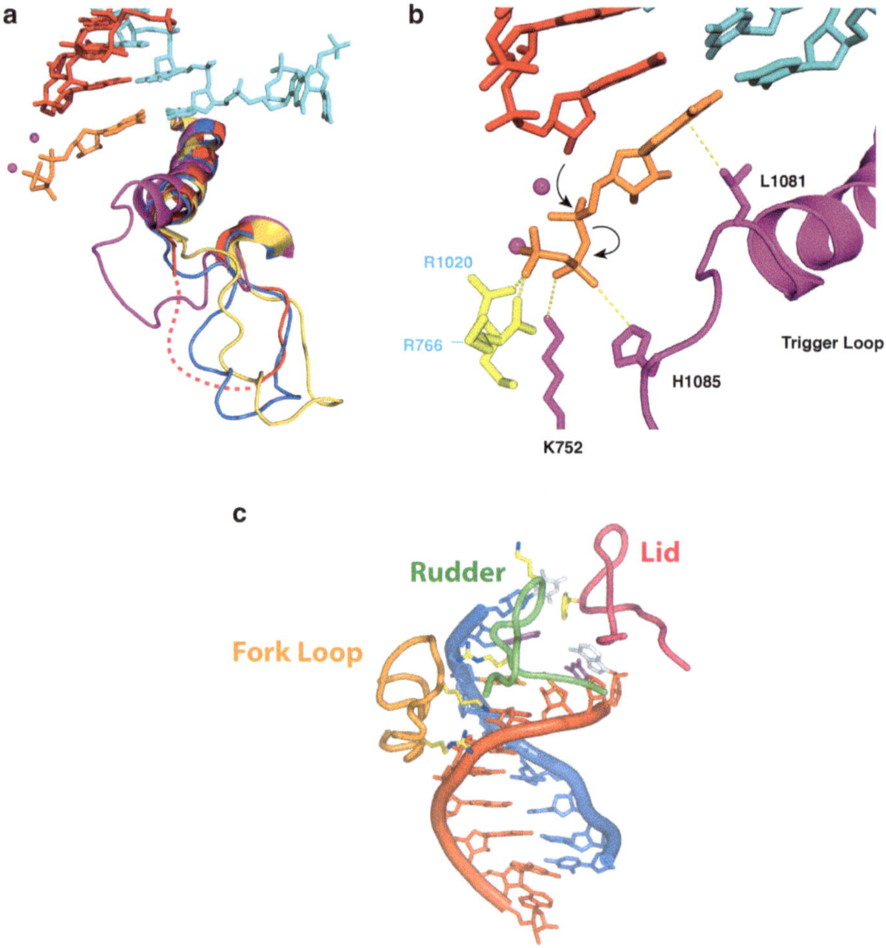

Fig. 12.3 (**a**) The trigger loop swings beneath NTP in the A site. View as in Fig. 12.2b, with multiple states of the trigger loop from various crystal structures in the absence of NTP in *red* (partially disordered), *blue*, and *yellow*. Only in the presence of NTP (*gold*) is the trigger loop (*purple*) directly beneath the A site, in position to contact the NTP (Wang et al. 2006). (**b**) Interactions of the trigger loop with NTP in the A site. Same as (**a**) with trigger loop in *purple*, with interactions indicated by *dashed yellow lines*, and the flow of electrons for phosphodiester bond formation and phosphoanhydride bond breakage indicated by *black arrows* (Wang et al. 2006). (**c**) Disruption of DNA–RNA hybrid in structure of pol II transcribing complex in post-translocation state. Same as *upper left panel* of Fig. 12.2a, except showing the DNA–RNA hybrid upstream of the A site, with protein loops in *green* (rudder), *purple* (lid), and *gold* (fork loop). A tyrosine side chain of the lid that appears to act as a wedge between DNA and RNA, and lysine and arginine side chains that interact with phosphates are shown in *yellow sticks* (Westover et al. 2004a)

12.2.3 Bridge Helix

An α-helix, spanning the cleft between Rpb1 and Rpb2 and therefore termed the "bridge helix," lies at the downstream edge of the transcription bubble in a transcribing complex (Fig. 12.1b). Alanine and threonine residues of the bridge helix interact with the coding base in the template DNA strand. The structure of bacterial RNA polymerase also contains a bridge helix, but in a partially unwound or bent state that would clash with the template DNA in the eukaryotic transcribing complex (Zhang et al. 1999). It was proposed that alternation of the bridge helix between straight and bent states underlies the translocation step in transcription (Gnatt et al. 2001). Biochemical studies have provided support for this idea and have further suggested a coupling of conformational changes of the bridge helix and trigger loop (Temiakov et al. 2005; Epshtein et al. 2002; Tuske et al. 2005).

12.2.4 Rudder, Lid, and Fork Loop

In contrast to the transcribing complex formed by initiation on a tailed template, in which a uniform DNA–RNA hybrid of nine base pairs is observed, the complex formed on a nucleic acid scaffold reveals disruption of the upstream end of the hybrid (Fig. 12.3c) (Westover et al. 2004a). The first 6 bp from the downstream end of the hybrid are coplanar with an appropriate separation for hydrogen bonding of the bases. The seventh, eight, and ninth base pairs are progressively noncoplanar, with an increasing separation of the bases. Protein loops termed "rudder" and "lid" interact with the separated DNA and RNA strands. A "fork loop" interacts through lysine side chains with the phosphates at positions 6 and 7 to stabilize the hybrid helix and prevent the separation of strands from extending downstream towards the active center. In this way, the enzyme achieves the disruption of the very stable DNA–RNA hybrid required for release of the RNA into solution.

12.2.5 Rpb4 and Rpb7

During logarithmic growth of yeast, the ratio of subunits Rpb4 and Rpb7 to other subunits of pol II is about 1:5, whereas during stress or in stationary phase, the ratio rises to near 1:1 (Kolodziej et al. 1990; Choder and Young 1993). Although Rpb4 and Rpb7 were originally removed by the deletion of the gene for Rpb4, to obtain homogeneous enzyme for crystallization, 12-subunit pol II containing Rpb4 and Rpb7 was later obtained in homogeneous form as well, either by purification of the enzyme with the use of an affinity tag on Rpb4 or by addition to the enzyme of Rpb4 and Rpb7 expressed in bacteria (Armache et al. 2003; Bushnell and Kornberg 2003). The structure of the 12-subunit enzyme at about 4 Å resolution reveals a heterodimer of Rpb4 and Rpb7 adjacent to the clamp, along a possible path of RNA exit from the enzyme. The heterodimer encloses part of the path, forming a tunnel large enough for RNA to pass through.

12.2.6 α-Amanitin

The death cap mushroom *Amanita phalloides* produces a specific inhibitor of pol II, α-amanitin (Wieland and Faulstich 1991). Upon soaking of pol II crystals with α-amanitin and difference diffraction analysis, the bound inhibitor could be seen beneath the bridge helix making contact primarily with Rpb1 (Bushnell et al. 2002). With improved computational methods, the inhibitor could be seen to contact the trigger loop and in particular His1085, the residue important for the role of the trigger loop in catalysis. Mutation of His1085 renders pol II resistant to inhibition by α-amanitin (Kaplan et al. 2008).

12.2.7 Pol II–TFIIB Complex

Biochemical studies have shown that general transcription factor TFIIB interacts with a promoter DNA–TFIID complex and with pol II (Buratowski and Zhou 1993; Ha et al. 1993). It has therefore been thought that TFIIB brings promoter DNA to the pol II surface. This idea was borne out by structure determination of a pol II–TFIIB cocrystal. The structure was solved at 4.5 Å by molecular replacement (Bushnell et al. 2004). A zinc anomalous difference map contained an extra peak beyond the eight normally found for pol II. The additional peak was located adjacent to the wall and the clamp domains of pol II, associated with a region of the pol II surface termed the "dock" domain. Electron density adjacent to the dock domain matched the NMR structure of the human IIB zinc ribbon (Fig. 12.4a). There was additional electron density between the wall and clamp domains, extending almost to the active center. A loop of polypeptide modeled into this density, termed the "B-finger," would appear to occupy the same location as the DNA–RNA hybrid in a transcribing complex. Superimposing the B-finger and DNA–RNA hybrid reveals no interference with the template DNA and no interference with RNA up to position 5 from the active center, but a steric clash with RNA at positions 6 and beyond (Fig. 12.4b). Consistent with this picture, biochemical studies show not only compatibility but a stabilizing effect of TFIIB on transcribing complexes containing RNA of five residues. Competition between TFIIB and RNA of length six residues or more could play a role in displacement of TFIIB, release of promoter DNA, and consequent "promoter escape."

The recent pol II–TFIIB cocrystal structure, obtained under different solution conditions from the original structure, reveals additional regions of TFIIB, including one of two C-terminal cyclin domains and a linker between the N- and

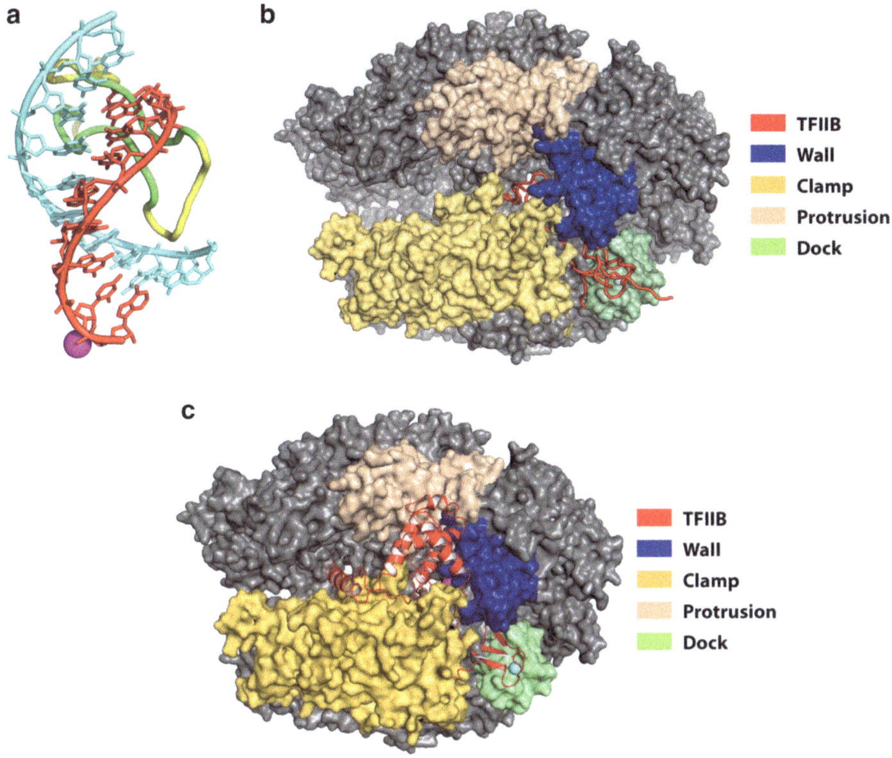

Fig. 12.4 (**a**) B-finger from pol II–TFIIB cocrystal structure superimposed on DNA–RNA hybrid from transcribing complex structure. View as in Fig. 12.3c, with B-finger in *yellow* and *green* (to indicate conserved regions) (Bushnell et al. 2004). (**b**) Pol II–TFIIB cocrystal structure showing the zinc ribbon domain and B-finger of TFIIB (*red chain trace*). View as in Fig. 12.1a except as surface representation. Dock, clamp, and wall domains of pol II are indicated, as well as "protrusion," a prominent feature of subunit Rpb2 (Bushnell et al. 2004). (**c**) Pol II–TFIIB cocrystal structure showing the linker and first cyclin domains of TFIIB (*red ribbon*). Otherwise the same as (**b**) (Liu et al. 2010)

C-terminal domains (Fig. 12.4c) (Kostrewa et al. 2009; Liu et al. 2010). The structure of a TFIIB cyclin domain in a complex with TATA-binding protein (TBP) and TATA box DNA, determined long ago by others, could be docked to the cocrystal structure to produce a model of a "closed" promoter complex (Fig. 12.5). Unwinding the promoter DNA to create an "open" complex, followed synthesis of RNA up to five residues, would cause a switch from the conformation represented by the recent cocrystal structure (Fig. 12.4c) to that of the original one (Fig. 12.4a). Contact between pol II and TFIIB would be disrupted in two stages: first, interaction of a short transcript with the B-finger and release of the cyclin domain and, second, ejection of the B-finger by a longer transcript to complete promoter escape.

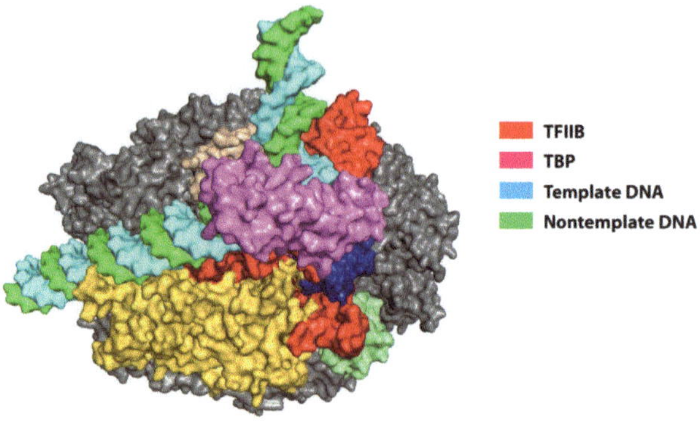

Fig. 12.5 Model of closed promoter complex, produced by docking X-ray structure of TFIIB–TBP (*pink*)–TATA box complex to structure of pol II–TFIIB complex, followed by extension of the ends of the TATA box DNA fragment with straight B-form DNA (*magenta* and *green*) (Liu et al. 2010)

References

Armache KJ, Kettenberger H, Cramer P (2003) Architecture of initiation-competent 12-subunit RNA polymerase II. Proc Natl Acad Sci USA 100(12):6964–6968

Buratowski S, Zhou H (1993) Functional domains of transcription factor TFIIB. Proc Natl Acad Sci USA 90(12):5633–5637

Bushnell DA, Kornberg RD (2003) Complete, 12-subunit RNA polymerase II at 4.1-A resolution: implications for the initiation of transcription. Proc Natl Acad Sci USA 100(12):6969–6973

Bushnell DA, Cramer P, Kornberg RD (2002) Structural basis of transcription: alpha-amanitin-RNA polymerase II cocrystal at 2.8 A resolution. Proc Natl Acad Sci USA 99(3):1218–1222

Bushnell DA, Westover KD, Davis RE, Kornberg RD (2004) Structural basis of transcription: an RNA polymerase II-TFIIB cocrystal at 4.5 Angstroms. Science 303(5660):983–988

Castro C, Smidansky ED, Arnold JJ, Maksimchuk KR, Moustafa I, Uchida A, Götte M, Konigsberg W, Cameron CE (2009) Nucleic acid polymerases use a general acid for nucleotidyl transfer. Nat Struct Mol Biol 16(2):212–218

Choder M, Young RA (1993) A portion of RNA polymerase II molecules has a component essential for stress responses and stress survival. Mol Cell Biol 13(11):6984–6991

Cramer P, Bushnell DA, Fu J, Gnatt AL, Maier-Davis B, Thompson NE, Burgess RR, Edwards AM, David PR, Kornberg RD (2000) Architecture of RNA polymerase II and implications for the transcription mechanism. Science 288(5466):640–649

Cramer P, Bushnell DA, Kornberg RD (2001) Structural basis of transcription: RNA polymerase II at 2.8 angstrom resolution. Science 292(5523):1863–1876

Darst SA, Ribi HO, Pierce DW, Kornberg RD (1988) Two-dimensional crystals of Escherichia coli RNA polymerase holoenzyme on positively charged lipid layers. J Mol Biol 203 (1):269–273

Darst SA, Kubalek EW, Kornberg RD (1989) Three-dimensional structure of Escherichia coli RNA polymerase holoenzyme determined by electron crystallography. Nature 340 (6236):730–732

Darst SA, Kubalek EW, Edwards AM, Kornberg RD (1991a) Two-dimensional and epitaxial crystallization of a mutant form of yeast RNA polymerase II. J Mol Biol 221(1):347–357

Darst SA, Edwards AM, Kubalek EW, Kornberg RD (1991b) Three-dimensional structure of yeast RNA polymerase II at 16 A resolution. Cell 66(1):121–128

Epshtein V, Mustaev A, Markovtsov V, Bereshchenko O, Nikiforov V, Goldfarb A (2002) Swing-gate model of nucleotide entry into the RNA polymerase active center. Mol Cell 10 (3):623–634

Fu J, Gnatt AL, Bushnell DA, Jensen GJ, Thompson NE, Burgess RR, David PR, Kornberg RD (1999) Yeast RNA polymerase II at 5 A resolution. Cell 98(6):799–810

Gnatt AL, Cramer P, Fu J, Bushnell DA, Kornberg RD (2001) Structural basis of transcription: an RNA polymerase II elongation complex at 3.3 A resolution. Science 292(5523):1876–1882
Ha I, Roberts S, Maldonado E, Sun X, Kim LU, Green M, Reinberg D (1993) Multiple functional domains of human transcription factor IIB: distinct interactions with two general transcription factors and RNA polymerase II. Genes Dev 7(6):1021–1032
Kaplan CD, Larsson K-M, Kornberg RD (2008) The RNA polymerase II trigger loop functions in substrate selection and is directly targeted by α-amanitin. Mol Cell 30(5):547–556
Kaplan CD, Jin H, Zhang IL, Belyanin A (2012) Dissection of pol II trigger loop function and pol II activity-dependent control of start site selection in vivo. PLoS Genet 8(4):e1002627
Kireeva ML, Komissarova N, Waugh DS, Kashlev M (2000) The 8-nucleotide-long RNA:DNA hybrid is a primary stability determinant of the RNA polymerase II elongation complex. J Biol Chem 275(9):6530–6536
Kolodziej PA, Woychik N, Liao SM, Young RA (1990) RNA polymerase II subunit composition, stoichiometry, and phosphorylation. Mol Cell Biol 10(5):1915–1920
Kostrewa D, Zeller ME, Armache KJ, Seizl M, Leike K, Thomm M, Cramer P (2009) RNA polymerase II-TFIIB structure and mechanism of transcription initiation. Nature 462 (7271):323–330
Liu X, Bushnell DA, Wang D, Calero G, Kornberg RD (2010) Structure of an RNA polymerase II-TFIIB complex and the transcription initiation mechanism. Science 327(5962):206–209
Poglitsch CL, Meredith GD, Gnatt AL, Jensen GJ, Chang WH, Fu J, Kornberg RD (1999) Electron crystal structure of an RNA polymerase II transcription elongation complex. Cell 98 (6):791–798
Sosunov V, Sosunova E, Mustaev A, Bass I, Nikiforov V, Goldfarb A (2003) Unified two-metal mechanism of RNA synthesis and degradation by RNA polymerase. EMBO J 22 (9):2234–2244
Temiakov D, Zenkin N, Vassylyeva MN, Perederina A, Tahirov TH, Kashkina E, Savkina M, Zorov S, Nikiforov V, Igarashi N et al (2005) Structural basis of transcription inhibition by antibiotic streptolydigin. Mol Cell 19(5):655–666
Tuske S, Sarafianos SG, Wang X, Hudson B, Sineva E, Mukhopadhyay J, Birktoft JJ, Leroy O, Ismail S, Clark AD Jr et al (2005) Inhibition of bacterial RNA polymerase by streptolydigin: stabilization of a straight-bridge-helix active-center conformation. Cell 122(4):541–552
Vassylyev DG, Vassylyeva MN, Zhang J, Palangat M, Artsimovitch I, Landick R (2007) Structural basis for substrate loading in bacterial RNA polymerase. Nature 448(7150):163–168
Wang D, Bushnell DA, Westover KD, Kaplan CD, Kornberg RD (2006) Structural basis of transcription: role of the trigger loop in substrate specificity and catalysis. Cell 127(5):941–954
Westover KD, Bushnell DA, Kornberg RD (2004a) Structural basis of transcription: separation of RNA from DNA by RNA polymerase II. Science 303(5660):1014–1016
Westover KD, Bushnell DA, Kornberg RD (2004b) Structural basis of transcription: nucleotide selection by rotation in the RNA polymerase II active center. Cell 119(4):481–489
Wieland T, Faulstich H (1991) Fifty years of amanitin. Experientia 47(11–12):1186–1193
Yuzenkova Y, Bochkareva A, Tadigotla VR, Roghanian M, Zorov S, Severinov K, Zenkin N (2010) Stepwise mechanism for transcription fidelity. BMC Biol 8(1):54
Zhang G, Campbell EA, Minakhin L, Richter C, Severinov K, Darst SA (1999) Crystal structure of Thermus aquaticus core RNA polymerase at 3.3 A resolution. Cell 98(6):811–824
Zhang J, Palangat M, Landick R (2009) Role of the RNA polymerase trigger loop in catalysis and pausing. Nat Struct Mol Biol 17(1):99–104

Chapter 13
Plant Multisubunit RNA Polymerases IV and V

Thomas S. Ream, Jeremy R. Haag, and Craig S. Pikaard

Abstract Plants are unique among eukaryotes in having five nuclear multisubunit RNA polymerases. These include RNA polymerases I, II, and III (Pols I, II, and III), which are ubiquitous among eukaryotes, plus two plant-specific RNA polymerases, Pol IV and Pol V, that are 12-subunit enzymes that evolved as specialized forms of Pol II. Pols IV and V are nonessential for viability but play important roles in RNA-mediated gene silencing pathways that tame transposons, defend against invading viruses, mediate cross talk among alleles, and influence development. Numerous amino acids that are invariant in the catalytic subunits of Pols I, II, and III are substituted, or even absent, in Pols IV or V, implying that Pols IV and V have fewer functional constraints on their evolution than other multisubunit RNA polymerases. In vitro, Pol IV and Pol V can extend an RNA primer hybridized to a DNA template, but the templates transcribed by Pol IV and Pol V in vivo are unclear. Likewise, the boundaries of Pol IV and Pol V transcription units and the characteristics of their primary transcripts remain undefined. In this chapter, the state of our understanding of Pol IV and Pol V subunit compositions and functions is discussed.

Keywords Pol IV • Pol V • *Arabidopsis* • Noncoding RNA • RNA silencing • siRNA • Heterochromatin • RNA-directed DNA methylation • Transcription

T.S. Ream
Department of Biochemistry, University of Wisconsin, Madison, WI 53706, USA

J.R. Haag • C.S. Pikaard (✉)
Department of Biology, Indiana University, Bloomington, IN 47405, USA

Department of Molecular and Cellular Biochemistry, Indiana University, Bloomington, IN 47405, USA

Howard Hughes Medical Institute, Indiana University, Bloomington, IN 47405, USA
e-mail: cpikaard@indiana.edu

K.S. Murakami and M.A. Trakselis (eds.), *Nucleic Acid Polymerases*, Nucleic Acids and Molecular Biology 30, DOI 10.1007/978-3-642-39796-7_13,

13.1 Introduction and Overview

All known eukaryotes require nuclear DNA-dependent RNA polymerases I, II, and III (Pols I, II, and III) for cell viability (see Chap. 12). In all but trypanosomes, Pol I specializes in the transcription of a single gene sequence, repeated in hundreds of copies, that encodes a primary transcript that is then processed to form the three largest RNAs of ribosomes, the protein synthetic machines of the cell. Pol II transcribes thousands of genes, including the genes that encode pre-messenger RNAs (mRNAs), pre-microRNAs (miRNAs), and a variety of small structural and regulatory RNAs. Pol III transcribes hundreds of genes, including those encoding pre-tRNAs, 5S ribosomal RNA, short interspersed nuclear elements (SINEs), and a collection of other relatively short (<400 nt) catalytic or regulatory RNAs.

Given the expectation of three, and only three, nuclear multisubunit RNA polymerases in plants, as in all other known eukaryotes, completion of the *Arabidopsis thaliana* genome sequence in 2000 (The Arabidopsis Genome Initiative 2000) revealed a surprise that plants have catalytic subunits for two additional nuclear RNA polymerases, now known as Pol IV and Pol V (Fig. 13.1a, b). The two genes encoding the largest subunits of nuclear Pols IV and V, *NRPD1* and *NRPE1*, respectively, are quite different in sequence compared to other multisubunit RNA polymerase largest subunits but have recognizable similarity to their conserved domains, including core sequences of the catalytic site (Fig. 13.1a, c) (Herr et al. 2005; Haag et al. 2009; Onodera et al. 2005; Pontier et al. 2005). Multiple intron–exon positions in the first half of *NRPD1* and *NRPE1* are identical to their positions in *NRPB1*, the gene encoding the largest subunit of Pol II in *Arabidopsis* (Luo and Hall 2007). However, the C-terminal domains (CTDs) of the NRPB1, NRPD1, and NRPE1 proteins are unrelated, with the heptad repeats typical of the Pol II largest subunit CTD missing. A domain related to the *DEFECTIVE CHLOROPLASTS AND LEAVES* gene implicated in chloroplast rRNA processing (Bellaoui et al. 2003) is present in the CTDs of both NRPD1 and NRPE1 (Fig. 13.1a). The CTD of the Pol V largest subunit, NRPE1, also contains a region composed of ten imperfect copies of a 16 amino acid sequence and a domain of repeating glutamine and serine amino acids, neither of which are present in the Pol II largest subunit, NRPB1 (Fig. 13.1a, e). Collectively, these observations suggest that duplication of the Pol II *NRPB1* gene, combined with rearrangements involving unrelated genes, gave rise to the *NRPD1* gene in an ancestor common to modern day algae and land plants (Luo and Hall 2007). Subsequent duplication of the *NRPD1* gene, occurring prior to the divergence of moss from vascular plants, is then thought to have given rise to the ancestral Pol V largest subunit gene (Luo and Hall 2007; Tucker et al. 2011).

In addition to *NRPD1* and *NRPE1* genes, the *Arabidopsis* genome sequence revealed two nearly identical genes (one of which is a nonfunctional pseudogene) encoding an atypical second-largest subunit, NRP(D/E)2 (Fig. 13.1b). The NRP (D/E)2 amino acid sequence is more similar to the NRPB2 subunit of Pol II than to

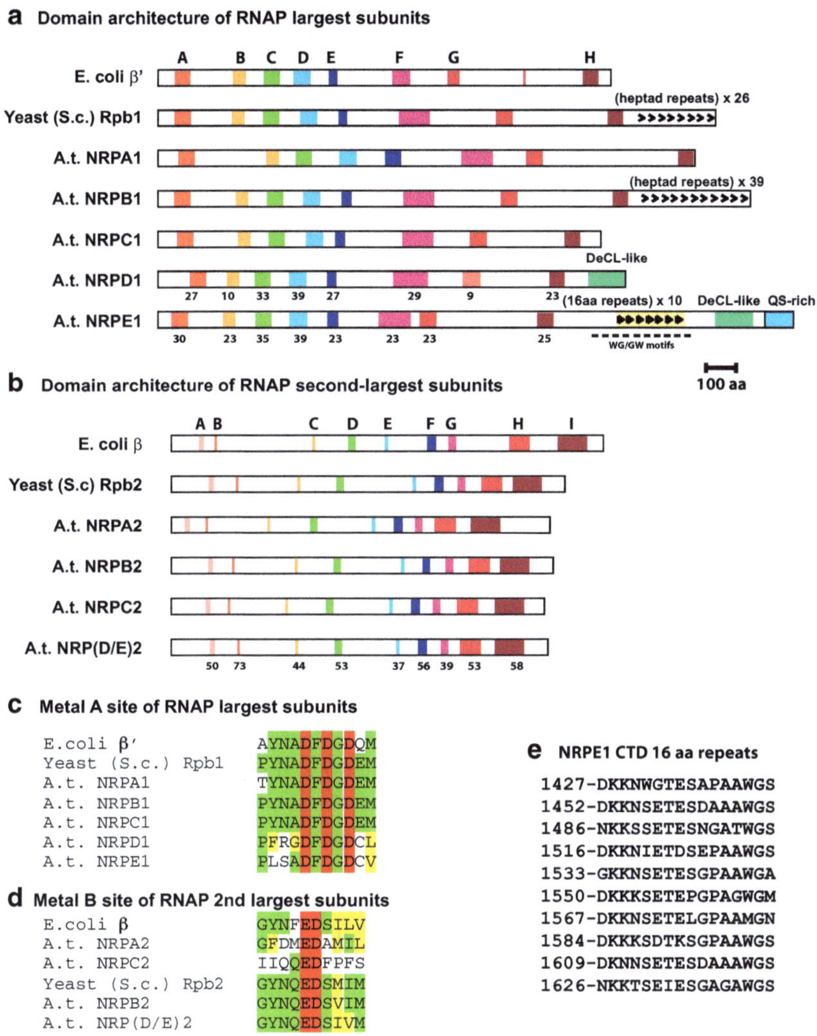

Fig. 13.1 Comparison of the catalytic subunits *Arabidopsis* nuclear RNA polymerases to those of *E. coli* RNA polymerase and yeast Pol II. (**a**) Diagrams for the largest subunits, with conserved domains A–H shown in different colors. Heptad repeats in the Pol II CTDs refer to tandemly arrayed seven amino acid sequences that are conserved in sequence, but not number, throughout eukaryotes. The DeCL-like domain of the Pol IV and Pol V subunit CTDs shows sequence similarity to the DEFECTIVE CHLOROPLASTS AND LEAVES protein in *Arabidopsis*. The QS rich domain is composed primarily of alternating glutamines and serine amino acids. The number below each of the Pol IV or Pol V domains shows the percent identity of that domain to the corresponding domain of the *Arabidopsis* Pol II subunit. (**b**) Diagrams for the second-largest subunits, with conserved domains A–I highlighted in different colors. The number below each of the Pol IV or Pol V domains is the percent identity between those domains and the corresponding domain of the *Arabidopsis* Pol II subunit. (**c**) Sequences surrounding the invariant aspartates of the magnesium-binding Metal A sites within conserved domain D of the largest subunits shown in panel **a**. (**d**) Sequences surrounding the invariant aspartate-glutamate duo of the magnesium-binding Metal B sites within the second-largest

the equivalent subunits of Pols I and III (Onodera et al. 2005), including sequences in the vicinity of the catalytic site (Fig. 13.1d). Genetic evidence indicated that *NRP(D/E)2* encodes the second subunit of both Pol IV and Pol V (Herr et al. 2005; Kanno et al. 2005; Onodera et al. 2005; Pontier et al. 2005).

The first insights into the functions of Pols IV and V (initially known as Pol IVa and Pol IVb) were revealed in 2005, based on molecular phenotypes of *nrpd1*, *nrpe1*, and *nrp(d/e)2* mutants (Herr et al. 2005; Kanno et al. 2005; Onodera et al. 2005; Pontier et al. 2005). These studies showed that Pols IV and V play nonredundant roles in a transcriptional silencing process known as RNA-directed DNA methylation (RdDM) that is important for silencing transposable elements, endogenous repeats, and transgenes (Zhang and Zhu 2011; Lahmy et al. 2010; Matzke et al. 2009). Roles for Pols IV and V in additional RNA-mediated silencing pathways have since been revealed. These include inter-allelic interactions responsible for the epigenetic phenomenon paramutation, pathogen resistance, and the spread of RNA silencing signals from cell to cell and via the vascular system (for reviews see Haag and Pikaard 2011; Molnar et al. 2011; Erhard and Hollick 2011; Arteaga-Vazquez and Chandler 2010; Brosnan and Voinnet 2011).

Subunit compositions of affinity-purified *Arabidopsis* Pols II, IV, and V, determined in 2009, showed that Pols IV and V are 12-subunit enzymes (Ream et al. 2009) that include subunits shared by Pol II (Ream et al. 2009; Huang et al. 2009; Lahmy et al. 2009; He et al. 2009a) and unique subunits that arose via stepwise duplication and sub-functionalization of Pol II subunit genes (Tucker et al. 2011; Ream et al. 2009) (Fig. 13.2). However, it remains unclear whether Pol IV and Pol V transcription is initiated in a Pol II-like fashion, making use of general transcription factors, activators, or coactivators that recognize promoters or enhancer sequences. Likewise, few details are known concerning Pol IV and Pol V activities as enzymes, including their templates in vivo, their modes of recruitment to specific loci, their transcription initiation or termination sites, the sizes of their primary transcripts, their processivity, or their fidelity.

Multiple subunits that are common to Pols II, IV, and V are encoded by two or more genes (Ream et al. 2009; Law et al. 2011), and one case study shows that alternative subunits can generate functionally distinct polymerase subtypes (Tan et al. 2012). New insights into the roles of the different subunits of Pols IV and V, and by inference, Pol II, may come from such studies, aided by the fact that null mutations eliminating Pol IV- or Pol V-specific subunits are not lethal. The recent demonstration that Pols IV and V can carry out RNA-primed transcription of DNA templates in vitro (Haag et al. 2012) also provides a starting point for learning more about their enzymatic capabilities and, in conjunction with genetic studies, the roles of catalytic versus non-catalytic subunits. These issues of general relevance to multisubunit RNA polymerase functions are explored in more detail in the subsequent sections of this chapter.

Fig. 13.1 (continued) subunits shown in panel **a**. (**e**) Sequence of the ten 16aa tandem repeats in the CTD of NRPE1. Numbers refer to amino acid positions within the full-length protein

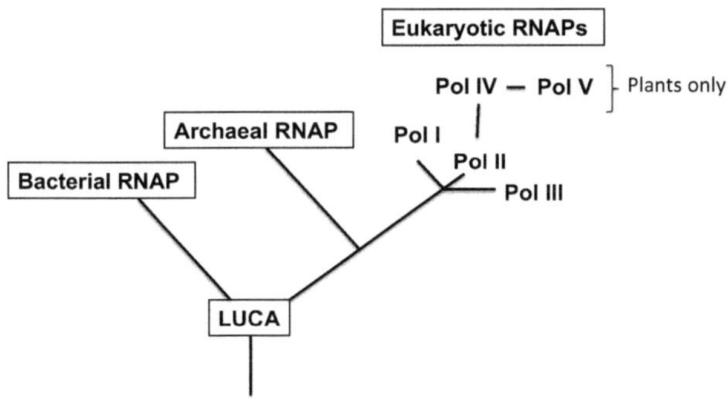

Fig. 13.2 Evolution of multisubunit RNA polymerases. LUCA refers to the last universal common ancestor of extant bacteria, archaea, and eukaryotes. The three essential polymerases of eukaryotes, Pols I, II, and III, must have existed in the common ancestor of all eukaryotes to explain their ubiquity

13.2 Pols IV and V Are Nonessential for Viability but Important for Development

In plants, as in all eukaryotes, Pols I, II, and III are essential. Heterozygous plants that have one wild-type allele and one null mutant allele for Pol I, II, or III catalytic subunits are viable, producing haploid gametophytes that are 50 % normal and 50 % mutant. Haploid female gametophytes that carry the mutant allele arrest early in development, failing to complete the three post-meiotic rounds of mitosis that are needed to form a mature eight-celled gametophyte that includes an egg cell. Thus, no transmission of the mutant alleles takes place through the female (Onodera et al. 2008). In contrast, haploid three-celled male gametophytes (pollen) bearing null alleles for essential Pol I, II, or III subunits are still able to complete development thanks to parentally supplied polymerase subunits (or holoenzymes) that persist from the post-meiotic pollen mother cell into the mature pollen. This parental contribution of Pols I, II, and III is sufficient to allow pollen germination upon landing on the stigma of a receptive female floral organ. There is also just enough parental Pol I, II, or III to allow for the growth of short pollen tubes and fertilization of the nearest (wild-type) ovules but not ovules deeper in the flower. In this way, some paternal transmission of mutant alleles to the next generation can take place, albeit at a low frequency (Onodera et al. 2008).

In contrast to null alleles for essential Pol I, II, or III subunits, null alleles of Pol IV or Pol V catalytic subunit genes (*nrpd1*, *nrpe1*, or *nrp(d/e)2*) are both male and female transmissible and homozygous mutants are viable and fertile. In *Arabidopsis thaliana*, the developmental consequences of being a Pol IV or Pol V mutant are subtle, at least under laboratory or greenhouse conditions; plants are normal in appearance but slower to flower, especially under short-day (long night) conditions. However, in maize, Pol IV null mutants display a number of developmental abnormalities, including altered cell fate in vegetative organs, stem outgrowths, and partial feminization of male floral organs, in addition to late flowering (Erhard et al. 2009; Parkinson et al. 2007). Collectively, these observations show that although Pols IV and V are nonessential for viability, they are nonetheless important for development. The differing severities of Pol IV mutant phenotypes in maize and *Arabidopsis* may be related to the different transposon contents of their genomes, with ~85 % of the maize genome composed of transposons, as opposed to ~15 % in *Arabidopsis*. Failure to tame transposons through Pol IV- and Pol V-mediated transcriptional silencing may lead to misregulation of adjacent genes.

13.3 Roles of Pols IV and V in RNA Silencing

In a wide variety of eukaryotes, small RNAs known as short-interfering RNA (siRNAs) or piRNAs direct the transcriptional silencing of homologous genes by bringing about DNA methylation and/or repressive histone modifications (Ishizu et al. 2012; Chen 2012; He et al. 2011; Law and Jacobsen 2010). In plants, this transcriptional silencing function is carried out by a specific class of siRNAs that are 24 nt in length and produced by DCL3, one of four Dicer endonucleases encoded by the *Arabidopsis thaliana* genome. The 24 nt siRNAs guide DNA methylation and heterochromatin formation to homologous loci, mostly retrotransposons or other repetitive nuclear elements, resulting in transcriptional silencing if promoters are methylated. This process, known as RdDM (Matzke et al. 2009; Zhang and Zhu 2011), is the process for which Pol IV and Pol V functions are best understood (Haag and Pikaard 2011). An abbreviated version of the RdDM process, emphasizing the steps involving Pols IV and V, is shown in Fig. 13.3. More complicated representations of the pathway involving additional activities can be found in recent reviews (Law and Jacobsen 2010; Haag and Pikaard 2011; Zhang and Zhu 2011).

RdDM has two major phases: the first involved in the biogenesis of 24 nt siRNAs and the second encompassing siRNA-programmed chromatin modifications. Pol IV is required in the first phase, and Pol V is critical for the second phase (Fig. 13.3). Mutations that knock out the catalytic subunits of Pol IV (*nrpd1* or *nrp(d/e)2*) essentially eliminate 24 nt siRNAs, as do mutations eliminating RDR2, one of six RNA-dependent RNA polymerases in *Arabidopsis*. These data provided early genetic evidence that Pol IV and RDR2 collaborate in the production of double-stranded RNA (dsRNA) precursors. Immunolocalization experiments showed that Pol IV colocalizes with repetitive loci that give rise to abundant 24 nt siRNAs and that RDR2 becomes mislocalized in *pol IV* mutants; in contrast, Pol IV localization is unaffected in *rdr2* mutants (Pontes et al. 2006). These observations suggested that Pol IV acts upstream of RDR2. More recently, affinity-purified Pol IV complexes analyzed by mass spectrometry were found to contain RDR2, with reciprocal immunoprecipitation experiments confirming that Pol IV and RDR2 associate in vivo (Haag et al. 2012; Law et al. 2011). The Pol IV–RDR2 interaction is dispensable for Pol IV transcription in vitro but appears to be required for RDR2

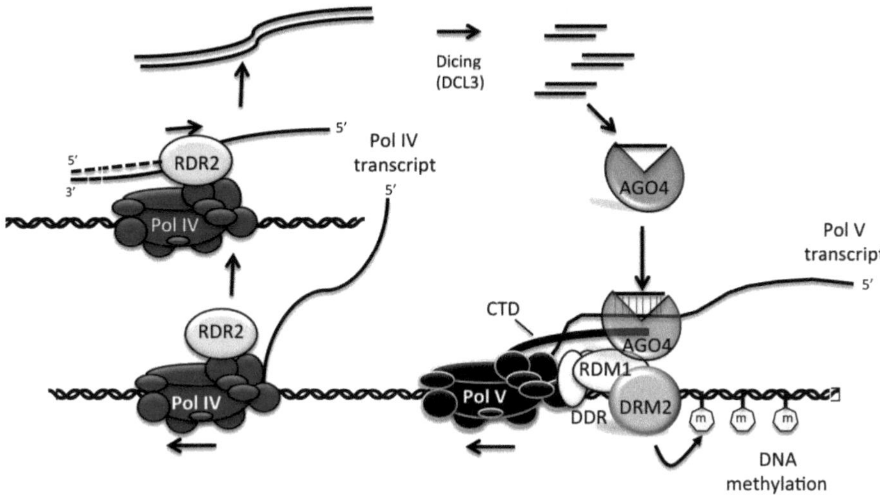

Fig. 13.3 Roles of Pols IV and V in RNA-directed DNA methylation in *Arabidopsis*. The diagram shows a simplified version of the pathway, focused on the roles of Pols IV and V. Double-stranded RNAs made by the Pol IV–RDR2 enzyme complex are diced by DCL3 and loaded into ARGONAUTE 4 (AGO4). The AGO4–siRNA complex is then recruited to target sites through binding to Pol V transcripts, and by physical interactions between AGO4 and the CTD of the Pol V largest subunit. The DDR complex assists in Pol V transcription and may also mediate interactions between AGO4 and the de novo DNA methyltransferase, DRM2

activity (Haag et al. 2012). This Pol IV-dependence of RDR2 activity suggests a mechanism in which Pol IV and RDR2 activities are coupled, with single-stranded Pol IV transcripts handed off to RDR2 to be used as templates, thus producing and channeling dsRNAs into the RdDM pathway (Fig. 13.3).

dsRNA products produced via Pol IV and RDR2 are diced by DCL3, and resulting 24 nt single-stranded siRNAs are loaded into the Argonaute family protein, AGO4, or its closest relatives, AGO6 or AGO9 (Zheng et al. 2007; Zilberman et al. 2003; Qi et al. 2006; Havecker et al. 2010). dsRNA synthesis, dicing, and AGO loading may be spatially coordinated given that RDR2, DCL3, and AGO4 partially colocalize, along with siRNAs and Pol V, within a nucleolus-associated Cajal body (Li et al. 2006; Pontes et al. 2006). Pol V has been shown to physically interact with AGO4 via WG or GW amino acid motifs (known as "AGO hooks") present in the Pol V CTD (see Fig. 13.1a) (El-Shami et al. 2007). Taken together, these observations suggested that Pol V might be guided to its sites of action by associating with AGO4–siRNA complexes within the Cajal bodies and then using the siRNAs to guide the complex to complementary target sites through base-pairing interactions (Li et al. 2006; Pontes et al. 2006). However, subsequent studies have argued against this hypothesis by showing that Pol V transcription at target loci is unaffected in *pol IV*, *rdr2*, or *dcl3* mutants defective for 24 nt siRNA biogenesis and is also unaffected in *ago4* mutants (Wierzbicki et al. 2008). Instead, siRNA binding to Pol V transcripts appears to be what recruits AGO4 to target loci.

Key evidence is that AGO4 can be chemically cross-linked to Pol V transcripts and that AGO4 associates with target loci (as shown using chromatin immunoprecipitation, or ChIP) in a Pol V-dependent manner (Wierzbicki et al. 2009). Moreover, AGO4's association with chromatin is abrogated in Pol V mutants bearing point mutations in the NRPE1 active site (Wierzbicki et al. 2009). These point mutations abolish Pol V transcriptional activity in vitro (Haag et al. 2012) and all known biological functions attributable to Pol V in vivo (Haag et al. 2009). Collectively, these findings indicate that AGO4 is primarily recruited to Pol V-transcribed loci via siRNA-mediated interactions with Pol V transcripts. AGO4 interactions with the Pol V CTD may further stabilize the complex or mediate downstream events (El-Shami et al. 2007). The functional significance of the Cajal bodies containing RDR2, DCL3, AGO4, siRNAs, and Pol V is not clear, but roles in RNA processing, chromatin modification, transport, or storage are possibilities (Li et al. 2008; Pontes and Pikaard 2008).

Several activities identified in genetic screens for disrupted RdDM are important for Pol V activity. Among these are DRD1, a putative SWI2/SNF2-family chromatin remodeling ATPase (Kanno et al. 2004); DMS3, a protein related to the hinge domains of cohesins and condensins (Kanno et al. 2008); and RDM1, a protein that binds methylated single-stranded DNA in vitro (Gao et al. 2010). Wierzbicki et al. showed that DRD1 and DMS3 are required for the production of Pol V transcripts and for stable Pol V association with chromatin (Wierzbicki et al. 2008, 2009). Subsequent studies showed that DRD1 and DMS3 associate with RDM1 to form a so-called DDR complex, named for the first initials of the three proteins (Law et al. 2010). Mass spectrometric analyses also identified Pol V subunits in affinity-purified DDR samples, suggesting that DDR and Pol V can physically associate (Law et al. 2010). The RDM1 protein of the DDR complex has also been shown to colocalize with AGO4 and the de novo DNA methyltransferase DRM2 (Gao et al. 2010), suggesting that RDM1 may serve as a bridge for recruitment of DRM2 to AGO4-siRNA- and Pol V-engaged loci (Zhang and Zhu 2011).

An SPT5-LIKE, KOW-domain transcription factor, SPT5L/KTF1, is involved in the downstream phase of the RdDM pathway, such that *ktf1/spt5l* mutants display a reduction in DNA methylation and reduced levels of siRNAs, similar to the effects of mutating the Pol V-specific subunit, NRPE5 (Bies-Etheve et al. 2009; Huang et al. 2009; Ream et al. 2009). In yeast and humans, Spt5 interacts with Spt4 to form the DSIF elongation factor complex, which is involved in mRNA capping, Pol II elongation, and prevention of premature transcription termination (Yamaguchi et al. 2013). One might expect SPT5L/KTF1 to similarly act as a Pol V elongation factor, consistent with the identification of SPT5L/KTF1 peptides in affinity-purified Pol V (Huang et al. 2009). However, Pol V transcripts are not decreased in abundance in a *spt5l/ktf1* loss-of-function mutant, but actually increase slightly (He et al. 2009c). A similar increase in Pol V transcript abundance is observed in *ago4* mutants (Wierzbicki et al. 2009) and SPT5L/KTF1 and AGO4 interact, via WG/GW "AGO-hook" motifs within SPT5L/KTF1 (He et al. 2009c; Bies-Etheve et al. 2009). These and other data have led to an alternative hypothesis that SPT5L/

KTF1 acts downstream of Pol V transcription possibly by helping AGO4 associate with Pol V transcripts or associated chromatin (Rowley et al. 2011). The increased abundance of Pol V transcripts observed in *ago4* and *ktf1* mutants may be a consequence of decreased slicing of Pol V transcripts by AGO4.

13.4 Subunit Compositions of *Arabidopsis* RNA Polymerases II, IV, and V

13.4.1 Subunits Common to Pols II, IV, and V

Affinity purification of *Arabidopsis* Pols II, IV, and V, followed by trypsin digestion and analysis of their peptides by LC-MS/MS mass spectrometry, revealed that Pols IV and V are composed of twelve subunits that are identical or homologous to the 12 core subunits of Pol II (Ream et al. 2009). Seven of the twelve subunits present in *Arabidopsis* Pols II, IV, and V are encoded by the same genes: the subunits homologous to yeast Rpb3, Rpb6, Rpb8, Rpb9, Rpb10, Rpb11, and Rpb12.

For all of the Pol II, IV, or V common subunits, except the 11th subunit (NRP (B/D/E)11), there are two genes in *Arabidopsis*. Mass spec analyses have revealed that the different variants can often be detected within the purified polymerases, sometimes equally and sometimes with only one variant being the major form (Ream et al. 2009; Law et al. 2011). Due to the high degree of similarity among most of these alternative subunits (~90 %, or higher, identity), one might expect the proteins to be functionally equivalent. However, a recent analysis of single or double mutants disrupting the two alternative forms of the ninth subunit [NRP (B/D/E)9a and NRP(B/D/E)9b] that are utilized by Pols II, IV, or V has forced a reconsideration of this assumption. This study revealed that the alternative ninth subunit proteins (and/or their genes) are only partially redundant with respect to Pol II functions and are nonredundant for Pol V-specific functions despite being 92 % identical (Tan et al. 2012). These data suggest that different Pol II, IV, and V subtypes exist in plants, resulting from their assembly using alternative subunits in various permutations. Consistent with this hypothesis, maize has three genes potentially encoding alternative NRP(D/E)2 subunits for Pol IV and/or V. These genes are not completely redundant, given that mutations in only one of the three genes disrupt the inter-allelic gene silencing phenomenon known as paramutation (Stonaker et al. 2009; Sidorenko et al. 2009). However, the developmental phenotypes of these mutants are not as severe as for mutants defective for the Pol IV largest subunit, NRPD1. Given that largest and second-largest subunits are equally important for forming the polymerase catalytic center, these observations suggest that there must be partial redundancy among the three second subunit variants with respect to functions affecting development (Pikaard and Tucker 2009).

13.4.2 *Pol V Makes Use of a Distinct 5th Subunit*

Rpb5, Rpb6, Rpb8, Rpb10, and Rpb12 are subunits common to Pols I, II, and III in yeast and humans; thus one might expect these subunits to be common to Pols I through V in *Arabidopsis*, as well. This is mostly true (as discussed above), but not entirely—although Pol IV uses the same fifth subunit as Pols I, II, and III, Pol V makes use of a distinct fifth subunit encoded by the *NRPE5* gene (Lahmy et al. 2009; Ream et al. 2009; Huang et al. 2009). Three additional Rpb5-like genes are also present in the *Arabidopsis* genome, at least one of which is used as an alternative minor subunit of Pol IV (Law et al. 2011). NRPE5 is distinguished from other eukaryotic Rpb5 subunits by the presence of an N-terminal extension, required for the protein's stability in vivo, and by divergence in its C-terminal assembly domain (Lahmy et al. 2009; Ream et al. 2009). In *nrpe5* mutants, normally silenced elements (such as retrotransposons) are derepressed as in *nrpe1* mutants, indicating that NRPE5 and the catalytic subunits are equally required for Pol V-dependent gene silencing (Huang et al. 2009; Lahmy et al. 2009; Ream et al. 2009; Douet et al. 2009). However, DNA methylation and siRNA abundance are less affected in *nrpe5* than *nrpe1* mutants (Ream et al. 2009; Lahmy et al. 2009), which is also the case for *nrp(d/e)4* mutants (He et al. 2009a). These results suggest that non-catalytic subunits of Pols IV and/or V may play roles in RNA silencing that are separable from the RNA synthesis function(s) of the catalytic subunits.

13.4.3 Arabidopsis *Pols II, IV, and V Have Unique Subunit 4/7 Sub-complexes*

In addition to their second-largest subunit, NRP(D/E)2, which is encoded by the same gene, Pol IV and Pol V have in common their fourth subunit, NRP(D/E)4, which is distinct from the NRPB4 subunit of Pol II (Ream et al. 2009). NRPD4 does not rescue a yeast *rpb4* mutation, whereas NRPB4 does (He et al. 2009a). Rpb4 and Rpb7 subunits interact, forming a dissociable sub-complex that in yeast has been shown to interact with the RNA transcripts of Pol II and regulate RNA processing, trafficking, and even translation (Choder 2004; Ujvari and Luse 2006; Runner et al. 2008; Harel-Sharvit et al. 2010). Interestingly, Pol II, Pol IV, and Pol V utilize distinct proteins as their major NRPB7, NPRD7, and NRPE7 subunits in *Arabidopsis* (Ream et al. 2009), with NRPE7 also serving as a minor form of the seventh subunit in Pol IV (Ream et al. 2009; Law et al. 2011). The *NRPD7* and *NRPE7* genes lack introns, suggesting that retrotransposition of an NRPB7 cDNA was the duplication event early in Pol IV/V evolution that gave rise to their seventh subunit genes (Tucker et al. 2011). The functional significance of having unique subunit-4/7 sub-complexes in Pols II, IV, and V is unclear but is likely to be important.

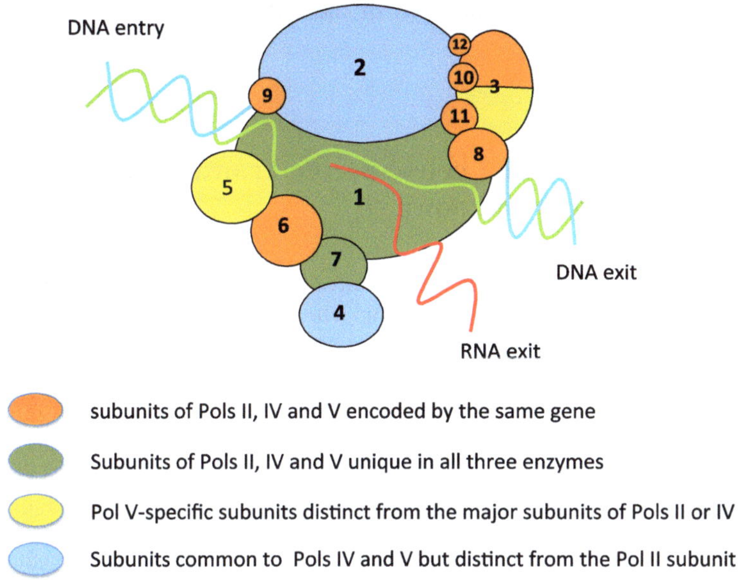

Fig. 13.4 Shared and unique subunits of *Arabidopsis* RNA polymerases II, IV, and V

To summarize, Pols II, IV, and V differ from one another in a subset of their subunits (Fig. 13.4). Pol IV differs from Pol II in four subunits (NRPD1, NRP(D/E)2, NRP(D/E)4, and NRPD7). Pol V differs from Pol II in five subunits (NRPE1, NRP(D/E)2, NRP(D/E)4, NRPE5, and NRPD7) and also makes nearly equal use of two alternative forms of the 3rd subunit, only one of which is primarily used by Pol II (or Pol IV). Pols IV and V differ in subunits 1, 5, and 7, and in the frequency of their use of the two alternative third subunits, with the major form of the Pol IV 3rd subunit being the same as for Pol II. Understanding how the different subunits contribute to the distinct functions of the three polymerases is a goal of ongoing research.

13.5 Pol V and Pol IV Largest Subunit CTD Functions Are Largely Unknown

As discussed previously, the Pol V largest subunit, NRPE1, has a long CTD that lacks the heptad repeats present in the Pol II CTD but has several unique domains, including a QS-rich domain, DeCL domain, and 10 imperfect tandem repeats of a 16 amino acid sequence (see Fig. 13.1a). Embedded within and adjacent to these 16 amino acid repeats are 18 WG/GW AGO-hook motifs implicated in interactions with AGO4 (El-Shami et al. 2007; He et al. 2009c). To date, this is the only function ascribed to the Pol V CTD. Unlike the Pol II largest subunit CTD, whose variable phosphorylation patterns confer regulatory meaning, nothing is known about potential posttranslation modifications of the NRPD1 or NRPE1 CTDs or their regulatory significance.

13.6 Rules Governing Locus-Specific Transcription by Pols IV and V Are Unknown

It is not clear how Pols IV or V are recruited to the thousands of loci where Pol IV-dependent siRNAs are generated or where RdDM occurs. Recent studies identified genomic sites of Pol V occupancy using ChIP-seq (Wierzbicki et al. 2012; Zhong et al. 2012), but motif finding algorithms failed to identify specific DNA sequences, such as putative or known promoter elements or transcription factor binding sites, that could explain the occurrence of Pol V in these regions. Pol IV ChIP-seq studies have not yet been reported nor have Pol IV primary transcripts been identified, but sites of 24 nt siRNA production are signatures of Pol IV (see Fig. 13.3), and motif finding at these siRNA loci has also failed to identify consensus sequences thus far.

Understanding more about the DDR complex may tell us more about Pol V recruitment or transcriptional regulation. In mutants such as *drd1* or *dms3* that disrupt the DDR complex, Pol V does not stably associate with chromatin, as shown by ChIP, and Pol V transcripts are no longer produced (Wierzbicki et al. 2008, 2009; Zhong et al. 2012). Whether the DDR complex plays a role in Pol V recruitment and initiation, or Pol V elongation and processivity is not yet known, but ChIP analyses of the DDR complex may reveal whether DDR localizes at specific subregions within Pol V-associated loci, perhaps revealing potential start sites.

Only two proteins with a known connection to Pol II regulatory mechanisms have emerged thus far in genetic screens for defects in RdDM. These are the SPT5-like protein, KTF1 discussed previously, and an *Arabidopsis* homolog of yeast IWR1, a protein which facilitates the nuclear import of assembled Pol II from the cytoplasm to the nucleus (Czeko et al. 2011), that was identified in two independent genetic screens (Kanno et al. 2010; He et al. 2009b). Mutant plants (*dms4*/*rdm4*) defective for the IWR1 homolog display pleiotropic phenotypic abnormalities in addition to disrupted RdDM, consistent with impaired functions of Pol II in addition to Pols IV and V.

13.7 RNAs Synthesized by Pols IV and V In Vivo

To date, no RNA corresponding to a primary transcript of Pol IV has been cloned and sequenced or detected in vivo by RT-PCR or other molecular assays. Most likely, this is due to the fleeting existence of these transcripts before being converted into dsRNA through the coupling of Pol IV and RDR2 activities, followed by dicing into siRNAs. As a result, nothing is known about Pol IV transcription start sites, termination sites (or regions), or potential transcript modifications, such as the possible addition of a 7-methylguanosine cap on the 5′ end. However, preliminary studies suggest that Pol V transcripts are not polyadenylated.

Unlike Pol IV transcripts, Pol V transcripts have been identified in vivo. Wierzbicki et al. examined genomic regions where no gene models existed and no significant transcription had been detected using DNA microarrays, yet where deep sequencing identified small RNAs that must have had precursors (Wierzbicki et al. 2008). This approach led to the identification of multiple loci where low-abundance transcripts detectable by RT-PCR are absent in *nrpe1* or *nrp(d/e)2* mutants, yet still detectable in *nrpd1* (Pol IV) mutants (Wierzbicki et al. 2008). ChIP analyses using anti-NRPE1 antibodies showed that Pol V physically associates with these loci. For several loci examined, transcripts appeared to initiate at multiple sites, based on 5′-RACE, were enriched among poly-A$^-$ RNA and had either 5′ caps or triphosphates—characteristic of newly initiated RNAs, as opposed to processed RNAs (Wierzbicki et al. 2008). Production of Pol V-dependent transcripts in vivo also requires the conserved aspartates of the NRPE1 Metal A site (see Fig. 13.1c), and these transcripts can be chemically cross-linked to NRPE1, allowing their immunoprecipitation using anti-NRPE1 antibodies. Collectively, these assays strongly suggested that these RNAs are Pol V transcripts.

13.8 Pol IV and Pol V Transcription In Vitro

RNA polymerase activity assays typically make use of sheared genomic DNA, or double-stranded DNA having 3′ overhangs, as a source of templates for promoter-independent transcription. Interestingly, Pols IV and V show no activity in such assays, unlike Pols I, II, or III (Onodera et al. 2005), thwarting early attempts by several laboratories, including ours, to detect Pol IV or Pol V biochemical activities. These negative results suggested that Pols IV and V might transcribe unconventional templates, or might even lack RNA polymerase activity altogether, consistent with the divergence, or absence, of more than 160 amino acids that are invariant in the catalytic subunits of Pols I, II, or III (Haag et al. 2009; Landick 2009). These substituted amino acids are clustered in the vicinity of the Metal A and Metal B sites of the catalytic center; in the bridge helix, trigger loop, cleft, and funnel domains of the largest subunits; and within the hybrid-binding domain of NRPD2 (Fig. 13.5; see also Fig. 13.1c, d). A number of these amino acids correspond to positions that interact with the incoming nucleotide triphosphate and the 3′ end of the growing RNA chain and are thus thought to be critical for catalysis. Among these are amino acids at the tip of the trigger loop and in the bridge helix that are not simply substituted but missing altogether in Pols IV and V (Landick 2009).

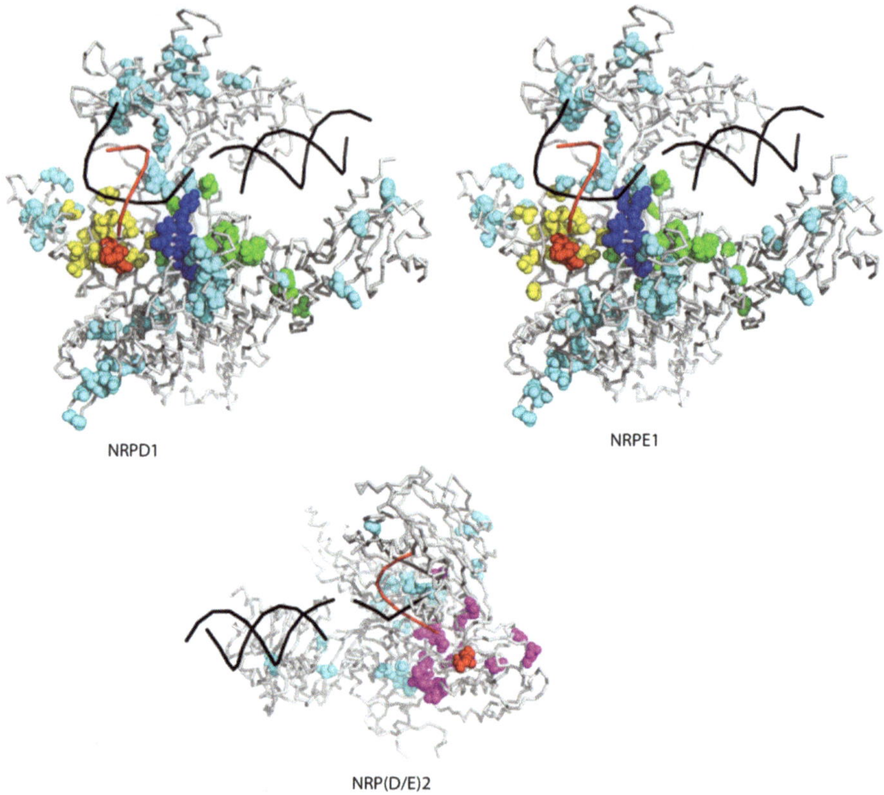

Fig. 13.5 Sequence divergence in Pols IV and V at amino acid positions that are invariant in Pols I, II, and III. The three images show these substituted positions as space-filling spheres mapped onto the Rpb1 and Rbp2 subunit structures of a yeast Pol II elongation complex determined in the Kornberg laboratory (Protein Data Bank structure 1R9T). The dsDNA substrate is shown in *black* and the RNA product in *red*. Amino acids colored *red mark* the positions of the invariant Metal A and Metal B sites in the largest and second-largest subunits, respectively. Substituted amino acids in the cleft, bridge helix, and active site domains of the largest subunits are colored *green*, *blue*, and *yellow*, respectively. Substituted amino acids in the hybrid-binding domain of the second-largest subunit are colored *magenta*. Substituted amino acids in the largest and second-largest subunits that are located outside of these domains are colored *cyan*. These images are reprinted from the open access article by Haag et al. (2009), as permitted under the Creative Commons Attribution License

Despite their lack of activity using conventional DNA templates, and their divergent amino acid sequences, *Arabidopsis* Pols IV and V are indeed functional RNA polymerases in vitro. Using templates in which an RNA primer is annealed to a DNA template oligonucleotide, thus forming an 8 bp RNA–DNA hybrid, Pols IV and V will elongate the RNA in a templated fashion (Haag et al. 2012), as shown previously for Pols I and II (Kuhn et al. 2007; Lehmann et al. 2007). Unlike Pol II, Pols IV and V are insensitive to the fungal toxin, alpha-amanitin, consistent with the divergence of multiple amino acids within the alpha-amanitin binding pocket of

Pol II (Haag et al. 2012). If double-stranded DNA is present downstream of the template DNA–RNA hybrid region, Pols II and IV are able to generate transcripts, but Pol V cannot, suggesting that Pol V cannot displace a non-template strand during transcription (Haag et al. 2012).

Our inability thus far to coax affinity-purified Pols IV and V to initiate transcription on DNA templates in the absence of an RNA primer may simply be due to the absence of unidentified helper proteins that do not copurify with the polymerases. However, cytological experiments have shown that Pols IV and V become mislocalized in RNase A-treated nuclei, unlike Pol II (Pontes et al. 2006), consistent with some form of RNA involvement, possibly as a template or primer.

13.9 Missing Pieces

Pols IV and V are remarkable enzymes given their evolution as specialized forms of Pol II but their substantial divergence at amino acid positions that are otherwise invariant among Pols I, II, and III. Approximately 120 of these normally invariant amino acids are substituted, or missing, in the largest subunits of *Arabidopsis* Pols IV and V, and another 40 are altered in the second-largest subunit (see Fig. 13.5). Moreover, comparison of the largest subunit sequences of Pols IV and V in a variety of plants reveals that they are evolving rapidly, with amino acid substitution rates occurring at 20 times the rate observed for the largest subunits of Pols I, II, or III (Luo and Hall 2007). Collectively, these observations indicate that there are fewer functional constraints on the evolution of Pols IV and V than for other polymerases.

DNA-dependent RNA polymerases perform a variety of functions in addition to RNA synthesis. These include pausing at specific sequences, backtracking along the template, RNA cleavage, and proofreading. These activities allow a polymerase to overcome barriers in the template or correct misincorporated nucleotides (Sydow and Cramer 2009). It is unknown if Pols IV and V possess any of these activities in vivo or in vitro. Given their involvement in noncoding RNA-mediated silencing processes, perhaps Pols IV and V do not need to be high-fidelity enzymes with proofreading or backtracking capabilities. In fact, error-prone transcription might actually be advantageous in the production of siRNAs that participate in the silencing of large transposon families, whose sequences are similar, but not identical. Perhaps by making use of RNA primers derived from Pol II (or other polymerase) transcripts, the catalytic centers of Pols IV and V have evolved to accommodate binding to DNA–RNA hybrid templates and only need to elongate RNAs rather than initiate their synthesis de novo from nucleoside triphosphates. This might also remove constraints on the evolution of the Pol IV and Pol V catalytic centers. Likewise, the possibility that Pols IV and V recognize methylated DNA, or chromatin templates bearing specific histone modifications, may contribute to their different amino acid sequences compared to Pol II. Such speculations should be possible to test upon further development of in vitro assays.

In vivo, identification of full-length Pol V or Pol IV transcripts through strand-specific RNA deep sequencing in appropriate wild-type and mutant backgrounds (coupled with comparative bioinformatics analyses) has not yet been accomplished. However, this basic information is critically important. Knowing where Pol IV and Pol V transcripts begin and end will define their transcription units and allow for more focused analyses of potential consensus motifs in the vicinity of these transcription units, possibly revealing promoters or other regulatory elements that can be tested. Moreover, knowing if Pol IV or Pol V transcripts are co- or post-transcriptionally modified or processed may reveal insights into their channeling, trafficking, and stability.

Understanding the functions of subunits that differ between Pols II, IV, and V is also likely to yield important new insights. Evidence that multiple structurally distinct subtypes of the different polymerases exist, based on alternative choices for many of their subunits, is intriguing. By exploiting the non-lethality of Pol IV- or Pol V-specific subunit mutations, the roles of different subunits in DNA methylation and chromatin modification in vivo, and transcription in vitro, can be explored, potentially revealing new insights, particularly into the functions of non-catalytic subunits. Last but not least, understanding the functions of the unique CTDs of the Pol IV and Pol V largest subunits should also be illuminating, potentially revealing mechanisms that couple transcription with noncoding RNA processing and chromatin modification.

Acknowledgments We apologize to the many researchers whose work was not cited for the sake of brevity. Pikaard lab research on Pols IV and V is supported by National Institutes of Health grant GM077590. T.R. is a Gordon and Betty Moore Foundation Fellow of the Life Sciences Research Foundation. C.S.P. is an Investigator of the Howard Hughes Medical Institute and the Gordon and Betty Moore Foundation. J.R.H. is supported by funds from the Howard Hughes Medical Institute. Opinions are those of the authors and do not necessarily reflect the views of our sponsors.

References

Arteaga-Vazquez MA, Chandler VL (2010) Paramutation in maize: RNA mediated trans-generational gene silencing. Curr Opin Genet Dev 20(2):156–163. doi:10.1016/j.gde.2010.01.008

Bellaoui M, Keddie JS, Gruissem W (2003) DCL is a plant-specific protein required for plastid ribosomal RNA processing and embryo development. Plant Mol Biol 53(4):531–543

Bies-Etheve N, Pontier D, Lahmy S, Picart C, Vega D, Cooke R, Lagrange T (2009) RNA-directed DNA methylation requires an AGO4-interacting member of the SPT5 elongation factor family. EMBO Rep 10(6):649–654. doi:10.1038/embor.2009.31, embor200931 [pii]

Brosnan CA, Voinnet O (2011) Cell-to-cell and long-distance siRNA movement in plants: mechanisms and biological implications. Curr Opin Plant Biol 14(5):580–587. doi:10.1016/j.pbi.2011.07.011

Chen X (2012) Small RNAs in development – insights from plants. Curr Opin Genet Dev 22 (4):361–367. doi:10.1016/j.gde.2012.04.004

Choder M (2004) Rpb4 and Rpb7: subunits of RNA polymerase II and beyond. Trends Biochem Sci 29(12):674–681. doi:10.1016/j.tibs.2004.10.007, S0968-0004(04)00272-5 [pii]
Czeko E, Seizl M, Augsberger C, Mielke T, Cramer P (2011) Iwr1 directs RNA polymerase II nuclear import. Mol Cell 42(2):261–266. doi:10.1016/j.molcel.2011.02.033
Douet J, Tutois S, Tourmente S (2009) A Pol V-mediated silencing, independent of RNA-directed DNA methylation, applies to 5S rDNA. PLoS Genet 5(10):e1000690. doi:10.1371/journal.pgen.1000690
El-Shami M, Pontier D, Lahmy S, Braun L, Picart C, Vega D, Hakimi MA, Jacobsen SE, Cooke R, Lagrange T (2007) Reiterated WG/GW motifs form functionally and evolutionarily conserved ARGONAUTE-binding platforms in RNAi-related components. Genes Dev 21 (20):2539–2544
Erhard KF Jr, Hollick JB (2011) Paramutation: a process for acquiring trans-generational regulatory states. Curr Opin Plant Biol 14(2):210–216. doi:10.1016/j.pbi.2011.02.005
Erhard KF Jr, Stonaker JL, Parkinson SE, Lim JP, Hale CJ, Hollick JB (2009) RNA polymerase IV functions in paramutation in Zea mays. Science 323(5918):1201–1205. doi:10.1126/science.1164508, 323/5918/1201 [pii]
Gao Z, Liu HL, Daxinger L, Pontes O, He X, Qian W, Lin H, Xie M, Lorkovic ZJ, Zhang S, Miki D, Zhan X, Pontier D, Lagrange T, Jin H, Matzke AJ, Matzke M, Pikaard CS, Zhu JK (2010) An RNA polymerase II- and AGO4-associated protein acts in RNA-directed DNA methylation. Nature 465(7294):106–109. doi:10.1038/nature09025, nature09025 [pii]
Haag JR, Pikaard CS (2011) Multisubunit RNA polymerases IV and V: purveyors of non-coding RNA for plant gene silencing. Nat Rev Mol Cell Biol 12(8):483–492. doi:10.1038/nrm3152
Haag JR, Pontes O, Pikaard CS (2009) Metal A and metal B sites of nuclear RNA polymerases Pol IV and Pol V are required for siRNA-dependent DNA methylation and gene silencing. PLoS One 4(1):e4110. doi:10.1371/journal.pone.0004110
Haag JR, Ream TS, Marasco M, Nicora CD, Norbeck AD, Pasa-Tolic L, Pikaard CS (2012) In vitro transcription activities of Pol IV, Pol V, and RDR2 reveal coupling of Pol IV and RDR2 for dsRNA synthesis in plant RNA silencing. Mol Cell 48(5):811–818. doi:10.1016/j.molcel.2012.09.027
Harel-Sharvit L, Eldad N, Haimovich G, Barkai O, Duek L, Choder M (2010) RNA polymerase II subunits link transcription and mRNA decay to translation. Cell 143(4):552–563. doi:10.1016/j.cell.2010.10.033, S0092-8674(10)01237-7 [pii]
Havecker ER, Wallbridge LM, Hardcastle TJ, Bush MS, Kelly KA, Dunn RM, Schwach F, Doonan JH, Baulcombe DC (2010) The Arabidopsis RNA-directed DNA methylation argonautes functionally diverge based on their expression and interaction with target loci. Plant Cell 22(2):321–334. doi:10.1105/tpc.109.072199, tpc.109.072199 [pii]
He XJ, Hsu YF, Pontes O, Zhu J, Lu J, Bressan RA, Pikaard C, Wang CS, Zhu JK (2009a) NRPD4, a protein related to the RPB4 subunit of RNA polymerase II, is a component of RNA polymerases IV and V and is required for RNA-directed DNA methylation. Genes Dev 23 (3):318–330. doi:10.1101/gad.1765209, 23/3/318 [pii]
He XJ, Hsu YF, Zhu S, Liu HL, Pontes O, Zhu J, Cui X, Wang CS, Zhu JK (2009b) A conserved transcriptional regulator is required for RNA-directed DNA methylation and plant development. Genes Dev 23(23):2717–2722. doi:10.1101/gad.1851809, gad.1851809 [pii]
He XJ, Hsu YF, Zhu S, Wierzbicki AT, Pontes O, Pikaard CS, Liu HL, Wang CS, Jin H, Zhu JK (2009c) An effector of RNA-directed DNA methylation in arabidopsis is an ARGONAUTE 4- and RNA-binding protein. Cell 137(3):498–508. doi:10.1016/j.cell.2009.04.028, S0092-8674 (09)00453-X [pii]
He XJ, Chen T, Zhu JK (2011) Regulation and function of DNA methylation in plants and animals. Cell Res 21(3):442–465. doi:10.1038/cr.2011.23
Herr AJ, Jensen MB, Dalmay T, Baulcombe DC (2005) RNA polymerase IV directs silencing of endogenous DNA. Science 308(5718):118–120

Huang L, Jones AM, Searle I, Patel K, Vogler H, Hubner NC, Baulcombe DC (2009) An atypical RNA polymerase involved in RNA silencing shares small subunits with RNA polymerase II. Nat Struct Mol Biol 16(1):91–93. doi:10.1038/nsmb.1539, nsmb.1539 [pii]

Ishizu H, Siomi H, Siomi MC (2012) Biology of PIWI-interacting RNAs: new insights into biogenesis and function inside and outside of germlines. Genes Dev 26(21):2361–2373. doi:10.1101/gad.203786.112

Kanno T, Mette MF, Kreil DP, Aufsatz W, Matzke M, Matzke AJ (2004) Involvement of putative SNF2 chromatin remodeling protein DRD1 in RNA-directed DNA methylation. Curr Biol 14 (9):801–805

Kanno T, Huettel B, Mette MF, Aufsatz W, Jaligot E, Daxinger L, Kreil DP, Matzke M, Matzke AJ (2005) Atypical RNA polymerase subunits required for RNA-directed DNA methylation. Nat Genet 37(7):761–765

Kanno T, Bucher E, Daxinger L, Huettel B, Bohmdorfer G, Gregor W, Kreil DP, Matzke M, Matzke AJ (2008) A structural-maintenance-of-chromosomes hinge domain-containing protein is required for RNA-directed DNA methylation. Nat Genet 40(5):670–675. doi:10.1038/ng.119, ng.119 [pii]

Kanno T, Bucher E, Daxinger L, Huettel B, Kreil DP, Breinig F, Lind M, Schmitt MJ, Simon SA, Gurazada SG, Meyers BC, Lorkovic ZJ, Matzke AJ, Matzke M (2010) RNA-directed DNA methylation and plant development require an IWR1-type transcription factor. EMBO Rep 11 (1):65–71. doi:10.1038/embor.2009.246, embor2009246 [pii]

Kuhn CD, Geiger SR, Baumli S, Gartmann M, Gerber J, Jennebach S, Mielke T, Tschochner H, Beckmann R, Cramer P (2007) Functional architecture of RNA polymerase I. Cell 131 (7):1260–1272

Lahmy S, Pontier D, Cavel E, Vega D, El-Shami M, Kanno T, Lagrange T (2009) PolV(PolIVb) function in RNA-directed DNA methylation requires the conserved active site and an additional plant-specific subunit. Proc Natl Acad Sci USA 106(3):941–946. doi:10.1073/pnas.0810310106, 0810310106 [pii]

Lahmy S, Bies-Etheve N, Lagrange T (2010) Plant-specific multisubunit RNA polymerase in gene silencing. Epigenetics 5(1):4–8

Landick R (2009) Functional divergence in the growing family of RNA polymerases. Structure 17 (3):323–325. doi:10.1016/j.str.2009.02.006, S0969-2126(09)00084-7 [pii]

Law JA, Jacobsen SE (2010) Establishing, maintaining and modifying DNA methylation patterns in plants and animals. Nat Rev Genet 11(3):204–220. doi:10.1038/nrg2719, nrg2719 [pii]

Law JA, Ausin I, Johnson LM, Vashisht AA, Zhu JK, Wohlschlegel JA, Jacobsen SE (2010) A protein complex required for polymerase V transcripts and RNA-directed DNA methylation in Arabidopsis. Curr Biol 20(10):951–956. doi:10.1016/j.cub.2010.03.062, S0960-9822(10)00388-X [pii]

Law JA, Vashisht AA, Wohlschlegel JA, Jacobsen SE (2011) SHH1, a homeodomain protein required for DNA methylation, as well as RDR2, RDM4, and chromatin remodeling factors, associate with RNA polymerase IV. PLoS Genet 7(7):e1002195. doi:10.1371/journal.pgen.1002195

Lehmann E, Brueckner F, Cramer P (2007) Molecular basis of RNA-dependent RNA polymerase II activity. Nature 450(7168):445–449

Li CF, Pontes O, El-Shami M, Henderson IR, Bernatavichute YV, Chan SW, Lagrange T, Pikaard CS, Jacobsen SE (2006) An ARGONAUTE4-containing nuclear processing center colocalized with Cajal bodies in Arabidopsis thaliana. Cell 126(1):93–106. doi:10.1016/j.cell.2006.05.032

Li CF, Henderson IR, Song L, Fedoroff N, Lagrange T, Jacobsen SE (2008) Dynamic regulation of ARGONAUTE4 within multiple nuclear bodies in Arabidopsis thaliana. PLoS Genet 4(2):e27. doi:10.1371/journal.pgen.0040027, 07-PLGE-RA-0748 [pii]

Luo J, Hall BD (2007) A multistep process gave rise to RNA polymerase IV of land plants. J Mol Evol 64(1):101–112

Matzke M, Kanno T, Daxinger L, Huettel B, Matzke AJ (2009) RNA-mediated chromatin-based silencing in plants. Curr Opin Cell Biol 21(3):367–376. doi:10.1016/j.ceb.2009.01.025

Molnar A, Melnyk C, Baulcombe DC (2011) Silencing signals in plants: a long journey for small RNAs. Genome Biol 12(1):215. doi:10.1186/gb-2010-11-12-219

Onodera Y, Haag JR, Ream T, Nunes PC, Pontes O, Pikaard CS (2005) Plant nuclear RNA polymerase IV mediates siRNA and DNA methylation-dependent heterochromatin formation. Cell 120(5):613–622. doi:10.1016/j.cell.2005.02.007, S0092-8674(05)00151-0 [pii]

Onodera Y, Nakagawa K, Haag JR, Pikaard D, Mikami T, Ream T, Ito Y, Pikaard CS (2008) Sex-biased lethality or transmission of defective transcription machinery in Arabidopsis. Genetics 180(1):207–218. doi:10.1534/genetics.108.090621, genetics.108.090621 [pii]

Parkinson SE, Gross SM, Hollick JB (2007) Maize sex determination and abaxial leaf fates are canalized by a factor that maintains repressed epigenetic states. Dev Biol 308(2):462–473. doi:10.1016/j.ydbio.2007.06.004, S0012-1606(07)01120-7 [pii]

Pikaard CS, Tucker S (2009) RNA-silencing enzymes Pol IV and Pol V in maize: more than one flavor? PLoS Genet 5(11):e1000736. doi:10.1371/journal.pgen.1000736

Pontes O, Pikaard CS (2008) siRNA and miRNA processing: new functions for Cajal bodies. Curr Opin Genet Dev 18(2):197–203. doi:10.1016/j.gde.2008.01.008

Pontes O, Li CF, Nunes PC, Haag J, Ream T, Vitins A, Jacobsen SE, Pikaard CS (2006) The Arabidopsis chromatin-modifying nuclear siRNA pathway involves a nucleolar RNA processing center. Cell 126(1):79–92. doi:10.1016/j.cell.2006.05.031

Pontier D, Yahubyan G, Vega D, Bulski A, Saez-Vasquez J, Hakimi MA, Lerbs-Mache S, Colot V, Lagrange T (2005) Reinforcement of silencing at transposons and highly repeated sequences requires the concerted action of two distinct RNA polymerases IV in Arabidopsis. Genes Dev 19(17):2030–2040

Qi Y, He X, Wang XJ, Kohany O, Jurka J, Hannon GJ (2006) Distinct catalytic and non-catalytic roles of ARGONAUTE4 in RNA-directed DNA methylation. Nature 443(7114):1008–1012

Ream TS, Haag JR, Wierzbicki AT, Nicora CD, Norbeck AD, Zhu JK, Hagen G, Guilfoyle TJ, Pasa-Tolic L, Pikaard CS (2009) Subunit compositions of the RNA-silencing enzymes Pol IV and Pol V reveal their origins as specialized forms of RNA polymerase II. Mol Cell 33 (2):192–203. doi:10.1016/j.molcel.2008.12.015, S1097-2765(08)00858-7 [pii]

Rowley MJ, Avrutsky MI, Sifuentes CJ, Pereira L, Wierzbicki AT (2011) Independent chromatin binding of ARGONAUTE4 and SPT5L/KTF1 mediates transcriptional gene silencing. PLoS Genet 7(6):e1002120. doi:10.1371/journal.pgen.1002120

Runner VM, Podolny V, Buratowski S (2008) The Rpb4 subunit of RNA polymerase II contributes to cotranscriptional recruitment of 3′ processing factors. Mol Cell Biol 28(6):1883–1891. doi:10.1128/MCB.01714-07, MCB.01714-07 [pii]

Sidorenko L, Dorweiler JE, Cigan AM, Arteaga-Vazquez M, Vyas M, Kermicle J, Jurcin D, Brzeski J, Cai Y, Chandler VL (2009) A dominant mutation in mediator of paramutation2, one of three second-largest subunits of a plant-specific RNA polymerase, disrupts multiple siRNA silencing processes. PLoS Genet 5(11):e1000725. doi:10.1371/journal.pgen.1000725

Stonaker JL, Lim JP, Erhard KF Jr, Hollick JB (2009) Diversity of Pol IV function is defined by mutations at the maize rmr7 locus. PLoS Genet 5(11):e1000706. doi:10.1371/journal.pgen.1000706

Sydow JF, Cramer P (2009) RNA polymerase fidelity and transcriptional proofreading. Curr Opin Struct Biol 19(6):732–739. doi:10.1016/j.sbi.2009.10.009

Tan EH, Blevins T, Ream T, Pikaard C (2012) Functional consequences of subunit diversity in RNA polymerases II and V. Cell Rep 1(3):208–214. doi:10.1016/j.celrep.2012.01.004

The Arabidopsis Genome Initiative (2000) Analysis of the genome sequence of the flowering plant Arabidopsis thaliana. Nature 408:796–815

Tucker SL, Reece J, Ream TS, Pikaard CS (2011) Evolutionary history of plant multisubunit RNA polymerases IV and V: subunit origins via genome-wide and segmental gene duplications, retrotransposition, and lineage-specific subfunctionalization. Cold Spring Harb Symp Quant Biol 75:285–297. doi:10.1101/sqb.2010.75.037

Ujvari A, Luse DS (2006) RNA emerging from the active site of RNA polymerase II interacts with the Rpb7 subunit. Nat Struct Mol Biol 13(1):49–54. doi:10.1038/nsmb1026, nsmb1026 [pii]

Wierzbicki AT, Haag JR, Pikaard CS (2008) Noncoding transcription by RNA polymerase Pol IVb/Pol V mediates transcriptional silencing of overlapping and adjacent genes. Cell 135 (4):635–648. doi:10.1016/j.cell.2008.09.035, S0092-8674(08)01192-6 [pii]

Wierzbicki AT, Ream TS, Haag JR, Pikaard CS (2009) RNA polymerase V transcription guides ARGONAUTE4 to chromatin. Nat Genet 41(5):630–634. doi:10.1038/ng.365

Wierzbicki AT, Cocklin R, Mayampurath A, Lister R, Rowley MJ, Gregory BD, Ecker JR, Tang H, Pikaard CS (2012) Spatial and functional relationships among Pol V-associated loci, Pol IV-dependent siRNAs, and cytosine methylation in the Arabidopsis epigenome. Genes Dev 26(16):1825–1836. doi:10.1101/gad.197772.112

Yamaguchi Y, Shibata H, Handa H (2013) Transcription elongation factors DSIF and NELF: promoter-proximal pausing and beyond. Biochim Biophys Acta 1829(1):98–104. doi:10.1016/j.bbagrm.2012.11.007

Zhang H, Zhu JK (2011) RNA-directed DNA methylation. Curr Opin Plant Biol 14(2):142–147. doi:10.1016/j.pbi.2011.02.003

Zheng X, Zhu J, Kapoor A, Zhu JK (2007) Role of Arabidopsis AGO6 in siRNA accumulation, DNA methylation and transcriptional gene silencing. EMBO J 26(6):1691–1701

Zhong X, Hale CJ, Law JA, Johnson LM, Feng S, Tu A, Jacobsen SE (2012) DDR complex facilitates global association of RNA polymerase V to promoters and evolutionarily young transposons. Nat Struct Mol Biol 19(9):870–875. doi:10.1038/nsmb.2354

Zilberman D, Cao X, Jacobsen SE (2003) ARGONAUTE4 control of locus-specific siRNA accumulation and DNA and histone methylation. Science 299(5607):716–719. doi:10.1126/science.1079695, 1079695 [pii]

Chapter 14
Structure, Dynamics, and Fidelity of RNA-Dependent RNA Polymerases

David D. Boehr, Jamie J. Arnold, Ibrahim M. Moustafa, and Craig E. Cameron

Abstract The RNA-dependent RNA polymerase (RdRp) is responsible for replicating the genomes of RNA viruses. The overall structure and function of RdRps is similar to other nucleic acid polymerases, although some RdRps employ unique initiation mechanisms. Recent biophysical studies indicate that the internal motions of RdRps, and other nucleic acid polymerases, are critical for their catalytic function and fidelity. In particular, these studies have suggested that the closing of the active site in preparation for catalysis involves the movement of the motif-D loop to help reposition a highly conserved lysine, enabling this residue to act as a general acid to protonate the pyrophosphate leaving group. Binding of incorrect nucleoside triphosphate does not induce the same structural changes in the motif-D loop, indicating a role for this loop in nucleotide discrimination. Indeed, substitution at the motif-D lysine increases polymerase fidelity and, intriguingly, decreases viral pathogenesis. The highly conserved nature of this lysine thus suggests a universal mechanism for rational vaccine design based on generating variants at this position. Moreover, substitutions elsewhere in the RdRp structure, including those remote from the active site, likewise lead to changes in polymerase fidelity and decrease viral pathogenesis. In these cases, the amino acid substitutions alter internal protein motions (including those in the motif-D loop) without substantially affecting the polymerase structure. A picture emerges in which RdRps and other nucleic acid polymerases can be viewed as "small world" networks of amino acids; communication pathways connect from the surface of the protein all the way to the catalytic center. These networks can be impacted by amino acid substitutions, inhibitor binding, and/or binding of accessory replication proteins to regulate RdRp catalysis and fidelity.

D.D. Boehr (✉)
Department of Chemistry, The Pennsylvania State University, University Park, PA 16802, USA
e-mail: ddb12@psu.edu

J.J. Arnold • I.M. Moustafa • C.E. Cameron
Department of Biochemistry and Molecular Biology, The Pennsylvania State University, University Park, PA 16802, USA

K.S. Murakami and M.A. Trakselis (eds.), *Nucleic Acid Polymerases*, Nucleic Acids and Molecular Biology 30, DOI 10.1007/978-3-642-39796-7_14,

Keywords RNA-dependent RNA polymerase • Fidelity • Pathogenesis • Vaccine • Protein structure • Protein dynamics

Abbreviations

CRE	Cis-acting element
DCCM	Dynamic cross correlation map
ds	Double stranded
FMDV	Foot-and-mouth disease virus
HCV	Hepatitis C virus
MD	Molecular dynamics
NMR	Nuclear magnetic resonance
NTP	Nucleoside triphosphate
NV	Norovirus
PCA	Principal component analysis
PV	Poliovirus
RdRp	RNA-dependent RNA polymerase
RT	Reverse transcriptase
SDKIE	Solvent deuterium kinetic isotope effect
ss	Single stranded
VPg	Virion protein genome linked
WNV	West Nile virus
WT	Wild-type

14.1 Introduction

The RNA-dependent RNA polymerase (RdRp) is responsible for the replication of viruses with RNA genomes. These viruses include some of the best known and most costly to human health and productivity, including rhinovirus (Gavala et al. 2011) and the influenza virus (Barik 2012) that are responsible for the common cold and "flu," respectively; hepatitis C virus (HCV) (Miller et al. 2012) that is a major cause of liver disease; and poliovirus (PV) (Aylward and Tangermann 2011) in which a decade-spanning effort has been directed towards its global eradication. The fast rates of replication, large virus yields, and high genetic diversity of these viruses make therapeutic interventions against these "moving targets" difficult. The catalytic efficiency and fidelity of their RdRps contribute to some extent to all of these factors.

Although RdRps have similar levels of nucleotide incorporation fidelity as the catalytic cores of other nucleic acid polymerases, their lack of a proofreading exonuclease activity leads to a higher mutation rate, approaching two nucleotide changes per RNA genome (Castro et al. 2005). Antiviral therapies may be compromised by these high mutation rates through resistant variants that arise from this vast genetic reservoir. The intrinsic genetic variation among these viruses is also critical for viral pathogenesis; it has been observed that viruses that express RdRp variants that are more faithful are less pathogenic (Pfeiffer and Kirkegaard 2003; Vignuzzi et al. 2006). Genetic diversity within the RNA virus population may allow these viruses to escape bottlenecks and defenses of the host that would otherwise force the virus into extinction.

Antiviral therapies can take advantage of the RdRp's lack of proofreading mechanism. For instance, the most successful drug combination against HCV involves the synthetic nucleotide analogue ribavirin, which becomes directly incorporated into viral RNA. Ribavirin can template equally well with cytidine and uracil, leading to increased mutation rates and the failure of the virus through "lethal mutagenesis" (Cameron and Castro 2001; Graci and Cameron 2002, 2008). Manipulating the fidelity of RdRps may also serve as critical starting points in the rational generation of new vaccine strains (Vignuzzi et al. 2006, 2008). Small animal studies have demonstrated that modification of the RdRp by changing its catalytic activity and/or nucleotide addition fidelity can lead to viral strains incapable of causing disease while providing immune protection to lethal dosages of "wild-type" (WT) virus (Van Slyke et al. 2012; Gnadig et al. 2012; Coffey et al. 2011; Coffey and Vignuzzi 2011; Vignuzzi et al. 2006, 2008; Pfeiffer and Kirkegaard 2003; Weeks et al. 2012). The importance of drug and vaccine developments based on knowledge of RdRps is further underscored by the finding that the virulence and pathogenicity of RNA viruses is directly linked to the function of their respective RdRps; increased virulence for influenza (Zhu et al. 2012; Leung et al. 2010) and West Nile virus (WNV) (Van Slyke et al. 2012) have been correlated to increased RdRp catalytic activity. Continued development of antiviral therapies will no doubt require a deeper understanding of RdRp catalytic function and fidelity.

An emerging theme in RdRp function, and nucleic acid polymerases in general, is the importance of internal protein motions for catalytic activity and regulation. As observed in other enzymes, internal motions may not be random, but rather they may have evolved to direct ligand binding, substrate selection, and chemical catalysis itself (Silva et al. 2011; Schramm 2011; Nashine et al. 2010; Villali and Kern 2010; Boehr et al. 2006a, b, 2009). With RdRps and other nucleic acid polymerases, it has long been recognized that conformational changes before and after nucleotide addition are critical for their function; nucleotide discrimination and phosphodiester bond formation may also depend on faster, more localized fluctuations. Computer simulations and biophysical techniques, such as nuclear magnetic resonance (NMR) spectroscopy, mass spectrometry, and single-molecule fluorescence (e.g., Moustafa et al. 2011; Yang et al. 2010; Seckler et al. 2009; Zhou et al. 2012), are just beginning to reveal the rich dynamic behavior of RdRps and other nucleic acid polymerases. Indeed, these techniques indicate that RdRps may be especially dynamic proteins. Thermodynamic melting data (Thompson et al. 2007) and NMR spectra (Yang and Boehr, unpublished results) are both consistent with PV RdRp behaving similar to a "molten globule," at least in their apo states. Such behavior would be consistent with the view that viral proteins possess unique biophysical characteristics that help them to compensate for high mutation rates and give them the ability to adapt to host defenses (Tokuriki et al. 2009).

14.2 Fundamentals of RdRp Structure and Function

14.2.1 Overall Structural Architecture of RdRps

RdRps are structurally similar to other template-directed nucleic acid polymerases, adopting the typical "cupped right-hand" structure with fingers, thumb, and palm subdomains (Hansen et al. 1997; Ago et al. 1999; Bressanelli et al. 1999; Lesburg et al. 1999; Butcher et al. 2001; Ng et al. 2002, 2008; Tao et al. 2002; Ferrer-Orta et al. 2006a, b, 2007, 2009; Choi et al. 2004, 2006; Yap et al. 2007; Malet et al. 2007; Campagnola et al. 2008; Zamyatkin et al. 2009; Gong and Peersen 2010; Lee et al. 2011; Mosley et al. 2012; Lescar and Canard 2009; McDonald et al. 2009). One major difference between RdRps and other nucleic polymerases is the extension of the fingers through the "fingertips" that helps to more fully enclose the active site (Fig. 14.1a). The full closure of the active site likely restricts large-scale conformational changes between the fingers and thumb subdomains and may have consequences in terms of protein stability (Thompson et al. 2007), their high processivity (Arnold and Cameron 2000; Rodriguez-Wells et al. 2001), and other features of RdRp function.

The fingers and thumb subdomains are primarily responsible for RNA primer and template binding, whereas the palm domain contains residues important for positioning the primer 3′-end, divalent cations, templating, and incoming nucleotide. Recent crystal structures of RdRps have identified channels necessary for template and NTP access and for the egress of newly synthesized RNA (Gong and Peersen 2010; Ferrer-Orta et al. 2007).

Conserved structural motifs A–G are also crucial for polymerase function (Table 14.1). Four of these motifs (A–D) are located in the palm subdomain and are conserved with other nucleic acid polymerases (Poch et al. 1989; O'Reilly and Kao 1998). Metal ion binding is mediated through the absolutely conserved Asp residues in motifs A and C. Other residues within motifs A and B are important for nucleobase and sugar selection (Gohara et al. 2000, 2004; Korneeva and Cameron 2007). Motifs D and F contain residues important for positioning of the nucleoside triphosphate moiety and may be important for stabilization of the pentaphosphate transition state (Castro et al. 2009; Gong and Peersen 2010; Iglesias et al. 2011; Yang et al. 2012).

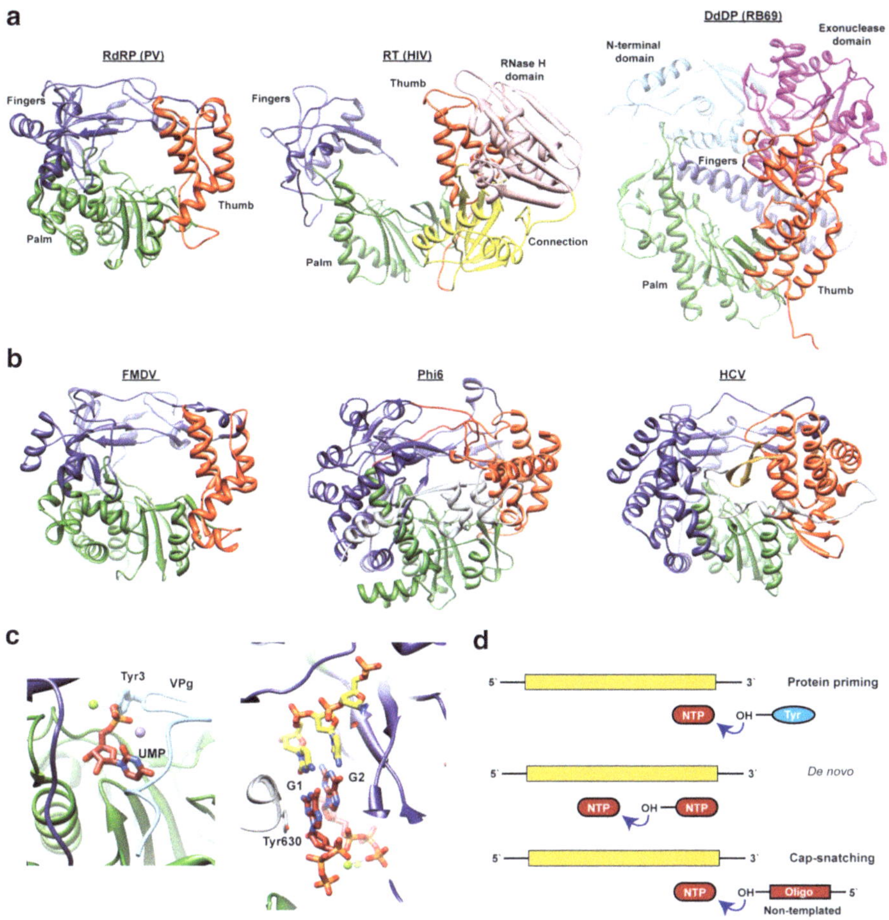

Fig. 14.1 Polymerase structures and initiation mechanisms. (**a**) Similarities and differences between the structures of nucleic acid polymerases. Structures of RdRp from PV (PDB code 1RA6) (Thompson and Peersen 2004), RT from HIV (PDB 1RTD) (Huang et al. 1998), and DdDp from RB69 (PDB 1IG9) (Franklin et al. 2001) are shown and labeled. The three polymerases revealed the canonical right-hand polymerase structure with the fingers (*blue*), palm (*green*), and thumb (*orange red*) subdomains. RdRps possess a unique feature compared to DdDps and RTs in that RdRp adopts an enclosed conformation caused by extensive interactions between the fingers and thumb subdomains, leading to a completely encircled active site. The structure of RB69 DdDp has two additional domains: the N-terminal domain (*cyan*) and the exonuclease domain (*purple*). The structure of HIV RT also has two additional domains: the RNase H domain (*pink*) and the connection domain (*yellow*), which connects the RNase H domain to the thumb. The metal-binding Asp residues are depicted as *sticks* to indicate the active site of the different polymerases. (**b–d**) RdRps from different RNA viruses employ different initiation mechanisms. (**b**) Crystal structures of RdRps from FMDV Picornaviridae (PDB 2F8E) (Ferrer-Orta et al. 2006a), bacteriophage Φ6 Cystoviridae (PDB 1HHS) (Butcher et al. 2001), and HCV Flaviviridae (PDB 1C2P) (Lesburg et al. 1999) are shown. The subdomains conserved among all polymerases are highlighted in different colors: fingers (*blue*), palm (*green*), and thumb (*orange red*). Φ6 RdRp and HCV RdRp have additional characteristic structural elements that are absent from FMDV RdRp. Φ6 RdRp has a C-terminal domain (*gray*) and two loops (*red*) protruding from the thumb interconnecting different parts of the fingers; HCV RdRp has C-terminal residues (*gray*) that

Table 14.1 Conserved sequence/structure motifs and their functional importance

Subdomain	Conserved structural/ functional regions	Sequence[a]	Function
Fingers	G	113 STSAGYPY 120	Template binding[b]
	F	153 PLVT**YVK**D**E**LRSKT**K**VEQ-GKS**R**LIEA 178	NTP and template binding
	I	107 LEALDL 112	Template binding
	II	184 SVAMRMAFGNLYAAFHK 200	Template binding
Palm	A	229 LFAF**D**YTGY**D**AS 240	2′-OH of NTP and metal-B binding
	B	293 TSIF**N**SMINNLIIRTLLLKT 312	Base of NTP and template binding
	C	323 MIAY**GDD**VIAS 333	Primer and metal-A binding and catalysis
	D	338 VDASLLAQSGK-DY**G**LTMTPAD**K**SAT 362	Triphosphate of NTP binding and catalysis
	E	363 FETVTWENVT**F**L**K**RF**F**RA 380	Nascent RNA 3′-end binding
Thumb	III	405 KDPRNTQDHVRSLCLL 420	Nascent RNA duplex binding

[a]The residues shown in double underlined and bold-faced type are completely conserved among RdRps; those residues representing signatures of the conserved sequence motifs are shown in black bold-faced type and single underlined. Residue number corresponds to PV RdRp
[b]The identification of residues' functions was based on the crystal structure of RdRp–RNA–NTP complex of FMDV RdRp (PDB 1WNE)

14.2.2 Structural Differences Among RdRps

RNA viruses can be divided into four basic classes: plus-sense single stranded (ss), minus-sense ss, double stranded (ds), and retroviruses; RdRps from these classes may differ structurally to accommodate the different replication challenges these

Fig. 14.1 (continued) constitute part of the polypeptide linking the polymerase to the membrane anchor and β-loop insertion (*gold*) within the thumb that protrudes into the polymerase active site. (**c**) The initiation complex of FMDV RdRP–VPg–UTP (PDB 2F8E), which employs the protein-priming initiation mechanism (*left panel*), and the initiation complex of Φ6 RdRp–DNA template–GTP (PDB 1HI0) (Butcher et al. 2001), which employs the de novo initiation mechanism (*right panel*), are shown. In the FMDV complex, the VPg primer protein (*cyan*) fits into the polymerase active site projecting Tyr3 (*cyan sticks*) that is covalently linked to the UMP molecule (*sticks* with *dark red* carbon atoms). In the Φ6 complex, two molecules of GTP substrate are bound, G1 and G2, and base paired to the thymine bases of the bound DNA oligomer (model of the preferred RNA template); Tyr630 stacks with the base of G1. The divalent catalytic ions are shown as *spheres*: magnesium (*green*) and manganese (*magenta*). (**d**) Schematic representations of the protein-priming and de novo initiation mechanisms shown in (**c**) and the cap-snatching initiation mechanism proposed for influenza polymerase

viruses face (Fig. 14.1b). Retroviral reverse transcriptases (RTs) are the subject of another chapter and will not be discussed in depth here, although it is important to point out the overall structural similarity between RdRps and RTs (Fig. 14.1a) and to note that mechanistic studies in one class are likely revealing for the other class of RNA-templated polymerases.

In plus-sense ssRNA viruses, the genomic RNA can be used directly as mRNA in infected cells to produce viral proteins. Only small amounts of minus-sense ssRNA are produced to guide the synthesis of multiple copies of plus-sense ssRNA (Uchil and Satchidanandam 2003; Choi 2012). The size of the thumb subdomain varies widely among RdRps from plus-sense ssRNA viruses and generally correlates with the polymerase initiation mechanism. RdRps from Flavivirus (also known as NS5B or NS5), including HCV and WNV, initiate replication de novo with the first two NTPs in sequence and generally have a larger thumb subdomain and/or additional structural elements that help to restrict access to the template-binding channel (Fig. 14.1b), only allowing ssRNA into the active site during de novo initiation (Ago et al. 1999; Bressanelli et al. 1999; Lesburg et al. 1999; Choi et al. 2004; Malet et al. 2007; Yap et al. 2007). These RdRps also have a GTP-binding site adjacent to the active site that helps to position the 3′-OH of the initiation NTP for nucleophilic attack by the second NTP (Choi et al. 2004; Ranjith-Kumar et al. 2002; Kao et al. 1999; Luo et al. 2000). Interestingly, widening the template-binding channel through removal of the C-terminal residues in HCV RdRp allows the use of a dsRNA template–primer complex for primer-dependent initiation (Hong et al. 2001).

In contrast, RdRps from Picornaviruses (also known as $3D^{pol}$), including those from PV (Hansen et al. 1997; Thompson and Peersen 2004; Thompson et al. 2007; Gong and Peersen 2010), foot-and-mouth disease virus (FMDV) (Ferrer-Orta et al. 2006a, 2007), and coxsackievirus (Campagnola et al. 2008), have smaller thumb subdomains, creating a larger template-binding channel that would accommodate the peptide primer necessary for initiation. These RdRps use a virally encoded 22 amino acid peptide called VPg (virion protein genome linked) to initiate replication. The cis-acting replication element (CRE) or ori-I is first used as a template for uridylation of VPg, and then the uridylated peptide primer is transferred to the 3′-end of the viral RNA template (Paul et al. 2000, 2003; Steil and Barton 2009; Ferrer-Orta et al. 2006a, 2009; Choi 2012) (Fig. 14.1c). RdRps from Caliciviruses use a similar protein-based initiation mechanism, but the details of the process are not well understood.

For minus-strand ssRNA viruses, no full crystal structure of any RdRp is currently available. This situation is unfortunate because minus-strand RNA viruses include some of the most prevalent and evasive viruses known, including influenza, measles and mumps viruses, rabies virus, and Ebola virus, and these structures could provide insight into some of the unique characteristics of RNA replication in these viruses. For instance, the functional influenza virus polymerase is a heterotrimer with one subunit containing endonuclease activity that is able to bind cellular mRNA cap and cleave it 10–15 nucleotides downstream from the 5′ end (Boivin et al. 2010). This action provides the RNA primer necessary for initiation through a mechanism known as "cap snatching" (Choi 2012) (Fig. 14.1d). The interactions between the RdRp and endonuclease may be important for coordinating initiation of RNA transcription. A structural description of these interactions may provide new avenues for antiviral therapies.

RdRps from dsRNA viruses, including reovirus and rotavirus (Tao et al. 2002; McDonald et al. 2009) and bacteriophage Φ6 (Butcher et al. 2001; Salgado et al. 2004; Poranen et al. 2008), are structurally similar to those characterized from plus-strand ssRNA viruses. One interesting RdRp variant comes from birnavirus. Here, the palm domain motifs are permuted such that the canonical A-B-C amino- to carboxyl-terminal order is changed to C-A-B (Garriga et al. 2007). Structural rearrangements to accommodate this new order lead to the motif-B loop occluding the active site in the ligand-free enzyme. Binding of the accessory protein VP3 leads to conformational changes in the motif-B loop to open the active site, which then allows RdRp catalysis (Garriga et al. 2007).

Structural differences among RdRps are generally related to differences in RNA synthesis initiation. In contrast, RNA elongation is predicted to occur through the same general mechanisms. As such, the rest of the chapter will focus mainly on PV RdRp, which has established kinetic and catalytic mechanisms, and crystal structures provide snapshots of most of the structural intermediates along its nucleotide addition cycle.

14.2.3 The Kinetic Mechanism of PV RdRp Highlights Fidelity Checkpoints

The first full description of the kinetic mechanism for a RdRp (Arnold and Cameron 2004) was enabled by the use of a symmetrical primer–template RNA substrate ("sym/sub") with a six base pair duplex and four nucleotide 5′ overhangs (Arnold and Cameron 2000). The use of this RNA substrate overcame previous difficulties in assembling stable RdRp–primer/template complexes and enabled single nucleotide incorporation events to be monitored and kinetic parameters, k_{pol} (i.e., the maximal rate constant for nucleotide incorporation) and $K_{d,app}$ (i.e., the apparent dissociation constant for incoming NTP), to be derived. These original studies set the stage for a more mechanistic understanding of RdRp catalysis and fidelity on par with what was known for DNA polymerases. Similar RNA substrates have since been used to characterize other viral RdRps (e.g., Arias et al. 2008; Jin et al. 2011, 2012). Newer RdRp assays have also been devised that are more amenable to high-throughput screening efforts and have the potential to provide new lead compounds for antiviral therapies (Mestas et al. 2007; Gong et al. 2009; Campagnola et al. 2011).

Pre-steady state kinetic experiments have established five key intermediates in the RdRp nucleotide addition cycle (Fig. 14.2a) (Arnold and Cameron 2004). The kinetic mechanism begins with enzyme bound with primer/template RNA

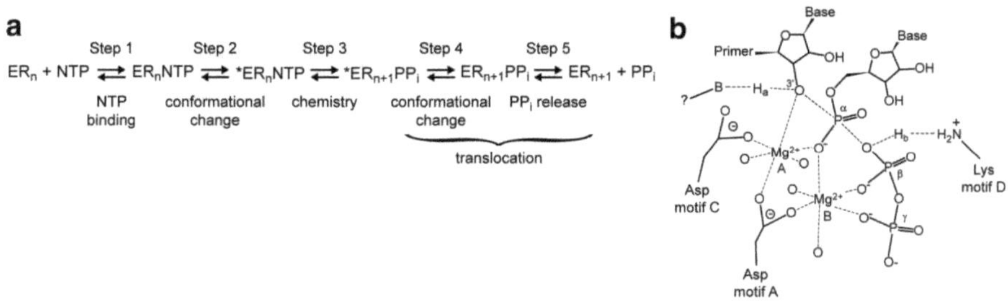

Fig. 14.2 Polymerase-catalyzed nucleotide incorporation. (**a**) Five-step kinetic mechanism for RdRp-catalyzed nucleotide incorporation. (**b**) Chemistry of nucleotidyl transfer. Initially, a nucleoside triphosphate bound by a divalent Mg^{2+} cation (metal B) enters the active site. Metal B is coordinated by the β- and γ-phosphates of the nucleotide, the conserved Asp residue located in structural motif A, and, likely, water molecules (indicated as oxygen ligands to metal B without specific designation). Metal B orients the triphosphate in the active site and contributes to charge neutralization during catalysis. Once the nucleotide is in place, a second divalent Mg^{2+} cation binds (metal A). Metal A is coordinated by the RNA primer 3′-OH, the α-phosphate of the nucleotide, and the Asp residues of structural motifs A and C. Metal A lowers the pK_a of the RNA primer terminus 3′-OH (denoted as H_a), facilitating nucleophilic attack on the nucleotide α-phosphorus atom. As the transition state of nucleotidyl transfer is approached (indicated by *dashed lines*), the primer 3′-hydroxyl proton, H_a, is transferred to a base, the identity of which is unknown, but that is hypothesized to be the conserved Asp residue in motif C. A conserved amino acid residue, a Lys, located in motif D of RdRps and RdDps, or helix O or helix P of DdDps or DdRps, respectively, serves as a general acid catalyst by donating a proton (denoted as H_b) to the pyrophosphate leaving group, which enhances the efficiency of nucleotidyl transfer (Castro et al. 2009)

(E:RNA); incoming NTP then binds to form a ternary complex (E:RNA:NTP) that induces a conformational change into a more active "closed" enzyme conformation (E*:RNA:NTP). Following phosphodiester bond formation (E*:RNA_{n+1}:PP_i), another conformational change occurs (E:RNA_{n+1}:PP_i) to prepare the enzyme for pyrophosphate release (E:RNA_{n+1}). Translocation along the RNA may be coupled to the post-chemistry conformational change and/or pyrophosphate release, before the cycle begins again. These kinetic steps are identical to those found for other nucleic acid polymerases (Joyce and Benkovic 2004).

In PV RdRp, both the pre-chemistry conformational change and the chemistry step itself are partially rate limiting when Mg^{2+} is used as the metal cofactor (Arnold and Cameron 2004). In the presence of Mn^{2+}, the chemistry step is fully rate determining, but the polymerase fidelity is substantially decreased (Arnold et al. 2004), presumably owing to the loss of the pre-chemistry fidelity checkpoint. The contributions of the pre-chemistry and chemistry steps to the overall rate can also be assessed through solvent deuterium kinetic isotope effects (SDKIE), given that the SDKIE can report on proton transfer steps during chemistry (see Fig. 14.2b). For PV RdRp, correct NMP incorporation (in the presence of Mg^{2+}) is associated with a significant SDKIE (~3) (Castro et al. 2007, 2009), consistent with the chemistry step making a substantial contribution to the overall rate. However, incorporation of NMP with incorrect nucleobase is associated with a much reduced SDKIE (~1.3) (Yang et al. 2012), indicating that the chemistry step is making a smaller contribution to the overall rate and suggesting that the pre-chemistry conformational change is becoming more rate limiting. Misincorporation further slows down the pre-chemistry conformational change for subsequent nucleotide addition as evidenced by the ~10-fold decrease in k_{pol} and the complete loss of the SDKIE (~1.0) (Yang et al. 2012). Misincorporation is also known to affect subsequent nucleotide addition in DNA polymerases (Kunkel and Bebenek 2000; Johnson and Beese 2004). These studies serve to highlight the importance of the pre-chemistry conformational change as a fidelity checkpoint.

Triggering the conformational changes necessary for catalysis likely depends on recognition of both the correct nucleobase and correct sugar (Castro et al. 2005). Disruption of a key hydrogen bond in PV RdRp between the 2′-OH of the incoming ribose and the carboxamide functional group of Asn297 through mutagenesis not only decreases sugar discrimination but also leads to ~30-fold decrease in k_{pol} for correct NMP incorporation (Gohara et al. 2004). Recognition of the 3′-OH is also critical; disruption of the interactions between Asp238 and the 3′-OH results in >2,000-fold decrease in k_{pol} (Gohara et al. 2004).

14.2.4 Crystal Structures Provide Molecular Snapshots of the Conformational Changes Along the Nucleotide Addition Cycle

RdRps catalyze phosphodiester bond formation through the two-metal ion mechanism, as previously suggested for other nucleic acid polymerases (Fig. 14.2b). Crystal structures of FMDV, NV, and PV RdRps have begun to reveal the structural changes that accompany NMP incorporation (Ferrer-Orta et al. 2007; Zamyatkin et al. 2009; Gong and Peersen 2010). The Peersen lab has been able to monitor structural changes to PV RdRp before and after NMP incorporation (Gong and Peersen 2010). Nucleotide addition resulted in local conformational changes in the palm region involving motifs A–C (Fig. 14.3a). Changes included a realignment of the motif-A backbone to allow for the formation of a complete three-stranded β-sheet with motif C and a repositioning of Asp233 in motif A to allow it to coordinate both Mg^{2+} metals in the active site. Upon NTP binding, there is also a repositioning of the side chain of Asp238 that induces a conformational change in Ser288, such that both Asp238 and Ser288 can form H-bonds to the ribose hydroxyls. Importantly, no significant changes in these active-site residues were observed in the presence of the corresponding 2′-dNTPs, suggesting a mechanism for correct sugar selection. Similar structural rearrangements have been observed for FMDV and NV RdRps (Ferrer-Orta et al. 2007; Zamyatkin et al. 2009).

One key question to address is what structural rearrangement(s) are important for the fidelity checkpoints highlighted by the kinetic experiments. It is unlikely that the pre-chemistry conformational change is reporting on a global repositioning of the fingers and thumb subdomains, considering that the fingertip interactions with the

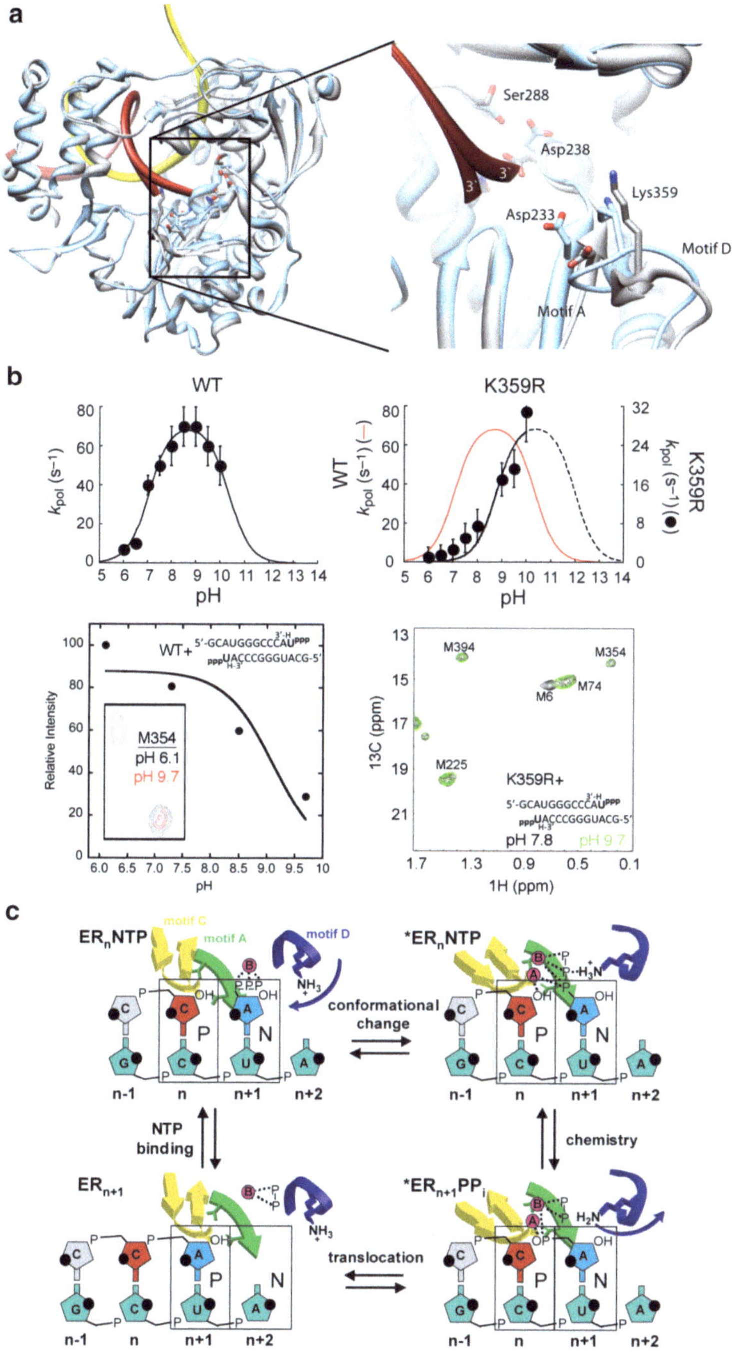

Fig. 14.3 Conformational changes of motif D and its functional role. (**a**) Conformational changes of RdRp revealed by X-ray crystallography. (*Left panel*) The PV RdRp binary (PDB 3OL6, *gray*) and ternary (PDB 3OL7, *cyan*) complexes are shown (Gong and Peersen 2010); the template–primer RNA are depicted as *yellow* and *dark red ribbons*, respectively. Subtle differences were observed between

thumb in RdRps would restrict such motions. Maintenance of the fingertip–thumb interactions in PV RdRp are also important for the protein's thermodynamic stability (Thompson et al. 2007), and even in molecular dynamics (MD) simulations, the interactions are kept intact while other regions undergo dynamic fluctuations (Moustafa et al. 2011).

The pre-chemistry conformational change is likely related to more subtle, local rearrangements in the palm subdomain, including the structural rearrangements in palm motifs A and C to bind magnesium and/or a reorientation of the triphosphate moiety of the incoming NTP into its catalytically active conformation; structural rearrangements within motifs D and F to reposition critical Lys and Arg residues may be important for proper alignment of the triphosphate for nucleophilic attack by the primer 3′-OH.

14.3 Newer Aspects in RdRp Structure and Function

14.3.1 The Catalytic Mechanism of RdRps Include General Acid Catalysis

One often overlooked aspect of the two-metal ion mechanism for DNA/RNA polymerases is the importance of proton transfer events that de-protonate the primer's 3′-OH to create a better nucleophile and protonation of the β-phosphate

Fig. 14.3 (continued) the binary and ternary complex structures; the protein chains in the two complex structures superimpose with an RMSD value of 0.53 Å. The key residues that revealed conformational changes in the ternary relative to the binary complex, resulting in closure of the polymerase active site, are shown as *sticks*. (*Right panel*) Zoom-in view of the region that showed conformational changes, enclosed by a *rectangle* in the *left panel*. The observed conformational changes involved motifs A and D, and repositioning of the side chains of Asp233, Asp238, Ser288, and Lys359, which are known to be important for the polymerase function. (**b**) Comparison of pH curves for WT and Lys359Arg PV RdRp from kinetics and NMR. Kinetics (*top panel*): pH rate profiles for WT and Lys359Arg PV RdRp obtained from evaluating the kinetics of nucleotide incorporation. WT PV RdRp yields two pK_a values of 7.0 ± 0.1 and 10.5 ± 0.1 (Castro et al. 2009). Lys359Arg RdRp yields a pK_a value of 8.8 ± 0.3 (*solid line*) with a second predicted pK_a value of 12.0 (*dashed line*). The *red line* shows the curve from WT PV RdRP for comparison. NMR (*bottom panel*): the open to closed transition of the WT PV RdRp–RNA-ternary complex is associated with a pK_a of 9.2, as obtained from the relative peak intensity of the Met354 resonance in the closed complex position. Met354 resonances for the RdRp–RNA-ternary complex at pH 6.1 (*black*) and 9.7 (*red*). Formation of the closed complex at high pH is restored by substitution of Lys359 with Arg, consistent with the higher pK_a of Arg (Yang et al. 2012). Met resonances for Lys359Arg RdRp–RNA–UTP ternary complexes at pH 7.8 are in *black* and at 9.7 are in *green*. The Met354 resonance remains unchanged between different pH regimes and is consistent with formation of the closed complex. (**c**) The stages of RNA synthesis and model for the role of motif D. During the conformational-change step, thought to be reorientation of the triphosphate for catalysis, Lys359 on motif D moves into a position both to serve as a proton donor and contribute to charge neutralization that develops during the transition state. After chemistry, loss of the proton on Lys359 triggers translocation to occur that facilitates PP_i release and reprotonation of Lys359 that would create a better leaving group. The pH rate profile for PV RdRp is consistent with acid/base catalysis (Fig. 14.3b), and proton inventory studies indicate that there are two proton transfer events in the rate-limiting step(s) (Castro et al. 2007, 2009). The identity of the general base is currently unknown, but recent studies indicate that the conserved motif-D Lys (Lys359 in PV RdRp) serves as the general acid. This residue is nearby the incoming NTP triphosphate, and substitutions at this position lead to changes in the SDKIE, proton inventory, and pH rate profile, all consistent with this Lys acting as a general acid (Fig 14.3b) (Castro et al. 2009).

These functional studies have been somewhat surprising given that in crystal structures of PV RdRp, the ε-amino group of Lys359 is out of reach (>5 Å away) of the β-phosphate (Gong and Peersen 2010). Nonetheless, additional structural rearrangements in motif D, such as those observed for NV RdRp, could reposition Lys359 within proper contact distance away, similar to the arrangement observed in telomerase (Gillis et al. 2008). Molecular dynamics (MD) simulations of HIV RT also demonstrate conformational fluctuations that would bring its motif-D Lys (i.e., Lys220) into proper catalytic register (Michielssens et al. 2011). Solution-state NMR experiments suggest that there are additional structural rearrangements in motif D, not observed in the currently available crystal structures (Yang et al. 2012). In bacteriophage Φ6 RdRp, NMR studies have demonstrated that Ile488 in motif D is undergoing conformational exchange on the same timescale as catalysis (Ren et al. 2010).

In PV RdRp, conformational rearrangements in motif D can be tracked by monitoring the chemical shift and relaxation behavior of the NMR peak for the nearby Met354, also located in motif D (Yang et al. 2012). The Met354 resonance shows a large, pH-dependent chemical shift change upon binding correct NTP to the RdRp/RNA complex (Yang et al. 2012). The pK_a associated with this conformational change closely matches the pK_a derived from the pH rate profile and assigned to Lys359 (Fig. 14.3b). An Arg substitution at this position leads to increases to both the kinetically and NMR-derived pK_as (Fig. 14.3b), indicating that the general acid Lys359 is intimately involved in the structural rearrangements in motif D that, in turn, help to position this residue for catalysis. The NMR studies also indicate that a similar conformational change does not occur when incorrect NTP binds, consistent with this step being kinetically slower for NMP misincorporation as demonstrated by slower overall rates and a decrease in the SDKIE mentioned above (Yang et al. 2012). De-protonation of the motif-D Lys and the loss of the electrostatic interaction with the pyrophosphate leaving group may trigger the post-chemistry conformational change, pyrophosphate release, and RNA translocation (see model in Fig. 14.3c).

General acid catalysis through a conserved Lys residue appears to be a common mechanism among DNA/RNA polymerases (Castro et al. 2009). In the case of A and B family DNA polymerases, the Lys is present on the O- and P-helices, respectively. These α-helices are repositioned only upon binding the correct NTP, and this serves as an important fidelity mechanism (Kiefer et al. 1998; Li et al. 1998; Tahirov et al. 2002; Temiakov et al. 2004; Yin and Steitz 2002); a similar helix is absent in RdRps and RTs. In an analogous manner to what has been observed with Lys359 in PV RdRp, Lys758 in the O-helix in *Escherichia coli* DNA polymerase I has been proposed to be involved in conformational changes in the ternary complexes before and after nucleotide addition (Kaushik et al. 1996). These studies indicate that the motif-D loop in RdRps and RTs plays an analogous role to the O/P "fidelity" helix in the DNA polymerases (Yang et al. 2012).

14.3.2 Polymerase Fidelity Is a Critical Determinant of Viral Pathogenesis

Alteration of the motif-D Lys may be a universal vaccine strategy for RNA viruses; a PV strain containing an Arg at this position was attenuated but provided immune protection to mice, along the same order of magnitude as the clinically used Sabin I vaccine strain (Weeks et al. 2012). The PV variant was delayed in growth and replicated with high fidelity, consistent with the in vitro kinetics of the Lys359Arg RdRp. As a side note, the Sabin I vaccine strain itself contains a mutation in motif D (Thr362Ile) that could also impact the function of RdRp and contribute to viral attenuation (Nomoto et al. 1982; Georgescu et al. 1995).

Previous studies had established RdRp fidelity as a determinant of viral pathogenesis and virulence. Serial passages of PV against the antiviral compound ribavirin led to the selection of a resistant variant of PV with a single amino acid change to the RdRp (i.e., Gly64 to Ser) (Pfeiffer and Kirkegaard 2003; Vignuzzi et al. 2006; Arnold et al. 2005). This variant also led to higher RdRp fidelity and decreased pathogenesis in mice (Arnold et al. 2005; Vignuzzi et al. 2006; Pfeiffer and Kirkegaard 2003). Other substitutions have since been discovered that also impact RdRp function and contribute to viral attenuation (Gnadig et al. 2012). Some of these changes have an opposite effect on polymerase fidelity compared to the Gly64Ser and Lys359 variants, i.e., they decrease polymerase fidelity, yet still decrease pathogenicity and virulence (Gnadig et al. 2012). Together, these studies suggest that there is a small window of genetic variation that is optimum for virus biology and which is finely tuned by RdRp fidelity (Fig. 14.4).

14.3.3 RdRp Dynamics: The Next Frontier in Understanding Polymerase Catalytic Function and Fidelity

Perhaps surprisingly, the Gly64 to Ser substitution that leads to ribavirin resistance in PV is not located in the active site and would not be predicted to make direct interactions with ribavirin. Instead, Gly64 is involved in a hydrogen-bonding network that includes Gly1 in the N-terminal β-strand, and Ala239 and Leu241 in motif A (Fig. 14.5a). It was originally predicted that the Gly64Ser amino acid

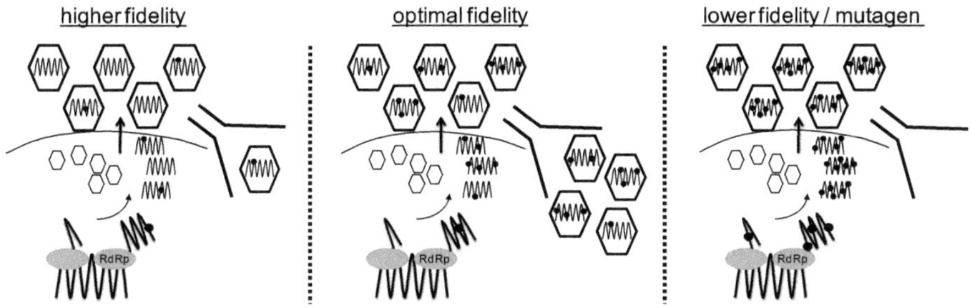

Fig. 14.4 Quasispecies nature of RNA viruses and effect of RdRp fidelity on virus biology. During viral replication, the number of mutations per genome is optimized and determined by the intrinsic error frequency, or fidelity, of the virus-encoded RdRp. A higher-fidelity RdRp leads to a decreased number of mutations per genome, limited diversity, bottleneck restriction, and lower viral fitness. A lower-fidelity RdRp or alternatively treatment with a viral mutagen, such as ribavirin, leads to an increased number of mutations per genome; this can exceed the tolerable threshold of mutations and reduce viral fitness

substitution would result in structural changes to the N-terminal β-strand and/or motif A that would change RdRp function and fidelity (Arnold et al. 2005). However, comparison of the X-ray crystal structures of WT and Gly64Ser RdRp did not reveal any substantial structural changes in these regions (Fig. 14.5b), although the modification to Ser enables the formation of additional hydrogen bond interactions with Glu2 (Marcotte et al. 2007). These results prompted the suggestion that static structures derived by X-ray crystallography were, by themselves, insufficient to explain the altered RdRp fidelity induced by the Gly64Ser substitution. Instead, changes to the dynamic excursions from the ground-state structure generated by the mutation may be responsible (Cameron et al. 2009). Consistent with this view, MD simulations have suggested that the Gly64Ser substitution leads to widespread ps–ns timescale dynamics in RdRp (Fig. 14.5c) (Moustafa et al. 2011). The observed dynamic changes with Gly64Ser RdRp appeared to be more substantial than dynamic differences between evolutionary distinct WT RdRps (Moustafa et al. 2011). Regions with altered flexibility were all associated with known functional regions of the enzyme, including motifs A, D, and F, and regions associated with the NTP- and RNA-binding channels (Fig. 14.5d). These results were consistent with dynamic NMR experiments that also showed that the Gly64Ser substitution induces global changes to the structure/dynamics of the enzyme (Yang et al. 2010).

The MD simulations may also be able to explain why other active-site remote mutations lead to changes to RdRp function; regions of flexibility were identified that had no previously known functional relevance to phosphodiester bond formation (Moustafa et al. 2011) but contained residues known to contribute to viral fitness (Diamond and Kirkegaard 1994) (Fig. 14.5e, f). One of these residues is Asp53, and it is worth noting that the Asp53Asn substitution is encoded by the Sabin I vaccine strain (Nomoto et al. 1982); the other RdRp substitutions are Tyr73His, Lys250Glu, and Thr362Ile. Tyr73 is next to Met74 that changes its chemical shift in response to RNA and NTP binding (Yang et al. 2010, 2012).

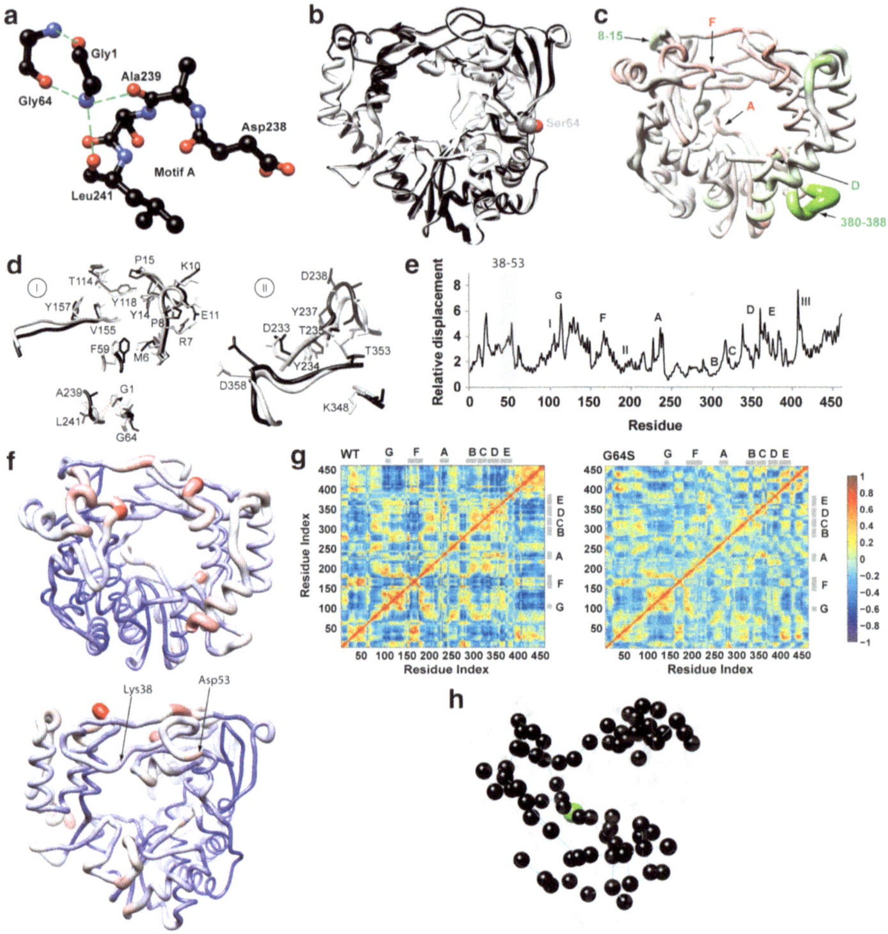

Fig. 14.5 Role of dynamics in RdRp function. (**a–d**) The high-fidelity derivative Gly64Ser revealed WT structure, but MD simulations showed global changes in dynamics of Gly64Ser compared to the WT enzyme. (**a**) Gly64 in PV RdRp (PDB 1RA6) plays a role in positioning the N-terminal Gly1 via H-bonding between the backbone atoms; Gly1 in turn stabilizes the motif-A conformation by forming H-bonding interactions with the backbone atoms of Ala239 and Leu241. Substitution of Gly64 by Ser was expected to alter motif-A conformation. (**b**) The crystal structure of Gly64Ser RdRp (PDB 2IJF, *gray*) (Marcotte et al. 2007) has been shown to be almost identical to that of the WT enzyme (*black*), RMSD value of 0.3 Å. (**c**) The differences between relative flexibility of Gly64Ser and WT RdRp obtained from MD simulations are mapped onto the WT structure. The differences are shown as color gradients of *green* (regions that are more flexible in Gly64Ser relative to WT RdRp) and *red* (regions that are less flexible in Gly64Ser relative to WT RdRp); the structure is rendered as a tube whose radius corresponds to the magnitude of flexibility differences between the two enzymes. The labeled *green* regions (motif D, residues 8–15, and residues 380–388) are much more flexible in Gly64Ser compared to WT RdRp, whereas the labeled *red* regions (motif A and motif F) are much less flexible in Gly64Ser compared to WT RdRp. (**d**) The average structures calculated from MD simulations for WT (*black*) and Gly64Ser (*gray*) RdRp are superimposed and regions that revealed structural perturbations are shown. Changes from position 64 to the N-terminal residues (8–15) and motif F are shown in **d-I**; changes in motif A and motif D are shown in **d-II**. Note that the hydrogen-bonding network involving residue at position 64 and the residues Gly1, Ala239, and Leu241 remain in both WT and

Lys250 is next to Met251, and NMR studies indicate that the Met251Ile substitution can have long-range structural and/or dynamic effects (Yang et al. 2010). Thus, it is tempting to speculate that the Sabin I substitutions may alter the structural dynamics of RdRp to impact catalytic function and fidelity.

Analysis of the MD simulations gives further insight into the functioning of the RdRp and, by extension, other nucleic acid polymerases. One important dynamic mode in RdRps involves the motions of the fingers and thumb subdomains relative to the palm subdomain such that the nascent RNA duplex channel expands and contracts (Moustafa et al. 2011). These motions appear to be "anticorrelated" to the opening/closing of the NTP-binding channel; when the RNA channel is open, the NTP channel is closed and vice versa. These coordinated motions may be important for binding of NTP, release of pyrophosphate, and/or translocation of the RNA.

14.3.4 RdRps Are "Small World" Networks of Amino Acid Residues

The MD simulations further suggest that there are long-range amino acid interaction networks important for RdRp function as demonstrated by the correlation of motions throughout the RdRp structure (Fig. 14.5g) (Moustafa et al. 2011).

Fig. 14.5 (continued) Gly64Ser RdRp. (**e**, **f**) Dynamics can explain biological effects caused by other remote-site mutations. (**e**) MD simulations revealed that functional motifs are largely contributing to the major motions observed during the course of simulations. The relative displacements of Cα atoms derived from principal component analysis (PCA) of the MD trajectory are plotted for each residue. Peaks correspond to regions that have large contributions to the major motions observed in the simulations. The region comprising residues 38–53 are highlighted by *gray shade*; mutations in this region have been shown to affect replication of PV. (**f**) Two views of PV RdRp structure (PDB 1RA6) are shown: (*top panel*) looking through the template–nascent RNA duplex channel and (*bottom panel*) looking through the NTP channel. The structure is rendered as a tube to indicate flexibility; the tube is colored in *blue-to-red* gradients with varying radii corresponding to the magnitude of the relative displacements shown in (**e**). The largest radius corresponds to the most flexible region, the smallest radius corresponds to the least flexible region. Positions of Lys38 and Asp53 are indicated in the *bottom view*. (**g**, **h**) Long-range network in RdRp revealed by MD simulations. (**g**) The color-coded dynamic cross correlation map (or DCCM), which measures the correlation between the displacements of Cα atoms, calculated from MD simulations of WT (*left panel*) and Gly64Ser (*right panel*) PV RdRp are shown. For each enzyme, the calculated matrix whose elements are the pairwise correlation scores between its residues is visualized as a colored map. The correlation scores are encoded with a color gradient from −1 (*blue*, completely anticorrelated) to +1 (*red*, completely correlated). Correlations between conserved structural motifs (*marked* with *gray bars*) can be observed in the maps. Interestingly, the single amino acid substitution in PV RdRp resulted in noticeable differences between the calculated DCCM maps of WT and Gly64Ser. (**h**) Residues shown to be correlated in the calculated DCCM are depicted as *black spheres* in the PV RdRp structure, shown as *transparent ribbon*. Of note, Gly64 (*green sphere*) falls along the inferred long-range network of polymerase residues

Additional evidence for these long-range interactions can be derived from NMR and mutational studies. In bacteriophage Φ6 RdRp, residues in motifs C, D, and E, along with residues in the outer edge of the template-binding channel, experience correlated dynamics on the μs–ms timescale as evidenced by their similar kinetic exchange parameters from NMR relaxation experiments (Ren and Ghose 2011; Ren et al. 2010). For PV RdRp, Met6 in the N-terminal β-strand and Met187 on a fingers subdomain helix are also conformationally coupled. These Met side chains occupy two conformations on the slow NMR chemical shift timescale, as evidenced by the presence of two resonances under some circumstances, and amino acid substitutions remote from either site have a similar effect on the dynamic equilibrium between these two conformations (Yang et al. 2010). This finding is intriguing considering that conformational changes in the N-terminal β-strand, mediated by Trp5, have been proposed to be involved in maintaining an elongation-competent complex (Campagnola et al. 2008; Hobdey et al. 2010).

RdRps, and other nucleic acid polymerases, can thus be viewed as having highly connected amino acid networks that provide communication pathways from the surface of the enzyme all the way to the catalytic center (Fig. 14.5h). Disruption of these networks through amino acid substitutions, as observed for Gly64Ser RdRp, or small molecule binding can have substantial impact on RdRp function. For the latter aspect, MD simulations for HCV RdRp have shown that inhibitor binding to an allosteric binding pocket influences internal fluctuations throughout the protein structure (Davis and Thorpe 2012). These networks also provide avenues for biological regulation; binding of accessory proteins in larger replication complexes and/or association of these replication complexes with the lipid membrane (Hsu et al. 2010) may alter RdRp structure, dynamics, and function. The evolution of the structure–function paradigm into a structure–dynamics–function paradigm may be the final frontier to a comprehensive understanding of nucleic acid polymerase function. Such studies have the potential to reveal new binding modes for RdRp inhibitors and give insight into how RdRp fidelity can be tuned through active-site remote mutations that may serve as the starting point for rational vaccine design.

Acknowledgments Research from the authors' labs is supported in part by a grant (AI45818) from the US National Institutes of Health to CEC and start-up funds from Pennsylvania State University and a US National Science Foundation Career Award (MCB1053993) to DDB.

References

Ago H, Adachi T, Yoshida A, Yamamoto M, Habuka N, Yatsunami K, Miyano M (1999) Crystal structure of the RNA-dependent RNA polymerase of hepatitis C virus. Structure 7(11): 1417–1426. doi:10.1016/S0969-2126(00)80031-3, st7b17 [pii]

Arias A, Arnold JJ, Sierra M, Smidansky ED, Domingo E, Cameron CE (2008) Determinants of RNA-dependent RNA polymerase (in)fidelity revealed by kinetic analysis of the polymerase encoded by a foot-and-mouth disease virus mutant with reduced sensitivity to ribavirin. J Virol 82(24):12346–12355. doi:10.1128/JVI.01297-08, JVI.01297-08 [pii]

Arnold JJ, Cameron CE (2000) Poliovirus RNA-dependent RNA polymerase (3D(pol)). Assembly of stable, elongation-competent complexes by using a symmetrical primer-template substrate (sym/sub). J Biol Chem 275(8):5329–5336

Arnold JJ, Cameron CE (2004) Poliovirus RNA-dependent RNA polymerase (3Dpol): pre-steady-state kinetic analysis of ribonucleotide incorporation in the presence of Mg2+. Biochemistry 43 (18):5126–5137. doi:10.1021/bi035212y

Arnold JJ, Gohara DW, Cameron CE (2004) Poliovirus RNA-dependent RNA polymerase (3Dpol): pre-steady-state kinetic analysis of ribonucleotide incorporation in the presence of Mn2+. Biochemistry 43(18):5138–5148. doi:10.1021/bi035213q

Arnold JJ, Vignuzzi M, Stone JK, Andino R, Cameron CE (2005) Remote site control of an active site fidelity checkpoint in a viral RNA-dependent RNA polymerase. J Biol Chem 280(27): 25706–25716. doi:10.1074/jbc.M503444200, M503444200 [pii]

Aylward B, Tangermann R (2011) The global polio eradication initiative: lessons learned and prospects for success. Vaccine 29(Suppl 4):D80–D85. doi:10.1016/j.vaccine.2011.10.005, S0264-410X(11)01599-4 [pii]

Barik S (2012) New treatments for influenza. BMC Med 10(1):104. doi:10.1186/1741-7015-10-104, 1741-7015-10-104 [pii]

Boehr DD, Dyson HJ, Wright PE (2006a) An NMR perspective on enzyme dynamics. Chem Rev 106(8):3055–3079. doi:10.1021/cr050312q

Boehr DD, McElheny D, Dyson HJ, Wright PE (2006b) The dynamic energy landscape of dihydrofolate reductase catalysis. Science 313(5793):1638–1642. doi:10.1126/science.1130258, 313/5793/1638 [pii]

Boehr DD, Nussinov R, Wright PE (2009) The role of dynamic conformational ensembles in biomolecular recognition. Nat Chem Biol 5(11):789–796. doi:10.1038/nchembio.232, nchembio.232 [pii]

Boivin S, Cusack S, Ruigrok RW, Hart DJ (2010) Influenza A virus polymerase: structural insights into replication and host adaptation mechanisms. J Biol Chem 285(37):28411–28417. doi:10.1074/jbc.R110.117531, R110.117531 [pii]

Bressanelli S, Tomei L, Roussel A, Incitti I, Vitale RL, Mathieu M, De Francesco R, Rey FA (1999) Crystal structure of the RNA-dependent RNA polymerase of hepatitis C virus. Proc Natl Acad Sci USA 96(23):13034–13039

Butcher SJ, Grimes JM, Makeyev EV, Bamford DH, Stuart DI (2001) A mechanism for initiating RNA-dependent RNA polymerization. Nature 410(6825):235–240. doi:10.1038/35065653, 35065653 [pii]

Cameron CE, Castro C (2001) The mechanism of action of ribavirin: lethal mutagenesis of RNA virus genomes mediated by the viral RNA-dependent RNA polymerase. Curr Opin Infect Dis 14(6):757–764

Cameron CE, Moustafa IM, Arnold JJ (2009) Dynamics: the missing link between structure and function of the viral RNA-dependent RNA polymerase? Curr Opin Struct Biol 19(6):768–774. doi:10.1016/j.sbi.2009.10.012, S0959-440X(09)00165-1 [pii]

Campagnola G, Weygandt M, Scoggin K, Peersen O (2008) Crystal structure of coxsackievirus B3 3Dpol highlights the functional importance of residue 5 in picornavirus polymerases. J Virol 82 (19):9458–9464. doi:10.1128/JVI.00647-08, JVI.00647-08 [pii]

Campagnola G, Gong P, Peersen OB (2011) High-throughput screening identification of poliovirus RNA-dependent RNA polymerase inhibitors. Antiviral Res 91(3):241–251. doi:10.1016/j.antiviral.2011.06.006, S0166-3542(11)00348-2 [pii]

Castro C, Arnold JJ, Cameron CE (2005) Incorporation fidelity of the viral RNA-dependent RNA polymerase: a kinetic, thermodynamic and structural perspective. Virus Res 107(2):141–149. doi:10.1016/j.virusres.2004.11.004, S0168-1702(04)00392-2 [pii]

Castro C, Smidansky E, Maksimchuk KR, Arnold JJ, Korneeva VS, Gotte M, Konigsberg W, Cameron CE (2007) Two proton transfers in the transition state for nucleotidyl transfer catalyzed by RNA- and DNA-dependent RNA and DNA polymerases. Proc Natl Acad Sci USA 104(11):4267–4272. doi:10.1073/pnas.0608952104, 0608952104 [pii]

Castro C, Smidansky ED, Arnold JJ, Maksimchuk KR, Moustafa I, Uchida A, Gotte M, Konigsberg W, Cameron CE (2009) Nucleic acid polymerases use a general acid for nucleotidyl transfer. Nat Struct Mol Biol 16(2):212–218. doi:10.1038/nsmb.1540, nsmb.1540 [pii]

Choi KH (2012) Viral polymerases. Adv Exp Med Biol 726:267–304. doi:10.1007/978-1-4614-0980-9_12

Choi KH, Groarke JM, Young DC, Kuhn RJ, Smith JL, Pevear DC, Rossmann MG (2004) The structure of the RNA-dependent RNA polymerase from bovine viral diarrhea virus establishes the role of GTP in de novo initiation. Proc Natl Acad Sci USA 101(13):4425–4430. doi:10.1073/pnas.0400660101, 0400660101 [pii]

Choi KH, Gallei A, Becher P, Rossmann MG (2006) The structure of bovine viral diarrhea virus RNA-dependent RNA polymerase and its amino-terminal domain. Structure 14(7):1107–1113. doi:10.1016/j.str.2006.05.020, S0969-2126(06)00259-0 [pii]

Coffey LL, Vignuzzi M (2011) Host alternation of chikungunya virus increases fitness while restricting population diversity and adaptability to novel selective pressures. J Virol 85 (2):1025–1035. doi:10.1128/JVI.01918-10, JVI.01918-10 [pii]

Coffey LL, Beeharry Y, Borderia AV, Blanc H, Vignuzzi M (2011) Arbovirus high fidelity variant loses fitness in mosquitoes and mice. Proc Natl Acad Sci USA 108(38):16038–16043. doi:10.1073/pnas.1111650108, 1111650108 [pii]

Davis BC, Thorpe IF (2012) Thumb inhibitor binding eliminates functionally important dynamics in the hepatitis C virus RNA polymerase. Proteins 81:40–52. doi:10.1002/prot.24154

Diamond SE, Kirkegaard K (1994) Clustered charged-to-alanine mutagenesis of poliovirus RNA-dependent RNA polymerase yields multiple temperature-sensitive mutants defective in RNA synthesis. J Virol 68(2):863–876

Ferrer-Orta C, Arias A, Agudo R, Perez-Luque R, Escarmis C, Domingo E, Verdaguer N (2006a) The structure of a protein primer-polymerase complex in the initiation of genome replication. EMBO J 25(4):880–888. doi:10.1038/sj.emboj.7600971, 7600971 [pii]

Ferrer-Orta C, Arias A, Escarmis C, Verdaguer N (2006b) A comparison of viral RNA-dependent RNA polymerases. Curr Opin Struct Biol 16(1):27–34. doi:10.1016/j.sbi.2005.12.002, S0959-440X(05)00221-6 [pii]

Ferrer-Orta C, Arias A, Perez-Luque R, Escarmis C, Domingo E, Verdaguer N (2007) Sequential structures provide insights into the fidelity of RNA replication. Proc Natl Acad Sci USA 104 (22):9463–9468. doi:10.1073/pnas.0700518104, 0700518104 [pii]

Ferrer-Orta C, Agudo R, Domingo E, Verdaguer N (2009) Structural insights into replication initiation and elongation processes by the FMDV RNA-dependent RNA polymerase. Curr Opin Struct Biol 19(6):752–758. doi:10.1016/j.sbi.2009.10.016, S0959-440X(09)00169-9 [pii]

Franklin MC, Wang J, Steitz TA (2001) Structure of the replicating complex of a pol alpha family DNA polymerase. Cell 105(5):657–667. doi:10.1016/S0092-8674(01)00367-1, S0092-8674(01)00367-1 [pii]

Garriga D, Navarro A, Querol-Audi J, Abaitua F, Rodriguez JF, Verdaguer N (2007) Activation mechanism of a noncanonical RNA-dependent RNA polymerase. Proc Natl Acad Sci USA 104 (51):20540–20545. doi:10.1073/pnas.0704447104, 0704447104 [pii]

Gavala ML, Bertics PJ, Gern JE (2011) Rhinoviruses, allergic inflammation, and asthma. Immunol Rev 242(1):69–90. doi:10.1111/j.1600-065X.2011.01031.x

Georgescu MM, Tardy-Panit M, Guillot S, Crainic R, Delpeyroux F (1995) Mapping of mutations contributing to the temperature sensitivity of the Sabin 1 vaccine strain of poliovirus. J Virol 69 (9):5278–5286

Gillis AJ, Schuller AP, Skordalakes E (2008) Structure of the Tribolium castaneum telomerase catalytic subunit TERT. Nature 455(7213):633–637. doi:10.1038/nature07283, nature07283 [pii]

Gnadig NF, Beaucourt S, Campagnola G, Borderia AV, Sanz-Ramos M, Gong P, Blanc H, Peersen OB, Vignuzzi M (2012) Coxsackievirus B3 mutator strains are attenuated in vivo. Proc Natl Acad Sci USA 109(34):E2294–E2303. doi:10.1073/pnas.1204022109, 1204022109 [pii]

Gohara DW, Crotty S, Arnold JJ, Yoder JD, Andino R, Cameron CE (2000) Poliovirus RNA-dependent RNA polymerase (3Dpol): structural, biochemical, and biological analysis of conserved structural motifs A and B. J Biol Chem 275(33):25523–25532. doi:10.1074/jbc.M002671200, M002671200 [pii]

Gohara DW, Arnold JJ, Cameron CE (2004) Poliovirus RNA-dependent RNA polymerase (3Dpol): kinetic, thermodynamic, and structural analysis of ribonucleotide selection. Biochemistry 43(18):5149–5158. doi:10.1021/bi035429s

Gong P, Peersen OB (2010) Structural basis for active site closure by the poliovirus RNA-dependent RNA polymerase. Proc Natl Acad Sci USA 107(52):22505–22510. doi:10.1073/pnas.1007626107, 1007626107 [pii]

Gong P, Campagnola G, Peersen OB (2009) A quantitative stopped-flow fluorescence assay for measuring polymerase elongation rates. Anal Biochem 391(1):45–55. doi:10.1016/j.ab.2009.04.035, S0003-2697(09)00285-1 [pii]

Graci JD, Cameron CE (2002) Quasispecies, error catastrophe, and the antiviral activity of ribavirin. Virology 298(2):175–180. doi:10.1006/viro.2002.1487, S004268220291487X [pii]

Graci JD, Cameron CE (2008) Therapeutically targeting RNA viruses via lethal mutagenesis. Future Virol 3(6):553–566. doi:10.2217/17460794.3.6.553

Hansen JL, Long AM, Schultz SC (1997) Structure of the RNA-dependent RNA polymerase of poliovirus. Structure 5(8):1109–1122

Hobdey SE, Kempf BJ, Steil BP, Barton DJ, Peersen OB (2010) Poliovirus polymerase residue 5 plays a critical role in elongation complex stability. J Virol 84(16):8072–8084. doi:10.1128/JVI.02147-09, JVI.02147-09 [pii]

Hong Z, Cameron CE, Walker MP, Castro C, Yao N, Lau JY, Zhong W (2001) A novel mechanism to ensure terminal initiation by hepatitis C virus NS5B polymerase. Virology 285(1):6–11. doi:10.1006/viro.2001.0948, S0042-6822(01)90948-1 [pii]

Hsu NY, Ilnytska O, Belov G, Santiana M, Chen YH, Takvorian PM, Pau C, van der Schaar H, Kaushik-Basu N, Balla T, Cameron CE, Ehrenfeld E, van Kuppeveld FJ, Altan-Bonnet N (2010) Viral reorganization of the secretory pathway generates distinct organelles for RNA replication. Cell 141(5):799–811. doi:10.1016/j.cell.2010.03.050, S0092-8674(10)00369-7 [pii]

Huang H, Chopra R, Verdine GL, Harrison SC (1998) Structure of a covalently trapped catalytic complex of HIV-1 reverse transcriptase: implications for drug resistance. Science 282(5394):1669–1675

Iglesias NG, Filomatori CV, Gamarnik AV (2011) The F1 motif of dengue virus polymerase NS5 is involved in promoter-dependent RNA synthesis. J Virol 85(12):5745–5756. doi:10.1128/JVI.02343-10, JVI.02343-10 [pii]

Jin Z, Deval J, Johnson KA, Swinney DC (2011) Characterization of the elongation complex of dengue virus RNA polymerase: assembly, kinetics of nucleotide incorporation, and fidelity. J Biol Chem 286(3):2067–2077. doi:10.1074/jbc.M110.162685, M110.162685 [pii]

Jin Z, Leveque V, Ma H, Johnson KA, Klumpp K (2012) Assembly, purification, and pre-steady-state kinetic analysis of active RNA-dependent RNA polymerase elongation complex. J Biol Chem 287(13):10674–10683. doi:10.1074/jbc.M111.325530, M111.325530 [pii]

Johnson SJ, Beese LS (2004) Structures of mismatch replication errors observed in a DNA polymerase. Cell 116(6):803–816. doi:10.1016:S0092-8674(04)00252-1, S0092867404002521 [pii]

Joyce CM, Benkovic SJ (2004) DNA polymerase fidelity: kinetics, structure, and checkpoints. Biochemistry 43(45):14317–14324. doi:10.1021/bi048422z

Kao CC, Del Vecchio AM, Zhong W (1999) De novo initiation of RNA synthesis by a recombinant flaviviridae RNA-dependent RNA polymerase. Virology 253(1):1–7. doi:10.1006/viro.1998.9517, S0042682298995174 [pii]

Kaushik N, Pandey VN, Modak MJ (1996) Significance of the O-helix residues of Escherichia coli DNA polymerase I in DNA synthesis: dynamics of the dNTP binding pocket. Biochemistry 35(22):7256–7266. doi:10.1021/bi960537i, bi960537i [pii]

Kiefer JR, Mao C, Braman JC, Beese LS (1998) Visualizing DNA replication in a catalytically active Bacillus DNA polymerase crystal. Nature 391(6664):304–307. doi:10.1038/34693

Korneeva VS, Cameron CE (2007) Structure-function relationships of the viral RNA-dependent RNA polymerase: fidelity, replication speed, and initiation mechanism determined by a residue in the ribose-binding pocket. J Biol Chem 282(22):16135–16145. doi:10.1074/jbc.M610090200, M610090200 [pii]

Kunkel TA, Bebenek K (2000) DNA replication fidelity. Annu Rev Biochem 69:497–529. doi:10.1146/annurev.biochem.69.1.497, 69/1/497 [pii]

Lee JH, Alam I, Han KR, Cho S, Shin S, Kang S, Yang JM, Kim KH (2011) Crystal structures of murine norovirus-1 RNA-dependent RNA polymerase. J Gen Virol 92(Pt 7):1607–1616. doi:10.1099/vir.0.031104-0, vir.0.031104-0 [pii]

Lesburg CA, Cable MB, Ferrari E, Hong Z, Mannarino AF, Weber PC (1999) Crystal structure of the RNA-dependent RNA polymerase from hepatitis C virus reveals a fully encircled active site. Nat Struct Biol 6(10):937–943. doi:10.1038/13305

Lescar J, Canard B (2009) RNA-dependent RNA polymerases from flaviviruses and Picornaviridae. Curr Opin Struct Biol 19(6):759–767. doi:10.1016/j.sbi.2009.10.011, S0959-440X(09)00164-X [pii]

Leung BW, Chen H, Brownlee GG (2010) Correlation between polymerase activity and pathogenicity in two duck H5N1 influenza viruses suggests that the polymerase contributes to pathogenicity. Virology 401(1):96–106. doi:10.1016/j.virol.2010.01.036, S0042-6822(10)00085-1 [pii]

Li Y, Korolev S, Waksman G (1998) Crystal structures of open and closed forms of binary and ternary complexes of the large fragment of Thermus aquaticus DNA polymerase I: structural basis for nucleotide incorporation. EMBO J 17(24):7514–7525. doi:10.1093/emboj/17.24.7514

Luo G, Hamatake RK, Mathis DM, Racela J, Rigat KL, Lemm J, Colonno RJ (2000) De novo initiation of RNA synthesis by the RNA-dependent RNA polymerase (NS5B) of hepatitis C virus. J Virol 74(2):851–863

Malet H, Egloff MP, Selisko B, Butcher RE, Wright PJ, Roberts M, Gruez A, Sulzenbacher G, Vonrhein C, Bricogne G, Mackenzie JM, Khromykh AA, Davidson AD, Canard B (2007) Crystal structure of the RNA polymerase domain of the West Nile virus non-structural protein 5. J Biol Chem 282(14):10678–10689. doi:10.1074/jbc.M607273200, M607273200 [pii]

Marcotte LL, Wass AB, Gohara DW, Pathak HB, Arnold JJ, Filman DJ, Cameron CE, Hogle JM (2007) Crystal structure of poliovirus 3CD protein: virally encoded protease and precursor to the RNA-dependent RNA polymerase. J Virol 81(7):3583–3596. doi:10.1128/JVI.02306-06, JVI.02306-06 [pii]

McDonald SM, Tao YJ, Patton JT (2009) The ins and outs of four-tunneled Reoviridae RNA-dependent RNA polymerases. Curr Opin Struct Biol 19(6):775–782. doi:10.1016/j.sbi.2009.10.007, S0959-440X(09)00160-2 [pii]

Mestas SP, Sholders AJ, Peersen OB (2007) A fluorescence polarization-based screening assay for nucleic acid polymerase elongation activity. Anal Biochem 365(2):194–200. doi:10.1016/j.ab.2007.03.039, S0003-2697(07)00189-3 [pii]

Michielssens S, Moors SL, Froeyen M, Herdewijn P, Ceulemans A (2011) Structural basis for the role of LYS220 as proton donor for nucleotidyl transfer in HIV-1 reverse transcriptase. Biophys Chem. doi:10.1016/j.bpc.2011.03.009, S0301-4622(11)00066-4 [pii]

Miller ER, McNally S, Wallace J, Schlichthorst M (2012) The ongoing impacts of hepatitis c – a systematic narrative review of the literature. BMC Public Health 12(1):672. doi:10.1186/1471-2458-12-672, 1471-2458-12-672 [pii]

Mosley RT, Edwards TE, Murakami E, Lam AM, Grice RL, Du J, Sofia MJ, Furman PA, Otto MJ (2012) Structure of hepatitis C virus polymerase in complex with primer-template RNA. J Virol 86(12):6503–6511. doi:10.1128/JVI.00386-12, JVI.00386-12 [pii]

Moustafa IM, Shen H, Morton B, Colina CM, Cameron CE (2011) Molecular dynamics simulations of viral RNA-dependent RNA polymerases link conserved and correlated motions of functional elements to fidelity. J Mol Biol 410:159–181

Nashine VC, Hammes-Schiffer S, Benkovic SJ (2010) Coupled motions in enzyme catalysis. Curr Opin Chem Biol 14(5):644–651. doi:10.1016/j.cbpa.2010.07.020, S1367-5931(10)00103-1 [pii]
Ng KK, Cherney MM, Vazquez AL, Machin A, Alonso JM, Parra F, James MN (2002) Crystal structures of active and inactive conformations of a caliciviral RNA-dependent RNA polymerase. J Biol Chem 277(2):1381–1387. doi:10.1074/jbc.M109261200, M109261200 [pii]
Ng KK, Arnold JJ, Cameron CE (2008) Structure-function relationships among RNA-dependent RNA polymerases. Curr Top Microbiol Immunol 320:137–156
Nomoto A, Omata T, Toyoda H, Kuge S, Horie H, Kataoka Y, Genba Y, Nakano Y, Imura N (1982) Complete nucleotide sequence of the attenuated poliovirus Sabin 1 strain genome. Proc Natl Acad Sci USA 79(19):5793–5797
O'Reilly EK, Kao CC (1998) Analysis of RNA-dependent RNA polymerase structure and function as guided by known polymerase structures and computer predictions of secondary structure. Virology 252(2):287–303. doi:10.1006/viro.1998.9463, S0042-6822(98)99463-6 [pii]
Paul AV, Rieder E, Kim DW, van Boom JH, Wimmer E (2000) Identification of an RNA hairpin in poliovirus RNA that serves as the primary template in the in vitro uridylylation of VPg. J Virol 74(22):10359–10370
Paul AV, Yin J, Mugavero J, Rieder E, Liu Y, Wimmer E (2003) A "slide-back" mechanism for the initiation of protein-primed RNA synthesis by the RNA polymerase of poliovirus. J Biol Chem 278(45):43951–43960. doi:10.1074/jbc.M307441200, M307441200 [pii]
Pfeiffer JK, Kirkegaard K (2003) A single mutation in poliovirus RNA-dependent RNA polymerase confers resistance to mutagenic nucleotide analogs via increased fidelity. Proc Natl Acad Sci USA 100(12):7289–7294. doi:10.1073/pnas.1232294100, 1232294100 [pii]
Poch O, Sauvaget I, Delarue M, Tordo N (1989) Identification of four conserved motifs among the RNA-dependent polymerase encoding elements. EMBO J 8(12):3867–3874
Poranen MM, Salgado PS, Koivunen MR, Wright S, Bamford DH, Stuart DI, Grimes JM (2008) Structural explanation for the role of Mn2+ in the activity of phi6 RNA-dependent RNA polymerase. Nucleic Acids Res 36(20):6633–6644. doi:10.1093/nar/gkn632, gkn632 [pii]
Ranjith-Kumar CT, Kim YC, Gutshall L, Silverman C, Khandekar S, Sarisky RT, Kao CC (2002) Mechanism of de novo initiation by the hepatitis C virus RNA-dependent RNA polymerase: role of divalent metals. J Virol 76(24):12513–12525
Ren Z, Ghose R (2011) Slow conformational dynamics in the cystoviral RNA-directed RNA polymerase P2: influence of substrate nucleotides and template RNA. Biochemistry 50(11): 1875–1884. doi:10.1021/bi101863g
Ren Z, Wang H, Ghose R (2010) Dynamics on multiple timescales in the RNA-directed RNA polymerase from the cystovirus phi6. Nucleic Acids Res 38(15):5105–5118. doi:10.1093/nar/gkq210, gkq210 [pii]
Rodriguez-Wells V, Plotch SJ, DeStefano JJ (2001) Primer-dependent synthesis by poliovirus RNA-dependent RNA polymerase (3D(pol)). Nucleic Acids Res 29(13):2715–2724
Salgado PS, Makeyev EV, Butcher SJ, Bamford DH, Stuart DI, Grimes JM (2004) The structural basis for RNA specificity and Ca^{2+} inhibition of an RNA-dependent RNA polymerase. Structure 12(2):307–316. doi:10.1016/j.str.2004.01.012, S0969212604000243 [pii]
Schramm VL (2011) Enzymatic transition states, transition-state analogs, dynamics, thermodynamics, and lifetimes. Annu Rev Biochem 80:703–732. doi:10.1146/annurev-biochem-061809-100742
Seckler JM, Howard KJ, Barkley MD, Wintrode PL (2009) Solution structural dynamics of HIV-1 reverse transcriptase heterodimer. Biochemistry 48(32):7646–7655. doi:10.1021/bi900790x
Silva RG, Murkin AS, Schramm VL (2011) Femtosecond dynamics coupled to chemical barrier crossing in a Born-Oppenheimer enzyme. Proc Natl Acad Sci USA 108(46):18661–18665. doi:10.1073/pnas.1114900108, 1114900108 [pii]
Steil BP, Barton DJ (2009) Cis-active RNA elements (CREs) and picornavirus RNA replication. Virus Res 139(2):240–252. doi:10.1016/j.virusres.2008.07.027, S0168-1702(08)00284-0 [pii]

Tahirov TH, Temiakov D, Anikin M, Patlan V, McAllister WT, Vassylyev DG, Yokoyama S (2002) Structure of a T7 RNA polymerase elongation complex at 2.9 A resolution. Nature 420 (6911):43–50. doi:10.1038/nature01129, nature01129 [pii]

Tao Y, Farsetta DL, Nibert ML, Harrison SC (2002) RNA synthesis in a cage – structural studies of reovirus polymerase lambda3. Cell 111(5):733–745. doi:10.1016/S0092-8674(02)01110-8, S0092867402011108 [pii]

Temiakov D, Patlan V, Anikin M, McAllister WT, Yokoyama S, Vassylyev DG (2004) Structural basis for substrate selection by t7 RNA polymerase. Cell 116(3):381–391. doi:10.1016/S0092-8674(04)00059-5, S0092867404000595 [pii]

Thompson AA, Peersen OB (2004) Structural basis for proteolysis-dependent activation of the poliovirus RNA-dependent RNA polymerase. EMBO J 23(17):3462–3471. doi:10.1038/sj.emboj.7600357, 7600357 [pii]

Thompson AA, Albertini RA, Peersen OB (2007) Stabilization of poliovirus polymerase by NTP binding and fingers-thumb interactions. J Mol Biol 366(5):1459–1474. doi:10.1016/j.jmb.2006.11.070, S0022-2836(06)01629-9 [pii]

Tokuriki N, Oldfield CJ, Uversky VN, Berezovsky IN, Tawfik DS (2009) Do viral proteins possess unique biophysical features? Trends Biochem Sci 34(2):53–59. doi:10.1016/j.tibs.2008.10.009, S0968-0004(08)00253-3 [pii]

Uchil PD, Satchidanandam V (2003) Architecture of the flaviviral replication complex. Protease, nuclease, and detergents reveal encasement within double-layered membrane compartments. J Biol Chem 278(27):24388–24398. doi:10.1074/jbc.M301717200, M301717200 [pii]

Van Slyke GA, Ciota AT, Willsey GG, Jaeger J, Shi PY, Kramer LD (2012) Point mutations in the West Nile virus (Flaviviridae; Flavivirus) RNA-dependent RNA polymerase alter viral fitness in a host-dependent manner in vitro and in vivo. Virology 427(1):18–24. doi:10.1016/j.virol.2012.01.036, S0042-6822(12)00083-9 [pii]

Vignuzzi M, Stone JK, Arnold JJ, Cameron CE, Andino R (2006) Quasispecies diversity determines pathogenesis through cooperative interactions in a viral population. Nature 439(7074):344–348. doi:10.1038/nature04388, nature04388 [pii]

Vignuzzi M, Wendt E, Andino R (2008) Engineering attenuated virus vaccines by controlling replication fidelity. Nat Med 14(2):154–161. doi:10.1038/nm1726, nm1726 [pii]

Villali J, Kern D (2010) Choreographing an enzyme's dance. Curr Opin Chem Biol 14(5):636–643. doi:10.1016/j.cbpa.2010.08.007, S1367-5931(10)00110-9 [pii]

Weeks SA, Lee CA, Zhao Y, Smidansky ED, August A, Arnold JJ, Cameron CE (2012) A polymerase mechanism-based strategy for viral attenuation and vaccine development. J Biol Chem 287(38):31618–31622. doi:10.1074/jbc.C112.401471, C112.401471 [pii]

Yang X, Welch JL, Arnold JJ, Boehr DD (2010) Long-range interaction networks in the function and fidelity of poliovirus RNA-dependent RNA polymerase studied by nuclear magnetic resonance. Biochemistry 49(43):9361–9371. doi:10.1021/bi100833r

Yang X, Smidansky ED, Maksimchuk KR, Lum D, Welch JL, Arnold JJ, Cameron CE, Boehr DD (2012) Motif D of viral RNA-dependent RNA polymerases determines efficiency and fidelity of nucleotide addition. Structure 29(9):1519–1527

Yap TL, Xu T, Chen YL, Malet H, Egloff MP, Canard B, Vasudevan SG, Lescar J (2007) Crystal structure of the dengue virus RNA-dependent RNA polymerase catalytic domain at 1.85-angstrom resolution. J Virol 81(9):4753–4765. doi:10.1128/JVI.02283-06, JVI.02283-06 [pii]

Yin YW, Steitz TA (2002) Structural basis for the transition from initiation to elongation transcription in T7 RNA polymerase. Science 298(5597):1387–1395. doi:10.1126/science.1077464, 1077464 [pii]

Zamyatkin DF, Parra F, Machin A, Grochulski P, Ng KK (2009) Binding of 2′-amino-2′-deoxycytidine-5′-triphosphate to norovirus polymerase induces rearrangement of the active site. J Mol Biol 390(1):10–16. doi:10.1016/j.jmb.2009.04.069, S0022-2836(09)00536-1 [pii]

Zhou J, Schweikhard V, Block SM (2012) Single-molecule studies of RNAPII elongation. Biochim Biophys Acta 1829:29–38. doi:10.1016/j.bbagrm.2012.08.006, S1874-9399(12)00149-6 [pii]

Zhu W, Zhu Y, Qin K, Yu Z, Gao R, Yu H, Zhou J, Shu Y (2012) Mutations in polymerase genes enhanced the virulence of 2009 pandemic H1N1 influenza virus in mice. PLoS One 7(3):e33383. doi:10.1371/journal.pone.0033383, PONE-D-11-15459 [pii]

Index

A
α-amanitin, 284
Archaeal DNA polymerases
  crenarchaeal B-family
    PolB1, 141–143
    PolB2, 143
    PolB3, 144
  D-family, 145–146
  vs. eukaryotes, 140
  euryarchaeal B-family, 144–145
  PriSL primase, 149
  proliferating cell nuclear antigen (PCNA)
    B-family polymerases, 150
    complexation, 150, 151
    D-family polymerases, 150
    lesion bypass polymerases, 151–152
    PCNA-interacting peptide (PIP) box, 150
  replication and repair, coordination
    oligomeric DNA polymerase complexes, 152–153
    participation, 155–156
    thermodynamic DNA polymerase selection, 153–154
    uracil read-ahead function, 155
  Y-family
    error rate, 146–147
    lesion bypass, 147–148
    pyrophosphorolysis, 148–149
    steric gate, 148
    structural similarity, 146–147
Archaeal Y-family lesion bypass polymerase, 146–149
Artificially expanded genetic information systems (AEGIS), 177–178

B
Bacteriophage RNA polymerases (RNAPs)
  N4 vRNAP
    characterization, 246
    nucleotidyl transfer reaction, 247–249
    structure, 246–247
    unique hairpin promoter DNA recognition, 247
  transcription elongation
    nucleotide addition cycle, 243–245
    promoter release and processivity, 241–243
    transition to elongation complex, 243
  T7 RNAP
    promoter binding, 238–240
    transcript initiation, 240–241
Base excision repair (BER) pathway
  backup DNA polymerases, 52–53
  damage sources, 47
  DNA polymerase β
    DNA damaging agents, 49
    mouse embryonic fibroblasts, 50
    mouse knockout, 51
    posttranslational modification, 50
    somatic/germline mutations, 51
    structural and functional details, 48, 49
  DNA polymerase γ
    MYH-mediated repair, 52
    nonhomologous end joining, 51–52
    oxidative damage, 52
  long-patch BER, 47, 48
  mitochondria, 258–259
  mitochondrial BER, 53–54
  short-patch BER, 47, 48

K.S. Murakami and M.A. Trakselis (eds.), *Nucleic Acid Polymerases*, Nucleic Acids and Molecular Biology 30, DOI 10.1007/978-3-642-39796-7,

B-family DNA replication polymerases
crenarchaeal
PolB1, 141–143
PolB2, 143
PolB3, 144
euryarchaeal, 144–145
BRCA1 C-terminal (BRCT) domain, 88, 89, 98
Bridge helix, 283

C
Cap snatching, 315–316
Cognate lesions, 86–87
Compartmentalized self-replication (CSR)
artificially expanded genetic information systems (AEGIS), 177–178
DNA polymerases, 179–181
history, 177
laboratory applications, 177
PCR amplification, 177, 178
Taq polymerase, heterotachous sites, 182, 183
Z:P pair, 178, 182
Crenarchaeal B-family DNA replication polymerases, 141–144
C-terminal extension (CTE), 219, 220

D
D-family polymerases (PolD), archaeal-specific, 145–146
Directed evolution
genetic diversity, 168, 169
genotype-phenotype linkage
compartmentalized self-replication (CSR), 177–182
phage display, 182–183
screening, 176
library creation and protein sequence space, 168–170
unguided library creation, 171–172
DNA interstrand cross-links (ICLs), 63–65
DNA lesions, 87
DNA polymerases
classification and function
RNA-dependent, 4–5
single catalytic subunit, 2–4
conserved structures, 7–9
domain, RTs
error rate, 199–200
HIV-1 RT, 197–199
low-processivity, 200
polymerization rate, 199
engineered (*see* Engineered DNA polymerases)
template-dependent DNA synthesis, 110, 111
template-independent DNA synthesis
double-strand breaks, 112
nonhomologous end joining, 113–114
pol β and λ, 112
pol μ, 112
primary amino acid sequence information, 123
TdT (*see* Terminal deoxynucleotidyl transferase (TdT))
DNA repair polymerases
BER pathway (*see* Base excision repair (BER) pathway)
double-strand break repair (*see* Double-strand break repair)
interstrand cross-links repair, 63–64
mismatch repair pathway
DNA metabolism, 62
DNA polymerase delta, 62–63
DNA replication fidelity improvement, 62
replicative polymerases δ and ε, 63
nucleotide excision repair pathway
Pol δ and Pol ε, 59–61
Pol κ, 61
sub-pathways, 59
repair pathways, 46
Double-strand break repair
homologous recombination pathway, 58–59
nonhomologous end joining (NHEJ) pathway
DNA polymerase μ, 56
DNA polymerase β, 57–58
DNA polymerase λ, 56–57
NHEJ functioning mechanism, 54, 55
terminal deoxynucleotidyltransferase (TdT), 57

E
Engineered DNA polymerases
direct design
fused polymerases, 166–167
guided modifications, 165–166
genotype–phenotype linkage
compartmentalized self-replication (CSR), 177–182
phage display, 182–183
screening, 176
literature, 183–184

polymerase chain reaction (PCR), 164
protein engineering methods
directed evolution, 168–172
gene shuffling/molecular breeding, 172–174
neutral drift libraries, 175–176
reconstructing evolutionary adaptive paths (REAP) approach, 174–175
Taq polymerase (*Taq* pol I), 164
Error-prone PCR (ePCR), 171
Eukaryotic replicative DNA polymerases
composition, 19
DNA polymerase δ
DNA substrates partitioning, 24
DNA synthesis, 23
high DNA synthesis fidelity, 24
homozygous mutations, 22
intrinsic kinetic properties, 24
mammalian Pol δ, 21–22
nucleotide misincorporation, 24–25
posttranslational regulation, 21–22
Schizosaccharomyces pombe, 22
structure, 22
DNA polymerase ε
base substitution error rates, 27
chromosomal replication, 26
high fidelity, 27
immunodepletion, 26
vs. Pol δ, 26
posttranslational regulation, 27–28
structure, 25–26
future aspects, 32
polymerase α-primase
catalytic activities, 19
chromosomal replication, 18
moderately accurate polymerase, 20–21
posttranslational regulation, 21
p49 subunit, 19, 20
short RNA-DNA primers synthesis, 20
structure, 18
replication fork
cell cycle checkpoint responses, 30–32
leading and lagging strands, 28–30
replication initiation, 28
Eukaryotic RNA polymerase II. *See* Yeast RNA pol II
Euryarchaeal B-family DNA replication polymerases, 144–145

F

Fialuridine (1-(2-deoxy-2-fluoro-b-D-arabinofuranosyl)-5-iodouracil (FIAU), 261–262
Fork loop, 283
Fused polymerases, 166–167

G

Gene shuffling
history, 172, 173
laboratory applications, 173–174
natural evolution, 173
Genotype–phenotype linkage
compartmentalized self-replication (CSR)
artificially expanded genetic information systems (AEGIS), 177–178
DNA polymerases, 179–181
history, 177
laboratory applications, 177
PCR amplification, 177, 178
Taq polymerase, heterotachous sites, 182, 183
Z:P pair, 178, 182
phage display, 182–183
screening, 176

H

HIV-1 RT
error rate, 199–200
nonnucleoside RT inhibitors (NNRTIs), 200
p51 subunit, 196
RNase H, 203–206
structure, 197–198
termination mechanism, 194

I

Insertion in fingers domain (IFD), 220

L

Library creation, directed evolution
protein sequence space, 168–170
unguided, 171–172
Lid loop, 283
Long-patch BER (LP-BER), 47, 48, 258

M

Minus-strand ssRNA viruses, 315–316
Mismatch repair (MMR) pathway
DNA metabolism, 62
DNA polymerase delta, 62–63
DNA replication fidelity improvement, 62
replicative polymerases δ and ε, 63

Mitochondria
DNA replication
asymmetrical synthesis mechanism, 252–254
conventional mode, 252
displacement mode, 252
fungi, 254
human mitochondrial DNA, 252, 253
pol γ (*see* Polmerase γ holoenzyme)
oxidative phosphorylation, 252
RNA transcription
gene expression, ATP, 269
gene structure, 264–265
HMG proteins, 266
Mtf1, 266
mtRNAP, 267, 268
oxidative mtDNA damage, 269
promoter-specific transcription, 267
Rpo41, 265–266, 268
TFAM, 266
TFB1M and TFBM2, 266, 267
T7 RNAP, 267
Molecular breeding. *See* Gene shuffling

N
Neutral drift libraries, 175–176
Nonhomologous end joining (NHEJ) pathway
base excision repair (BER) pathway, 51–52
double-strand break repair
DNA polymerase μ, 56
DNA polymerase β, 57–58
DNA polymerase λ, 56–57
NHEJ functioning mechanism, 54, 55
terminal deoxynucleotidyltransferase (TdT), 57
template-independent DNA synthesis, 113–114
Non-LTR retrotransposition mechanism, 195
Nucleic acid polymerases
classification and function
DNA polymerase, 2–5
RNA polymerase, 5–6
conserved structures
DNA polymerase, 7–9
RNA polymerase, 10
discovery, 2
family members, 2, 3
future applications, 12–13
implications, disease/therapy, 11–12
Nucleoside/nucleotide reverse transcriptase inhibitors (NRTIs)
clinical manifestations, 259
drug toxicity, 260, 261
HIV and HBV, 260–262
Nucleotide addition cycle
nucleoside triphosphates (NTPs), 243
substrate loading, catalytic site, 244
substrate selection, pre-insertion site, 244
translocation, 245
Nucleotide excision repair (NER) pathway
DNA polymerases
Pol δ and Pol ε, 59–61
Pol κ, 61
sub-pathways, 59
N4 vRNAP
characterization, 246
nucleotidyl transfer reaction, 247–249
structure, 246–247
unique hairpin promoter DNA recognition, 247

P
PCNA-interacting peptide (PIP) box, 150
Phage display, 182–183
Plus-sense single stranded (ss) RNA viruses, 315
Poliovirus (PV) RdRp
catalytic mechanism, 320–322
crystal structures, 318–320
dynamics
Gly64Ser substitution, 322–325
MD simulations, 323, 325
NMR and mutational studies, 326
role of, 324
structure–function paradigm, 326
X-ray crystallography, 323, 324
kinetic mechanism
chemistry, 317–318
nucleotide addition cycle, 316–317
pre-chemistry conformational change, 317–320
symmetrical primer–template RNA substrate, 316
structural differences, 316
viral pathogenesis and virulence, 322
Polmerase γ holoenzyme
in mitochondrial DNA repair, 258–259
mutations
Alpers syndrome, 263–264
MELAS, 264

oxidative damage, 264
progressive external ophthalmoplegia, 263
NRTIs
clinical manifestations, 259
drug toxicity, 260, 261
HIV and HBV, 260–262
pol γA
AID subdomain, 254–255
vs. bacteriophage T7 DNA polymerase, 255–256
enzymatic activities, 253
IP subdomain, 254
vs. pol γB, 257
pol γB interaction, 255
spacer domain, 254
pol γB
functions, 253–254
Pol γA–distal Pol γB monomer, 257–258
processivity, 256–257
species-dependent variation, 257
Polymerase-associated domain (PAD), 87, 88, 91, 92
Posttranslational regulation
DNA polymerase δ, 21–22
DNA polymerase ε, 27–28
polymerase α-primase, 21
PriSL primase, 149
Proliferating cell nuclear antigen (PCNA)
B-family polymerases, 150
complexation, 150, 151
D-family polymerases, 150
DNA polymerase δ, 23
lesion bypass polymerases, 151–152
PCNA-interacting peptide (PIP) box, 150
Y-family polymerase interaction
ubiquitin-modified PCNA, 98–99
umodified PCNA, 96–98
Protein engineering method
directed evolution
library creation and protein sequence space, 168–170
unguided library creation, 171–172
gene shuffling/molecular breeding
history, 172, 173
laboratory applications, 173–174
natural evolution, 173
neutral drift libraries, 175–176
reconstructing evolutionary adaptive paths (REAP) approach, 174–175

R
Reconstructing evolutionary adaptive paths (REAP) approach, 174–175
Replication factor C (RFC), 23
Retrotransposition mechanism, non-LTR elements. *See* Non-LTR retrotransposition mechanism
Retrotransposons, 190
Reverse transcriptases (RTs)
connection and RNase H domains
cellular types, 201
metal ion-assisted catalysis, 203
nucleic acid substrate, 203–204
phosphate-binding pocket, 202–203
definition, 190
DNA polymerase domain
error rate, 199–200
HIV-1 RT, 197–199
low-processivity, 200
polymerization rate, 199
non-LTR retrotransposition mechanism, 195
retrotransposons, 190
retroviruses, 190
substrate binding and coordination, 204–206
subunit organization
alpharetroviruses, 196
monomeric and dimeric, 196–197
TERT, 219–220
viral DNA synthesis, LTR elements
central termination, 194–195
polypurine tract-primed (+) strand DNA synthesis, 193
(-) strand strong-stop DNA synthesis and strand transfer, 193
tRNA-primed (-) DNA synthesis, 191, 193
tRNA primer removal and (+) strand DNA transfer, 193–194
Rev1-interacting region (RIR), 89, 101
Ribonucleoprotein (RNP)
core composition, 217
evolutionary aspects, 224–225
RNA-dependent DNA polymerase, 4–5
RNA-dependent RNA polymerase (RdRp)
antiviral therapies, 311
biophysical techniques, 311–312
catalytic mechanism, 320–322
conformational changes, 311
dynamics

Gly64Ser substitution, 322–325
MD simulations, 323, 325
NMR and mutational studies, 326
role of, 324
structure–function paradigm, 326
X-ray crystallography, 323, 324
internal protein motions, 311
intrinsic genetic variation, 311
kinetic mechanism
chemistry, 317–318
nucleotide addition cycle, 316–317
pre-chemistry conformational change, 317–320
symmetrical primer–template RNA substrate, 316
nucleotide incorporation fidelity, 310–311
structural architecture
domains, 312
initiation mechanisms, 312–314
motifs and functional importance, 312, 314
structural differences
dsRNA viruses, 316
minus-strand ssRNA viruses, 315–316
plus-sense single stranded (ss), 315
viral pathogenesis and virulence, 322, 323
RNA polymerase
classification and function, 5–6
conserved structures, 10
RNase H domains
cellular types, 201
metal ion-assisted catalysis, 203
nucleic acid substrate, 203–204
phosphate-binding pocket, 202–203
Rudder loop, 283

S
Short-patch BER, 47, 48
Single-nucleotide BER (SN-BER), 258
Solvent deuterium kinetic isotope effects (SDKIE), 317–318

T
Taq polymerase (*Taq* pol I)
DNA polymerase I, 164, 165
fused polymerase, 166–167
heterotachous sites, 182, 183
TATA-binding protein 214 (TBP), 285, 286
Telomerase
anticancer therapeutics, 229–230
catalytic cycle, 217, 218
mechanism
duplex binding, 229
duplex dissociation, 227
nucleotide addition, 225–227
strand separation, 228
template realignment, 229
ribonucleoprotein (RNP)
core composition, 217
evolutionary aspects, 224–225
telomerase reverse transcriptase (TERT)
C-terminal extension (CTE), 219, 220
reverse transcriptase (RT), 219–220
telomerase essential N-terminal (TEN), 218, 219
telomerase RNA-binding domain (TRBD), 219
telomerase RNA (TR)
conserved core, 221, 222
H/ACA domain, 223
pseudoknot and stem-loop moiety, 221–222
template boundary element (TBE), 223
yeast and filamentous fungi, 222–224
telomeres, 216
Telomerase essential N-terminal (TEN), 218, 219
Telomerase reverse transcriptase (TERT)
C-terminal extension (CTE), 219, 220
reverse transcriptases (RTs), 219–220
telomerase essential N-terminal (TEN), 218, 219
telomerase RNA-binding domain (TRBD), 219
Telomerase RNA (TR)
conserved core, 221, 222
H/ACA domain, 223
pseudoknot and stem-loop moiety, 221–222
template boundary element (TBE), 223
yeast and filamentous fungus, 222–224
Telomerase RNA-binding domain (TRBD), 219
Telomeres, 216
Template-dependent DNA synthesis, 110, 111
Template-independent DNA synthesis
DNA polymerases
double-strand breaks, 112
nonhomologous end joining, 113–114
pol β and λ, 112
pol μ, 112
primary amino acid sequence information, 123
TdT (*see* Terminal deoxynucleotidyl transferase (TdT))

double-strand breaks, 112
primary amino acid sequence information, 123
Terminal deoxynucleotidyl transferase (TdT)
biochemical applications, 130
cancer
acute lymphocytic leukemia, 127
anticancer agents, 127–128
chronic lymphocytic leukemia, 127
3-Eth-5-NIdR, 128–129
merkel cell carcinoma, 129–130
5-nitroindolyl-2′-deoxynucleoside triphosphate, 128
prognosis and survival, 127
enzymatic properties of, 119–120
kinetic mechanism
conformational changes, 121
ordered substrate binding, 120
product release, 120–121
regulation of
PCNA interaction, 127
posttranslational regulation, 126
protein–protein interactions, 126–127
sources and purification, 118–119
tertiary structure
amino acids, 126
crystal structure, 123, 124
metal ions, 123
subdomains, 124, 125
Transcription elongation, RNAPs
nucleotide addition cycle
nucleoside triphosphates (NTPs), 243
substrate loading, catalytic site, 244
substrate selection, pre-insertion site, 244
translocation, 245
promoter release and processivity
N-terminal domain (NTD), 241
RNA exit channel formation, 241, 243
transition to elongation complex, 243
Translesion synthesis (TLS), 86
T7 RNAP
promoter binding, 238–240
transcript initiation, 240–241

U
Ubiquitin-binding motif (UBM), 89, 98
Ubiquitin-binding zinc finger (UBZ), 89, 99
Ubiquitin-modified PCNA, 98–99
Unmodified PCNA, 96–98

V
Viral DNA synthesis and reverse transcription
central termination, 194–195
polypurine tract-primed (+) strand DNA synthesis, 193
(−) strand strong-stop DNA synthesis and strand transfer, 193
tRNA-primed (−) DNA synthesis, 191, 193
tRNA primer removal and (+) strand DNA transfer, 193–194
Virion protein genome (VPg), 315

W
Werner syndrome protein (WRN), 25

X
X-family DNA polymerases
template-independent DNA synthesis
nonhomologous end joining, 113–114
pol β and λ, 112
pol μ, 112
primary amino acid sequence information, 123
TdT (*see* Terminal deoxynucleotidyl transferase (TdT))
V(D)J recombination, 115–117

Y
Yeast RNA pol II
electron density map, 278
synchrotrons, 278
ten-subunit pol II, yeast, 278, 279
transcribing complex
α-amanitin, 284
bridge helix, 283
DNA-RNA hybrid, 279, 280
fork loop, 283
magnesium ions and NTP entry site, 280
pol II-TFIIB complex, 284–286
post-translocation complex, 280
pre-translocation complex, 280
Rpb4 and Rpb7 subunits, 283–284
rudder and lid loop, 283
structure, 279
tailed template, 278
trigger loop, 280–282

Y-family polymerases
catalytic activity
core regions and active sites, 92
DNA polymerase η, 90–91
DNA polymerase ι, 91, 93
DNA polymerase κ, 93–94
Rev1, 95–96
cognate lesions, 86–87
DNA lesions, 87
interactions
DNA polymerase η, 100–101
protein–protein interactions, 100
Rev1, 101–102
ubiquitin-modified PCNA, 98–99
unmodified PCNA, 96–98
lesion bypass
error rate, 146–147
lesion bypass, 147–148
PCNA, 151–152
pyrophosphorolysis, 148–149
steric gate, 148
structural similarity, 146–147
structures of
BRCA1 C-terminal domain, 88
C-terminal regions, 87–88
domains, 87
full-length Y-family polymerases, 89–90
translesion synthesis, 86

MIX
Papier aus verantwortungsvollen Quellen
Paper from responsible sources
FSC® C105338

If you have any concerns about our products,
you can contact us on
ProductSafety@springernature.com

In case Publisher is established outside the EU,
the EU authorized representative is:
Springer Nature Customer Service Center GmbH
Europaplatz 3, 69115 Heidelberg, Germany

Printed by Libri Plureos GmbH
in Hamburg, Germany